# Talking about Leaving Revisited

Elaine Seymour • Anne-Barrie Hunter
Editors

# Talking about Leaving Revisited

## Persistence, Relocation, and Loss in Undergraduate STEM Education

Heather Thiry, Timothy J. Weston, Raquel P. Harper,
Dana G. Holland, Andrew K. Koch, Brent M. Drake,
Anne-Barrie Hunter, and Elaine Seymour
Authors

Springer

*Editors*
Elaine Seymour
Ethnography & Evaluation Research (E&ER)
University of Colorado Boulder
Boulder, CO, USA

Anne-Barrie Hunter
Ethnography & Evaluation Research (E&ER)
University of Colorado Boulder
Boulder, CO, USA

Center for STEM Learning
University of Colorado Boulder
Boulder, CO, USA

ISBN 978-3-030-25306-6 ISBN 978-3-030-25304-2 (eBook)
https://doi.org/10.1007/978-3-030-25304-2



This Springer imprint is published by the registered company Springer Nature Switzerland AG
The registered company address is: Gewerbestrasse 11, 6330 Cham, Switzerland

# Foreword

## *Talking About Leaving* Revisited: A Return to the Scene of the "Crime"

*Talking about Leaving* took many of us by storm when it was published in 1997. Unusual for its time, the study did not indulge in "victim blaming" and instead pointed us toward strategies that went beyond "fixing the students" as a way to diversify the STEM (science, technology, engineering, mathematics and medical) community.

We were motivated to join the overall undergraduate STEM education reform movement by the publication of *Talking about Leaving: Why Undergraduates Leave the Sciences*. This study opened the eyes and minds of many who were wondering about the exodus of students from STEM fields, especially the leave-taking of women and students of color. The work was important for many reasons, not the least of which were the clues offered about how and where things were going awry and what might be done to get them back on track. But unlike the prevailing presumptions that the students who were leaving didn't have "the right stuff" for STEM, the evidence that emerged pointed to a much more complex story that included complicity of poor teaching by faculty and, at least for me, neglect by institutions to declare the losses as unacceptable—to act and to provide needed support.

I had personal reasons to want to know what happened to everyone; as a Black woman scientist, I have spent my entire career wondering where everyone had gone and how to address their leaving. In part this is related to my own pathway, from Birmingham, Alabama, the Jim Crow South, to years as "the only," in my class or in my major or in my lab group, on faculty or on a board or committee. I was drawn to science after the launch of Sputnik because of the compelling vision and opportunities, even for a little girl from Birmingham, for understanding the world, making a difference in the world, for earning a living and making a life. I knew there are many more people out there, from all backgrounds and experiences, who were drawn to and interested in STEM, who needed to see the pathway to turn interest into outcomes. So, where did they go?

## The Mystery Is Revealed

My career at AAAS, working to improve the quality and increase access to education and careers in STEM, aligns completely with my love for mystery stories. Follow the clues; look for the evidence; talk to the victims, survivors, witnesses; identify the suspects. Though it's okay to develop hypotheses about what happened, make no assumptions about whodunit; keep your mind open, your eyes peeled, and be ever suspicious. P.D. James, the great mystery writer, is said to have noted that when she first heard about Humpty Dumpty she wondered if he fell or if he was pushed. How many of the students who left STEM were pushed?

For those of us who work on diversity, equity, and inclusion in STEM, *Talking about Leaving Revisited* (TALR) reads like a mystery story. While some of the actions, such as poor teaching, have impacts on everyone, they have differentially negative effects on women and male students of color in prompting "relocation" or leaving. But the story is not all bad news; for example, there have been some improvements in teaching over the 20 years between studies. The reform movement had real impact in calling attention to the relationship between poor teaching and student leaving.

But a new threat to degree completion has been added: the significant number of students who just leave school without a degree. When returning to the scene of the crime and retracing the steps from the previous volume, we find that some new elements have been added, changes in the landscape of higher education that must be accounted for to understand why students leave STEM. Is it money? Motivation? Lack of clear connection to careers? A hot job market? Too much complexity involved in navigating the pathway to the degree?

I chaired the panel of National Academies exploring two-year and four-year STEM degree completion in 2014–2016. During that study, we had a chance to look at the "headline topics," and larger contours that surround leaving, the details of which are amplified in *TALR*. A diverse group of people enter higher education with interest in STEM, and you look around a year or two later and many of them are gone. What happened? The committee followed a number of different evidence threads: the culture of STEM; the institutional supports; the policy environment; and the challenges of organizational change. I only wish we had had *TALR* at that time since it provides many of the missing pieces of evidence needed to reconstruct the "crime scene."

## *The Culture of STEM*

Through the richness of the data and students' own voices in *TALR*, we gain a more fulsome picture of the dynamics that operate. Leavers and persisters alike experience challenges; the differences in outcome emerge from how they react, respond, and/or adapt to these.

If scientists enter their fields and undertake their research because of the satisfaction and enjoyment they derive from this work, one might expect that they would want to be ambassadors for their disciplines, sharing knowledge and excitement and engaging in practices that would make students feel welcome. Yet the discussions in *TALR*, especially around the so-called introductory, weed-out courses, lead us to the exact opposite conclusion. In these courses, there is content overload, incoherent presentation, curve grading, with material pitched too high and inappropriately abstract, a focus on rote learning, boring delivery—in other words, mind-numbing, something to be endured rather than enjoyed—the exact opposite of what you get with inclusive pedagogy and active learning.

Why might students feel "pushed out," and why might they actually *be* pushed out? There is evidence to support the idea that the normative culture of science includes the view that natural ability determines the capacity for STEM learning; many of the faculty within STEM exhibit a "fixed ability mindset." This would mean that as early as possible their role would be perceived as that of "identifying the best and weeding out the rest"; or as noted in Chap. 7, the *structured intentionality of the weed-out system* [is] *to get rid of a higher proportion of them* [students] *rather than teach them.* The evidence presented here is damning: where faculty teach these introductory weed-out courses in entirely different ways than they teach upper-level courses, for example. And the "weed-out" works—in promoting relocation and/or switching, but often with the loss of many of those who would be seen as *most* able, especially among young women, rather than those who might be considered "unworthy." Students who persist do so in spite of the negative messaging! Where women and minority students might face loss of confidence, white males might call upon explanations of institutional complicity in not addressing poor teaching. Where one's identity as a good student is in jeopardy, able students may choose to switch rather than fight.

Science loses talent—those who might contribute to its advancement and its diversity—in, as *TALR* describes, *a normalized process of structured wastage.*

Other aspects of the negative impact of the culture of STEM include gender stereotyping and bias against students of color. I found it sad that women of color still felt safer approaching graduate TAs for assistance (rather than faculty) 50 years after I made that same choice following a similar assessment. And they likely made that decision for the reason that I did—to avoid experiencing the bias associated with not belonging.

## *Institutional Supports*

What assumptions are made at the institutional and department level about who the STEM students are and what they might need to be successful? There are often major disconnections between what the faculty and institutions assume to be students' knowledge and prior experiences and what those really are. There are often also misalignments of expectations on all sides.

Fewer students arrive at age 18, fresh out of high school, with adequate preparation that will allow them, without supports, to successfully navigate an STEM course of study. Students may be older; they may be employed, working a significant number of hours that competes with their time for study. Students of color especially may arrive from under-resourced K-12 systems, where adequate labs, challenging courses, and career guidance are lacking; first-generation students may lack the social capital to understand what to expect at college and how to navigate a complicated system. Even when arriving with strong K-12 preparation, students of color may still encounter isolation and other barriers to success. Students may not understand the need to seek help or the level of effort required for college coursework compared with how they managed in high school.

*TALR* situates the entangled nature of these barriers within the culture of STEM: *The popular meritocratic narrative of individual effort and intelligence absolves faculty and institutions of any responsibility for student learning and success.* Persistence involves students' finding peers, using learning centers, graduate assistants, minority and women's STEM organizations, any help that feels safe, to navigate a complicated structure that is not transparent, flexible, forgiving, or welcoming.

Failure to provide clear information and examples of jobs available to and work done by individuals with discipline-specific preparation seems to be a challenge for students and faculty alike. It's hard to imagine possibilities without this. In this issue, there may be a role for professional societies as well as institutions, such as sharing the career outcomes of alumni of programs.

## *The Policy Environment*

Many different aspects of navigating a pathway to STEM involve issues of institutional and/or departmental policies, including issues such as the adequacy and form of financial support for education; how much students must work to support their education; their ability to move and transfer credits across institutions. Savvy students obtain information about courses and faculty, avoiding situations which threaten to result in poor outcomes, taking courses with more adept teaching faculty or at other institutions and transferring in those credits. The less savvy suffer the consequences. Would policies aimed at monitoring courses (and the faculty teaching them) for the numbers and patterns of DFWI (D's, F's, withdrawals, and incompletes) reduce the tolerance for and incidence of these outcomes? Are there expectations that faculty who are hired will need to demonstrate knowledge and practice of effective teaching strategies? Are graduate students and postdoctoral scholars expected to participate in professional development around effective teaching? Unless institutions and departments are willing to understand and assume responsibilities for their roles in student loss, nothing will change.

## *Challenge of Organizational Change*

Are we satisfied with the loss of talent that has been documented in *TALR*? Do we see how the traditional STEM narratives and "standard operating procedures" serve to weed rather than cultivate diversity in STEM? I found the barriers and quicksand described in this volume all too familiar. Fifty years ago, I survived the hazing of introductory courses but only by seeking help and convincing the only African-American graduate student in one of the departments that I was underprepared from my under-resourced high school. I was not dumb, but I needed help to persist. I knew how to learn; I just needed the time to catch up while I was trying to keep up. I had expected my grades to fall and was determined not to be put off by that. But we cannot address the challenge of stemming loss by imbuing each and every student with stubbornness and resilience.

Two years ago AAAS, inspired by Athena SWAN, a gender equity program in the UK, launched SEA Change (STEM Equity Achievement) as an initiative to provide positive incentives for institutional transformation that supports diversity, equity, and inclusion (DEI) in STEM, especially for women and people of color as undergrads, graduate students, and faculty. The initiative involves working with colleges and universities as they undertake self-assessment of their policies, programs, processes, procedures, and practices to align those with research-based, "best in class" efforts that are DEI affirming. Driven by the institutions' own data, a self-assessment team uses the SEA Change "scaffold" to answer questions about their own structures that may serve as barriers to or catalysts for DEI in STEM. An action plan to close the gaps between current and desired states is proposed; and this application is subjected to peer review. A successful applicant receives an institutional SEA Change Bronze Award, a public, positive, and hopefully desirable incentive to move toward structural transformation. SEA Change is designed to support continuous improvement through progression to higher award levels as the goals of ambitious action plans are achieved and as institutions become beacons for DEI in STEM.

This effort has been launched in Australia and Canada as well as the United States. The UK has a 13-year history of accomplishment with Athena SWAN, the impact of which was enhanced when funders announced their intention to use departmental ratings in determining who might apply to certain solicitations.

While colleges and universities bear outsize responsibility for addressing losses stemming from their immediate actions, challenges lie in other areas that, at first glance, seem to lie beyond their boundaries; or do they? Take for example the issue of K-12 education which is found to have failed so many students of color. K-12 teachers of STEM receive their content education in the same institutions, from the same STEM faculty. Many leave without real conceptual understanding and, unfortunately, many teach as they are taught.

## *Larger Societal Concerns*

Careers in STEM have become a destination of choice, especially with increased visibility of tech-related careers, messaging from parents and teachers about lucrative salaries, low unemployment, satisfying jobs, and the opportunities to help others, which study in these fields can offer. All of this is true to some extent. But students who enter colleges or universities with little information about other disciplines can be attracted to consider other fields once exposed to these options and especially should they be disillusioned by their experiences with STEM classes. We can all support students' leaving when associated with an affirmative decision to go toward something better or more desirable; we recoil at the idea of students being pushed out by a system designed to "thin the herd."

There is often the confluence of many different pressure points for students who go into STEM, those who leave and those who persist, making it seem very hard to navigate successfully through all the barriers. So we really should not be surprised at the level of the losses, especially for women and students of color. *TALR* is an explication of those factors that constitute Robert Merton's idea of cumulative disadvantage in science that results in members of some groups being lost to science. To a fan of mystery stories, it feels like in Agatha Christie's *Murder on the Orient Express* (spoiler alert!!) no one person is responsible because everyone shares the guilt!

*TALR* appropriately indicts many. And, seeing the evidence, I concur in those indictments. And were I a juror, I would vote to convict. But what happens after that? The only reasonable way forward would be to recommend a sentence of "restorative justice." We must use the evidence here to activate the next movement—one that drives us toward institutional transformation, where each of us owns our part in the crime, where we work together to lay out a plan for reconciliation and rehabilitation. At the same time, there remain unanswered questions and unquestioned "answers."

The constellation of factors that contribute to loss are broad and interconnected. Likelihood of poor K-12 schooling which so disadvantages students of color links to socioeconomic status and lack of social capital, lack of knowledge about careers or about navigating college. No single intervention can address this—only systemic approaches.

The harmful actions may not have been done with malice, but there has been and continues to be harm. *TAL1* provided us with hope and direction and focused attention that helped launch the undergraduate STEM education reform movement. *TALR* makes the case for systemic change, pointing us to actions that can positively affect STEM for all. This is an opportunity to use the research to guide our actions, to reject the traditions born of a different time, place, and talent pool, letting *TALR* guide the path forward.

Shirley M. Malcom
Senior Advisor, American Association for the Advancement of Science
Washington, DC, USA
Director, SEA Change, Washington, DC, USA

# Acknowledgments

The authors and editors wish to thank the Alfred P. Sloan Foundation and the National Science Foundation for their generous financial support of this project. We offer special thanks to our Program Officers, Dr. Elizabeth Boylan at the Sloan Foundation and Dr. Myles Boylan at the National Science Foundation's Division of Undergraduate Education for their encouragement, advice, and critique of our work throughout its planning, research, and development into this volume. The Sloan Foundation supported the original study which produced the volume *Talking about Leaving: Why Undergraduates Leave the Sciences* (Westview Press, 1997) and has generously embraced the current undertaking. We also thank NSF's Dr. Boylan for his active and long interest in our work, and we acknowledge his considerable contributions to the STEM education reform movement of which our work has been an ongoing part.

We thank the many people in the six universities and colleges that, after a gap of two decades, once again agreed to take part in this study. They include: the presidents and deans of these institutions who supported and enabled the work; the departmental chairs, faculty, and administrative staff who hosted the research team and arranged on-site logistics; and the directors and staff of the institutional research and records offices who provided the student records from which our analyses were conducted, and the student samples and courses were drawn.

We thank the members of our advisory board whose contributions in the planning and early stages of the work were especially valuable, and from whom we have received feedback and advice on an individual basis as the work progressed: Drs. Robin Adams, Angela Byers Winston, Kevin Eagan, Mary Frank Fox, Angela Ginorio, Charles Henderson, Jillian Kinzie, and Marcia Baxter Magolda.

Our work has also been supported by our host organization at the University of Colorado Boulder, the Center for STEM Learning, its Directors, Drs. Noah Finkelstein (Department of Physics) and Valerie Otero (School or Education), and Business Manager, William Tarantino, and by our colleagues in our home organization, Ethnography & Evaluation Research, Center to Advance Research and Teaching in the Social Sciences.

We thank our research colleagues who have offered insights and contributions from their own concurrent work in this field, particularly, Drs. Kevin Eagan, Joseph J. Ferrare, You-Geon Lee, Mark Connolly, Ross J. Benbow, Erika Vivyan, and Christine Pfund. We also thank our long-time collaborator, Dr. Susan B. Millar for her editorial suggestions.

To all the students who talked to us about their experiences and gave us their insights in interviews, focus groups, and class surveys, we thank you for your generosity and candour. We hope that those who read our distillations of what we learned from you will recognize them as faithful accounts. Many students told us that they accepted our interview invitations in hopes of contributing, though our work, to the enhancement of education in the STEM disciplines. We too hope that their accounts will inform policy and practice as STEM education moves forward.

This material is based upon work supported by the Alfred P. Sloan Foundation Award No. 2012-6-05 and the National Science Foundation Award No. DUE 1224637. Any opinions, findings, and conclusions or recommendations expressed in this material are those of the authors and do not necessarily reflect the views of the Alfred P. Sloan Foundation or the National Science Foundation.

# Contents

# Authors

**Heather Thiry** Ethnography and Evaluation Research (E&ER), University of Colorado Boulder, Boulder, CO, USA

**Timothy J. Weston** Ethnography and Evaluation Research (E&ER), University of Colorado Boulder, Boulder, CO, USA

National Center for Women and Information Technology (NCWIT), University of Colorado Boulder, Boulder, CO, USA

**Raquel P. Harper** Ethnography and Evaluation Research (E&ER), University of Colorado Boulder, Boulder, CO, USA

Center for STEM Learning, University of Colorado Boulder, Boulder, CO, USA

**Dana G. Holland** Ethnography and Evaluation Research (E&ER), University of Colorado Boulder, Boulder, CO, USA

Center for STEM Learning, University of Colorado Boulder, Boulder, CO, USA

**Andrew K. Koch** John N. Gardner Institute for Excellence in Undergraduate Education, Brevard, NC, USA

**Brent M. Drake** John N. Gardner Institute for Excellence in Undergraduate Education, Brevard, NC, USA

**Anne-Barrie Hunter** Ethnography and Evaluation Research (E&ER), University of Colorado Boulder, Boulder, CO, USA

Center for STEM Learning, University of Colorado Boulder, Boulder, CO, USA

**Elaine Seymour** Ethnography and Evaluation Research (E&ER), University of Colorado Boulder, Boulder, CO, USA

# About the Editors

**Elaine Seymour** is co-founder and director emerita of Ethnography & Evaluation Research (E&ER) at the University of Colorado at Boulder (UCB). Her work has focused on change in STEM education and career paths, including efforts to improve quality, access, and diversity. Her best-known work, co-authored with Nancy M. Hewitt, *Talking about Leaving: Why Undergraduates Leave the Sciences*, (1997) which the current volume revisits and extends, is widely cited for its contribution to the nationwide effort to improve undergraduate STEM education. In addition to her many articles, published volumes by Seymour and E&ER members based on their research include: *Talking about Disability: The Education and Work Experiences of Undergraduates with Disabilities in Science, Mathematics, and Engineering Majors* (1998); *Partners in Innovation: Teaching Assistants in College Science Courses* (2005) which draws on E&ER's evaluation studies of science education reform initiatives; and *Undergraduate Research in the Sciences: Engaging Students in Real Science* (2010), a comprehensive study of the benefits and costs of undergraduate research and the processes whereby these arise. She also contributed chapters on the processes of change and resistance to change to *Transforming Institutions: Undergraduate STEM Education for the 21st Century* (2015). In recognition of her research on women in science and engineering, WEPAN awarded her their 2002 Betty Vetter Award for Research, and, in 2006, she testified before Congress on trends and needs in STEM education reform. To meet the needs of classroom innovators for course evaluation methods relevant to their learning objectives, she co-developed, and continues to expand the scope of, the widely used Student Assessment of their Learning Gains (SALG) online instrument that focuses exclusively on what students gain from their courses. She is a sociologist and a British-American whose education and career have been conducted on both sides of the Atlantic.

**Anne-Barrie Hunter** is co-director of Ethnography & Evaluation Research and program manager for the Center for STEM Learning at the University of Colorado Boulder. Hunter worked with Seymour and Hewitt on the original *Talking about Leaving* study. Since then, she has collaborated on many E&ER research and evaluation projects focused on postsecondary STEM education improvement, specializing in those aimed at persistence to graduation of students historically underrepresented in STEM fields. She and Seymour have worked closely on multiple projects, notably in co-authoring *Talking about Disability: The Education and Work Experience of Graduates and Undergraduates with Disabilities in Science, Mathematics, and Engineering Majors* (1998) and *Undergraduate Research in the Sciences: Engaging Students in Real Science* (2010), to which Hunter contributed; a qualitative meta-analysis of evidence on the outcomes of undergraduate research, identification of the specific benefits of "authentic" research experiences and of the costs and benefits to faculty of UR engagement. Grounded in this research, she co-developed E&ER's Undergraduate Research Student Self-Assessment (URSSA) survey, a free, online, validated evaluation instrument for the assessment of UR programs in STEM fields. Hunter has acted as PI, co-PI, external evaluator, and consultant for numerous STEM education improvement projects, including the Howard Hughes Medical Institute's Undergraduate Science Education Program and Inclusive Excellence grants, as well as multiple NSF projects, notably awards that support the advancement of women physicists, climate change education initiatives, and institutional STEM transformation efforts. She was the PI for the five-year, multiple methods study on which the present volume is based; is co-PI for an NSF grant to align administrative systems to enable easier two-year college transfers and is an external evaluator for the Science Technology Center at CU Boulder. Her interests focus on institutional change to support quality STEM education and on faculty professional development in pedagogy that promotes and sustains broader participation in college STEM education.

# About the Authors

**Heather Thiry** is research faculty member at the University of Colorado, Boulder. Her research interests focus on the educational and career pathways and identities of students from groups traditionally underrepresented in scientific and technological fields, including low-income, first-generation college, and underrepresented minority students. She is especially interested in systems-level change efforts to support the STEM pathways of diverse students in higher education institutions. She is currently one of the evaluators of the NSF INCLUDES Alliance, the Computing Alliance of Hispanic-Serving Institutions (CAHSI). She also leads a five-year research study of underrepresented students' STEM transfer pathways from two-year to four-year institutions. Her evaluation and research interests also focus on the influence of co-curricular activities and experiences on the persistence in STEM majors, including the role of undergraduate research and student organizations on students' development as STEM professionals. She recently served as a member of the National Academy of Sciences committee that produced the report, *Undergraduate Research Experiences for STEM Students: Successes, Challenges and Opportunities*.

**Timothy J. Weston** is a research associate at the University of Colorado's National Center for Women and Information Technology (NCWIT), where he has conducted evaluation and research on National Science Foundation, Department of Education, NASA and private foundation grants for 20 years. Projects have included large-scale experimental evaluations of virtual tutors for the Institute of Education Sciences (IES), assessing new STEM curricula and teaching methods, and designing and implementing tests and surveys, some used nationally. Additionally, he has developed and analyzed quantitative observational protocols in comparative studies of research-based teaching, participated in studies of female representation in academic positions, and evaluated the use of digital libraries to help teach computer science. Currently, he is working on efforts related to student persistence in higher education with a longitudinal study of high school girls moving into college and the workplace. He is also studying the pathways taken by students who are transferred to four-year universities from community college, and the teaching practices of new mathematics professors.

**Raquel P. Harper** is a research associate in Ethnography & Evaluation Research (E&ER) at the University of Colorado Boulder. At CU, she conducts research on undergraduate students' experiences in science, technology, engineering, and mathematics (STEM) programs. Her current research interests include students' learning experiences in STEM courses, scientific career paths, the relationship of institutional and departmental climate on students' persistence, and the underrepresentation of women and minorities in STEM.

She has a diverse academic background in both arts and sciences. She earned a Ph.D. in Public Communication and Technology from Colorado State University in Fort Collins. She also has a B.S. in Physical Therapy from Western Washington University in Bellingham and an M.S. in Technical and Scientific Communication from Colorado State. She has expertise in mixed methods, qualitative research, science education and communication, and training in public health. Her research has been published in science-, health-, and communication-related publications.

Prior to working at CU, Harper was an assistant professor of science, health, and technical communication at Massey University in New Zealand. She primarily worked with students in science, engineering, and healthcare-related disciplines. She has also taught science and technical communication courses at both the University of Colorado and Colorado State University. And she has worked in science and health communication roles for the Department of Energy, the National Cancer Institute, and for the magazine *Biophotonics International.*

**Dana G. Holland** is a social scientist with over 20 years of experience in higher education and education policy research. She currently holds a research faculty position at the University of Colorado, Boulder. Dr. Holland has abiding research interest in understanding institutional practices and organizational changes that contribute to student success, particularly for traditionally disadvantaged and low-income students. She serves as a co-principal investigator on the five-year *STEM Transfer Pathways* study that investigates both student experiences and promising institutional practices in two- to four-year transfer STEM pathways in relatively open access institutions. She also has projects that focused on institutional practices contributing to degree attainment for Native American students, nontraditional engineering students, and Hispanic students beginning STEM studies in community college. Dr. Holland's academic research appears in *Anthropology & Education Quarterly*, *Comparative Education Review*, *Higher Education*, and *Science, Technology & Human Values*. She has served as a proposal reviewer for the National Science Foundation and several journals, including *Science Education.*

Dr. Holland holds a joint Ph.D. in Education and Sociology from the University of Pennsylvania and an M.A. in Applied Anthropology from the University of Maryland. For 4 years, Dr. Holland was an Assistant Professor of Social Science at the American University of Afghanistan in Kabul, where she served as Chair of the University Curriculum Committee and member of the Accreditation Steering

Committee. She also taught as an adjunct faculty member at Arcadia University in Philadelphia and as a visiting Lecturer at the University of Malawi in Southern Africa. In these positions, she developed and taught courses in social research methods, education foundations, public policy, and anthropology at both undergraduate and graduate levels and gained insights into the opportunities and constraints to change in higher education.

**Andrew K. Koch** is the President and Chief Operating Officer for the non-profit John N. Gardner Institute for Excellence in Undergraduate Education which he joined in 2010. In his role, he provides strategic leadership and operations oversight for the Institute in its efforts to help colleges improve teaching, learning, student success and, in the process of doing so, mitigate inequitable outcomes and advance social justice. Prior to coming to Gardner Institute, Drew spent nearly 20 years working in both independent and public postsecondary institutions on student enrollment, access, success, accreditation, learning, and completion efforts with a particular emphasis on first-generation, low-income, and historically underrepresented students. He holds a B.A. in History and German from the University of Richmond, an M.A. in History from the University of Richmond, an M.A. in Higher Education Administration from the University of South Carolina, and a Ph.D. in American Studies from Purdue University. His scholarship focuses on critical university studies and the role of colleges and universities in shaping culture, equity, and democracy in the United States. He has served as the principal investigator or co-principal investigator on more than two dozen grant-funded research projects with support coming from sources such as the Bill & Melinda Gates Foundation, GEAR UP, Kresge Foundation, Lilly Endowment, Lumina Foundation, the National Science Foundation, and the Schusterman Family Foundation. He has published widely on student access and success topics, with a particular emphasis on historically underrepresented and underserved populations, the first-year experience, gateway courses, and redesign of unjust education systems.

**Brent M. Drake** currently serves as the Vice Provost of Decision Support at the University of Nevada, Las Vegas, where he focuses on data related to student attainment and learning, overall institutional effectiveness, institutional reporting, faculty activity, and data analytics. Through his work at UNLV, Brent advocates for increased access to tools and research for university faculty, administration, and students that can guide better-informed university decision-making. These include policies and practices that have consequences for attainment of institutional goals such as student achievement, retention, and completion.

Prior to his appointment at UNLV, Brent worked for 16 years at Purdue University, where his final position as Chief Data Officer was oversight of the Office of Institutional Research, Assessment, and Effectiveness. He began his career in 2001 as the assessment coordinator for the Lily Endowment retention initiatives at Purdue University. Brent earned all of his degrees at Purdue: a B.S. in Athletic Training in 1995, an M.S. in Sports Psychology in 1997, and, in 2009, a Ph.D. in Educational Psychology, with an emphasis on both motivation theory and psychometrics.

He presents and publishes on topics in higher education, including motivational models related to student success, retention-enhancing programs, business intelligence and data analytics, enrollment modeling, recruitment, enrollment trends, and student success efforts. Brent also serves as a Fellow for Research and Innovation with the John N. Gardner Institute for Excellence in Undergraduate Education.

**Anne-Barrie Hunter**: see About the Editors

**Elaine Seymour**: see About the Editors

# Chapter 1
# Why We Are Still Talking About Leaving

**Elaine Seymour, Anne-Barrie Hunter, and Timothy J. Weston**

## What May Be Amiss in STEM Education

We began the account of our original study *Talking about Leaving: Why Undergraduates Leave the Sciences* (TAL) with the following observation:

> Whenever traditional practices are called into question and new practices are proposed, it is always worth asking: Why at this time? Who seeks these changes? Who resists them? And by what rhetoric do they support these positions? (Seymour & Hewitt, 1997, p. 1).

When we began this research in Spring 1990, concerns had just begun to be expressed that there might be something seriously awry with what has now come to be referred to as "science, technology, engineering, and mathematics (STEM) education." By the time we published our findings in 1997, efforts to improve quality

E. Seymour (✉)
Ethnography and Evaluation Research (E&ER), University of Colorado Boulder, Boulder, CO, USA
e-mail: Elaine.Seymour@colorado.edu

A.-B. Hunter
Ethnography and Evaluation Research (E&ER), University of Colorado Boulder, Boulder, CO, USA

Center for STEM Learning, University of Colorado Boulder, Boulder, CO, USA
e-mail: abhunter@colorado.edu

T. J. Weston
Ethnography and Evaluation Research (E&ER), University of Colorado Boulder, Boulder, CO, USA

National Center for Women and Information Technology (NCWIT), University of Colorado Boulder, Boulder, CO, USA
e-mail: westont@colorado.edu



E. Seymour, A.-B. Hunter (eds.), *Talking about Leaving Revisited*,
https://doi.org/10.1007/978-3-030-25304-2_1

and access in STEM education had already begun. The arguments that launched what was to become a national movement for STEM education improvement have changed over time and are still evolving. The question of whether too few STEM graduates are being recruited and retained to meet the nation's future needs continues to be debated. However, from the outset, key questions prompting change initiatives were (and still are): How may the practices of STEM education be redesigned so that they more effectively foster interest, competence, and persistence in the sciences, and secure a growing, more diverse, population of STEM-qualified graduates?

An influential contributor to recent debates has been the 2012 report of the President's Council of Advisors on Science and Technology (Olson & Riordan, 2012), *Engage to Excel: Producing One Million Additional College Graduates with Degrees in Science and Technology*. This, and many other reports and articles, argue for improvements in STEM undergraduate education because a workforce of well-prepared STEM graduates is critical, "to ensure the economic strength, national security, global competitiveness, environment, and health of the United States" (NASEM, 2016, p. 7). The PCAST Report proposes that if only 10% of students who left STEM majors were retained, the USA could meet its future national workforce needs. Whether the nation will achieve this is argued out from a number of perspectives that hinge on how the STEM workforce is defined (e.g., Carnevale, Smith, & Melton, 2011; Rothwell, 2013; Salzman, 2013). However, a somewhat different rationale for securing increased STEM participation and graduation rates cuts across the debate about whether we face a shortage or a shortfall in graduates to pursue STEM careers. This perspective rests on the projection that, by 2020, almost two-thirds of all jobs that require knowledge and skills in STEM will require a bachelor's degree or higher (Bureau of Labor Statistics, 2012; Carnevale et al., 2011). Thus, the debate about the under- or over-supply of graduates for STEM *careers* is reframed in terms of an ever-expanding need for workforce entrants that are mathematically, scientifically, and technically *competent*. Seen in this light, the critical problem is *diversion*, a countervailing process by which students, college graduates, and workers with demonstrated capability in STEM are not reaching jobs where these skills are needed. As Carnevale and his colleagues point out, "Diversion, coupled with the observation that the market for STEM competencies is broader than the market for STEM workers, illuminates why we look like we're producing enough STEM workers—but we're actually not" (p. 42).

Thus, the debate has partially cycled back to the observation that prompted our original study; namely, that the nation can only produce sufficient competent STEM graduates if we can attract and retain more students in these majors. The added concern is whether and how best can we place them in careers where their skills are needed. Clearly, we also need to understand what gets in the way of our achieving these objectives. Our progress is undercut because we still have a revolving door problem: although increasing numbers of students—including those from underrepresented minority groups (URMs) —enter STEM disciplines, losses from these majors remain persistently high. As we further discuss in Chap. 2, of all students who enter college intending to major in a STEM field, recent studies estimate that only 40–50% (varying by discipline) complete a degree in a STEM major (Carnevale et al., 2011; Chen, 2015; Eagan, Hurtado, Figueroa, & Hughes, 2014; Gates & Mirkin, 2012; Hurtado, Eagan, & Hughes, 2012; Lee & Ferrare, 2019; National Science Foundation, 2012; Whalen & Shelley, 2010). Assessments of the extent of these losses include students who enter STEM majors, then either switch to non-STEM majors, or leave college without a degree in any major.

How does this picture compare with the rates of loss that we reported in TAL two decades ago? The best national source then available was the Cooperative Institutional Research Program (CIRP) data for 1991 provided to us by UCLA's Higher Education Research Institute (HERI). In the CIRP analysis, for students who entered college in 1987, there was, by 1991, a continuum of stability to instability from original majors into majors in a completely different disciplinary group. Beginning in English, where the switching rate was only 15%, it progressed through the social sciences, fine arts, education, history, and political science with rates ranging from 28% to 35%; thence to engineering and business, where rates were 38% to 4%. Finally, in the sciences, computer science, and mathematics switching rates varied between 47% and 63%. The most stable STEM major was engineering, which is also the most selective in its screening of entrants. In mathematics and the physical and biological sciences, between 51% and 53% of all entrants switched to non-STEM majors.

The further question is, which students were lost? Twenty years ago, the rates at which men of color and women of all races and ethnicities switched out of STEM majors were higher than those for white males. When we began the TAL study, most of the research on STEM losses focused on high loss rates among able, well-prepared women (cf., Astin & Astin, 1993; Hall & Sandler, 1984; Manis, Sloat, Thomas, & Davis, 1989; Rosser, 1990; Tobias, 1990, 1992). From the CIRP analysis, we concurred that STEM switching rates in the sciences and mathematics were higher overall for women (52%) than for men (41%). The loss rates by sex were comparable in engineering (37% for women; 38% for men), but with much smaller numbers of women entrants. We also reported that women left STEM majors with higher average PGPAs than those of men who persisted (McLelland, 1993). That women of high ability are still lost is well attested in a review of the evidence by Lindberg, Hyde, Petersen, and Linn (2010), and recent studies by Bressoud, Mesa, and Rasmussen (2015), and Islam and Al-Ghassani (2015) which we discuss later in this chapter.

Although national data on STEM losses among students of color were less available in the early 1990s, the National Action Council for Minorities in Engineering (NACME) surveys provided information about engineering majors. Among these, Morrison and Williams (1993) reported that the graduation rate for students of color in engineering was about half that of white students: only 34% of students of color who entered engineering majors completed them, compared with 68% of white students. In sciences or mathematics, by junior year, 65% of students of color who entered these majors had left, compared with 37% of white students (cf., *Science*, 1992), and half of the students of who left engineering majors dropped out of college altogether (Campbell, 1993). While the numbers of women and students of color entering STEM majors have increased over time, they are still underrepresented in these majors in relation to their proportions in the US population and both groups earn proportionately fewer STEM bachelor's degrees than white men (Eagan et al., 2014; Hurtado et al., 2012; Rodriguez et al., 2012). Growing awareness of this substantial loss of natural talent underlies an ongoing national effort, supported by both public and private foundations, to attract and retain more students of color in STEM majors (Carnevale et al., 2011; Griffith, 2010; Rodriguez et al., 2012).

To increase the retention of able women, Ellis, Fosdick, and Rasmussen (2016) propose a strategy that focuses on the impact of a single course, Calculus 1, on STEM losses. Because of its gateway role in all or most STEM majors, they argue that, were women to proceed to Calculus 2 at the same rate as men, the number of women entering the STEM workforce would increase by 75% and thereby bring an additional 20% of new graduates into the STEM workforce. What prevents this is not, they argue, lack of ability or effort. Rather, it is the loss of incoming confidence that teaching and assessment methods designed to weed out students engender among women, especially women of color. The challenge to the professoriate that the study authors present is fundamentally the same as that made by the researchers in the early 1990s; namely, that it is unacceptable to discard high proportions of students who enter with the interest and ability to undertake an undergraduate science education.

Why, we may ask, are these losses still occurring, given a major national effort over the last two decades to improve access, quality, and outcomes of science and mathematics education at all levels—from kindergarten to graduate school? Despite evidence of progress (reviewed in Seymour & Fry, 2016), research continues to point to rates of switching that prevent us reaching national workforce goals, and to disparities by gender, race, and ethnicity in the extent of those losses. The more serious consequences of STEM switching evident in the TAL data, however, may be wastage of talent, compromise, or distortion of career aspirations, time and money wasted, debts increased, lost confidence, pride, and a sense of direction—all of which also affect switchers' families and communities. In Chen's (2013) analysis of National Center for Education Statistics data for 2004–2009, for entrants to STEM majors in this cohort, 28% switched to a non-STEM field and 20% left college without any degree or certificate—losses that we must consider as a form of permanent wastage. Lee and Ferrare (2019) clarify that students at particular risk are those who switch early in their academic trajectory and those whose parents do not have a bachelor's degree. These students are less likely to graduate on time and more likely to drop out. Lee and Ferrare also propose that switching is a risk that some students are more able to surmount because their cultural and economic advantages mitigate switching's degree completion risks. The TAL findings also raise the question of whether losses from the sciences are (as conventional wisdom implies) the net result of myriad individual decisions based on students' personal and intellectual limitations. Alternatively, does the group patterning evident in STEM attrition data reflect an accumulation of structured disadvantages?

## Seeking Explanations

When we started the TAL study in 1990, it had already begun to be argued that STEM attrition could not be viewed solely as a natural consequence of differential levels of ability, and that high school preparation, and the climate and activities of STEM classrooms, likely played a part in determining which students did and did not persist in these majors. However, the work of teasing out the whole range of factors that contribute to high STEM diversion rates had not been attempted; neither had the relative hierarchy, or possible interrelationship, of explanatory factors been explored. It was this combination of tasks that we set out to accomplish. Our goals were, "to discover, and establish the relative importance of, factors with greatest bearing upon the decisions of undergraduates at 4-year colleges and universities to switch from science, engineering, and mathematics majors into disciplines that are not science-based" (Seymour & Hewitt, 1997, p. 13). We think it is useful to review what we discovered so that we may make comparisons with what we and later researchers have found, and assess what has, or has not, changed over the intervening years.

### *How the Original TAL Study Was Done*

We focused our inquiry on students at a sample of seven types of 4-year colleges and universities that, taken together, contribute most to the national supply of baccalaureate STEM graduates.[1] This decision well predated the growth of 2-year STEM degrees and the multiple pathways to 4-year STEM degrees now taken by

---

[1] The sample comprised four public IHEs:

- a large urban, mid-western university with prestige ranking for its STEM research and high production of STEM undergraduates and graduates;
- a "flagship" western state university with high reputation for its engineering school and several science departments;
- a comprehensive, urban north-eastern university with large and diverse STEM undergraduate enrollment;
- a western state (originally land-grant) university serving a large rural population, with a prestigious engineering research program, and applied science specialties.

And three private IHEs:

- a large west coast university with selective admission and high STEM research prestige;
- a western liberal arts college with a strong reputation for its science teaching (engineering is not offered);
- a small western city university offering masters' degrees and doctorates in the sciences and undergraduate engineering.

The selection of four western state institutions was made because the study was funded in two stages—the first of which focused on four regional IHEs; the second stage added three additional IHE types and wider geographic scope.

many students, especially first-generation undergraduates (NASEM, 2016). Because evidence about the causes of STEM diversion was lacking, we took an ethnographic approach to data collection and analysis that made no presumptions about what kinds of contributory causes might emerge, nor about their relative significance.[2] Also, in line with ethnographic traditions (Agar, 1996; Clifford, 1988; Geertz, 1973; Spradley, 1979; Van Maanen, 1988), we posited that students who had recently switched out of STEM majors into non-STEM majors, and seniors in STEM majors who were poised to graduate, would be expert informants about the experiences and considerations that had shaped their decisions. We selected only those students who had entered college with demonstrated mathematical ability. Based on the advice of a sample of STEM faculty, we drew this line at mathematics scores at or above 650 SAT or 28 ACT. This allowed us to exclude the possibility that those in our sample who had switched out of STEM majors simply lacked the requisite ability to complete them.

We included in our disciplinary sample, the physical sciences, life sciences, and earth sciences, plus mathematics and applied mathematics, and all engineering specialties. Computer science, which had not at this date developed as an independent discipline, was not included, so our original pneumonic was SME, not STEM. From across these disciplines, we interviewed 335 students at the seven institutions over a 3-year period (1990–1993), and, subsequently, an additional 125 students at a further six institutions (making a total of 460) in order to check the validity of tentative hypotheses emerging from the analysis. The student sample comprised 55% juniors who had switched and 45% persisting seniors. Within each of these groups, we over-sampled (i.e., in terms of their representation in STEM majors) by gender (52% of white students in our sample were women), and by race/ethnicity (26% of the whole sample were students of color—with almost equal numbers of women and men). This was done in order to get a clear understanding of what best explained higher switching rates in these groups. The authors conducted individual interviews with switchers and all students of color, and single-gender focus groups with graduating seniors. The length of interviews requested of participants was 1 h, but this varied according to how much time participants chose to spend in discussing the questions that we raised.

All interviews and focus groups were conducted as semi-structured, open-ended, conversational explorations with the order of protocol questions guided by students' responses and by issues that they spontaneously raised. We used the same protocol (cf., TAL, pp. 401–402) for all informants, making adjustments for gender, race/ethnicity, and academic decisions. New lines of inquiry emerged as informants introduced issues, experiences, and examples that they saw as relevant in helping us to understand the rationales behind their academic and career decisions. These were always followed up with the original speaker and were checked out with subsequent interviewees. Thus, we gradually built up a framework of explanations for student decisions.

[2] Ethnographic research explores cultural phenomena from the point of view of the subjects of the study—as expert witnesses to events of which they are a part.

Interviews were recorded and transcribed *verbatim*; then coded using "The Ethnograph" (Seidel, Kjolseth, & Seymour, 1988). Computer-assisted qualitative research—then in its infancy—made possible coding and domain analysis of our text data to a high degree of complexity, notwithstanding the unusually large sample for an ethnographic study. The information that we were seeking about the factors bearing upon students' decisions to move or stay was typically embedded in their narrative accounts; rarely in abstract or summary statements. There were no preconceived codes; code names referenced discrete ideas encountered in students' narratives, illustrations, and comments rather than our questions. Where several points were made in the same statement, each was coded. Codes that clustered around particular themes were gathered into domains. These were gradually built into a set of issues, each of which made a distinctive contribution to switching decisions for some students, but (as we discuss) also created problems for many of those who persisted. An overview of the results of our study follows next.

## *What Contributed Both to STEM Switching Decisions and to Problems for Those Students Who Remained?*

We discovered that three kinds of processes accounted for decisions to leave STEM majors. They were: (1) "Push" factors—problems in students' precollege and college experiences that made it difficult for them to persist with their original choices of majors and career aspirations; (2) "Pull" factors—perceived attractions or advantages that drew students to alternative majors and career possibilities—often while they struggled with problems in their original STEM majors; and (3) Pragmatic or instrumental considerations that made students' original choices seem less feasible or promising than the alternatives they were considering. Tracking these broad groupings and bundling together groups of factors that fell within them gave us an understanding of how switching decisions are reached by processes that are simultaneously underway over time. There was often (although not always) a "last straw" event or realization that triggered diversion decisions, but these were not "causes" per se.[3]

We used the term "persistence" to describe a spectrum of student thinking and behavior focused on the effort to continue in the major that they originally chose. Like switching, we found that this effort included complex "push–pull" decision-making processes. Students who continued in their STEM majors to graduation developed coping strategies and discovered and drew upon sources of support to help them persevere. Thus, persistence represents a range of student experiences whereby some students succeed in surmounting the same problems and situations that students who decide to switch do not.

---

[3]This explains why exit interviews may not provide useful explanations for a student's decisions to leave. The interviewers tend to hear about "last straw" incidents but may not learn about the substantive problems that led to them.

We found a hierarchy of 23 factors that contributed to students' decisions to switch out of their STEM majors and to problems for other students. Dominant among them were push factors, but there were also pull factors that offered alternative paths, and pragmatic considerations that prompted rethinking, especially of career options. Unexpectedly, we discovered an even larger problem among STEM majors who had persisted despite their experiences of the same problems that prompted some of their peers to leave. We used the metaphor of a "problem iceberg" to describe this "above and below the waterline" finding. We also found that no student left their major for a single reason. Indeed, the average number of factors contributing to each switching decision was 4.2. However, STEM switchers reported more of the same problems that were reported as troublesome to all students. Persisters reported an average of 5.4 concerns; switchers averaged 8.6. Thus, switchers could be defined as students who experienced more of the same issues that also troubled persisters.

Contrary to the common assumptions that STEM switching was largely caused by students' intellectual or personal inadequacies in face of academic challenges, a strong finding was the high proportion of contributors to switching decisions that arise from problems in course design, the poor quality of teaching, negative classroom culture, and difficulties in securing help with academic difficulties. In short, issues with aspects of STEM students' learning experiences were the dominant contributors to STEM majors' switching decisions. Criticisms of faculty pedagogy contributed to one-third (36%) of all switching decisions. However, complaints about poor teaching were cited as a near-universal concern for switchers overall (90%) and were the most commonly cited problems of persisting seniors (74%). Highly ranked factors contributing to switching decisions related to aspects of teaching or made invidious comparisons between the quality of the learning experienced in STEM classes and that offered by former high school science teachers, or by instructors in non-STEM disciplines. Indeed, concerns about pedagogical effectiveness, assessment practices, and curriculum structure pervaded all but seven of the 23 factors driving switching decisions. We found a strong similarity between the concerns of switchers and persisters in almost half of the 23 issues raised. The four factors contributing most to switching were two "push" factors—the effects of poor teaching by STEM instructors and overwhelm created by the heavy pace and load of course demands, and two pull factors—consequential loss of incoming interest in the STEM major while assessing a non-STEM alternative as offering more interest and a better education. All four of these issues were also highly cited by persisters. Seven issues were cited as shared concerns by over one-third of both switchers and persisters. In addition to the four concerns listed above, they were: choosing an STEM major for reasons that proved inappropriate; difficulties in securing academic help or advising; and inadequate high school preparation. An additional four concerns were shared by 20–30% of all switchers and persisters: financial problems experienced in completing STEM majors; and conceptual difficulties with one or more STEM courses. Factors that proved to have little or no significance for switching were: language difficulties with foreign instructors or teaching assistants, poor teaching by teaching assistants, poor lab or computer facilities, and large class size.[4] All of these variables were commonly suggested in 1990s reports as likely explanations for STEM losses. Thus, we asked students about them to explore their explanatory significance.

These findings did not vary by type of institution, varied only a little by discipline, and, as noted, the same problems were identified both by switchers and persisters. Indeed, we found no evidence to support the theory that students who leave and those who persist can be distinguished by individual characteristics, such as intellectual potential, moral attributes (such as diligence or effort), or motivation. Nor could we explain persistence difficulties for both groups of students in terms of the intrinsic "hardness" of the conceptual material. Students reported that material could be "made difficult" by poor teaching methods but was comprehensible when taught well. In six of the seven participating institutions, graduating seniors ranked "poor faculty teaching" as their highest educational concern; in the seventh school (with a high minority student population) the seniors ranked poor high school preparation first and poor undergraduate teaching second.

## *Dimensions of Problems with STEM Learning Experiences*

Here we describe the nature of the issues that contributed to switching decisions and that also created persistence problems for students overall as reported in the original TAL volume.

### Problems with Teaching and Learning

Several "iceberg" items referenced STEM learning experiences and the comparisons that students made with their class experiences in other majors. The three most frequently cited contributors to switching decisions were:

- Loss or dissipation of disciplinary interest in their STEM majors: ranked first (43%) of all reasons for switching; of concern to 60% of all switchers and cited by 36% of persisters (a push factor).
- A non-STEM major assessed as offering more interest or a better education: ranked second (40%) of all reasons for switching; of concern to 38% of all switchers and cited by 32% of persisters (a pull factor).
- Problems with STEM instructor pedagogy: ranked third (36%) of all reasons for switching; of concern to 90% of all switchers and cited by 74% of persisters (a push factor).

A related set of push issues derived from problems with curriculum design or structure. Most frequently cited were:

[4] How instructors taught large classes did contribute to switching. Class size, by itself, did not.

- Overload from the high volume of work and high speed at which this had to be completed in foundational courses ranked fourth (35%) for all reasons for switching; and was of concern to similar proportions of all switchers (45%) and persisters (41%) (a push factor).

Two less frequently cited problems related to learning experiences were:

- Lack of peer study support (a push factor that was more commonly cited by men than by women);
- Preference for the teaching approach used in particular non-STEM courses (a pull factor).

## Problems with Instructor Pedagogy

As noted, accounts of "poor teaching" in STEM classes were by far the commonest complaint of all switchers (90%). Problems with instructor pedagogy were also mentioned by far more persisters (74%) than any other issue and ranked first or second in persisters' concerns at all seven institutions. Issues with course structure, curricula, pacing of courses, assessment systems, and student workloads were also of concern to both groups, but were rated as less serious than concerns about the quality of STEM instructor pedagogy.

Students were very specific about what was wrong with much of the teaching they had experienced. Their most common complaints were that lessons lacked preparation, logical sequencing, or coherence, and that little attempt was made to check that students were understanding class content. Students were frustrated by instructors who seemed unable to explain their material sequentially, coherently, or break it down into sequences that would enable conceptual grasp. "Poor" teachers did not appear to understand the relationship between the amount of material which can be presented in a single class and the level of comprehension and retention which they could expect from students. Nor did they pitch their class materials or test questions at a level which was appropriate for students at their stage of conceptual understanding. Students looked for, and mostly did not find, illustration, application, and discussion of the implications of material being taught. They also found it hard to retain their interest in the subject where their instructors failed to present the material in a stimulating manner. STEM classes were often faulted for their dullness of presentation—predominantly straight lecture—and over-focus on memorization.

However, by far the most effective technique for dissipating student interest was the widely reported practice of reading materials from text books and what students referred to as "silent teaching." Here, the instructor writes on the board with his or her back to the class and addresses the students infrequently and minimally. Reports of these teaching methods came from every STEM discipline and were reported on every campus. The message that students inferred from these behaviors was that these instructors took no responsibility for their students' learning. Unfortunately, what many STEM instructors also communicated was an apparent disinterest in the class subject matter. This was a commonly offered and strongly stressed reason for dissipation of interest in a chosen STEM discipline and accounts for its top ranking as a contributor to switching decisions. Again, we were made aware of the "push-and-pull" nature of decisions to leave: poor teaching and the dullness of classes made it hard (even for students with a strong liking for science and mathematics) not to feel drawn towards disciplines where they experienced the excitement of intellectual exploration and debate. Unfortunately, students who most often reported that they were "bored out of the sciences" by the teaching in foundational courses were high-performing, multi-talented students who moved to non-STEM majors with greater intellectual appeal. As we discuss in Chap. 7, this is one of the contradictory effects of "weed-out class" teaching methods that we encountered.

Another set of criticisms focused on instructors' limited understanding of how students learn. The absence of apparent structure in the selection of class materials, the order, and logic of their presentation; and lack of fit between class materials, homework, and the content of tests, all suggested that instructors knew little about organizing their teaching around learning objectives that were shared with their students. Students rejected the argument that some STEM subjects matter is inherently tedious, and that learning it is just part of the hard grind to be expected in STEM majors. Rather, they viewed material as interesting when it is taught in an interesting manner. Some students who repeated a course offered contrasting examples of instructors who taught the same material, but, unlike the original instructor, structured their presentation and assessments so as to build comprehension. These students reported a marked difference in how much they had learned and how well they did.

Important elements in what students also saw as "good teaching" were openness, respect for students, encouragement of discussion, and a sense of discovering things together. Comparing teaching styles encountered in STEM courses with those experienced in courses outside of STEM disciplines were marked by dichotomies: coldness versus warmth, elitism versus democracy, aloofness versus openness, and rejection versus support. Both switchers and seniors recounted experiences from which they inferred that STEM instructors avoid contact with undergraduates, are indifferent to their academic problems, dislike teaching, and lack motivation to teach well. Some accounts of distancing behavior by STEM faculty involved sarcasm or ridicule. These behaviors also created classroom atmospheres in which students were afraid to ask questions or to "say something stupid." Dismissive or rejecting attitudes towards students who approached instructors with questions were interpreted as indications that instructors placed all responsibility for learning squarely on their students' shoulders or saw it as a matter for delegation to teaching assistants.

Seniors with undergraduate research experiences described the pleasant, open way in which faculty advisors treated students in a research relationship and contrasted this with their apparent indifference to their learning in the classroom. Not only did lack of student–teacher dialogue convey indifference, it also meant that instructors lacked feedback about how much their students were, and were not, learning. Again, students made invidious comparisons with what they experienced in particular non-STEM classes. Not only did students feel more comfortable talking about academic difficulties to their non-STEM professors, they often found the atmosphere in their classes intellectually stimulating. Again, the push–pull character of switching decisions is evident in these accounts.

The distancing behavior of STEM instructors in foundational courses had particular consequences for students who had learned their high school science and math in interactive settings that included both peer–peer and teacher–student dialogue. Learning to learn in supportive relationships left students especially vulnerable to culture shock in early STEM courses. Inability to evoke a supportive interaction from instructors prompted many students to doubt their ability and interest and undermined their confidence. These effects were particularly marked among women and first-generation students, including students of color and students from small high schools.

## Issues of Curriculum Design and Structure

When students explained to us what was "hard" about STEM majors, they commonly referred to the large volume of work required and the high speed at which it had to be completed. The challenge was physical and moral as well as intellectual: "To survive the constant round of assignments, problem sets, tests, lab work and reports required by several courses simultaneously, classwork had to take precedence over all other educational interests, personal relationships, athletic commitments, social life, paid employment, leisure, and sleep" (TAL, p. 93). Engineering students were at greater risk of switching for reasons related to pace and load issues than other STEM majors: they were a factor in 45% of engineering majors' switching decisions, compared with 25% for science and math switchers. Indeed, engineers complained that representing their majors as 4-year degrees was tacitly dishonest.

The rigid curriculum structure of STEM majors often did not accommodate errors in the choice of classes, and made no allowance for illness, accident, or family crises. Although seniors accepted that some of the high volume of work arose from the nature of particular disciplines, there was widespread suspicion that the pace of introductory classes was made deliberately faster than necessary. Courses were also defined as unnecessarily hard where material was not presented in a logical sequence.

Another aspect of design that commonly created learning problems was the misalignment of course elements. These included: an unbalanced selection of course content, tests that did not reflect the content presented, failure to distinguish between content of greater or lesser significance, or to coordinate with colleagues in content sequencing. Students' commonest complaints about misalignment of labs described mis-fit between classes and lab syllabi, labs that were out of sync with class content, and the discrepancy between lab credits and the amount of lab work required.

The contribution of curriculum design issues to the process of switching often began with feeling overwhelmed—experienced both by students who were well-prepared, and more acutely, by those who were not. This was commonly the start of a downward spiral in which students fell behind and attempted to repeat classes. Often, it also prompted a growing awareness that persistence involved an unpalatable choice about the kind of education that students wanted and whether they were prepared to give up whatever else in life they valued.

## The Significance of Grades

Almost one-quarter (23%) of reasons for switching involved problems with grades, mostly in the first two semesters. One-third of all switchers and 13% of persisting seniors cited grades' issues as having been, or continuing to be, problematic. Conventional wisdom might predict that grade-related problems would figure more highly in switching decisions, and rather less in the accounts of persisters. However, predicting which students are more or less likely to leave STEM majors by establishing relationships between performance scores and persistence outcomes is difficult because this does not take into account how people respond to the grades they receive. Grades are not objective, neutral, facts about people; they are labels to which people react emotionally and that may prompt behavioral and identity adjustments. We observed that students who are apparently competent to complete STEM majors are often lost because they interpret particular grades as indications that they are unfit to continue. Such losses include some highly talented students, as exemplified by the finding at one institution that women left STEM majors with grades that were higher than the average grades both of men who left and those who remained (McLelland, 1993).

Some students do abandon their original intention to major in a STEM discipline for grade-related reasons: their preparation is not as good as their high school grades would suggest; they get poor grades because they do not settle down to college work quickly enough, or simply do not work hard enough. Our data provide examples of students for whom one or more of the above statements would, by their own accounts, be accurate. However, at this level of demonstrated mathematical ability, we did not expect, and did not find, these to be dominant factors in our informants' relationships with grades.

By far the most common way in which grades affected switching decisions was by the shock of introductory course grades that were far lower than those to which students were accustomed to receiving in high school. What, then, allowed some students to persist despite discouraging grades while others, whose grades were similar or even better, did not? A key problem was that, throughout their precollege education, students had been encouraged to use grades as extrinsic measures of comprehension, progress, and self-worth. Over time, grades became defining elements in identity and self-esteem. As with all extrinsic measures of self-worth, they made the self-image vulnerable to external definition. As one senior explained, "If I get As, I must be an A kind of person. If I get a C, I am a C." A tough grading policy for first-year students and sophomores may have been intended to force "lazy" high school graduates to deal with the hard realities of academic work. However, it also had the potential to undermine the sense of self-worth of students with good ability and study habits. As seniors explained, in order to survive in the major, students must rethink the personal significance of grades.

"Curve grading," which is widely practiced in foundational STEM (and many later) courses, tends to reify grades and detach them from their pedagogical purpose. This makes it more difficult for students to break with old ways of thinking. Indeed, it tends to reinforce them. Curve grading also raised concerns about misalignments between the grade awarded and the level of comprehension students thought they had attained, the standard of work they had demonstrated, or the amount of effort they had expended. Seniors expressed concerns about the contrary situation whereby curved grades allowed students who knew they did not really understand the material to end up with a reasonable grade. By forcing separations between comprehension, performance, and reward, curve grading can make the grades received seem totally arbitrary to students. As one student observed, "Grades are like weather. Sometimes it rains; sometimes it doesn't."

Students questioned the purpose and ethics of grade manipulation which made many first-year students feel useless, hopeless, and incompetent. It also reinforced the tendency to focus on tests and how to pass them. First-year students developed resentments about "unfairness" because trying hard did not necessarily lead to success and went unrecognized by instructors. In the face of poor early test grades, many first-year students fell back on the same strategies that they learned in high school: they tried harder, they crammed, tried to work out what the teacher "really wants to know," and they cheated. In classes where work demands seemed unreasonable, and the assessment system implied that the main objective was to beat classmates to a good grade, cheating was discussed by seniors in focus groups as logical and predictable. It was not, however, legitimated, because, in a curve-grading situation, it gave one's competitors an edge.

We found that students' responses to grades in their early classes that were much lower than those to which they were accustomed were critical to their persistence in the major. This was as true for students who were doing reasonably well as it was for students who were academically marginal. As seniors attested, unless students were able to change their study strategies and maintain a sense of their own worth in face of disappointment and frustration, a plummeting sense of self-esteem accelerated a downward spiral out of the major. Typically, this process began with discovery of being underprepared for the level of work demanded in early STEM classes; it continued with the shock of grades by which students who had thought of themselves as competent redefined themselves as incompetent. When old high school strategies proved less effective than hitherto in securing higher grades, self-doubt, panic, and depression increased. Some students sought help; many did not because (young men especially) believed they were supposed to "hack it" alone. Some students found guidance and support from senior peers, dorm mates, TAs, advisers, and sometimes faculty, but many did not. As their sense of hopelessness increased, students began to skip classes, isolate from peers, do less work, understand less and less, perform even worse in exams, fail, and leave. Switching thus allowed some students to escape from a situation in which they continually "felt bad" about themselves.

Many seniors recounted how close they came to switching and how a timely intervention by faculty, senior students, or peers was critical to their decision to stay. For those who found academic or personal help at the right time, the downward spiral could be reversed. Seniors described it as vital to survival to accept an average grade and trust their own judgment about how well they understood the material.

## The Competitive Culture and the "Weed-Out" Tradition

Dislike of the competitive culture of STEM classes was a factor in 15% of reasons for switching and a source of complaint for 28% of switchers overall. However, it was mentioned as a problem by only 9% of seniors. This was consistent with seniors' observations that the uncomfortable competitive atmosphere is largely confined to the first two years. Early in junior year, seniors reported a dramatic change in faculty's approach to teaching, including their use of collaborative learning methods, and more personal treatment of their surviving majors. They concluded that the competitive atmosphere of their introductory classes was deliberately engineered. Many switchers also came to the conclusion that students were set up to fight each other for grades.

Competitiveness was reported to be greatest in majors leading to professional careers—mainly the health professions and engineering. For students with these career aspirations the competitive climate was often a factor in their change of career plans. However, it also diminished the educational experience of many students with other career plans. Curve grading was portrayed as the engine driving competition because it forces students to compete with each other by exaggerating fine degrees of difference in performance. Seniors criticized forced competition as counter-productive. It perpetuates the high school habit of focusing on grades rather than on understanding. It also distorts normal social interactions, creating isolation and mutual suspicion, and it perverts what seniors regarded as a preference for cooperative, collegial work that they expected to find in professional settings. More women than men found the competitive ethos of STEM classes alien or offensive. They avoided those classes known to be very competitive and offset their effects by forming study groups.

The competitive culture of foundational STEM courses and indeed many of the problems cited by students in the structure, teaching, and assessment of these courses are characteristics of what students described as "weed-out" courses. Weeding students out is a common, long-established practice in a number of academic disciplines. However, it is dominant in STEM majors, where it has a semi-legitimate status and is part of what gives STEM majors their reputation for hardness. Pedagogical strategies, deployed as tests for ability and character, are a mechanism by which STEM disciplines seek to identify students who are presumed to be the most able and interested to continue in their majors. However, there are no references to weed-out courses in official literature, and deans and faculty may be evasive about their existence. As with studies of hazing (Ambrose, 1966; Festinger, 1961; Nuwer, 2001, 2004), practices that are overlooked or denied may be perpetuated because they are perceived as serving valued ends that are thought difficult to achieve by other means.

## Finding Timely Help

As indicated earlier, in face of these and other difficulties, survival in a STEM major often hinged on whether a student knew that it was appropriate to seek help, knew what help was available, and found an appropriate source in a timely manner. What students sought were: help in understanding particular academic material; accurate information about required courses and course sequencing for their degree requirements; and practical help or advice with problems that impinged on their academic performance (e.g., problems with finances, employment, and health). Among all of the factors contributing to attrition, student difficulties in getting appropriate help is the one that most clearly derives from problems with institutional provisions. We found that most students had experienced some problems with support systems at all seven institutions. However, the perceived effectiveness of advice and counseling services varied as much within campuses as between them. On every campus, we found gaps, overlaps, and confusion in the division of responsibility between departments (or colleges), central advising services, and advising programs for underrepresented groups. One of the most difficult problems for first-year students was to learn quickly enough how the campus system of advising, counseling, and tutorial services worked in order to prevent small problems from becoming large ones.

Failure to find adequate advice, counseling, or tutorial help contributed to one-quarter (24%) of all switching decisions was a source of frustration for three-quarters (75%) of all switchers. It was an issue raised by half (52%) of all persisters for whom it was the second most commonly cited concern. First-year students are at particular risk because they often need several kinds of help simultaneously—accurate information about choice and sequencing of classes, setting up finance and work-study plans, tutorial help with early academic difficulties, and personal encouragement. They tend to see their problems as interrelated and become frustrated with systems that assign each kind of problem to a different agency. It would be easier to dismiss their difficulty in finding appropriate sources of help as a temporary problem were it not for the testimony of half of the seniors who reported that confusion and gaps in provision of support had continued throughout their academic careers.

The search for an advisor with whom to discuss their progress, problems, and options was of special importance to students thinking about switching. They need to know whether the problems they had experienced are to be expected at this stage in their major, and should be tolerated as such, or whether they need to rethink their career plans. They found that instructors who can give advice about their discipline may know little about alternative majors and careers. Students who raised the possibility of leaving a STEM major with advisors were often encouraged to switch, whether or not this was the right decision.

Students understood that instructors probably get no training for their advisory role and appreciated that professional advisors work with a complex requirements system. However, what students wanted from advisors, above all else, was personal attention. Some complained that instructors were negligent about keeping office hours, and, for some students, the system of assigned advisors had completely broken down. Indeed, some of the problems they experienced with erroneous course advice arose from the cursory nature of advisory interviews. The difficulties of under-classmen in persuading instructors to take an interest in them may arise because of a tension between their roles as academic advisors and that of disciplinary gatekeepers. Students' accounts illustrated a fundamental clash of perspectives: students do not perceive the need to distinguish between advising, counseling, and tutoring functions, while faculty tend to resist all but the most formal of these functions—course advising. A painful experience with a professor at an academic crisis point was often the "last straw" incident in the process leading to switching. Seniors also recounted how close they had come to switching following a discouraging encounter with their faculty advisor. However, some switchers faulted themselves for not seeking help from instructors, advisors, or tutors, and speculated whether they might have survived in the major had they done so.

Deans and faculty also ask why students do not make better use of the support systems available to them, and we found a number of reasons for this. First-year students, especially, were intimidated by the unapproachable demeanor that some instructors customarily project toward under-classmen. They quickly picked up the message that it is more appropriate to approach TAs or tutors for help with academic problems. They also feared humiliation if they asked, "dumb questions." Many seniors remained wary of personal contact with faculty and, rather than trying to guess what professors might consider an appropriate matter for consultation, preferred not to approach them at all. Students who work may also not be unable to attend scheduled office hours. Notwithstanding these discouragements, to get the best out of campus support systems, seniors advocated: learning to be assertive and persistent; making contact with instructors; double-checking all information about graduation requirements; soliciting first-hand knowledge from more experienced students; and making full use of TAs, tutoring services, and refusing to be brushed off with unsatisfactory answers. In short, they advocated a proactive consumer approach to STEM education.

## *Contributors to Switching That Arise Outside of STEM Education Experiences*

Not all the issues that contributed to switching or to the ongoing problems of STEM seniors derived from aspects of their experiences with STEM education. Some predated college; others were external to it. However, these variables also intersected with problems within college.

## Reasons for Choosing STEM Majors That Did and Did Not Enable Persistence

Choosing a major for reasons that subsequently proved inappropriate or insufficient was not (at 14%) a major contributor to switching decisions. However, ill-considered choices created problems for 83% of switchers, and also for 39% of seniors who described themselves as making the best of choices that, with hindsight, were poorly grounded. Students usually offered several reasons for their choice of a STEM major, with some combinations proving more durable than others. Persisting seniors were more likely than switchers to have chosen their major (and/or their career path) for reasons that included intrinsic interest. Switchers were more likely to have chosen for reasons that included: the influence of significant others; materialistic considerations; and doing well in mathematics and/or science in high school. Their choices were often less well informed about the nature of the major and related careers. That said, many persisters' choices were also prompted by materialistic considerations, pressures from people close to them, compromises, and limited understanding of what a major might entail. Persisters, however, were less driven than were switchers by factors other than intrinsic interest. It seemed to matter less what other reasons students had for their choices, so long as one strong element in their decision was an intrinsic interest in the disciplines that comprised the major and the careers to which it might lead.

A marked difference between the sexes found in this study lay in the reasons for their choice of majors. Women were about twice as likely as men to have chosen a STEM major through the active influence of people significant to them (especially parents and teachers) or by following family career traditions (notably, in medicine and engineering). Among switching women, these influences were 25% of reasons for their choices, but were also evident among women who persisted. For both groups of women, the influence of significant others was greater than for both switching and persisting men. By contrast, men were almost twice as likely as women to cite "being good at mathematics and/or science in high school" as a reason to choose a STEM major. Given the limitations of our data, we did not know whether they were actually more able or better prepared than women entrants. However, it was clear that far more young men than young women felt confident in their readiness to undertake higher level science and mathematics. Switching men stood out, both against non-switching men and against all women, as more apt to make this the basis for their choice. However, choices based on perceptions of aptitude for science or mathematics proved particularly unreliable if coupled with inadequate prior understanding of the nature of the chosen major.

The contrasts between switching and non-switching women were also of interest. The choices of switching women were the most materialistic of any group, again reflecting strong family promptings about good career opportunities and a sound financial future. They also showed less evidence of intrinsic interest in the sciences than women who persisted. By contrast, female persisters stood out as more highly motivated by interest than any other group in our sample, and as the least driven by material considerations.

For some students, intrinsic interest was coupled with a desire to enter a particular field because of commitment to a wider social purpose. Attaching one's career goals to a clear altruistic purpose in some practical form appeared to sustain interest and momentum through periods of difficulty. Certain themes recur in the kinds of socially directed careers that students aspired to: service to others (such as adaptation of particular skills or technologies to help groups with special needs); protection of the environment (wildlife, eco-systems, climate, etc.); and the promotion of international peace. Ninety-one percent of altruistic reasons for choosing STEM majors were expressed by men of color and by women of all races and ethnicities. Persisting senior women were also more likely than their male peers to rank materialistic goals lower than a desire to work at something they cared about—whether as a matter of personal fulfillment, or in pursuit of a valued social cause. The goals of students of color often included making a long-term contribution to their families and communities.

## Inadequacies in High School Preparation

Given the apparently adequate SAT and ACT scores of our sample, we did not expect problems with the quality of high school preparation or conceptual difficulties sufficient to prompt switching. Both proved to be the case: inadequate preparation accounted for 15% and conceptual difficulties 13% of contributions to switching. However, the effects of inadequate high school preparation were the most common contributor to *early* switching decisions. They were cited as problems by approximately 40% of all switchers and persisters, and more often experienced by students of color.

Accounts of under-preparation were of two types: deficiencies of curriculum content and subject depth, and failure to acquire appropriate study skills and habits. Some students had received no high school calculus and/or described their science or mathematics knowledge as insufficient for their first college classes. Other deficiencies were: no introductions to theoretical material or analytic thinking; no opportunity to take college preparation classes (including Advanced Placement); and lack of laboratory experience. A subtle form of under-preparation was having to unlearn a tendency to see material in modular form, without transfer, connections, or framework. These students were shaken to discover how poor their high school preparation had been in comparison to other first-year students. Their above-average performances in standardized tests gave them little indication of their insufficient readiness for college work until their first gateway classes.

Students who attended poorly endowed schools in working-class areas (including small rural schools, large inner-city schools, and reservation schools) were those who most often cited insufficient resources and limited access to well-qualified teachers as salient features in their high school science and mathematics education. Even with good teachers, an able student at a school with multiple social, financial, and educational limitations was still at risk. These students knew they had received a substandard education and many of them knew this while they were in high school. However, they described their parents and communities as often unaware of their schools' inadequacies.

Many of the students of color that we interviewed who had been outstanding students in high schools serving predominantly minority populations faced an uphill battle in the competitive culture of the university. They were shocked to discover that they had overestimated their readiness for college work. Treated as special by their high school teachers, they entered STEM classes both underprepared and overconfident in their ability to undertake them. They were at a loss to comprehend how, in a single semester, they could have gone from the top of their high school class to the bottom of college mathematics or physics classes: Although many white students also suffered from inadequate academic preparation or entered STEM courses with an inflated view of their readiness for the level of work required, these problems were much more common among students of color. Among students of color who switched, these problems were almost universal.

However, some underprepared students came from well-resourced high schools. They included students with a good natural ability in mathematics who had failed to learn the study skills and discipline needed for college-level work. Students who found mathematics and science easy described how elements in the culture of school, home, and peer group mitigated against their acquiring these critical aspects of college preparation. They scored high grades with minimal effort and were accustomed to praise from teachers, family, and peers. Teachers made limited demands and set achievement targets below their capabilities, so that they were neither stretched nor challenged. They did little homework, or did it at the last minute, and were often left to their own devices while teachers worked with others. In college, they quickly discovered a gap between their incoming expectations and their ability to perform as required. Viewing grades as an acknowledgement of their talent rather than their efforts, and with little experience of coping with frustration or set-backs (compared with peers who always had to work hard) they had no psychological defenses against lower grades. Thus, mathematically able students who had learned to underachieve in high school were often early casualties of weed-out courses.

Among all switchers, 34% reported a sharp drop in confidence caused by their expectation of high (or easy) grades, and shock at receiving the lower grades that are traditional in introductory STEM courses. Twenty-three percent of switchers reported that this had contributed to their decision not to continue, and 13% of seniors also described their struggles to overcome the initial blow to ego. Learning not to interpret grades as personal criticism was essential to surviving the discovery that they were not adequately prepared for college. Overall, 40% of both switchers and persisters described struggles created by under-preparation, usually from the first week in college. Those who dropped out or failed classes because of underpreparation were in the first wave of attrition from STEM majors.

### Career and Lifestyle: Time and Money

The remaining factors contributing to STEM switching reflected students' perspectives on their university education as a preparation for life and work. They comprised: rejection of careers based on STEM majors or of the lifestyles they were presumed to imply; choice of non-STEM careers that seemed more appealing; doubts that the rewards of an undergraduate STEM degree would compensate the effort required to complete it; and financial difficulties in completing a STEM major in the length of time required. These considerations again illustrate the "push and pull" nature of the processes leading to switching.

***Rejection of STEM Careers and Lifestyles*** Twenty-nine percent of contributions to switching decisions, and 43% of all switchers' concerns, reflected doubts about the kinds of jobs that would be available on graduation and the lifestyles they were perceived to involve. One-fifth (21%) of STEM seniors expressed similar anxieties. Dominant concerns were that work available to graduates—particularly without a higher degree—would be unfulfilling, offer little responsibility or autonomy, and would make demands at the expense of valued life interests. These thoughts often emerged from internships, contacts with working professionals, or observation of academics at work.

Many engineering majors entered with scant idea of what professional engineers do. As they developed a picture of engineering, some students questioned whether they would like the work. Internships in junior year often settled the matter. Conversely, some seniors reported that their first professional experiences were critical in their decision to remain. Clearly, hands-on experience enabled appropriate career choices. Some science and mathematics switchers who left, in part, because they rejected the prospect of graduate school and the academic life were uncertain what else they might do. Mathematics switchers had least sense of what careers might be open to them: on every campus, we encountered high-achieving, predominantly female, mathematics switchers who had been unable to develop clear career goals. Science and mathematics switchers gave more complex reasons for their career-related decisions than did engineering switchers. These included: wanting a balanced life, valuing work for its intrinsic satisfactions and social purposes more than its material rewards, and rejecting careers and projected lifestyles in corporate science. Some students anticipated that STEM careers would require adoption of an unacceptable *persona*. This presumption was grounded in their experience that, in order to survive in the major, they and their peers were already developing a *persona* that was introverted, single-minded, perfectionist, socially inept, and alien to them.

***The Appeal of Alternative Careers*** About one-quarter (27%) of switching decisions were made with non-STEM alternative career directions in mind, and 17% of STEM seniors were also considering work in non-STEM fields beyond graduation. These students had clear ideas about the kinds of work or work contexts that appealed to them. Common preferences were work that was intrinsically interesting or served a social purpose that they cared about—for example, work in environmental protection and teaching. Such aspirations were most often expressed by white women and students of color. Some were seeking careers that combined STEM knowledge and skills with those of other disciplines. This was a marked trend among high ability students who undertook double majors, including cross-disciplinary blends. Twenty-one percent of switchers and 19% of persisters who were considering teaching science or mathematics described this as a deviant ambition that they pursued despite the disapproval of STEM faculty, families, and peers.

***Weighing the Hedonistic Calculus*** By contrast, one-third of switchers (31%) rejected their STEM major partly because it did not lead to financial rewards commensurate with the effort required to complete the degree. One-fifth of all seniors also raised this issue. Switchers explained that, had the STEM educational experience been more fulfilling, they could have tolerated its discomforts. However, where their interest had not been engaged, they looked for post-graduate rewards as compensation for "sticking it out." Where such rewards seemed uncertain, they considered other majors and careers. Seniors familiar with this logic described it as weighing the "profit-to-grief" ratio. The perception that the career opportunities and material rewards of completing STEM degrees were not worth the considerable effort required to acquire them contributed to 40% of switching decisions by science and mathematics switchers and 31% of engineering switchers.

From the outset, engineers expected more in terms of material rewards than other STEM majors and were willing to put up with the discomforts of engineering majors because they seemed to promise that all the effort would pay off in the end. Engineers who entered with expectations of good material rewards also spoke of their undergraduate education in return-on-investment terms. Reflecting on the employment difficulties of recently graduated friends, they saw uncertainty about jobs and incomes in the early 1990s as a reason to switch to career paths with better prospects of employment and rewards. Science and mathematics majors expected much less in material terms than engineers, but also wanted work commensurate with the considerable effort which their majors had demanded.

Concerns about the job market or finding employment without an advanced degree were a major topic of discussion among seniors. Seniors with concerns about an uncertain future sought to increase their chances of profitable work by raising their GPA found inside tracks with prospective employers through internships, developed flexible career plans, and looked for new market niches. Some seniors were considering graduate school as one way to delay their job search. Concern about the limited job availability without an advanced degree also prompted the intention of switching disciplines after graduation.

### Financial Problems and Their Consequences

Our study was undertaken at the end of a period (1979–1992) in which both the federal government and state legislatures had substantially reduced their contributions to higher education. Mortenson (1995) estimated the average annual increase in college fees averaged two to three times the annual rate of inflation and reported opportunities for higher education in this period becoming increasingly unequal. Our findings illustrated the direct, personal consequences of this trend for many of our informants. Students experienced fewer grants, more loans, and more time spent working while universities and colleges were increasing tuition and fees. The resulting fall in enrollments triggered more increases in costs to students. Over one-quarter (27%) of all the students in our sample reported financial problems that were serious enough to influence their academic progress or career plans. Financial difficulties were a factor in 17% of switching decisions and were of concern to 30% of all switchers and 23% of seniors. Both engineering switchers and seniors reported more financial difficulties than science and mathematics seniors. Many engineering first-year students and their families had expected that engineering degrees would take 4 years rather than the 5 years that was becoming the norm. Financial plans made in accordance with this expectation often fell apart.

Less than one-third of all participants were funded by scholarships, sponsors, or private resources. Approximately two-thirds (63%) had taken out loans and over half (56%) were meeting some of their education and personal expenses through work. The average was 18 h per week, although some students at the four public institutions worked 30–45 h a week. Financial aid was universally reported to be difficult to get, with an application process that was overly complex, and with many limitations or exclusions. In the opinion of seniors, getting any kind of financial assistance, including loans, had become increasingly difficult at a time when tuition, fees, books, and the cost of living were all increasing. Some students who were excluded from financial aid programs, or limited in what they could get because of their family's income level, reported little or no financial support from their families.

Students were aware of, and broadly supported, national efforts to recruit more students of color into STEM majors. However, against the background of steep increases in higher education costs that students were meeting themselves, focus group discussions (especially at state institutions) revealed feelings of resentment toward students of color who were believed to be receiving public funds to which white students in financial need had less access.

Being employed commonly lengthened the time taken to graduate. This was compounded where financial aid was refused in the final year(s). Problems with degrees that took more than 4 years to complete figured in only 8% of switching decisions but created difficulties for 20% of seniors. STEM degrees were thought particularly hard to reconcile with student employment because they make greater time demands than other degrees: they often require more credits; their curriculum structure is more intense; and laboratory courses require large time commitments. Working students also felt at great disadvantage in courses with intense competition time spent at work and travelling to and from work reduced the time available for study. Those who worked more hours than was consistent with good scholarship were at constant risk of failing classes or earning poor grades. Working students were constantly forced to choose between academic commitments and their need for financial survival. Students who paid most of their own way through school had less margin for error than did those with scholarships or family assistance. Mistakes in the choice of classes or course failures had disproportionately negative effects because they could not afford to repeat classes and risk subsequent failures. The constant strain of juggling time and energy between the demands of work and school was cited by 17% of students as a major factor in leaving STEM majors.

Financial problems had other consequences: some students chose a particular major or institution largely because they were offered financial help; some stayed in majors in which they had lost interest rather than lose a scholarship; and the debt burden accumulated by those who chose borrowing over employment had profound consequences for some career choices. Notwithstanding the belief of some white students that students of color had an unfair advantage in getting scholarships or financial aid, black and Hispanic students were overrepresented among those whose decision to switch was directly related to their financial difficulties. It became clear that students of color from poorer communities were not only more at risk of switching majors, they were, as both Porter (1990) and Rotberg (1990) also reported, at greater risk than white students of dropping out of university altogether.

To a marked degree, students accepted the responsibility of contributing financially to their own education. What they found hard to accept was the inadequacy, inequity, and complexity of the financial aid system. They also expressed anger at what they saw as political and institutional insensitivity to recent increases in the proportion of higher education costs which students must meet by working and borrowing, and about the academic, career, and personal consequences of this situation.

## Explaining the Loss of Able Women

Beginning in the 1980s, the earliest studies of losses from STEM majors reported lack or loss of confidence and lowered self-esteem as significant contributors to the failure of qualified women to enter or complete STEM majors.[5] The TAL study also explored why women of high demonstrated ability might lose confidence after entering STEM majors, and what else might explain their greater vulnerability to switching.

A gender breakout of the "iceberg table" results yielded some clues. The strongest difference between men and women was found, not in their reasons for leaving STEM majors, but in their reasons for entering them. As recounted, women differed sharply from men in that the influence of significant adults was a more important factor in their choice of STEM majors than it was for men. Women also entered with

[5] Reviewed in *Talking about Leaving*, pp. 235–236 and 239.

less well-developed views of what they wanted out of college, less clear ideas about what drew them intellectually to any major and what they wanted from it in career terms. Women were also more often altruistic than men in their career goals and were more likely than men to switch in order to pursue careers offering more humanitarian or personally satisfying work. Indeed, the educational focus of male STEM majors was more instrumental than that of women in that men were more willing to place career goals above considerations of personal satisfaction. Women showed greater concern than did men to make their education, career goals, and personal priorities fit coherently together. More women than men switched, in part, because another major offered greater intrinsic interest, a better overall educational experience, or because the career options or lifestyle which STEM majors appeared to offer were less appealing. Similarly, more senior women than men felt that another major might have given them a better education or expressed doubts about the personal and job satisfactions they could expect from STEM careers.

In the interviews, many women described being held back by a low self-assessment of their abilities and their difficulty in knowing that they were "doing okay" without the reassurance they had been accustomed to from high school teachers. There was little gender difference in switchers' accounts of the contribution "poor teaching" had made to their switching decisions. However, their definitions of "good teaching" diverged around what they expected of instructor–student interactions and the consequences of their unmet expectations. More women than men entered college with an expectation of personal relationships with instructors. This was embedded in gendered definitions of "good" and "bad" teachers: women more often stressed the importance of a teacher's personal behavior toward them, and defined "bad" teachers as "unapproachable," "impersonal," and "intimidating." "Good" teachers were "approachable," "nice," "friendly," "patient," "interested in how you respond," and "present the subject in a friendly manner." Men were more concerned with instructors' effectiveness in presenting course material: the "good" professor is "enthusiastic," "interesting," can "explain well," "uses good analogies," "stresses application," "allows questions," and "knows whether the students comprehend." More often for women than for men, engaging the teacher in a personal dialogue determined the ease with which they could learn and become confident in the adequacy of their performance. Failed attempts to establish a personal relationship with instructors was a major loss to many women, and also to those men whose high school teachers gave them personal attention and fostered their potential. These students consistently used the word "discouragement" to describe their reaction to the experience of weed-out classes, especially where instructors refused to interact with them as individual learners: Women who were looking for encouragement to bolster their self-confidence, but who could not evoke it from instructors tended to feel discouraged (TAL, p. 270). Senior women testified that developing an independent sense of their own ability and progress had been vital to their survival.

We noted that the process of losing confidence often began before college entry. Notwithstanding the encouragement of teachers, women described difficulties in giving themselves permission to choose STEM majors, referencing the dampening effect of cultural messages that suggested that women couldn't or shouldn't do science. On entry, while in the process of considering possible majors, some women reported overt discouragement by faculty advisors against attempting particular STEM courses. Many women described a process in which they had learned to set their aspirations at a level lower than their abilities, and to wonder whether they "belonged." This was far more rarely an issue for their male peers. Women's doubts manifested as: less assertiveness in asking for what they needed, less inner-strength to cope with set-backs, and more dependence on others for reassurance. Another common precollege deficit was limited hands-on technical or laboratory experience. It meant that women approached technical tasks more tentatively than they subsequently found was warranted. However, it gave male peers a psychological advantage and was another source of women's fears about incompetence and doubts about belonging. Male appraisals of their academic worth often had negative consequences for their identity as women. The problems of belonging and identity are linked, because the qualities that women felt that they must demonstrate in order to win recognition for their right to belong (especially "smartness," assertiveness, and competitiveness) raised the anxiety that recognition could only be won at the expense of (then prevailing) notions of "femininity." In this double-bind situation, women felt they could only win male acceptance, in academic terms, by losing it in personal terms.

It was important to establish whether switching decisions or the problems of women who remained in STEM majors were related to discriminatory behavior by STEM faculty, TAs, or fellow students. Out of the 173 women interviewed, eight women (four switchers and four seniors) reported direct experience of discriminatory behavior, rudeness to all the women in a class, or sexually inappropriate behavior. More common were an array of more subtle experiences by which some instructors conveyed the message that women were unwelcome in their major. Women felt excluded from some class activities, and some instructors set a misogynist tone by encouraging, ignoring, or failing to check the rudeness of young men towards the much smaller number of women in their classes. The degree to which instructors did (or did not) tolerate rude classroom behavior toward women was transmitted to their teaching and laboratory assistants, who then repeated these patterns. Indeed, bad behavior by male TAs and class peers had a direct impact on women's discomfort levels. Rude behavior by male classmates was a constant, daily source of stress for many women, especially in engineering, physics, chemistry, and the applied sciences. We learned both from women and sympathetic men that women were subjected, on a daily basis, to unkind and sexually suggestive remarks and jokes intended to make women feel uncomfortable and unwelcome. High grades earned by women were explained in ways that did not concede intellectual merit, such as by hard work rather than "smartness," as "freak" occurrences, or by flirting with instructors and TAs. Women were angry when men behaved in ways intended to make them feel unwelcome and devalued—both as women and intellectual competitors. Although they rarely described these experiences as a direct contributor to decisions to switching decisions, in classes where they were prevalent, they were a background factor that made decisions to leave easier when other issues came into play.

During TAL data collection in the 1990s, women who were numerically isolated in most STEM majors experienced male peer hostility on a daily basis. Whether these damaging behaviors continue two decades later is among our research questions for the present study. However, in 1997, we concluded that, "the prospect of four years of isolation and male hostility on the one hand, and the abrupt withdrawal of familiar sources of praise, encouragement and reassurance by faculty on the other were the most common reason for the loss of self-confidence that makes women peculiarly vulnerable to switching," including women of high ability (p. 271).

## Persistence Difficulties for Students of Color

We analyzed the situations of students of color by racial or ethnic group to avoid global generalizations about "minorities" (cf., Museus, Agbayani, & Chang, 2016). That said, four of the risk factors in the iceberg table were more likely to be cited by students of color overall than by their white peers:

- Inappropriate reasons for choice of a STEM major (35% vs. 6%).
- Conceptual difficulty with one or more STEM subject(s) (31% vs. 5%).
- Inadequate high school preparation in mathematics, science, and/or study skills (25% vs. 11%).
- Shift to a more appealing career option (33% vs. 23%).

By contrast, white students more commonly cited three other factors in their switching decisions than did students of color overall:

- Lack or loss of interest in their STEM major (49% vs. 29%).
- Poor teaching by STEM instructors (42% vs. 21%).
- Curriculum overload and over-fast pace (41% vs. 19%).

We turned to the text data to explore the reasons for these differences. However, it is important to highlight three other distinctions. First, students of color tended to blame themselves rather than instructors or institutions for their difficulties. Second, as with women (with the exception of black women), the decision to leave a STEM major was often preceded by loss of confidence. Third, the process of switching, which was especially painful for students of color, often had long-term consequences that were more serious for them than for their white peers.

Choosing a STEM major for reasons that subsequently proved inappropriate was mentioned as a problem by 94% of all students of color, by 35% as a reason for switching, and was one of the strongest overall differences between white and non-white STEM students. Some had been encouraged to enter majors for which they had insufficient interest or preparation through the active influence of others, including initiatives to recruit more students of color into STEM fields by offering scholarships. Families and communities also encouraged students of color to make choices that reflected social rather than personal career goals. Some parents and teachers were mistaken about students' actual interest and abilities or had limited understanding of what levels of ability and preparation were required. Hispanic students reported that, because engineering was seen in their community as synonymous with success, they had been encouraged to pursue engineering to the exclusion of other options. Few Asian-American students chose their major without reference to family priorities: job security and following parental occupational paths were stressed over personal interest. Indeed, respect for parents' wishes and a strong desire to realize them were major reasons why Asian-American STEM majors were less likely to switch than any other group of students. However, the dominance of family over individual choice was a major contributor to high levels of dissatisfaction among Asian-American seniors.

Although many white students also chose their majors inappropriately, suffered from inadequate preparation, and entered with inflated views of their readiness to undertake the level of work required, these problems were far more common among students of color. Among students of color who switched, they were almost universal. However, we found that students of color also experienced a set of unique problems that made it harder to persist even where students *were* well prepared, and their field appropriately chosen. As summarized in Table 1.1, whether students in any particular racial or ethnic group had to contend with each issue is indicated as a "Yes" or "No."

In order to succeed in STEM majors, students of color often found it necessary to alter or override cultural values that were important to themselves, their families, and their communities. Those unable to ignore or discard cultural values that hindered their academic success were vulnerable to switching majors or abandoning the attempt to attain any degree. Interviewees of color reported that white instructors and students appeared to be unaware of the extra layers of difficulty with which they had to contend. Table 1.1 also clarifies another reason why it is an error to treat all non-white students as if they are a homogenous group.

**Table 1.1** Comparison by racial/ethnic group of cultural values expressed by non-white students

| | African-Americans | | | | Native Americans | |
|---|---|---|---|---|---|---|
| Value | Inner-city | All others | Hispanics | Asian-Americans | Reservation | All others |
| Obligation to serve community | Yes | No | Yes | No | Yes | No |
| Obligation to be a role model | Yes | Yes | Yes | No | Yes | No |
| Conflict between student/family roles | Yes | No | Yes | No | Yes | No |
| Educational goals defined by parents | No | No | No | Yes | No | No |
| Encouraged to be self-assertive | No | Yes | No | No | No | No |
| Encouraged to be self-reliant/autonomous | Yes | Yes | No | No | No | Yes |
| Supportive, effective peer group culture | No | No | Yes | No | Yes | No |

## Testing the *Talking about Leaving* Hypotheses

Ethnographic studies are used to generate rather than test hypotheses. Although the capacity to work with large samples enabled by text analysis software has increased the strength of ethnographic findings, it broadly falls to subsequent research to confirm, augment, or refute them. Since TAL's publication, studies of discontinuation from STEM majors have tended to concentrate on particular problem areas, on single disciplines, or on the efficacy of particular forms of remedial intervention. How does this subsequent body of work clarify the contributions to losses from STEM majors of single or combined variables identified in the original study?

Many of the studies reviewed below explore STEM attrition among students of color, and, increasingly, all first-generation students. This is not surprising, given the seeming intractability of high losses from these groups despite considerable nationwide efforts to reduce them. In the following review, we include findings bearing on lower persistence rates of students of color within each section.

### *Studies of Risk and Persistence in Undergraduate STEM Education*

A limited number of studies of factors contributing to losses from STEM majors report the significance of a combination of variables similar to those described in TAL. They report that STEM switchers are more likely than persisters to report loss of interest in their major as a result of poor teaching in introductory courses, rigid curricula, and negative classroom climates (Biggers, Brauer, & Yilmaz, 2008; DeAngelo, Franke, Hurtado, Pryor, & Tran, 2011; Suresh, 2007). Researchers have also identified particular STEM courses that act as barriers to progression in STEM majors and careers. They include, sequences in general chemistry, physics and biology, calculus and differential equations, computer programming, and several courses in engineering (cf., Alexander, Chen, & Grumbach, 2009; Chang, Cerna, Han, & Saenz, 2008; Gasiewski, Eagan, Garcia, Hurtado, & Chang, 2012; NASEM, 2016; Suresh, 2007). Other studies highlight the ways that the behavior and attitudes of faculty and peers can significantly reduce the likelihood that women or students of color in STEM majors will persist towards graduation (Gayles & Ampaw, 2014; Price, 2010).

Researchers have also addressed the obverse question of what changes in instructional practices in gateway courses enable persistence (e.g., Freeman et al., 2014; Graham, Frederick, Byars-Winston, Hunter, & Handelsman, 2013; Henderson, Beach & Finkelstein, 2011). By studying the outcomes of STEM education reform initiatives, these studies point to the value of pedagogies that actively engage students (e.g., Olson & Riordan, 2012; Wieman, Perkins, & Gilbert, 2010). A further group of studies focuses on classroom strategies found to enable STEM persistence, namely academically oriented peer interactions such as study groups, discussion, and group projects (Callahan, 2009; Hyde & Gess-Newsome et al., 2000; Ost, 2010; Packard, 2005), and student–faculty interactions that provide academic assistance and support (Cole & Espinoza, 2008; Hyde & Gess-Newsome et al., 2000). As reported in TAL, failure to establish, or limited opportunity to develop, such relationships leaves many STEM majors at risk. Academically oriented peer support increases STEM persistence and degree completion among students of color who join STEM-related undergraduate societies and programs (Chang et al., 2008; Espinosa, 2011; Herrara & Hurtado, 2011; Hurtado, Eagan, & Chang, 2010; Palmer, Maramba, & Dancy, 2011). As also illustrated in TAL, engagement in a lively, inclusive community of STEM students (such as sociable departmental research seminars and women's profession-oriented societies) increases the chances of persistence for both white women and students of color (Chang, Sharkness, Hurtado, & Newman, 2014; Espinosa, 2011; Ong, Wright, Espinosa, & Orfield, 2011).

One particular source of culture shock reported in TAL that often (though not exclusively) affects women, and especially women of color, is the loss of teacher–learner interaction. In high schools and community colleges, learning in dialogue with teachers is an important factor in enabling students' learning and confidence in their ability to do math and science. However, it is more rarely encountered in introductory STEM courses (Reyes, 2011). As reported in the TAL study, finding instructors and advisors who are academically helpful and personally encouraging is a key element in many women's persistence (Jackson, 2013; Packard, 2005). Zhang and Allen's (2015) study of students transferring from community college into engineering majors cites support from instructors as a significant factor in overcoming academic difficulties, Also, as reported in TAL, instructors may not appreciate the unintended consequences of teaching styles that limit interaction with students, especially for students of color, and most particularly for women of color (Johnson, 2007).

## *The Effects of Negative Classroom Climates*

The findings of early studies (reviewed and reconfirmed in TAL), and explored in more recent work, explain how negative racial and gender climates experienced in STEM classrooms—or more widely in the institution—undermine the chances of success of men of color and women of all races and ethnicities (Gayles & Ampaw, 2014; Hurtado et al., 2010; Museus & Liverman, 2010; Ong et al., 2011). Students of color often report feeling that they do not belong in STEM departments (Strayhorn, Long, Kitchen, Williams, & Stentz, 2013; Tate & Linn, 2005). Welcoming and supportive cultural climates established within STEM classrooms, departments, and other disciplinary settings are found to enhance commitment to STEM majors and increase retention, especially for students of color (Callahan, 2009; Espinosa, 2011; Garcia & Hurtado, 2011; Gayles & Ampaw, 2014; Hurtado et al., 2010; Litzler & Young, 2012; Museus & Liverman, 2010; Ong et al., 2011).

## *Social Integration: The Role of Cocurricular Experiences*

Many lines of inquiry follow Tinto's (1975) model of social and academic integration based on the seminal work of Emile Durkheim published in 1897 (translated, Spaulding and Simpson, 2005). Tinto proposes that dropping out of majors (or out of school altogether) occurs as the result of students' failing to integrate into social and academic relationships that primarily develop in the classroom. In addition to their considerable value in the learning process, classroom interactions, small group discussions, and study groups shape students' academic and social integration, and, thereby, their likelihood of persisting (Braxton, Shaw Sullivan, & Johnson Jr., 1997). Similarly, cocurricular experiences, such as co-ops, internships, and undergraduate research, promote students' identification with their major, support STEM-related career aspirations, and increase degree-completion rates, including that of students of color (Eagan, Hurtado, & Chang, 2010; Jaeger, Eagan, & Wirt, 2008; Jones, Barlow, & Villarejo, 2010; Kim, Rhoades, & Woodard, 2003; Laursen, Hunter, Seymour, Thiry, & Melton, 2010; Nagda, Gregerman, Jonides, von Hippel, & Lerner, 1998; Nasr, Pennington, & Andres, 2004). Undergraduate research (UR) experiences do this by inducting students into the processes of "thinking and working like a scientist" (Seymour, Hunter, Laursen, & DeAntoni, 2004), adviser-protégé mentoring, students' presenting and discussing their work with others in their field—all of which promote the bonding of novices to the collaborative scientific enterprise (Hunter, Laursen, & Seymour, 2007; Laursen et al., 2010). The resulting increase in STEM graduation rates for UR participants is documented in a substantial number of studies, that include: Chang et al. (2014), Graham et al. (2013), Herrara and Hurtado (2011), Espinosa (2011), Hurtado et al. (2012), Jones et al. (2010), Perna et al. (2009), Chang et al. (2008), and Clewell et al. (2006).

The body of work that, in various ways, draws on Tinto's model of social and academic integration, underwrites TAL findings that students are more at risk of switching if they fail to: establish academically and personally supportive relationships; find and use academic help in timely fashion; and make mentoring connections with their instructors. As we also reported, failure to integrate into college life leaves all students who enter with limited know-how about how to navigate college (particularly first-generation students and those from recent immigrant families) at greater risk than peers who have acquired such social capital from family members with experience of higher education.

## *Inadequacies of High School Preparation*

Other studies validate the TAL finding that high school preparation (reflected in AP course enrollment, high GPA, high SAT, or ACT scores) is critical in predicting both success and loss in STEM majors (Bonous-Hammarth, 2000; Eagan et al., 2010;

Kokkelenberg & Sinha, 2010; Rask, 2010; Whalen & Shelley, 2010). Early interest in STEM fields and high-school achievement are also shown to predict success in the sciences (Chang et al., 2014; Crisp, Nora, & Taggart, 2009; LeBeau et al., 2012; Maltese & Tai, 2011; Nicholls, Wolfe, Besterfield-Sacre, & Shuman, 2010; Riegle-Crumb & King, 2010; Tai, Liu, Maltese, & Fan, 2006). Also reported in TAL (and in early studies by Hills, 1965; Cejda, Kaylor, & Rewey, 1998), under-preparation, both in terms of academic readiness for college-level work and knowledge of how to navigate college systems, is a problem for all first-generation students. They more often attend high schools that insufficiently prepare them for higher education, have problems paying for college, and have to adjust to a bewildering new environment (Goldrick-Rab, 2010). These students often struggle to surmount the abrupt decline in good grades to which they had been accustomed in high school. Again, as reported in TAL, even for students from better resourced high schools, grade shock in foundational courses is a severe blow to confidence and raises questions about whether they "belong" in these majors.

## *The Significance of Grades*

A large number of studies have explored the effects and significance of performance scores, both from high school and within college. Rather than reviewing these here, we refer readers to Chaps. 2, 6, 7, and 8 where this work is considered in the context of the present study.

## *The Significance of Calculus Courses to STEM Persistence*

While the present study has been underway (2012–2017), a linked set of nationwide studies with broad significance for STEM persistence has explored the critical role of college calculus in determining persistence beyond freshman year. Described in Bressoud et al. (2015) the Mathematical Association of America (MAA) survey was administered to a large, random, and stratified sample of first-year calculus students. The researchers surveyed 213 colleges and universities, 502 instructors, and more than 14,000 students to learn, "who takes Calculus 1 in college, why they take it, their preparation for this class, and their experiences in this class" (p. 179). As with other studies, they underscore the importance of good high school math preparation but cite the findings of the Office for Civil Rights report (US Department of Education, 2014) that only half of all high schools offer calculus. Bressoud and colleagues found that, "although black and Latino students make up 37% of all high school students, they constitute only 20% of the students who take the AB Calculus exam and only 11% of those who take the BC exam" (p. 180). Racial and social-class disadvantages that limit access to AB Calculus are also reflected in the types of institutions of higher education that students attend.

The researchers also quantify which students are "weeded out" in Calculus 1 courses: students who received grades of C, D, or F collectively averaged 57% across the institutional sample—numbers that are the more disturbing because they are averages. Higher rates of failure—D, F, or withdrawal (DFW)—were found in master's universities and 2-year colleges. Other losses predominantly occurred among women who received As and Bs, notwithstanding the presumption that these grades indicate an ability to continue in the calculus sequence. As the authors point out, their findings have particular significance for losses among "serviced" STEM majors who account for 94% of students enrolling in Calculus 1. Highest losses are from engineering and computer science (48% of whom are men) and the life sciences or teaching (53% of whom are women). Only 6% of Calculus 1 enrollees planned majors in mathematics or the physical sciences—many of whom place directly into higher calculus courses. Thus, Calculus 1 alone makes a profound contribution to rates of freshman switching out of other STEM majors.

Bressoud and his colleagues also explore the contribution to success in Calculus 1 made by students' levels of confidence in their mathematical abilities, their enjoyment and understanding of high school mathematics, and their resulting sense of readiness for college calculus. They clarify the characteristics that most strongly correlate with maintaining student confidence, enjoyment of, and the desire to continue in, mathematics. As with studies cited earlier, they found these attributes were best enabled by a good rapport between students and instructors in the learning experience. However, their strongest finding about the impact of Calculus 1 courses across the US colleges and universities surveyed is, "how effective this course is in destroying (incoming) confidence" (p. 182). They discovered that students' belief that they have the knowledge and ability to succeed in this course dropped from 80% to 50% between the start and end of the term: confidence in their perceived ability to do mathematics fell by half a standard deviation, and enjoyment of mathematics dropped by one-third. This effect was particularly marked among women. Even with final grades of As and Bs, twice as many women as men with the same grades abandoned the idea of taking Calculus 2. When asked why, 18% of these high-scoring women (compared with 4% of men with the same high grades) explained that they did not understand calculus well enough to proceed to Calculus 2, and 7% of these women (but none of the men) thought that their grade was "not good enough" to continue. These findings align with those reported in TAL, Chap. 5 (pp. 234–235 and 274).

## Explaining Gender Differences in Persistence

Since the TAL study, a number of studies have endorsed and expanded on its findings about the nature of the link between low confidence and low STEM persistence among women (Good, Rattan, & Dweck, 2012; Holmegaard, Madsen, & Ulriksen, 2014). In a recent study that explores the causes of gender disparities in STEM switching patterns, Hardin and Longhurst (2016) observed changes in critical social–cognitive variables near the middle and at the end of the first semester of an introductory chemistry course. Even after controlling for course performance, they report that women show lower STEM self-efficacy, coping self-efficacy, and STEM interest than do men. Men experienced a small but significant increase in their perceived support for pursuing a STEM degree, whereas women did not.

As one of the NSF-funded studies of the role of college calculus in STEM persistence, Ellis et al. (2016) also approach the question of why women and men still do not complete STEM degrees at comparable rates. Adding to Bressoud et al.'s (2015) findings, they report that, while controlling for other factors, women fail to progress from Calculus 1 to Calculus 2 at a rate 1.5 times greater than men. They discard the hypothesis that women understand the material less well than men—citing Lindberg et al.' (2010) meta-analysis that found no evidence of gender differences in mathematics ability; and a study of Calculus 1 in which women out-performed men (Islam & Al-Ghassani, 2015). Comparing women and men with above-average mathematical abilities and preparation, Ellis and her colleagues found that women started and ended the term with significantly lower mathematical confidence than men and stated significantly more often than men that they did not understand the course material well enough to continue. The researchers conclude that Calculus 1 courses contribute significantly to the STEM "gender filter" (c.f., Blickenstaff, 2005) by weeding out women through lack of confidence in their own ability rather than actual mathematical capability. As noted at the beginning of this chapter, Ellis and colleagues projected substantial gains in the graduation rates of all STEM majors were the issues raised for women in Calculus 1 courses effectively addressed.

## Why We Are Revisiting *Talking about Leaving*

Since the 1990s, a growing community of education reformers has drawn on the findings of TAL and related research to address the improvement of STEM undergraduate education. The NSF and many private foundations have funded STEM educational innovations to undertake this work. They have collectively produced a body of research-grounded and classroom-tested learning materials, interactive and active pedagogies that engage students in their own learning, and assessment methods that probe students' depth of understanding and ability to apply, extend, and transfer their knowledge. This expanding body of tested and adapted materials and methods has been disseminated by workshops that offer hands-on exposure to learning theories, research findings, and their classroom applications. Universities have developed teaching and learning centers that offer practical help for instructors and TAs to incorporate research-based instructions strategies (RBIS) into their courses. Disciplinary societies have added education sections to their meetings that disseminate research findings, support a growing scholarship of teaching and learning among practitioners, have developed websites that offer information about RBIS, provide connections to networks of expertise, and offer online discussion. Both traditional and new (often online) journals publish articles on the theory and methods of what is sometimes referred to as "scientific teaching" (Handelsman et al., 2004).

In their review of studies on uptake or scale-up of scientific teaching methods, Seymour and Fry (2016) found indications of success in institutional uptake of particular pedagogies and their spread among individual instructors (Beichner, 2008; Beichner et al., 2007; Beichner & Saul, 2003; Dori & Belcher, 2005; Ege, Coppola, & Lawton, 1997; Elizondo-Montemayor, Hernández-Escobar, Ayala-Aguirre, & Aguilar, 2008; Merton, Froyd, Clark, & Richardson, 2009). The UCLA-HERI faculty survey also reports an increase in instructors using student-centered pedagogies (DeAngelo, Hurtado, Pryor, Kelly, & Santos, 2009). Other studies offer mixed reports for uptake including problems with discontinuation of methods initially tried (Borrego, Froyd, & Hall, 2010; Henderson & Dancy, 2007; Henderson & Dancy, 2008; Dancy & Henderson, 2010; Seymour & DeWelde, 2016; Walczyk, Ramsey, & Zha, 2007). Overall, this body of work suggests increased awareness of alternative ways to teach and some inclination to try them.

In Chap. 8, we discuss the scholarship underlying efforts to encourage uptake of research-grounded teaching methods in STEM education. Many colleagues engaged in these initiatives have asked us to revisit our original study. Both they and we wished to establish whether STEM switching rates have declined nationally and in our institutional sample, and to learn from student accounts whether their STEM learning experiences reflect RBIS in course design and classroom practice, and whether such changes increase perseverance in STEM majors. We also wanted to learn to what extent the conditions that undermined STEM persistence two decades ago have abated and whether there are new problems that pose persistence risks.

In this chapter, we have also reviewed studies of particular variables that influence student persistence and attrition in STEM fields. We have also stressed that multiple factors intersect in contributing to STEM attrition. Some early studies have also considered an array of variables that combine to undermine STEM persistence (Kuh, Schuh, Whitt, & Associates, 1991, 2005; Pascarella & Terenzini, 1991, 2005). However, there has been no recent work that explores the relative weight and interrelationship of the array of factors that contribute to both STEM switching and persistence. Thus, we also wanted to learn what has, and has not, changed over the last 20 years in the interconnected spectrum of problems encountered by students in STEM majors that we originally identified. We also wanted to discover what new issues may have emerged in the changing socioeconomic context in which higher education now operates. We were not at all sure that we could accomplish all of this in one study, but, as before, we proceeded in the expectation that other researchers will test, augment, and refine our work.

Finally, after a gap of years, it is perhaps inevitable that the findings of older studies are forgotten. If new work is not undertaken, commentators are apt to resort to conjecture, unsupported by data, in order to account for STEM switching rates—as is the case in this selection of opinion articles (e.g., Berrett, 2011; Drew, 2011; Reich, 2011; Shi, 2011; Taylor, 2011).

For all of these reasons, it seemed timely to revisit *Talking about Leaving: Why Undergraduates Leave the Sciences*. In 2012, with the financial support of the National Science Foundation and the Alfred P. Sloan Foundation (who also funded the original study) we began a new 5-year study whose findings are explored and discussed in the rest of this book.

## Design and Methods of the New Study

### *A Multi-Component, Mixed-Methods Research Study*

This mixed-methods research both replicated and augmented the original TAL study. Building upon a qualitative interview study with students who switched from, and who persisted in, STEM majors, we added component studies that explored the dimensions of persistence risks from different sources of data. This also offered the possibility of triangulating component study findings. The whole 5-year project comprised:

- *A comparative review of evidence from two national data sets* to estimate the current national rates of persistence, relocation, and loss (whether by switching or dropping out of college) in STEM majors. (a) The 2013 National Center of Educational Statistics (NCES) report *STEM Attrition: College Students' Paths Into and Out of STEM Fields* (Chen, 2013) that examined data from the 2004–2009 Beginning Postsecondary Students Longitudinal Study and the 2009 Postsecondary Education Transcript Study; (b) The Higher Education Research Institute (HERI) provided us with recent estimates of STEM persistence, relocation and switching through its Cooperative Institutional Research Program (CIRP) data. These represent an update of the CIRP data that HERI provided for the original 1997 study;
- *Analyses of institutional student transcript and attribute data* using multiple statistical methods, including logistical regression, to explore the switching and persistence patterns of STEM majors at the six institutions in our study;
- *Administration of the online Student Assessment of their Learning Gains (SALG) end-of-course survey* in a matched set of STEM foundation courses across the six sample institutions. This yielded numeric estimates of the extent of several types of student learning gains, the nature of benefits and problems experienced in these courses augmented by written responses to open-ended questions, and information about other aspects of students' experiences (e.g., students' choice of majors and careers);
- *In-depth interviews and focus groups with structured samples of "switchers" and "persisters"* This study replicates the original research, using ethnographic analyses of verbatim transcripts from interviews and focus groups with a structured sample of 346 students across the six participating institutions. Its broad

purposes were to explore what has and has not changed in what causes switching and relocation and what enables persistence, and to discern what new sources of influence on student decisions have arisen in the intervening years. Are switchers leaving STEM majors for the same reasons as those found in *TAL*? Our research questions were:

1. What role is played in persistence-related problems of both switchers and non-switchers by learning experiences that students define as unsatisfactory? Do these factors play a role in student persistence comparable to that documented in *TAL*?
2. Has the relative significance of particular factors shifted over time? Do other factors prompt field switching that we did not previously identify?
3. Do graduating STEM seniors continue to experience the same educational problems as students who switched out of their majors? Are there differences between these groups?
4. What are the variations in student answers to all of these questions for sample subgroups (i.e., students of color or ethnicity, all women, all men, students with above and below 650 mathematics SAT entry scores, disciplinary groupings, and whole institutions)?
5. Why did students choose or aspire to STEM majors or careers? How have reasons for choice of major changed since our original study and how do these relate to persistence?
6. Into which disciplines do STEM field-switchers go, and why?
7. Is there evidence that changes in instructors' teaching methods have had a beneficial effect on persistence?
8. Have other problems identified in *TAL* been ameliorated?

Additionally, we reference studies and findings from *two concurrent collaborating studies* by colleagues working in the same field. These are:

- A classroom observation study led by. J. J. Ferrare (University of Kentucky) in the same 71 gateway courses as those to which the SALG survey was administered using the Teaching Dimensions Observation Protocol (TDOP) instrument.
- A study by Andrew K. Koch and Brent M. Drake of "DFWI" rates in four foundation-level STEM courses located in each of 36 institutions participating in the Gardner Institute's "Gateways to Completion" project. This study aligns with our institutional data analysis of DFWI rates at the six study sites, and the two studies triangulate in presenting the impact of "severe foundation courses" on persistence for particular student demographics.

All data (except the Gardner Institute study) were collected at six of the seven original participating site institutions: one institution declined to participate in the current study. The six participating institutions are four public universities with high research activity and two private institutions, one with high research activity and one with a science focus, according to the Carnegie Classifications.

- *PB1R1*: A western state (originally land-grant) university serving a large rural population, with a prestigious engineering research program, and applied science specialties;
- *PB2R1*: A comprehensive, urban north-eastern university with large and diverse STEM undergraduate enrollment;
- *PB3R1*: A large urban, mid-western university with prestige ranking for its STEM research and high production of STEM undergraduates and graduates;
- *PB4R1*: A "flagship" western state university with high reputation for its engineering school and several science departments;
- *PV1R3*: A western liberal arts college with a strong reputation for its science teaching (engineering is not offered);
- *PV2R1*: A small western city university offering engineering, and masters' and doctoral degrees in the sciences.

Institutions comprising our study sample were selected to represent the array of US colleges and universities and that typify most undergraduates' STEM education. Enrollment ranged from ~2000 (PV1R3) to over 34,000 (PB3R1) undergraduates and, compared to national averages, most of the institutions in our study had a greater representation of white students, and fewer Hispanic and African-American students. Representation of women was also lower than is seen nationally. However, the baccalaureate graduation rates were higher than the national average at our sample of institutions, as were averages for baccalaureate graduation rates for racial/ethnic groups at each of our participating institutions. Although our schools have higher graduation rates than the national average, most of our schools are drawn from R1 institutions where graduation rates are higher overall. (See Appendix A for detailed demographic data of the student populations of institutions in our study sample.)

We next provide an overview of methods guiding our work. We have placed a more detailed discussion of our quantitative methods and protocols in appendices for readers who are interested in reviewing the instruments, protocols, and specific aspects of the analyses used in this research.

## National and Institutional Transcript Study of STEM Field-Switching Patterns

An analysis of the national Beginning Postsecondary Students Longitudinal Study (BPS) data set was undertaken to identify current rates of switching across STEM majors, both overall and by disciplinary groupings and sex, race/ethnicity. In addition, as in TAL, we collaborated with the Higher Education Research Institute (HERI) at University of California Los Angeles to construct a current national portrait of STEM field-switching, also overall and by disciplinary groupings and sex, race/ethnicity.

Additionally, we undertook an analysis of student transcript records from our six participating institutions to learn how many students switched out of STEM, who switched, and when. We collected current and retrospective transcripts for all juniors and seniors for 2013 and 2014, and in the same years we conducted site visits at each institution. Student transcripts for students' academic careers came from the years 2007 to 2014. The records from the six institutions included both STEM and non-STEM students. We analyzed 45,565 records from individual students with demographic and academic information, such as gender, ACT/SAT scores, and declared majors. Of these students, 14,626 started college in STEM disciplines, with 2132 students switching out of STEM majors. We also collected 1,437,806 transcript records of the same students' grades, and descriptions of each class they took over multiple terms. These records included term-by-term reports of current major, and cumulative and term GPAs.

Several operational definitions guided the transcript study. We categorized STEM majors based on the *Categorization of Instructional Study* (CIP) code of the major provided by National Center for Educational Statistics (NCES, 2019). STEM disciplines include agriculture, environmental studies, computer science, biology/life sciences, mathematics, engineering, and physical sciences. We called students *switchers* if they started in a declared STEM major, but then shifted to, and persisted in, any major in the social sciences, art, humanities, business, education, or any other non-STEM major.

Several smaller subgroups of switchers also emerged during our analysis. *Switch-in* students started college in non-STEM disciplines and switched into and persisted in STEM majors, while *switch-around* students entered and left STEM majors multiple times during their academic careers. *STEM-relocator* students changed from one STEM major to another. Our analyses also included students who entered and persisted in STEM majors (*persisters*), and students who had never entered STEM (*non-STEM*). Our transcript records did not include students who left or transferred out of the institution. An in-depth description of quantitative methods for the institutional transcript records study is provided in Appendix B.

We note that switching rates varied considerably by institution. However, while much of our analysis aggregates across schools and treats students as "one big group," we did adhere to a rule of thumb in our analyses in testing each pattern of findings to learn if the pattern seen for the whole was also present at each school. For instance, we found that women switched more than men at each institution, also finding that at *all* schools this was true. If findings did not generalize, we either noted this in the analysis or did not present the finding.

Our sample, while very large in terms of students and student records, is restricted to six institutions, and thus limits generalization of results. The records analysis therefore is descriptive of these students and schools and may or may not be representative of a wider population. However, we do know that the large public universities in our sample and the smaller private schools are not outliers in terms of student demographic representation, gender balance, socioeconomic representation, or academic selectivity compared with similar schools in the USA.

## *The Gateway Course-Taking Study*

Because instructors are particularly influential on the trajectory of student performance early in students' academic careers, this component study focused on current undergraduate experiences of instruction in these courses. At each study site, we identified 10 "foundational" courses considered by our site informants to play important roles in the early stages of their STEM degree programs. We selected courses offered across study sites in order to facilitate cross-institutional comparisons of instructor teaching and student experiences. Because foundational courses often include more than one section and instructor, for two courses at each site we included two instructors, making a total of 12 instructors per research site ($n = 84$). The student sample was potentially all students in these classes. These students were asked by their instructor to consider participating in a focus group and to complete the online Student Assessment of Learning Gains (SALG) survey, which offers systematic assessment of what students are gaining from their educational experiences and detailed feedback on precisely which pedagogical elements of the course are contributing to students' learning gains. The SALG survey thus provides instructors with detailed information about what progress their students are making toward course learning goals. Quantitative methods used for analyzing TALR-SALG survey responses are provided in Appendix B. A copy of the TALR-SALG survey is provided in Appendix C.

To understand what particular aspects of each instructor's teaching methods contributed to students' growth as learners, interest in the discipline, and motivation to remain in their majors, we examined student responses to the SALG survey. The survey asks students about their commitment to their major, and to comment on open-ended questions on any potential instructional or institutional factors related to field-switching. The SALG survey was administered to all students enrolled in classes included in our introductory course sample.

We received 1427 full responses from students in 52 classrooms at the six institutions in our study. The survey was administered online using the Surveymonkey survey software with the help of participating instructors who sent emails with a web-link to students. Students received three reminder emails to participate in the study. Nearly 40% of respondents were Sophomores; first-year students and juniors each represented about one-quarter of respondents; and ~8% were seniors. Students filling out the survey represented a range of STEM fields, including agriculture, computer science, engineering, life sciences, mathematics, and physical sciences. Some students who filled out the survey were majoring in non-STEM fields; we included their responses to assess classroom teaching and institutional climate.

The average response rate to the survey was 24%, although rates varied substantially by class and among institutions. We were able to assess the representativeness of survey responses at each institution given the wider gender and race/ethnicity percentages of each school. All samples were within plus or minus three percentage points of the school population for gender. At two institutions, African-American students were underrepresented by 3% and 4% respectively. All other racial-ethnic groups were within 2% of school populations. While samples were generally representative by gender and race-ethnicity, they were not random given that students had a choice to answer, or not answer the survey.

## *In-Depth Interviews and Focus Groups with Structured Samples of "Switchers" and "Persisters"*

The qualitative analyses presented in this volume focus on students who entered college as STEM majors and switched to non-STEM majors and on seniors who remained in their original STEM majors. Students who relocated into other STEM majors, and students who undertook multiple majors and minors that included a STEM discipline were also encountered both in the interview study and in the national and institutional data analyses. The design of this part of the project replicates and augments the original TAL interview study. Its results are grounded in ethnographic analyses of the verbatim transcripts of 96 individual interviews with switchers (largely juniors) and focus groups with 250 persisting seniors. We did individual interviews with all students of color, whether switchers or persisters, and also with the low-math seniors. The focus groups with seniors were all-female or all-male in order to offer opportunities for frank discussion of gender-related issues. In these protocol-based conversations, we explored what has and has not changed, and what new factors have arisen, in the contributory causes of switching from STEM majors since *Talking about Leaving* was published in 1997. Interview protocols were developed in consultation with salient members of the project's advisory board and were designed to solicit answers to our research questions in semi-structured conversations that encouraged exploration both of original and new lines of inquiry. Protocols used for the Persistence study interviews and focus groups are provided in Appendix D.

As in the original study, we have identified "switchers" as students who enter the university as STEM majors and then switch to a non-STEM discipline. We defined "persisters" as seniors who were persisting to graduation in their original STEM major.

The STEM majors included in the study were physical sciences, life and agricultural sciences, engineering, mathematics, computer science, and information technology. The latter were not included in the original study because computer science and information technology had not yet emerged as formal majors. We defined non-STEM majors as: social sciences; arts; humanities; and certain applied STEM related fields, such as architecture, nursing, and landscape design. We classified students who switched to other majors within STEM as "relocators" not as switchers and gave them special attention in our analyses (cf., Chaps. 3, 9, and 10).

To select our interview sample of juniors who had switched and seniors who were persisting in their original STEM majors, we built an intentional sample frame from student transcripts at each institution. The interview sample that we sought was stratified by race/ethnicity, gender, discipline, and indicators of "math readiness" (by math SAT or ACT scores) with sufficient numbers in each cell to provide clear patterns of response. Switching largely occurs in the first 2 years of a STEM degree, so most switchers were interviewed during their junior year—close enough to their decision to be able to recall what had led to it and also able to reflect on its subsequent impact on their lives.

We operationalized "high-math ability" as having at least a 650 Math SAT or 28 Math ACT score. This decision was based on surveys of faculty beliefs (that were solicited both for the TAL study and again at the start of this study) about what level of demonstrated math competence was likely to predict students' success in STEM majors. Though faculty identified a range of metrics that they deemed important, the majority of responses focused on SAT and ACT indicators. As in the original study, selecting students with this level of math readiness allowed us to focus on students whom faculty might, prima facie, wish to keep in their major. It was, thus, a means to rule out switching caused largely by difficulties with the level of math required in STEM majors. We also wanted to understand why (as we also found in TAL) some high-achieving students, including women, leave STEM majors. Most of our persister interviewees entered with high math scores. However, we also selected a subsample of persisters with low-math readiness indicators (<650 SAT; <27 ACT) to better understand what enables these students to persist despite their incoming disadvantages.

A total of 3750 interview invitations were sent out by email in four waves at each site. This yielded 346 interviewees who met our selection criteria for individual or focus group interviews. Response rates varied by institution and targeted group: the response rates for persisters varied from 3% to 18% per institution and for switchers from 5% to 23% per institution—a 9% overall response rate. Follow-up recruitment emails and phone calls were used to secure adequate participation from members of key groups such as students of color or students from underrepresented disciplines. All switchers and all students of color (both switchers and persisters) and persisters who entered with low-math readiness scores were interviewed individually. In a private setting, we hoped to capture as fully as interviewees were comfortable with the nuances of their experiences and educational choices. Aside from the low-math persisters, white persisting seniors (all of whom had high-math readiness indicators) were interviewed in all male and all female focus groups. We found in the original study that this single sex configuration allows for more open discussion of issues about which we particularly wanted to learn, including issues of gender.

## Student Samples

***Switchers*** As in the original study, we deliberately over-sampled some groups the better to understand particular patterns in their persistence rates. Thus, we interviewed 96 switchers at six institutions of whom 64% were women ($n = 61$) and 36%

were men ($n = 35$). This proportion reflects our concern to understand why (as discussed in our review of recent research findings) women are still switching disproportionately from STEM disciplines. In our sample of talented students, 69% of women and 37% of men left their STEM major with a GPA of 3.5 (B+) or better (i.e., 42 of 61 women and 3 of 35 men). We also over-sampled students of color to get a clear picture of their experiences and concerns. Half of the high-performing women in our sample who switched were women of color ($N = 21$), with a smaller number who were men of color (i.e., 38%, $N = 5$). Additionally, 73% of our switcher sample were white ($n = 70$) and 27% were students of color ($n = 26$). Of these, nine were Latino/a, three who self-identified as multi-racial, five were African-American, and nine were Asian or Pacific Islander.[6] Of all students of color, 12 were the first in their family to attend college and 7 students were immigrants.

Students switched out of a variety of disciplines: 43 from life sciences (e.g., microbiology, evolutionary biology, animal sciences, etc.); 25 from engineering (including civil, biomedical, aerospace, mechanical, computer, environmental, chemical, and electrical engineering); 16 from the physical sciences (e.g., physics, chemistry, geology); eight from mathematics; and four from computer science. Switchers were highly qualified by commonly used math-readiness indicators. In high school, almost all had taken precalculus and 68% had taken AP or IB Calculus, their average SAT and ACT math scores were 680 and 28.5, respectively, and their average cumulative college GPA was 3.21. By the definition described above, 76% of the switcher sample had "high-math readiness" ($n = 72$) and 23% were categorized by "low-math readiness" (plus two for which no SAT/ACT scores were available).

***Persisters*** The persister sample comprised 250 students: 41% were men ($n = 102$) and 59% women ($n = 148$); 18% of women and 14% of men qualified as high-performing persisters. Sixty percent of persisters were white ($n = 150$), 40% were students of color ($n = 100$), including 41 Asian/Pacific Islanders, 19 African-Americans, 18 Hispanics/Latinos, 18 who were identified as multi-racial, and four Native Americans. It was not possible to determine how many persisters were first-generation college students, as only three institutions provided this information.

---

[6] Asian-American students are commonly omitted from STEM education studies that focus on underrepresented minorities (URMs). To do this is to treat as a single category students from a wide array of national and linguistic groups and fails to distinguish those communities that are long established in the USA from those of recent immigration. In addition, as we discovered in the original study (and recounted earlier in this chapter), their exclusion misses problems that are distinctive to some Asian groups. A 2018 study by the Center for Law and Social Policy (CLASP) highlights one important persistent-related issue: "Across the board, Asian-American students have the greatest amount of unmet need, regardless of the institution they attend. We offer a few explanations for this. Across all races/ethnicities, Asian-Americans are the most income-stratified; while some Asian-American subpopulations are as financially secure as whites, many others live in deep poverty." Kochhar and Cilluffo (2018) also report that, income inequality in the U.S. is rising most rapidly among Asians.

Forty-five percent of persisters were majoring in the life sciences ($n = 113$); 26% in engineering; 15% in the physical sciences; 7% in mathematics; and 6% in computer science. Eighty-six percent of persisters had taken precalculus and 50% had taken AP or IB Calculus in high school. Their average SAT and ACT exam math scores were 668 and 28.8, respectively. The persisters' average cumulative college GPA was 3.36, 66% were categorized by "high-math readiness" ($n = 165$) and our subsample of low-math-ready persisters were 33% ($n = 83$) of all persisters. Figures showing TALR study student samples are given in Appendix E.

Interviews were conducted during 2-week site visits to each of the six institutions between Spring 2013 and Fall 2014. Semi-structured individual interviews were conducted in private study rooms and typically ranged from 45 to 90 min. Focus groups ran from 90 min to two-and-a-half hours, depending on participants' levels of engagement with the issues explored in the protocols. To identify changes of speakers in focus groups while ensuring anonymity, each focus group participant chose a card with the name of a famed STEM scholar and used that pseudonym to identify themselves and each other during their discussion. Following interviews and focus groups, each student was provided $20 as a small "thank you" for their participation. Interviews and focus groups were digitally recorded and labeled with a letter-and-date filing convention to further ensure anonymity of institutions. The interviews and focus groups were transcribed verbatim by a professional transcription service and entered into server-based *NVivo 10.0* (QSR International, 2016), which allows simultaneous use of the qualitative coding software by team members. All research was conducted with the approval of the Institutional Review Board at the University of Colorado Boulder and in compliance with rules and regulations governing human subjects research.

Our ethnographic analysis of the text data followed procedures originally described by Spradley (1979). Some codes were generated deductively, based on themes from the original study for then–now comparison, or to reflect their prominence in the STEM retention literature (e.g., experiences of teaching and learning, class climate, advising or support systems, and high school preparation). Codes were also generated inductively, based on new issues that were raised by participants in describing the sources of their educational decisions (e.g., concerns about career opportunities or the current economic climate, parental influence in choice of majors or careers, and the association of identity with grades). Codes were organized hierarchically within domains that represent the larger categories of interest in our study (e.g., reasons for students' choice of major or career, college transition issues, financial problems, STEM and non-STEM teaching, and learning experiences). Like individual codes, some domains were created deductively with reference to prior research findings, but almost all codes within each domain were inductively grounded in issues spontaneously raised by interviewees.

Ethnographic analysis proceeds by coding each segment of textual information on any matter that speakers identified as significant to their decisions. Although the protocol questions (also deductively derived from TAL and other research findings) initially framed the topics of discussion, it was always what interviewees raised as relevant in response to any question that determined into what existing code category any response was placed or whether a new code was needed to capture an idea raised by the interviewee. Thus, more than is the case in content analysis, ethnographic analysis proceeds by coding each segment of text as a piece of evidence that the speaker has identified as significant to their decisions. Because normal conversations often reference different matters in the same few sentences and may be layered by observations of different kinds, any segment of interview text is apt to contain several distinct ideas, each of which is coded with a separate code name and may be placed in different domains from other utterances in the same segment. For instance, a question like "How did you come to choose your major?" is apt to solicit a response that references teachers, school prowess, parental expectations or knowledge, perceived career opportunities, prestige of various careers, and so on: these are each captured and categorized in relevant domains.

Five team members shared the conduct, coding, and analysis of student interviews. The text files of each interview or focus group and the emergent codebook that the team was collectively evolving were electronically available to all coders. The team met weekly to discuss the development of new codes, combine or expand existing codes, change code names, and other adjustments. Code and domain changes were always available for discussion and resolution. When the coding process was completed, and all codes had been checked for accuracy, frequencies were run for code clusters across the whole data set and for particular subsets of participants. We analyzed both individual codes and broader domains grounded in themes that had emerged in the interview data for patterns by switchers and persisters, gender, race/ethnicity, discipline, low-math readiness, or other characteristics (Boyatzis, 1998; Creswell, 2012). To get a sense of the size of the resulting final electronic code book, our ethnographic analysis of verbatim transcripts using *NVivo 10.0* generated ~25 major domains, ~95 sub-domains, and ~1500 discrete codes. The numbers generated in frequency tables offer a sense of the relative weight of particular student observations on any topic. However, single observations can be very important in providing explanations and insights into matters that other informants have simply described or illustrated. Thus, the analysis that is reflected in our accounts of findings throughout this book is built from frequency tables, the researchers' observation of patterns and themes captured in the coding process, and the insights of individual participants that we offer as direct citations from the transcribed interviews. The coding and analysis process described here was also used for students' written responses to open-ended questions in the SALG instrument.

Overall, our set of component studies and findings from the two collaborating studies provide rich and detailed information. Each study, designed to answer different, though related, research questions, produced a wealth of data. Taken together, they provided opportunities for triangulation of qualitative and quantitative data that offered a nuanced picture of the attitudes, beliefs, rationales, behaviors, interpretations, and decisions of students and, by inference, their families, peers, instructors, departments, and institutions. These analyses offer an updated understanding of factors affecting students' decisions to persist, relocate, or leave STEM majors and the processes whereby these decisions are taken.

## Scope and Structure of This Volume

From its inception, this has been a highly collaborative project, and so we have written a highly collaborative book about it. The whole team worked together in the design, data collection, processing, and primary analyses. However, each chapter reflects the final analytic work of team members, variously combined, with contributions also from some of our colleagues in the field.

The first three chapters place the study in context. Chaps. 2 and 3 offer an overview of major patterns in our findings from the two main study components at our six study sites—the institutional records analyses, and the overall results from the interview study with targeted student samples. Thereafter, we write in depth about strong themes in our findings, commenting as we go along on student groups at particular risk of switching, including white women and women and men of color. In each chapter, we also compare and contrast new findings with those of the original study.

Chapters 4 and 5 focus on what students bring into college, why they choose particular majors, how well they adjust to the STEM majors they have chosen, and the import of all these variables for their persistence. Chapters 6, 7, 8, and 9 focus on facets of students' educational experiences in STEM majors—both in early foundational courses and across the whole 4 years of academic study—and the bearing these experiences have for persistence, relocation, and loss. Chapter 8 also includes some evidence in student accounts of improvements in the effectiveness of teaching methods. Chapter 10 discusses variations in the processes whereby decisions to switch or relocate are reached, and the negative personal consequences that switching often involves. In a series of vignettes, it also explores several reasons why high-performing STEM majors—particularly women—switch to non-STEM fields.

Not all influences on decisions to move or leave arise from educational experiences in college. In Chap. 11, we discuss the consequences of working while in school, parental influence on academic and career choices, and students' perceptions of the career opportunities available in the current economy. Chapter 12 addresses the complex, interrelated factors that contribute to students' persistence in STEM majors. Like switching, persistence is a process that unfolds over time and involves many adjustments to prior identities, practices, and habits. We describe how students must draw on an array of individual, social, cultural, and institutional resources to persist and graduate with a STEM degree. In the concluding chapter, we review what particular study findings suggest as remedies of different kinds and levels to address the diagnoses we have offered for the loss or diversion of able students from STEM majors.

## References

Agar, M. (1996). *The professional stranger: An informal introduction to ethnography*. San Diego, CA: Academic Press.

Alexander, C., Chen, E., & Grumbach, K. (2009). How leaky is the health career pipeline? Minority student achievement in college gateway courses. *Academic Medicine, 84*(6), 797–802.

Ambrose, S. (1966). *Duty, honor, country: A history of west point*. Baltimore, MD: John Hopkins University Press.

Astin, A., & Astin, A. S. (1993). *Undergraduate science education: The impact of different college environments on the educational research pipeline in the sciences*. Los Angeles, CA: Higher Education Research Institute, UCLA.

Beichner, R. J. (2008). *The SCALE-UP project: A student-centered active learning environment for undergraduate programs*. Commissioned paper for the national academies workshop: Evidence on promising practices in undergraduate science, technology, engineering, and mathematics (STEM) education. Retrieved from http://www7.nationalacademies.org/bose/PP_Commissioned_Papers.html

Beichner, R. J., & Saul, J. M. (2003). Introduction to the SCALE-UP (student-centered activities for large enrollment undergraduate programs) project. In *Proceedings of the International School of Physics "Enrico Fermi"* (pp. 1–17). Varenna, Italy.

Beichner, R. J., Saul, J. M., Abbott, D. S., Morse, J. J., Deardorff, D. L., Allain, R. J., … Risley, J. S. (2007). The student-centered activities for large enrollment undergraduate programs (SCALE-UP). In E. F. Redish & P. J. Cooney (Eds.), *Research-based reform of university physics*. Retrieved from http://www.compadre.org/PER/per_reviews/volume1.cfm

Berrett, D. (2011, November 6). *What spurs students to stay in college and learn? Good teaching practices and diversity*. The Chronicle of Higher Education. Retrieved from http://chronicle.com/article/What-Spurs-Students-to-Stay-in/129670/

Biggers, M., Brauer, A. & Yilmaz, T. (2008, March). *Student perceptions of computer science: A retention study comparing graduating seniors vs. CS leavers*. Paper presented at the 39th SIGCSE technical symposium on computer science education, Portland, OR. https://doi.org/10.1145/1352135.1352274

Blickenstaff, J. C. (2005). Women and science careers: Leaky pipeline or gender filter? *Gender and Education, 17*(4), 369–386. https://doi.org/10.1080/09540250500145072

Bonous-Hammarth, M. (2000). Pathways to success: Affirming opportunities for science, mathematics, and engineering majors. *Journal of Negro Education, 69*(1/2), 92–111.

Borrego, M., Froyd, J. E., & Hall, T. S. (2010). Diffusion of engineering education innovations: A survey of awareness and adoption rates in U.S. engineering departments. *Journal of Engineering Education, 99*(3), 185–207.

Boyatzis, R. E. (1998). *Transforming qualitative information: Thematic analysis and code development*. Thousand Oaks, CA: Sage.

Braxton, J. M., Shaw Sullivan, A. V., & Johnson, R. M., Jr. (1997). Appraising Tinto's theory of college student departure. In J. C. Smart (Ed.), *Higher education: Handbook of theory and research*. New York, NY: Agathon.

Bressoud, D., Mesa, V., & Rasmussen, C. (2015). *Insights and recommendations from the MAA national study of college calculus* (MAA notes). Washington, DC: The Mathematical Association of America.

Bureau of Labor Statistics, U.S. Department of Labor. *Occupational outlook handbook*, 2012–2013 ed. Retrieved from http://www.bls.gov/ooh

Callahan, K. M. (2009). Academic-centered peer interactions and retention in undergraduate mathematics programs. *Journal of College Student Retention, 10*(3), 361–389.

Campbell, G. (1993, March 31). *Visions of engineering education in Century 11*. The Porth Distinguished Lecture, University of Missouri at Rolla, Rolla, MO.

Carnevale, A. P., Smith, N., & Melton, M. (2011). *STEM: Science, technology, engineering, mathematics*. Georgetown University, Center on Education and the Workforce. Retrieved from http://www9.georgetown.edu/grad/gppi/hpi/cew/pdfs/stem-complete.pdf

Cejda, B. D., Kaylor, A. J., & Rewey, K. L. (1998). Transfer shock in an academic discipline: The relationship between students' majors and their academic performance. *Community College Review., 26*(3), 1–14.

Center for Law and Social Policy (CLASP). (2018). *When financial aid falls short: New data reveal students face thousands in unmet needs*. Washington, DC.

Chang, M. J., Cerna, O., Han, J., & Saenz, V. (2008). The contradictory roles of institutional status in retaining underrepresented minorities in biomedical and behavioral science majors. *The Review of Higher Education, 31*(4), 433–464.

Chang, M. J., Sharkness, J., Hurtado, S., & Newman, C. (2014). What matters in college for retaining aspiring scientists and engineers from underrepresented racial groups. *Journal of Research in Science Teaching., 51*(5), 555–580.

Chen, X. (2013). *STEM attrition: College students' paths into and out of STEM fields (NCES 2014-001)*. Washington, DC: National Center for Education Statistics, Institute of Education Sciences, U.S. Department of Education.

Chen, X. (2015). STEM attrition among high-performing college students in the United States: Scope and potential causes. *Journal of Technology and Science Education, 5*(1), 41–59.

Clewell, B. C., de Cohen, C. C., Tsui, L., Forcier, L., Gao, E., Young, N., … West, C. (2006). *Final report on the evaluation of the National Science Foundation Louis Stokes Alliance for Minority Participation Program*. Washington DC: The Urban Institute.

Clifford, J. (1988). *The predicament of culture: Twentieth century ethnography, literature, and art*. Cambridge, MA: Harvard University Press.

Cole, D., & Espinoza, A. (2008). Examining the academic success of Latino students in science technology engineering and mathematics (STEM) majors. *Journal of College Student Development, 49*(4), 285–300.

Creswell, J. W. (2012). *Qualitative inquiry and research design: Choosing among five approaches*. London, England: Sage.

Crisp, G., Nora, A., & Taggart, A. (2009). Student characteristics, pre-college, college, and environmental factors as predictors of majoring in and earning a STEM degree: An analysis of students attending a hispanic-serving institution. *American Educational Research Journal, 46*(4), 924–942.

Dancy, M., & Henderson, C. (2010). Pedagogical practices and instructional change of physics faculty. *American Journal of Physics, 78*(10), 1056–1063.

DeAngelo, L., Franke, R., Hurtado, S., Pryor, J. H., & Tran, S. (2011). *The American freshman: National norms for fall 2009*. Los Angeles, CA: Higher Education Research Institute, UCLA.

DeAngelo, L., Hurtado, S., Pryor, J. H., Kelly, K. R., & Santos, J. L. (2009). *The American college teacher: National norms for the 2007–2008 HERI faculty survey*. Los Angeles, CA: Higher Education Research Institute, UCLA.

Dori, Y. J., & Belcher, J. (2005). How does technology-enabled active learning affect undergraduate students' understanding of electromagnetism concepts? *Technology, 14*(2), 243–279.

Drew, C. (2011, November 4). Why science majors change their minds (it's just so darn hard). *The New York Times*. Retrieved from http://www.nytimes.com/2011/11/06/education/edlife/why-science-majors-change-their-mind-its-just-so-darn-hard.html?_r=1&scp=1&sq=christopher%20drew%20science%20majors&st=cse

Durkheim, E. (2005) *Suicide: A study in sociology*. In J. A. Spaulding & G. Simpson (Trans.) London and New York, NY: Routledge and Kegan Paul.

Eagan, K., Hurtado, H., Figueroa, T. & Hughes, B. (2014). *Examining STEM pathways among students who began college at four-year institutions*. Commissioned paper prepared for the Committee on Barriers and Opportunities in completing 2-year and 4-year STEM degrees, National Academy of Sciences, Washington, DC. Retrieved April, 2015, from: http://sites.nationalacademies.org/cs/groups/dbassesite/documents/webpage/dbasse_088834.pdf

Eagan, M. K., Hurtado, S., & Chang, M. J. (2010, November). *What matters in STEM: Institutional contexts that influence bachelor's degree completion rates*. Paper presented at the annual meeting of the Association for the Study of Higher Education, Indianapolis, IN.

Eagan, M. K., Hurtado, S., Figueroa, T., & Hughes, B. (2012). *Examining STEM pathways among students who begin college at four-year institutions*. Los Angeles, CA: Higher Education Research Institute, UCLA.

Ege, S. N., Coppola, B. P., & Lawton, R. G. (1997). The University of Michigan undergraduate chemistry curriculum. 1. Philosophy, curriculum, and the nature of change. *Journal of Chemical Education, 74*(1), 74–83.

Elizondo-Montemayor, L., Hernández-Escobar, C., Ayala-Aguirre, F., & Aguilar, G. M. (2008). Building a sense of ownership to facilitate change: The new curriculum. *International Journal of Leadership in Education, 11*(1), 83–102.
Ellis, J., Fosdick, B. K., & Rasmussen, C. (2016). Women 1.5 times more likely to leave STEM pipeline after Calculus compared to men: Lack of mathematical confidence a potential culprit. *PLoS One, 11*(7), e0157447. https://doi.org/10.1371/journal.pone.0157447
Espinosa, L. L. (2011). Pipelines and pathways: Women of color in undergraduate STEM majors and the college experiences that contribute to persistence. *Harvard Educational Review, 81*(2), 209–240.
Festinger, L. (1961). The psychological effects of insufficient rewards. *American Psychologist, 16*(1), 1–11.
Freeman, S., Eddy, S. L., McDonough, M., Smith, M. K., Okoroafor, N., Jordt, H., & Wenderoth, M. P. (2014). Active learning increases student performance in science, engineering and math. *Proceedings of the National Academies of Science., 111*(23), 8410–8415.
Garcia, G. A., & Hurtado, S. (2011). *Predicting Latina/o STEM persistence at HSIs and non-HSIs*. Paper presented at the American Educational Research Association annual meeting, New Orleans, LA.
Gasiewski, J. A., Eagan, M. K., Garcia, G. A., Hurtado, S., & Chang, M. (2012). From gatekeeping to engagement: A multi-contextual, mixed method study of student academic engagement in introductory STEM courses. *Research in Higher Education, 53*(2), 229–261.
Gates, S. J., & Mirkin, C. (2012). Engage to excel. *Science, 335*(6076), 1545.
Gayles, J. G., & Ampaw, F. (2014). The impact of college experiences on degree completion in STEM fields at four-year institutions: Does gender matter? *The Journal of Higher Education, 85*(4), 439–468.
Geertz, C. (1973). Thick description: Toward an interpretive theory of culture. In *The interpretation of cultures: Selected essays* (pp. 3–30). New York, NY: Basic Books.
Goldrick-Rab, S. (2010). Challenges and opportunities for improving community college student success. *Review of Educational Research, 80*(3), 437–469.
Good, C., Rattan, A., & Dweck, L. S. (2012). Why do women opt out? Sense of belonging and women's representation in mathematics. *Journal of Personality and Social Psychology, 102*(4), 700.
Graham, M. J., Frederick, J., Byars-Winston, A., Hunter, A.-B., & Handelsman, J. (2013). Increasing persistence of college STEM majors. *Science, 341*, 1455–1456.
Griffith, A. L. (2010). Persistence of women and minorities in STEM field majors: Is it the school that matters? *Economics of Education Review, 29*(6), 911–922. https://doi.org/10.1016/j.econedurev.2010.06.010
Hall, R. M., & Sandler, B. (1984). *Out of the classroom: A chilly campus climate for women? Project on the status and education of women*. Washington, DC: Association of American College.
Handelsman, J., Ebert-May, D., Beichner, R., Bruns, P., Chang, A., DeHaan, R., … Wood, W. B. (2004). Scientific teaching. *Science, 304*(5670), 521–522. https://doi.org/10.1126/science.1096022
Hardin, E. E., & Longhurst, M. O. (2016). Understanding the gender gap: Social cognitive changes during an introductory STEM course. *Journal of Counseling Psychology, 63*(2), 233–239. https://doi.org/10.1037/cou0000119
Henderson, C., & Dancy, M. H. (2007). Barriers to the use of research-based instructional strategies: The influence of both individual and situational characteristics. *Physical Review Special Topics–Physics Education Research, 3*(2), 020102.
Henderson, C., & Dancy, M. H. (2008). Physics faculty and educational researchers: Divergent expectations as barriers to the diffusion of innovations. *American Journal of Physics, 76*(1), 79–91.
Henderson, C., Beach, A., & Finkelstein, N. (2011). Facilitating change in undergraduate STEM instructional practices: An analytic review of literature. Wiley Online Library. https://doi.org/10.1002/tea.20439

Herrara, F.A. & Hurtado, S. (2011). *Maintaining initial interests: Developing science, technology, engineering, and mathematics (STEM) career interests among underrepresented racial minority students*. Paper presented at the American Educational Research Association annual meeting, New Orleans, LA.

Hills, J. (1965). Transfer shock: The academic performance of the transfer student. *The Journal of Experimental Education., 33*(3), 201–215.

Holmegaard, H. T., Madsen, L. M., & Ulriksen, L. (2014). To choose sciences: Constructions of desirable identities among young people considering a STEM higher education programs. *Journal of Science Education, 36*(2), 186–215.

Hunter, A.-B., Laursen, S. L., & Seymour, E. (2007). Becoming a scientist: The role of undergraduate research in students' cognitive, personal, and professional development. *Science Education, 91*(1), 36–74. https://doi.org/10.1002/sce.20173. Retrieved 10/13/2006 from http://www3.interscience.wiley.com/cgi-bin/jissue/108069556

Hurtado, S., Eagan, K., & Chang, M. (2010). *Degrees of success: Bachelor's degree completion rates among initial STEM majors*. Los Angeles, CA: Higher Education Research Institute, University of California.

Hurtado, S., Eagan, M. K., & Hughes, B. (2012). *Priming the pump or the sieve: Institutional contexts and URM STEM degree attainments*. Los Angeles, CA: Higher Education Research Institute, UCLA.

Hyde, M. S., & Gess-Newsome, J. (2000). Adjusting education practice to increase female persistence in the sciences. *Journal of College Student Retention, 1*(4), 335–355.

Islam, M. M., & Al-Ghassani, A. (2015). Predicting college math success: Do high school performance and gender matter? Evidence from Sultan Qaboos University in Oman. *International Journal of Higher Education, 4*(2), 67. https://doi.org/10.5430/ijhe.v4n2p67

Jackson, D. L. (2013). Making the connection: The impact of support systems on female transfer students in science, technology, engineering, and mathematics (STEM). *The Community College Enterprise., 19*(1), 19–33.

Jaeger, A. J., Eagan, M. K., & Wirt, L. G. (2008). Retaining students in science, math, and engineering majors: Rediscovering cooperative education. *Journal of Cooperative Education and Internships, 42*(1), 20–32.

Johnson, A. C. (2007). Unintended consequences: How science professors discourage women of color. *Science Education, 91*, 805–821.

Jones, M. T., Barlow, A. E., & Villarejo, M. (2010). The importance of undergraduate research for minority persistence and achievement in biology. *The Journal of Higher Education, 81*(1), 82–115.

Kim, M. M., Rhoades, G., & Woodard, D. B. (2003). Sponsored research versus graduating students? Intervening variables and unanticipated findings in public research universities. *Research in Higher Education, 44*(1), 51–81.

Kochhar, R., & Cilluffo, A. (2018). *Income inequality in the U.S. is rising most rapidly among Asians*. http://www.pewsocialtrends.org/2018/07/12/income-inequality-in-the-u-s-isrising-most-rapidly-among-asians/

Kokkelenberg, E. C., & Sinha, E. (2010). Who succeeds in STEM studies? An analysis of Binghamton University undergraduate students. *Economics of Education Review, 29*(6), 935–946. https://doi.org/10.1016/j.econedurev.2010.06.016

Kuh, G. D., Kinzie, J., Schuh, J. H., Whitt, E. J., & Associates. (2005). *Student success in college: Creating conditions that matter*. San Francisco, CA: Jossey-Bass.

Kuh, G. D., Schuh, J., Whitt, E. J., & Associates. (1991). *Involving colleges: Successful approaches to fostering student learning and development outside the classroom*. San Francisco, CA: Jossey-Bass.

Laursen, S. L., Hunter, A.-B., Seymour, E., Thiry, H., & Melton, G. (2010). *Undergraduate research in the sciences: Engaging students in real science*. San Francisco, CA: Jossey Bass.

LeBeau, B., Harwell, M., Monson, D., Dupuis, D., Medhanie, A., & LeBeau, T. R. (2012). Student and high school characteristics related to completing a science, technology, engineering or mathematics (STEM) major in college. *Research in Science & Technological Education, 30*(1), 17–28.

Lee, Y.-G., & Ferrare, J. J. (2019). Finding one's place or losing the race? The consequences of STEM departure for college dropout and degree completion. *The Review of Higher Education.*, *43*(1), 221–261.

Lindberg, S. M., Hyde, J. S., Petersen, J. L., & Linn, M. C. (2010). New trends in gender and mathematics performance: A meta-analysis. *Psychological Bulletin, 136*(6), 1123. https://doi.org/10.1037/a0021276

Litzler, E., & Young, J. (2012). Understanding the risk of attrition in undergraduate engineering: Results from the project to assess climate in engineering. *Journal of Engineering Education, 102*(2), 319–345.

Maltese, A. V., & Tai, R. H. (2011). Pipeline persistence: Examining the association of educational experiences with earned degrees in STEM among U.S. students. *Science Education, 95*(5), 877–907.

Manis, J., Sloat, B. F., Thomas, N. G., & Davis, C.-S. (1989). *An analysis of factors affecting choices of majors in science, mathematics, and engineering at the University of Michigan.* Michigan, MI: Center for Continuing Education of Women, University of Michigan.

McLelland, L. (1993). *Students entering science, mathematics and engineering majors as fall freshmen, 1980–1988.* Unpublished data provided by L. McLelland, Director, Office of Research and information, University of Colorado at Boulder.

Merton, P., Froyd, J. E., Clark, M. C., & Richardson, J. (2009). A case study of relationships between organizational culture and curricular change in engineering education. *Innovative Higher Education, 34*(4), 219–233.

Morrison, C., & Williams, L. E. (1993). Minority engineering programs: A case for institutional support. *NACME Research Letter, 4*(1), 1–11.

Mortenson, T. (1995, January 26–29). *Financial opportunity for post-secondary education.* Paper to the NACME Research and Policy Conference: Minorities in mathematics, science and engineering. Wake Forest University, Winston-Salem, NC.

Museus, S. D., Agbayani, A., & Chang, D. M. (Eds.) (2016). *Focusing on the underserved: Immigrant, refugee, and indigenous Asian-American and Pacific Islanders in Higher Education.* University of Hawai'i, Information Age Publishing

Museus, S. D., & Liverman, D. (2010). High-performing institutions and their implications for studying underrepresented minority students in STEM. *New Directions for Institutional Research, 148*, 17–27.

Nagda, B. A., Gregerman, S. R., Jonides, J., von Hippel, W., & Lerner, J. S. (1998). Undergraduate student-faculty research partnerships affect student retention. *The Review of Higher Education, 22*(1), 55–72.

Nasr, K., Pennington, J., & Andres, C. (2004). A study of students' assessments of cooperative education outcomes. *Journal of Cooperative Education, 38*(1), 13–21.

National Academies of Sciences, Engineering, and Medicine. (2016). *Barriers and opportunities for 2-year and 4-year STEM degrees: Systemic change to support students' diverse pathways.* National Academies Press.

National Science Foundation. (2012). *Science and engineering indicators 2012.* Washington, DC: National Science Board.

National Center for Educational Statistics CIP User Site. (2019, May 1). Retrieved from: https://nces.ed.gov/ipeds/cipcode/browse.aspx?y=55

Nicholls, G. M., Wolfe, H., Besterfield-Sacre, M., & Shuman, L. J. (2010). Predicting STEM degree outcomes based on $8^{th}$ grade data and standard test scores. *Journal of Engineering Education, 99*(3), 209–223.

Nuwer, H. (2001). *Wrongs of passage.* Bloomington, IN: Indiana University Press.

Nuwer, H. (2004). *The hazing reader.* Bloomington, IN: Indiana University Press.

Olson, S., & Riordan, D. G. (2012). *Engage to excel: Producing one million additional college graduates with degrees in science, technology, engineering, and mathematics* (Report to the President) (130 p.). Executive Office of the President.

Ong, M., Wright, C., Espinosa, L., & Orfield, G. (2011). Inside the double bind: A synthesis of empirical research on undergraduate and graduate women of color in science, technology, engineering, and mathematics. *Harvard Educational Review, 81*(2), 172–208.

Ost, B. (2010). The role of peers and grades in determining major persistence in the sciences. *Economics of Education Review, 29*(6), 923–934. https://doi.org/10.1016/j.econedurev.2010.06.011
Packard, B. W. (2005). Mentoring and retention in college science: Reflections on the sophomore year. *Journal of College Student Retention, 6*(3), 289–300.
Palmer, R. T., Maramba, D. C., & Dancy, T. E. (2011). A qualitative investigation of factors promoting the retention and persistence of students of color in STEM. *The Journal of Negro Education, 81*(4), 491–504.
Pascarella, E., & Terenzini, P. (2005). *How college affects students: A third decade of research* (Vol. 2). San Francisco, CA: Jossey-Bass.
Pascarella, E., & Terenzini, P. T. (1991). *How college affects students: Findings and insights from twenty years of research.* San Francisco, CA: Jossey-Bass.
Perna, L., Lundy-Wagner, V., Drezner, N. D., Gasman, M., Yoon, S., Bose, E., & Gary, S. (2009). The contributions of HBCUs to the preparation of African-American women for STEM careers: A case study. *Research in Higher Education, 50*, 1–23.
Porter, O. (1990). *Undergraduate completion and persistence at four-year colleges and universities: Completers, persisters, stop-outs and drop-outs*. Washington, DC: National Institute of Independent Colleges and Universities.
Price, J. (2010). The effect of instructor race and gender on student persistence in STEM fields. *Economics of Education Review, 29*(6), 901–910.
QSR International (2016). *NVivo 10.0*. https://www.qsrinternational.com/nvivo/what-is-nvivo
Rask, K. (2010). Attrition in STEM fields at a liberal arts college: The importance of grades and pre-collegiate preferences. *Economics of Education Review, 29*(6), 892–900. https://doi.org/10.1016/j.econedurev.2010.06.013
Reich, D. (2011, November 9). Why engineering majors change their minds. *Forbes*. Retrieved from http://www.forbes.com/sites/danreich/2011/11/09/why-engineering-majors-change-their-minds/
Reyes, M. (2011). Unique challenges for women of color in STEM transferring from community colleges to universities. *Harvard Educational Review, 81*(2), 241–262.
Riegle-Crumb, C., & King, B. (2010). Questioning a white male advantage in STEM: Examining disparities in college major by gender and race/ethnicity. *Educational Researcher, 39*(9), 656–664.
Rodriguez, C., Kirshstein, R., Amos, L., Jones, W., Espinosa, L., & Watnick, D. (2012). *Broadening participation IN STEM: A call to action*. Washington, DC: American Institutes for Research.
Rosser, S. V. (1990). *Female-friendly science: Applying women's study methods and theories to attract students*. New York, NY: Pergamon.
Rotberg, I. (1990). Sources and reality: The participation of minorities in science and engineering education. *Phi Delta Kappan, 71*, 358–361.
Rothwell, J. (2013). *The hidden STEM economy: Metropolitan policy program at brookings institution*. Retrieved April, 2015, from http//www.brookings.edu/~/media/research/files/reports/2013/06/10-stem-economy-rothwell/thehiddenstemeconomy610.pdf
Salzman, H. (2013). What shortage? The real evidence about the STEM workforce. *Issues in Science and Technology, 29*(4). Retrieved from http://issues.org/29-4/what-shortgae-th-real-evidence-about-the-stem-workforce/
Science editorial. (1992). Minorities in science: The pipeline problem. *Science, 285*(5085), 1175.
Seidel, J. V., Kjolseth, J. R., & Seymour, E. (1988). *The ethnograph: A user's guide*. Littleton, CO: Qualis Research Associates.
Seymour, E., & DeWelde, C. (2016). Why doesn't knowing change anything? Constraints and resistance, leverage and sustainability (with K. De Welde). In G. C. Weaver, W. D. Burgess, A. L. Childress, & L. Slakey (Eds.), *Transforming institutions: Undergraduate STEM education for the 21st century*. Purdue, IN: Purdue University Press.
Seymour, E., & Fry, C. F. (2016). The Reformers' Tale: Determining progress in improving undergraduate STEM education. In G. C. Weaver, W. D. Burgess, A. L. Childress, & L. Slakey (Eds.), *Transforming institutions: Undergraduate STEM education for the 21st century*. Purdue, IN: Purdue University Press.

Seymour, E., & Hewitt, N. M. (1997). *Talking about leaving: Why undergraduates leave the sciences*. Boulder, CO: Westview Press.
Seymour, E., Hunter, A.-B., Laursen, S., & DeAntoni, T. (2004). Establishing the benefits of research experiences for undergraduates in the sciences: First findings from a three-year study. *Science Education, 88*(4), 493–534.
Shi, R. (2011, November 16). An alternative to grade inflation. *The Daily Texan*. Retrieved from http://www.dailytexanonline.com/opinion/2011/11/16/alternative-grade-inflation
Spradley, J. P. (1979). *The ethnographic interview*. Belmont, CA: Wadsworth Group/Thomson Learning.
Strayhorn, T. L., Long, L. L., Kitchen, J. A., Williams, M. S., & Stentz, M. E. (2013). *Academic and social barriers to Black and Latino male collegians' success in engineering and related STEM fields*. Atlanta, GA: American Society for Engineering Education.
Suresh, R. (2007). The relationship between barrier courses and persistence in engineering. *Journal of College Student Retention: Research, Theory, and Practice, 8*(2), 215–239.
Tai, R. H., Liu, C. Q., Maltese, A. V., & Fan, X. (2006). Planning early for careers in science. *Science, 312*(5777), 1143–1144.
Tate, E. D., & Linn, M. C. (2005). How does identity shape the experiences of women of color engineering students? *Journal of Science Education and Technology., 14*(5/6), 483–493.
Taylor, T. (2011, November 14). *Grade inflation and choice of major* [Web log post: The conversable economist]. Retrieved from http://conversableeconomist.blogspot.com/2011/11/grade-inflation-and-choice-of-major.html
Tinto, V. (1975, Winter). Dropout from higher education: A theoretical synthesis of recent research. *Review of Educational Research, 45*(1), 89–125.
Tobias, S. (1990). *They're not dumb, they're different: Stalking the second tier*. Tucson, AZ: Research Corporation.
Tobias, S. (1992). *Revitalizing undergraduate science: Why some things work and most don't*. Tucson, AZ: Research Corporation.
U S Department of Education. (2014). *2013–2014 civil rights data collection: Key data insights on equity and opportunity gaps in our nation's public schools*. Washington DC: Office for Civil Rights.
Van Maanen, J. (1988). *Tales of the field: On writing ethnography*. Chicago, IL: University of Chicago Press.
Walczyk, J. J., Ramsey, L. L., & Zha, P. (2007). Obstacles to instructional innovation according to college science and mathematics faculty. *Journal of Research in Science Teaching, 44*(1), 85–106.
Whalen, D. F., & Shelley, M. C. (2010). Academic success for STEM and non-STEM majors. *Journal of STEM Education, 11*(1–2), 45–60.
Wieman, C., Perkins, K., & Gilbert, S. (2010). Transforming science education at large research universities s: A case study in progress. *Change: The Magazine of Higher Learning, 42*(2), 7–14.
Zhang, Y., & Allen, T. O. (2015). Challenges and support: Transfer experiences of community college engineering students. *Journal of Applied Research in Community College, 22*(1), 43–50.

# Chapter 2
# Patterns of Switching and Relocation

Timothy J. Weston

## What Is the Rate of Switching Nationally?

The 2013 National Center of Educational Statistics (NCES) report *STEM Attrition: College Students' Paths Into and Out of STEM Fields* (Chen, 2013) examined data from the 2004–2009 Beginning Postsecondary Students Longitudinal Study and the 2009 Postsecondary Education Transcript Study. These studies used a large (7800 students) random sample of college students and collected information at three points during the 6 years of the research. The NCES study found that 48% of students pursuing a bachelors' degree entering STEM majors either left school before graduating or switched to a non-STEM major. Twenty-eight percent of students stayed in college, but switched out of their STEM major to a non-STEM major. It is this group of students who are the central focus of this book.

The NCES study also presented descriptive analyses about which students in which majors switch at what rates, and where students move to after they leave STEM. The discipline with the highest switching rates was biology/life sciences (30%), the lowest engineering (21%). Business was the most popular destination non-STEM major for these students (22%), and more women switched out of STEM majors than men (32% vs 26%). The NCES study also identified students with higher rates of switching as those with poorer pre-college preparation, especially in

T. J. Weston (✉)
Ethnography and Evaluation Research (E&ER), University of Colorado Boulder, Boulder, CO, USA

National Center for Women and Information Technology (NCWIT), University of Colorado Boulder, Boulder, CO, USA
e-mail: westont@colorado.edu



E. Seymour, A.-B. Hunter (eds.), *Talking about Leaving Revisited*,
https://doi.org/10.1007/978-3-030-25304-2_2

mathematics. During their first year of college, STEM switchers earned fewer science credits and were enrolled in fewer higher-level math courses than STEM-persisters. They also experienced more "academic duress," defined as failure or withdrawal from courses and achieved lower overall GPAs.

The NCES researchers used regression models to weigh the individual contribution of different factors associated with STEM switching. Their statistical models showed that students who took a low credit load of STEM courses tended to switch, as well as students who did not take advanced math courses. Lower GPA's in STEM courses and lower overall GPA were also associated with switching, and women switched at higher rates than men. While both African-American and Hispanic students switched more than other groups, in the NCES regression analysis, differences between groups were explained better by other variables such as GPA. Poor high school preparation, very low high school GPA, and lower high school mathematics credits all predicted switching.

The Higher Education Research Institute (HERI) also provides estimates of undergraduate students switching out of STEM through its Cooperative Institutional Research Program (CIRP) data (Eagan, Hurtado, Figueroa, & Hughes, 2014). This study matches data from HERI's annual Freshman Survey (TFS) given to hundreds of thousands of students with transcript data from the National Student Clearinghouse (NSC) for the same years 2004–2010 as the NCES study (Eagan et al., 2014). In the latest version of the survey, 28.1% of students who answered the survey after 4 years had switched to another non-STEM major. However, this figure does not take into account those students who left college before the second survey. If we assume (as is stated in the NCES report) that 20% of STEM students leave the university altogether, the percentage of all students starting college in STEM and then switching to a non-STEM major would be closer to 23% in the CIRP data.

Some patterns in switching were consistent between the CIRP and NCES analyses. The switching rate for women was higher than for men (30.6% vs. 25.8%), relative differences in switching rates between biology (30.6%) and engineering (22.6%) were similar, as were much higher switching rates for African-American (41.7%) and Hispanic (40.6%) students than for white students (27.8%). The number of students who switched out of STEM to business majors (17% vs. 22%) was also comparable. As we will show later in this chapter, most of these relative comparisons are also reflected in our non-random sample of six institutions.

## *Have Switching Rates Changed Since the Publication of Talking About Leaving?*

Gauging whether switching rates have changed in the 20 years since publication of *Talking about Leaving* is complicated by changes over time in how students were sampled by the CIRP survey, and by differences between CIRP and NCES sampling methods. Between years 1991 and 2011, the student survey administered by HERI switched from administering the survey for free to charging for their service. This changed the composition of institutions who participated with private institutions making up a greater part of the sample. On average, students at private institutions leave STEM majors at lower rates than those students at public institutions and the gap between private and public institutions is approximately 6% in the NCES study.

Comparisons between current CIRP and NCES estimates are also confounded by other issues. Because students complete the CIRP Freshman Survey during summer orientation or in the first few weeks of their first fall term, students report their intended majors as opposed to actual declared majors. It is possible that some students intending to become STEM majors changed their minds before declaring their major and chose a major in another non-STEM field, or vice versa. The CIRP analyses comprise only five disciplinary STEM groupings: engineering, biological sciences, physical sciences, mathematics and statistics, and agriculture. However, because it treats computer science and related technical majors separately as "other technical" majors, they are not included in overall counts of STEM persistence, relocation, and switching. The NCES surveys include students who leave college altogether and covers a 6-year timeframe, while CIRP does neither. Thus, the primary difference seen between the STEM persistence rates relate to the fact that the HERI College Senior Survey is neither designed nor intended to collect data from students who have left college prior to their fourth year. By contrast, the NCES survey is designed as a 6-year longitudinal study of the 2003 entering first-year cohort, regardless of whether or how long they persist as a college student. Thus, the NCES data include a large number of students who left college prior to earning a degree 6 years after first entering, all of whom are classified as STEM leavers if they originally reported a STEM major.

It is our view, therefore, that the best recent national estimates of losses from STEM majors are those provided by the 2004–2009 NCES survey, namely, that 28% of students who enter STEM majors leave them for non-STEM majors and another 20% leave college altogether.

Nevertheless, bearing these caveats in mind, we can examine differences in persistence and switching rates between the 1991 CIRP survey presented in the original study and the most recent, 2011 CIRP survey. Some differences stand out:

- Switching from STEM majors into non-STEM majors declined dramatically from 44.1% to 28.1%, a 16% improvement.
- Persistence in original STEM majors improved by 13.9% from 46.0% to 59.9%.
- Relocation within STEM majors increased slightly from 10% to 11.9%.
- For women, persistence increased substantially by 23.6% from 37% to 61.3%.

As indicated in Table 2.1, persistence for some disciplines has changed much more than for others. Persistence in mathematics and biology increased by 27.1% and 19.3% respectively, while persistence in computer science decreased by 5.3%.

Changes were also evident by gender within specific disciplines. As presented in Table 2.2, persistence for women increased by 34.7% in both mathematics and physical sciences, and by 25.4% in biology. Increases for men ranged between 19.8% in mathematics to a decrease of 4% in computer science.

**Table 2.1** Comparing CIRP Freshman Survey data results: Persistence rates by discipline, 1991–2011. (Corresponds to TAL Table 1.1 "Stayed in same major")

| Discipline | 1991 (%) | 2011 (%) | Difference (%) |
|---|---|---|---|
| Mathematics | 34.1 | 61.2 | 27.1 |
| Biology | 42.0 | 61.3 | 19.3 |
| Engineering | 51.4 | 65.1 | 13.7 |
| Physical science | 29.9 | 45.0 | 15.1 |
| Agriculture[a] | – | 36.4 | – |
| Computer science (technical) | 46.4 | 41.1 | −5.3 |
| All STEM | 46.0 | 59.9 | 13.9 |

[a]Results not reported for men. Reliable figure for total group missing

**Table 2.2** Comparing CIRP Freshman Survey data results: Persistence rates by discipline and gender, 1991–2011

| Discipline | Women 1991 | Women 2011 | Difference (Women) | Men 1991 | Men 2011 | Difference (Men) |
|---|---|---|---|---|---|---|
| Mathematics | 27.8 | 62.5 | 34.7 | 39.9 | 59.7 | 19.8 |
| Biology | 35.8 | 61.2 | 25.4 | 47.1 | 61.5 | 14.4 |
| Engineering | 50.8 | 62.5 | 11.7 | 51.5 | 66.1 | 14.6 |
| Physical science | 27.7 | 61.7 | 34.0 | 30.6 | 49.5 | 18.9 |
| Agriculture | 20.9 | 23.2 | 2.3 | n/a | 23.5 | – |
| Computer science (technical) | 30.8 | 44.1 | 13.3 | 54.2 | 50.2 | −4.0 |
| All STEM | 37.7 | 61.3 | 23.6% | 49.9 | 61.3 | 11.4% |

The rates of switching mirrored increases in persistence. As indicated in Table 2.3, the greatest decreases in switching between years were found in mathematics, physical sciences, and biology. The percentage of students switching within STEM disciplines ("STEM relocators") remained fairly constant, except in mathematics, where fewer students switched to another STEM major in 2011 than in 1991.

As illustrated in Table 2.4 switching rates by gender also varied substantially. The rate in mathematics fell precipitously from 72.3% in 1997 to 29.5% in 2017, a 42.8% difference. Rates in computer science (37.7%) and agriculture (42.1%) also fell steeply for women. Rates for men also fell, ranging from 27.1% in physical science, but gained 9.6% in computer science.

Given the differences in sampling methods between years, many of the changes in persistence and switching may be overstated. However, even if these findings underestimate the true percentages, they nevertheless indicate a trend that students are switching out of STEM majors at a much lower rate than they did 20 years ago. Reviewing all the estimates for switching and persistence that we discuss in this chapter, and, notwithstanding variations created by sampling and data collection methods, all sources (including those from our own study) converge on the conclusion that the rate of switching losses from STEM majors have substantially dropped since the CIRP estimates of rates of 38–63% across STEM majors (44% overall) that were cited in the original 1997 study.

**Table 2.3** Switching rates by discipline, 1991 and 2011

| Discipline | 1997 | 2017 | Difference | STEM relocator 1997 | STEM relocator 2017 | Difference |
|---|---|---|---|---|---|---|
| Mathematics | 62.7 | 30.2 | 32.5 | 11.4 | 8.6 | −2.8 |
| Biology | 51.0 | 30.6 | 20.4 | 7.1 | 8.1 | 1.0 |
| Engineering | 38.1 | 22.6 | 15.5 | 10.5 | 12.2 | 1.7 |
| Physical science | 51.2 | 28.5 | 22.7 | 18.9 | 26.4 | 7.5 |
| Agriculture | 47.2 | 36.4 | 10.8 | 0 | | |
| Computer science (technical) | 53.6 | 39.7 | 13.9 | 0 | | |
| All STEM | 44.1% | 28.1% | 16.0% | 10.0% | 11.9% | 1.9% |

*Note:* STEM relocator category corresponds to "Moved to major in same group" in TAL Table 1.1

**Table 2.4** Switching rates by discipline and gender, 1991 and 2011

| Discipline | Women 1991 | Women 2011 | Difference (Women) | Men 1991 | Men 2011 | Difference (Men) |
|---|---|---|---|---|---|---|
| Mathematics | 72.3 | 29.5 | 42.8 | 53.9 | 32.3 | 21.6 |
| Biology | 56.7 | 32.0 | 24.7 | 46.2 | 27.8 | 18.4 |
| Engineering | 37.1 | 22.7 | 14.4 | 38.3 | 22.7 | 15.6 |
| Physical science | 44.3 | 30.6 | 13.7 | 53.5 | 26.4 | 27.1 |
| Agriculture | 79.1 | 37.0 | 42.1 | n/a[a] | 35.3 | – |
| Computer science (technical) | 69.2 | 31.5 | 37.7 | 45.8 | 55.4 | −9.6 |
| All STEM | 52.4% | 30.2% | 22.2% | 41.2% | 25.8% | 15.4% |

Note: Switching difference computed 2011 minus 1991
[a]Missing in 1991 data

## Patterns of Switching in the Institutional Records Analyses

In this section we discuss the extent of and variations in the switching and persistence patterns of students found in the sample of six institutions in our study.

### *What Were the Rates of STEM Switching, Persistence, and Relocation in Our Sample?*

At the six institutions in our study, 12,565 students did not switch out of their STEM major, 2020 students switched out of a declared STEM major, and 1726 relocated to another STEM major. As percentages of all students who declared a STEM major, these are 86.2% persisters and 13.8% switchers. Of those counted as persisters, 13.7% were STEM-relocators who switched majors into a different STEM discipline. Another 16.4% relocated within their STEM discipline, sometimes as a result of administrative reclassification of major. We also included non-STEM majors who formed 55% of individuals in our total dataset, and "switch-in" students who started

**Table 2.5** Rates of switching at six participating institutions

| Institution | % Switched | Total *N* | *N* of Switchers | *N* of Persisters |
|---|---|---|---|---|
| PB4R1 | 10 | 2640 | 256 | 2383 |
| PV2R1 | 28 | 568 | 161 | 406 |
| PB3R1 | 9 | 4250 | 369 | 3881 |
| PB2R1 | 13 | 1847 | 249 | 1598 |
| PV1R3 | 7 | 115 | 8 | 107 |
| PB1R1 | 19 | 5168 | 977 | 4190 |
| | 13.8 | 14,626 | 2020 | 12,565 |

college in non-STEM disciplines and then switched into STEM, another 9% of all persisters.

Our figures for switchers are considerably lower than the 28% reported by NCES in 2013 that, as we discussed in the prior section, are currently the most accurate national estimates available. The far more conservative figure in our findings reflects the characteristics and limitations of our dataset:

- The transcript records provided did not include students who left or transferred out of the institution after starting a STEM major. These were students who switched out of STEM and then left college directly from STEM majors;
- We also had no way of accounting for STEM majors who had switched before they transferred into our six institutions;
- We received current records for juniors, seniors (including fifth-year seniors and graduates), thus, a portion of these students had not yet finished school and still might decide to switch. Although this number is small among the records of graduates (e.g. 3%), given the low rates of switching among seniors found in our data for graduates, our analyses only include "survivors"—those who graduated or were still enrolled as of their junior or senior years;
- Some of the institutions in our (historically-derived) institutional sample also proved to have lower switching rates than the national average. We were made aware during site visits of ongoing efforts by all six schools to improve STEM retention—a response, in part, to their particular awareness of findings from our original study.

Notwithstanding these caveats, our total sample of 14,585 STEM majors proved more than adequate to the task of discovering patterns that enable our understanding of what institutional and disciplinary contexts and student characteristics are highly connected with the risks of switching and relocation. It is these patterns that we address in the balance of this chapter.

## Variations by Institution and Discipline

As shown below in Table 2.5, switching rates varied considerably by institution. However, while much of our analysis aggregates across schools and treats students as one big group, we adhered to a rule of thumb in our analyses by testing each set

**Table 2.6** Rates of STEM switching and persistence by STEM discipline at six participating institutions

| Discipline | % Switched | Total *N* | *N* of Switchers | *N* of Persisters |
|---|---|---|---|---|
| Agriculture | 12 | 1091 | 132 | 959 |
| Environmental science | 13 | 703 | 92 | 611 |
| Computer science | 18 | 638 | 116 | 522 |
| Engineering | 8 | 6421 | 525 | 5896 |
| Biology/life sciences | 20 | 4564 | 928 | 3636 |
| Mathematics | 24 | 414 | 100 | 314 |
| Physical sciences | 17 | 742 | 126 | 616 |
| | 13.8 | 14,573 | 2020 | 12,553 |

of findings to learn if the pattern seen for the whole institutional sample was also present at each school. For instance, we found that women switched more than men both at each institution and at all schools taken together. If findings did not generalize, we either note this in the analysis or do not present the findings.

We are also aware that our sample, while very large in terms of students and student records, is limited to six institutions. The analysis, as such, is descriptive of these students and schools and may or may not be representative of a wider population. However, we do know that the large public universities in our sample and the smaller private schools are not outliers in terms of student demographic representation, gender balance, socio-economic representation, or academic selectivity compared with similar schools in the USA.

As shown in Table 2.5, the number of switchers varied substantially across institutions with a high of 28% at PV2R1 and a low of 7% at PV1R3, both of which are smaller, private schools. Overall, 13.8% of students in our data switched out of STEM majors.

Students also switched at different rates, depending on the discipline of their declared STEM major. As outlines in Table 2.6, we saw the highest rates of switching in biology (20%), mathematics (24%), and computer science (18%), and the lowest rate in engineering (9%).

## *Which Students Are at Greater Risk of Switching?*

We also examined switching rates by the demographic and academic characteristics of students, and combinations of these characteristics. We often found switching patterns to be complex with unexpected interactions between factors such as gender and mathematics score. As reflected in Table 2.7, we found that women switch at higher rates than men, with 18.3% of women and 11% of men switching. The gap between genders is 7.3%. As indicated in Fig. 2.1, the difference between men and women varies little across institutions with women switching more than men at every college in our sample.

**Table 2.7** Rates of STEM switching and persistence by gender

| Gender | % switch | *N* | Switchers | Persisters |
|---|---|---|---|---|
| Women | 18.3 | 5696 | 1041 | 4654 |
| Men | 11.0 | 8864 | 979 | 7885 |

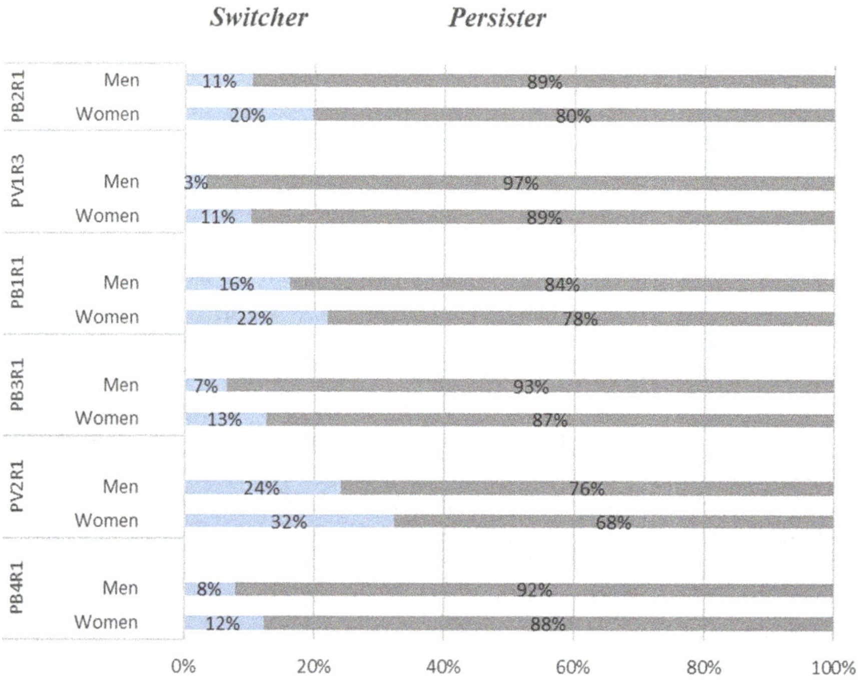


**Fig. 2.1** Rates of switching and persistence by gender at each institution

We used the Underrepresented Minority (URM) designation for many of our analyses. This group comprises students who are African-American, Native American, Hispanic/Latino, or Native Hawaiian/Pacific Islander. The switching rate for URM students taken together was 20% compared with 13.4% for all other students. As indicated in Table 2.8, African-American, and Hispanic students switched at higher rates than White and other student groups.

As shown in Fig. 2.2, the gap in switching rates for URM and non-URM students was not uniform across institutions and switching rates between groups were almost the same at two universities.

As explained in the research methods discussion, based on the expectations of STEM faculty that standardized math scores at, or above, certain levels were a good predictor of a student's ability to progress from foundational to more advanced STEM courses, we used these faculty-based cut-off scores as way of grouping our interview sample. We also found that switching rates varied substantially for

**Table 2.8** Switching and persistence rates by racial/ethnic groups

| Race/ethnicity | % Switch | *N* | Switchers | Persisters |
|---|---|---|---|---|
| American Indian or Alaska Native | 13 | 53 | 7 | 46 |
| Asian | 12 | 1201 | 145 | 1056 |
| African-American | 22 | 359 | 78 | 281 |
| Hispanic/Latino | 19 | 782 | 154 | 628 |
| Multi-racial | 26 | 240 | 62 | 178 |
| Native Hawaiian/Other Pacific Islander | 18 | 11 | 2 | 9 |
| White (non-hispanic) | 14 | 10759 | 1501 | 9258 |
| Non-resident alien | 4 | 643 | 23 | 620 |
| | 13.8 | 14,426 | 2020 | 12,406 |

*Note:* Small numbers of Native Hawaiian and Native-American students make percent estimates of switching unreliable
*Note:* Total includes unknown race/ethnicity

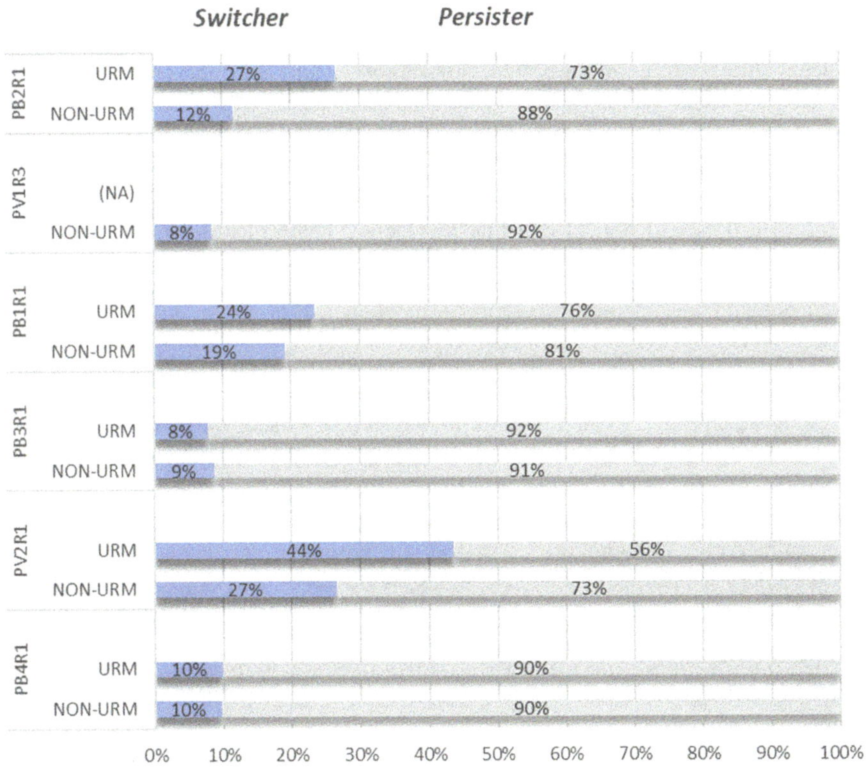


**Fig. 2.2** Percentage of switchers and persisters for URM students at each institution. *Note*: URM group at PV1R3 too small to make stable comparison

**Table 2.9** Standardized math blocks corresponding to ACT and SAT math score blocks

| Block | ACT-M | SAT-M |
|---|---|---|
| Lowest (L) | Lowest–24 | Lowest–560 |
| Low-medium (LM) | 25–27 | 561–620 |
| High-medium (HM) | 28–30 | 621–680 |
| Highest (H) | 31–highest | 681–highest |

**Table 2.10** Percentage of students switching by standardized math score block

| Standardized math (block) | % Switch | *N* | *N* of switchers | *N* of persisters |
|---|---|---|---|---|
| Lowest | 24 | 3739 | 883 | 2856 |
| Low | 16 | 2226 | 408 | 1818 |
| High | 12 | 3624 | 418 | 3206 |
| Highest | 7 | 2870 | 192 | 2678 |
| | 13.8 | 12,859 | 1900 | 10,959 |
| (No score reported by institution) | | 1729 | 120 | 1609 |

students entering the university with different ACT and SAT math scores. Indeed, standardized math scores proved to be one predictor of student switching, with lower-scoring students switching more often than their higher-scoring peers. For the purposes of our institutional records analyses, we combined the scores from the ACT and SAT math tests available in the institutional records and divided the distribution of scores into four blocks that represented math quartiles (as indicated in Table 2.9).

As is evident in Table 2.10 and Fig. 2.3, we saw a strong linear relationship between mathematics scores and switching rates with 24% of our students in the lower-math block switching out of STEM, and only 7% of students switching in highest block.

As shown in Fig. 2.4, the relationship between math scores and switching rates was constant across institutions in our sample with students with lower math scores switching at higher rates.

The patterns by gender, URM status, and math score block became clearer when we examined rates of switching for student sub-groups comprising combinations of gender, URM status, and math score block. For example, by comparing men and women at each math score block, we found that:

- Across math score blocks, women switched from 3% to 8% more than men.
- URM students switched at almost identical rates to non-URM students in the medium math score blocks,
- But *both* highest and lowest scoring URM students switched out at higher rates than non-URM students: 4% higher for top scoring students and 6% higher for lowest.
- Gaps between switching rates for men and women remained almost uniformly constant across math score blocks at each institution.

**Fig. 2.3** Percentage of switchers by mathematics score block, highest to lowest

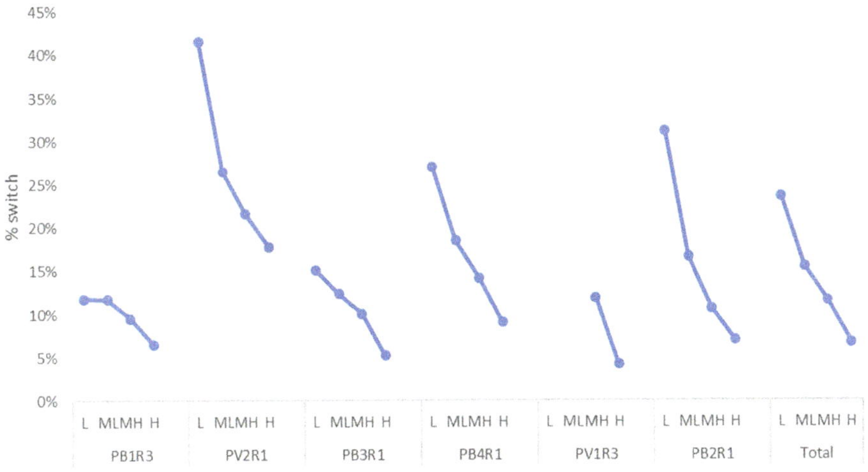


**Fig. 2.4** Percent of students switching by mathematics score blocks at each institution. *Note*: Two lowest math score categories at PV1R3 too small to make stable comparison

The patterns of variation in STEM switching that result from examining combinations of math score blocks with gender and with URM status are represented in Figs. 2.5 and 2.6.

Differences in rates of switching were constant between URM and non-URM students for both men and women. In Fig. 2.7 women in both URM and non-URM categories switched more often than men.

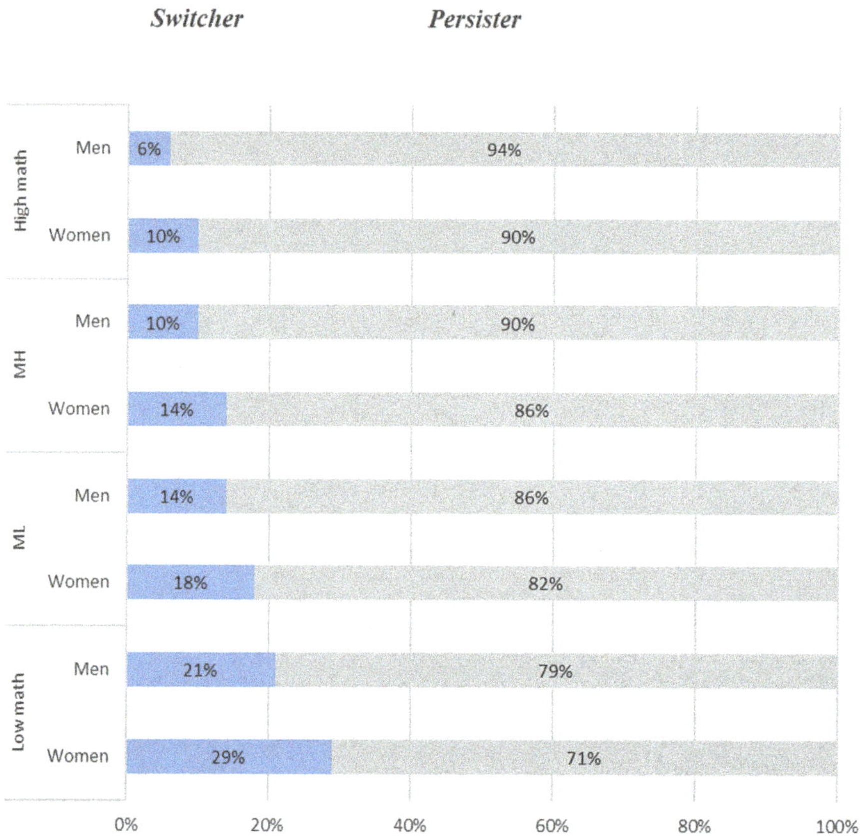


**Fig. 2.5** Percentage of switchers by math score block and gender. *Note*: Uses available math scores $n = 12{,}859$

Analysis of the three-way combinations of math score, gender, and URM status also revealed an unexpected pattern that was masked in the two-way combinations. As indicated in Fig. 2.8:

- URM women in the lowest math block switched out of STEM at a much higher rate (41%) than their peers.
- By contrast, higher-math-ability URM women switched at about the same rate as non-URM women (14% vs. 12%). URM men switched at higher rates than non-URM men in the highest math block (12% vs. 7%).
- At four of the six institutions, switching rates for URM students followed the same pattern of low and high math URM students switching at greater rates than non-URM students.

We also examined Pell grant status for students where these data were available. Pell grant recipients typically come from lower-income families. In our data, this

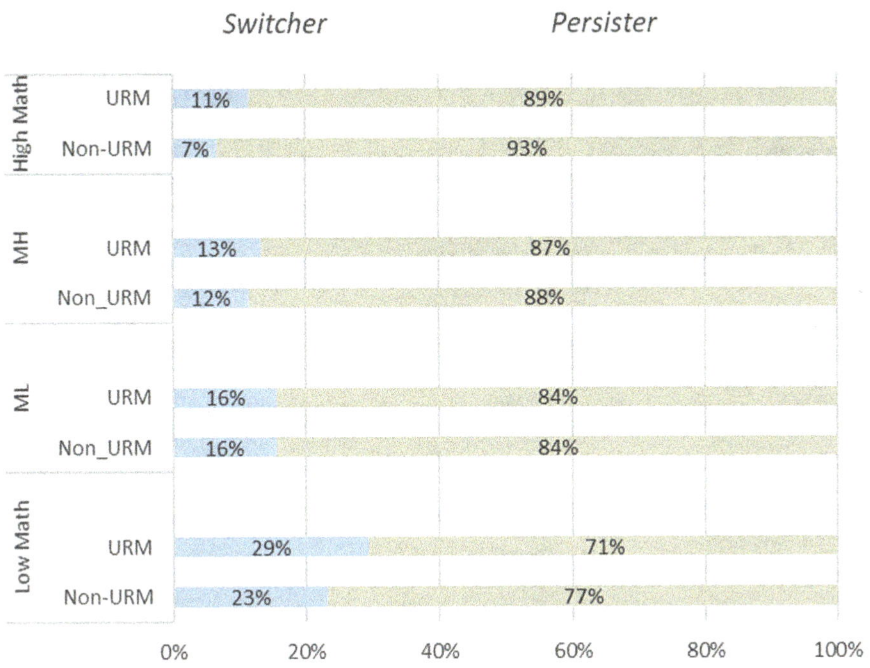


**Fig. 2.6** Percentage of switchers by math score block and URM status

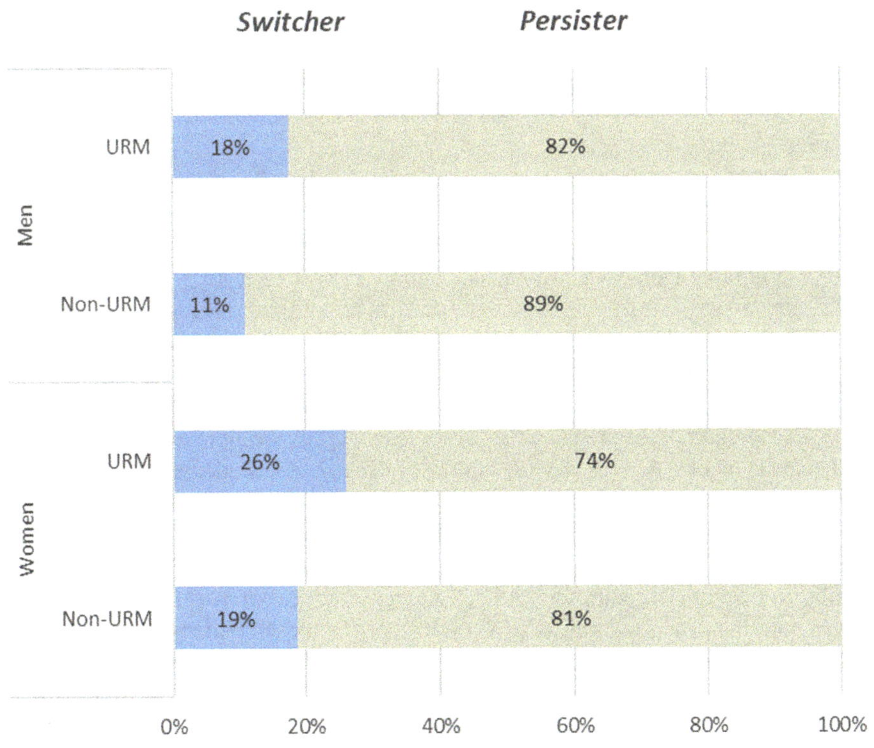


**Fig. 2.7** Percentage of switchers by gender and URM status

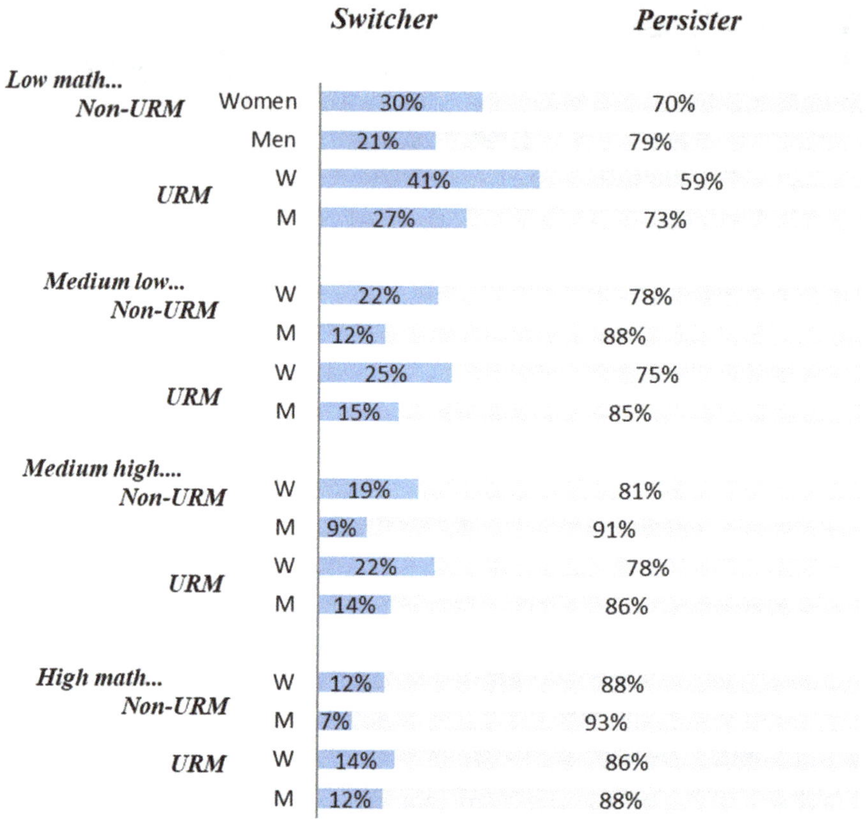


**Fig. 2.8** Percentage of switchers by Gender, URM status and Math Block

proxy for family income was somewhat confounded with URM status, with 50% of URM students receiving Pell grants versus 22% of non-URM students. Consequently, their rates of switching were similar to those of all URM students with male and female students showing very different patterns of switching (c.f., Fig. 2.9).

We also looked at first semester Grade Point Average (GPA) which is, in some ways, an indicator of the preparedness of students. Again, we divided the distribution into quartile blocks. As was also the case with math scores, students with lower first semester GPAs switched at higher rates than their peers who received better grades (c.f., Fig. 2.10). This pattern held at four of the six institutions.

As shown in Fig. 2.9, the pattern of women switching at higher rates than men is similar to the pattern found with ACT/SAT math scores. Women switch at an 8% higher rate than men regardless of GPA except for those students with very low GPA's where the difference is 5%. The gap between men and women for each GPA block remained uniform across institutions (Fig. 2.11).

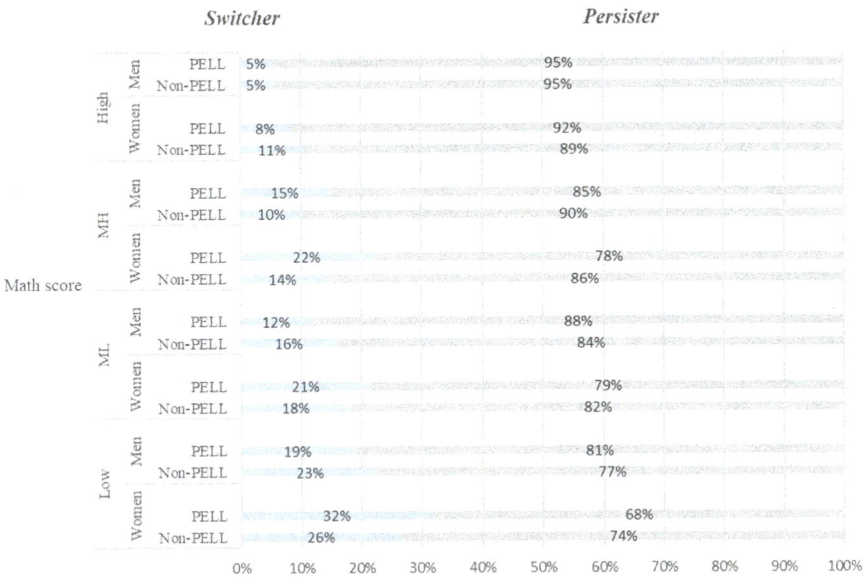


**Fig. 2.9** Percentage of switchers by math score block, gender and Pell grant status

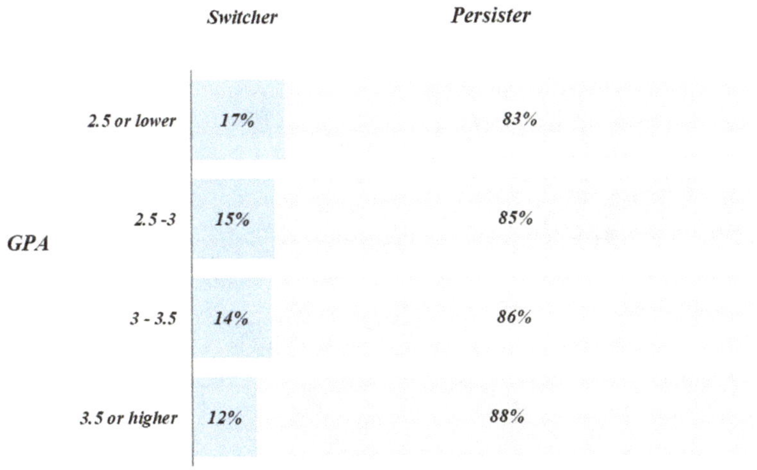


**Fig. 2.10** Percentage of switchers by GPA block

Underrepresented minority students also switched at higher rates within each GPA block. It is noticeable (as represented in Fig. 2.12) that the differences between URM and non-URM students widened in the lower GPA blocks.

Again, we found that the combination of GPA level, gender, and URM status revealed wide gaps between the switching rates of URM women and men across

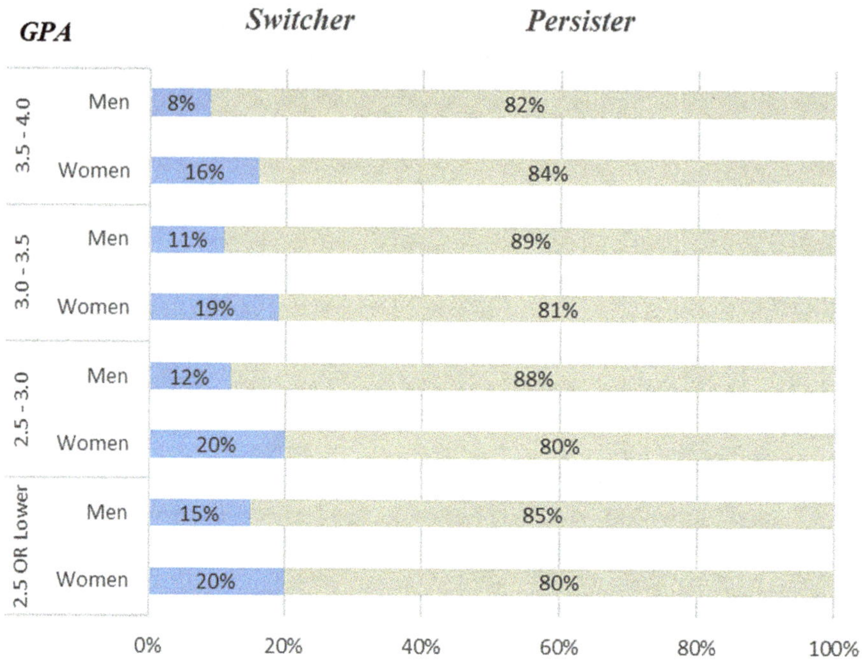


**Fig. 2.11** Percentage of switchers by GPA block and gender

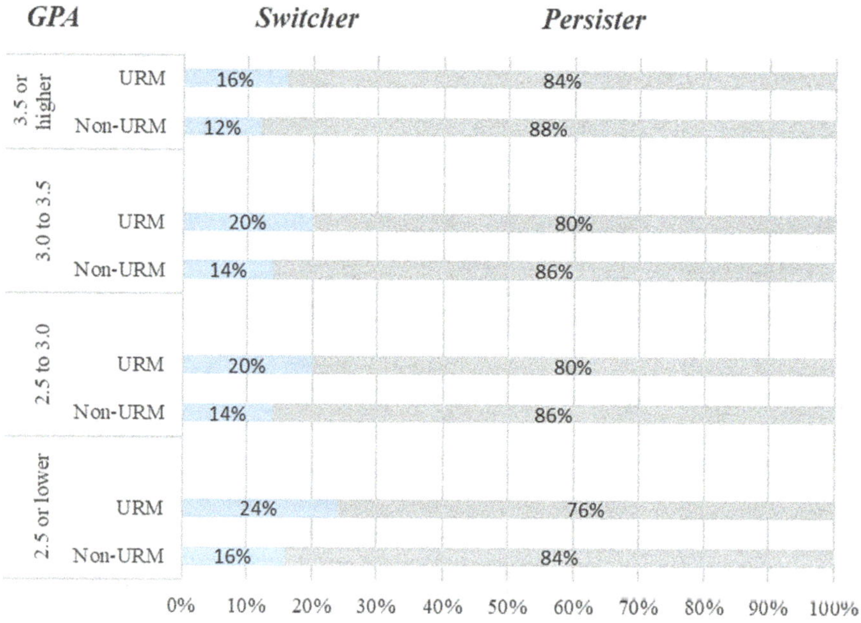


**Fig. 2.12** Percentage of switchers by GPA block and URM/non-URM status

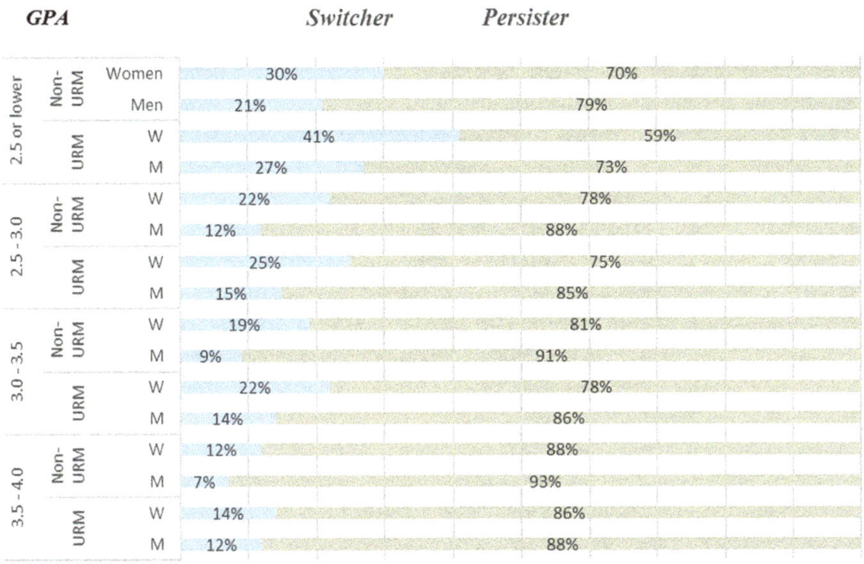


**Fig. 2.13** Percentage of switchers by gender, URM and GPA

the GPA range. As is indicated in Fig. 2.13, URM women with lower GPA's switched at higher rates than any other group. The same pattern was evident at three out of five schools, with the sixth school not having enough cases for disaggregation.

## *What Factors in the Institutional Records Data Best Predict Switching?*

The previous section examined differences in rates of STEM switching by institutions, discipline, gender, underrepresented minority status, ACT/SAT math scores, Pell grant-eligibility, and first-generation status. Assessing how much each factor contributes to predicting switching is possible using a logistic regression model. We used statistical covariates to control differences in the length of time students had been in school, and differences in the difficulty of courses experienced by students. Variables for average course difficulty experienced, class level, and terms enrolled all entered the model suggesting that switchers and persisters differed on these underlying factors. (See Chap. 1 and Appendix B for full explanations of the logistic regression model and description of variables.)

We found that academic factors such as first-term GPA, ACT/SAT math scores, and the number of incomplete grades were strongly associated with students switching out of STEM majors. Students with lower math scores and lower grade point

**Table 2.11** Logistic regression model for switching

| | *B* | SE | Sig. | Exp(B) |
|---|---|---|---|---|
| GPA (first term) | −0.77 | 0.050 | <0.001** | 0.46 |
| Number of IW's | 0.22 | 0.017 | <0.001** | 1.24 |
| Number of DF's | 0.09 | 0.014 | <0.001** | 1.09 |
| Math score (ACT/SAT) | −0.23 | 0.037 | <0.001** | 0.79 |
| Number of repeated courses | −0.18 | 0.017 | <0.001** | 0.83 |
| Gender (1 = male) | −0.36 | 0.062 | <0.001** | 0.70 |
| URM (1 = URM) | 0.11 | 0.095 | 0.16 | 1.12 |
| URM*women*low math | 0.29 | 0.14 | 0.042* | 1.34 |
| Number of terms | 0.15 | 0.015 | <0.001** | 1.16 |
| Class level | −0.41 | 0.06 | <0.001** | 0.66 |
| Average course difficulty | −1.07 | 0.050 | <0.001** | 0.34 |
| Discipline[a] | 0.87 | 0.16 | <0.001** | 2.4, 0.48 |
| Institution[a] | 0.77 | 0.11 | <0.001** | 1.3, 0.5 |
| Constant | 1.061 | 0.28 | <0.001** | 2.88 |

Dependent variable is Switcher: Switched = 1, Persisted = 0

[a]Coefficients are average of absolute value of individual discipline and institutional effects, Exp(B) is average of values below and above 1 for individual discipline and individual effects

$**p < 0.01$, $*p < 0.05$. $R^2 = 0.26$

averages, and those who withdrew from courses, left STEM at higher rates. ACT/SAT Verbal/Reading did not significantly predict switching (Table 2.11).

As shown in group comparisons of switching rates (c.f, Table 2.7), the analysis confirmed that women switched at higher rates than men. However, underrepresented minority status did not (by itself) predict switching, although female URM students with low math scores did switch at statistically significantly higher rates than other students. The non-significant result for URM students is due to the overrepresentation of URM students in the lower quartiles of the standardized math score distribution and the higher association of math scores with switching.

As reported earlier in this chapter, students in different disciplines and institutions leave STEM at differing rates. While the logistic model controls for this variation, enrollment in particular disciplines is one of the strongest predictors of switching, with very low rates for engineering, and higher rates for biology and math. Variation among institutions is lower, but this also accounts for substantial variability in switching rates. Thus, which major students enter in which school has important consequences for their switching risk.

## *When Do Students Switch Out of STEM Majors?*

As illustrated in Fig. 2.14, 50% of students in our sample who switched did so during their first college year, 25% left after the first semester or quarter, and an additional 30% of students switched after the second year. Only a small percentage of students (13%) switched in the third year or later.

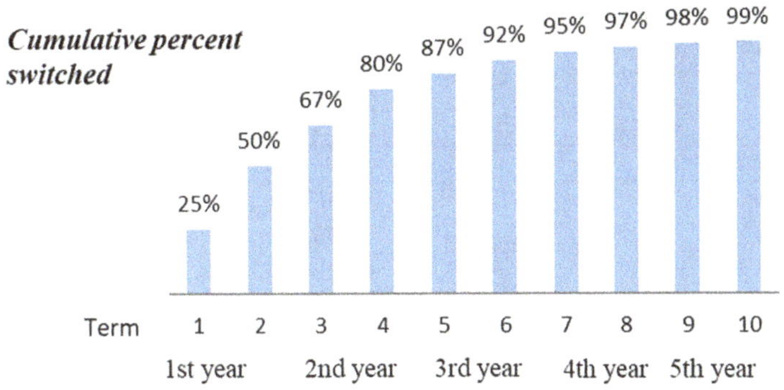


**Fig. 2.14** Cumulative percent of students switching by term. Terms are semester or quarter

**Table 2.12** Average switching times for students at each institution in semester/quarter units

| Institution | Average | *N* | SD |
|---|---|---|---|
| PB4R1 | 3.6 | 256 | 2.54 |
| PV2R1 | 3.7 | 161 | 1.97 |
| PB3R1 | 3.3 | 369 | 1.95 |
| PB2R1 | 3.5 | 249 | 2.49 |
| PB1R1 | 2.7 | 976 | 1.98 |
| Total | 3.0 | 2010 | 2.13 |

*Note*: PV1R3 had too few cases to make reliable comparison

We make comparisons for the average switching time between groups, although average numbers cannot be interpreted literally given that semesters and quarters are discrete numbers. Men tend to switch slightly later than women (mean of 3.2 terms compared to 3.0), and URM students tend to switch later (mean of 3.5) than non-URM students (mean of 3.0). Pell grant recipients left later than non-Pell grant recipients (3.2–2.9). Average times for leaving STEM majors also mapped onto math scores, with students with higher math scores leaving later than those in lower math blocks (low to high: 3.0, 3.1, 3.2, 3.4).

We also wanted to know if the time of switching varied substantially by institutions or discipline. As indicted in Table 2.12, we found that most variation occurred between institutions, with students at PB4R1 (the Western state university) who switched significantly earlier than the other institutions. In subsequent chapters, we will discuss what features of STEM educational practices may account for institutional variations in switching patterns.

No obvious patterns were found for switching between disciplines (c.f., Table 2.13) although students in mathematics tended to switch earlier and engineering students later than students in other STEM disciplines.

**Table 2.13** Average and median switching times for students for each major discipline in semester/quarter units

| Discipline | Average | *N* | SD | Median |
|---|---|---|---|---|
| Agriculture | 3.0 | 123 | 2.04 | 2.5 |
| Environmental science | 2.7 | 94 | 2.28 | 2.0 |
| Computer science | 3.0 | 118 | 1.79 | 2.7 |
| Engineering | 3.2 | 573 | 2.16 | 2.8 |
| Biology/life sciences | 2.9 | 1012 | 2.19 | 2.4 |
| Math | 2.5 | 90 | 1.76 | 2.0 |
| Physical sciences | 2.8 | 122 | 1.85 | 2.4 |
| Total | 3.0 | 2020 | 2.13 | 2.5 |

## What Are the Switching Patterns for STEM Majors with Higher and Lower Academic Performances?

We compared switchers with better grades—those in the highest quartile of cumulative GPAs before switching—to those with switchers in the lowest quartile of cumulative GPAs before switching. We found contrasts between these two groups by discipline, gender, URM status and destination majors chosen.

In biology, higher academic switchers were over-represented with 51% of all higher academic switchers coming from this group compared to 44% of lower academic switchers coming from the life sciences. Highest and lowest academic performers were also distinguished by their destination majors. We found the most dramatic difference in the choice of business majors such as accounting, marketing or finance: 16% of high performing switchers (primarily from engineering and biology) moved into these majors while only 6% of lower performing students did so. The lower academic group also switched more often to non-psychology majors in the social sciences than did students with better grades (21% vs. 13%). Some of this difference is attributable to lower performing students switching from engineering and math to economics. This destination was chosen rarely by switchers from these disciplines who had achieved higher grades.

As we discuss further in the following section, undecided majors are a frequent choice of students who leave STEM majors. The undeclared major has different titles: in some schools it is called, "undecided", "open-option", or "unclassified." Slightly higher numbers of students from the lowest than the highest performing group chose undecided majors (22% low vs. 18% high). However, nearly 10% of switchers stay in undecided majors for more than one semester. Higher proportions of lower performing students (15% low vs. 6% high) stay in these majors, and some switchers stayed in undecided majors up to five terms.

High- and low-performing students also differed by their gender representation. In the highest performing group, 59% were women and 41% were men. These proportions were (coincidentally) reversed in the lowest performing group where 41% were women and 59% were men. This finding again supports the contention made

**Table 2.14** Comparison of groups of switchers with highest and lowest cumulative GPAs, percentage of group

| S group | Highest Cum. GPA group[a] (*n* = 532) (%) | Lowest Cum. GPA group (*n* = 456) (%) |
|---|---|---|
| Switched from biology | 51 | 44 |
| Switched from engineering | 22 | 26 |
| Switched to business | 16 | 6 |
| Switched to social sciences | 13 | 21 |
| Switched to undeclared | 18 | 22 |
| Switched to undeclared and stay more than one term | 6 | 15 |
| Women in GPA group | 59 | 41 |
| URM switchers | 6 | 19 |

[a]*Percentage of group* should be read as "51% of high GPA switchers were biology majors"

both in this study and the original that STEM majors lose high-performing women who, before switching were outperforming the men who remain.

The two groups were also differentiated by race/ethnicity (as indicated by URM and non-URM status). URM students made up 19% of switchers in the lowest performing group and 6% of switchers in the highest group.

On average, switchers in both the highest and lowest performing groups left their STEM majors at roughly the same time, that is, after the second term of their first year.

A summary of these distinctions between the highest and lowest performing groups of switchers is offered in Table 2.14.

## Which Majors Do STEM Switchers Move into?

We analyzed the destination majors that switchers chose after they left their STEM majors. Rather than being individual decisions, we found that students' subsequent pathways reflect a patterned relationship to the STEM majors that they left. Describing these pathways is useful because they can illuminate some of the institutional and disciplinary norms that shape them. Some pathways suggest affiliation between the content and skills studied in STEM and non-STEM majors, as in the case of students who leave the life sciences for public health majors, or math majors pursuing finance degrees. Institutional course requirements may also drive switching choices when STEM and non-STEM majors share particular pre-requisite courses.

In the description of our study methods in the prior chapter, we explained our methods for finding the expected frequencies for students switching from STEM to non-STEM disciplines (cf., Chap. 1 and Appendix B). Underlying all calculations is the comparison between the relative proportion of students in any given STEM discipline at an institution compared with the proportion of students in the destination non-STEM discipline. We quantify this proportion with the *expected frequency ratio* which is the number of students observed switching divided by the expected number. Ratios greater than one indicate that greater than expected numbers of students switched; ratios less than one indicate the opposite. This comparison provides a framework for assessing the probability of switching from one discipline to another. In reality, we know that students rarely switch exactly at rates dictated by representation of students in each discipline, and that most pathways are more or less frequently traveled than others.

When we conducted comparisons at each institution we also wanted to know if expected frequency ratios generalized across the schools in our study, or whether the size of particular switching pathways might be better explained by more idiosyncratic factors at particular institutions. We considered that pathways were generalized if four out of the six institutions shared ratios above or below one. Because very small numbers exaggerate the size of proportions, we only reported pathways with more than 20 students. (A more detailed explanation of decision rules, dealing with small numbers, and missing data can be found in the methods section of Chap. 1 and Appendix B.)

## *Do Some Disciplines Switch at Higher or Lower than Expected Rates?*

First, we simply wanted to know if students switched out of STEM majors in numbers proportional to a discipline's representation on campuses. That is, if 5% of all students are math majors, are 5% of all STEM switchers also math majors?

In five of the seven STEM disciplines in our study, more students than expected switched into non-STEM majors. These disciplines included;

- Biology/Life sciences (O/E = 1.5),
- Computer Science (O/E = 1.35), and
- Mathematics (O/E = 1.58).

Both the environmental sciences and physical science students switched at roughly expected rates, while engineering (O/E = 0.61) and agriculture (O/E = 0.73) had fewer students than expected switching to non-STEM majors.

Ratios generalized across institutions for all disciplines except for agriculture where the pattern varied by institution. Institution PB4R1 had significantly lower than expected rates of switching (O/E = 0.59) from agriculture majors, while the other two institutions with agriculture schools showed larger than expected proportions of switching loss.

## What Are the Most Populous Pathways for Students Switching Out of STEM Majors?

In terms of sheer numbers, the most frequently taken paths out of STEM majors are:

- Biology to psychology (167);
- Biology to social science (131);
- Engineering to social science (103);
- Undeclared majors into which a sizable number of students switched from almost all STEM majors with higher numbers from engineering (137), life sciences (173), and computer science (42).

There were other larger paths that attracted greater-than-expected numbers of students. Notably, large expected frequency ratios were found for pathways from:

- Life sciences to psychology (O/E = 4.8), and
- Life sciences to non-psychology social science (O/E = 2.6).

While relatively large numbers of students follow individual, non-patterned pathways, some pathways show a pattern of lower- than-expected student numbers given the relative size of majors on each campus, Students who take pathways at lower-than-expected rates are:

- Engineers who switch to a business path (76 students; O/E = 0.69) and,
- Life science majors (58) who switch to humanities at a much lower than expected number (O/E = 0.23).

The path diagrams (Figs. 2.15 and 2.16) show expectancy ratios and the number of students for the five largest paths out of engineering and life sciences. (All reported paths pass the four-out-of-six rule for generalization across institutions.)

The most striking result of this analysis (represented in Figs. 2.17) may be the large number—and in much greater-than-expected numbers—of students from all STEM disciplines who switch to undeclared majors. (This outcome is represented in Fig. 2.16.) We also found that switchers go to undeclared majors at much higher rates than non-STEM majors. In Chap. 10 we explore what happens to these students, many who continue in undeclared majors for more than one semester after switching.

While some of the discipline-to-discipline pathways are too small to calculate stable expectancy ratios, it is still of interest to learn what are the most popular destination majors for switchers from each discipline. Table 2.15 presents the number of students leaving each discipline and the number in the most popular destination majors. For the largest disciplines, life sciences, engineering and computer science, the most frequently-chosen destination was an undeclared major; in all other disciplines the most populous major was social science.

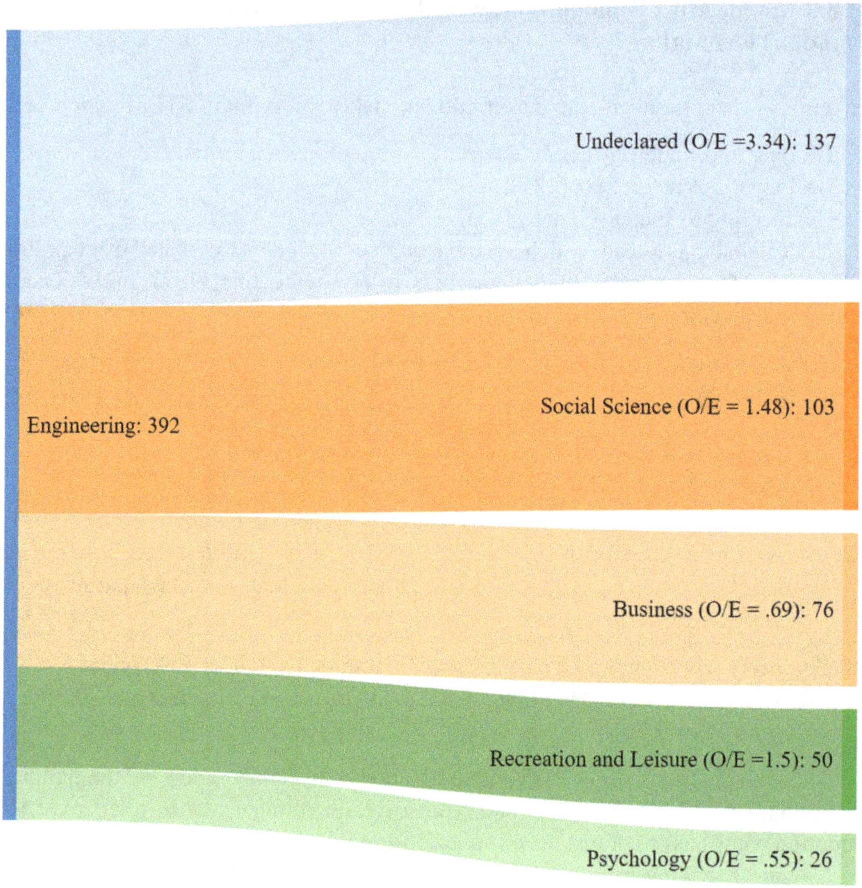


**Fig. 2.15** Sankey diagram of expectancy ratios and student numbers for pathways from engineering to non-STEM majors

### Within Discipline-to-Discipline Pathways, What Are the Most Frequently Destination Chosen Majors?

We wanted to know which specific majors within destination disciplines were most often chosen by STEM switchers. As indicated in Table 2.16, for most pathways, a single major was chosen by at least one-third to one-half of the students who switched. Many of these choices suggested an affiliation between the original STEM major and the destination major, for example, moves from life sciences to community health education, computer science and math to economics, and engineering to finance. Other paths were not dominated by a single major, notably, moves from life sciences into business where students went to a long list of majors.

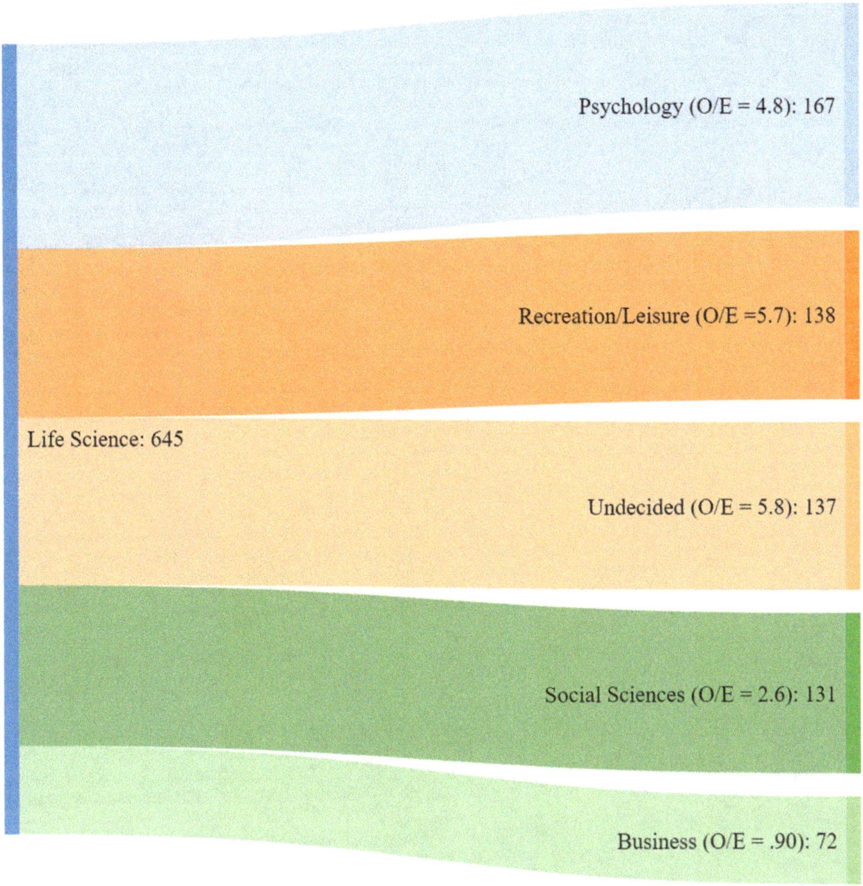


**Fig. 2.16** Sankey diagram of expectancy ratios and student numbers for pathways from life sciences/biology to non-STEM majors

## *What Differences Are Evident in Destination Pathways by Gender and GPA?*

As indicated in Table 2.17, both gender and GPA were associated with distinctive patterns in switching pathways. Large differences in proportions were evident for students switching from life sciences to psychology, with female majors much more likely than male majors to take this path. The same gender-related pattern was true for women switching more than men from life sciences to either education or undeclared majors. Male engineers were more likely than female engineers to switch to business or to undeclared majors, and men more likely than women to change from math majors to social science.

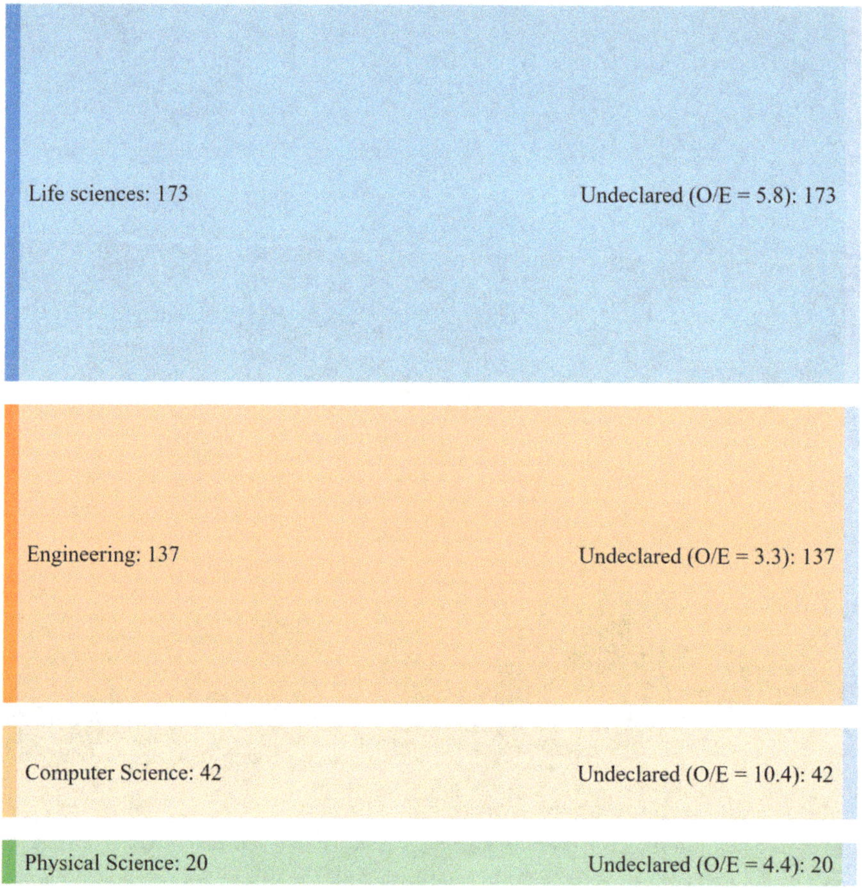


**Fig. 2.17** Sankey diagram of expectancy ratios and student numbers for pathways from stem to undeclared majors

In engineering, women tended to switch to infrequent pathways, such as art, that were not included in the analysis because of small pathway size; while men accounted for proportionally more of the switchers who took more 'traditional' paths into social science and business.

We also divided students into two groups that were above or below average GPA. As shown in Table 2.18, these two groups showed differences across destination pathways. Students with lower GPAs went from life sciences to psychology in greater numbers than those with higher GPAs. Higher GPA students in both engineering and life sciences switched to business at higher rates than students with

**Table 2.15** Number of students in most populous destination majors for STEM switchers

| Pathway | Number in pathway | % of switchers in destination discipline[a] | Total number of switchers in original STEM discipline |
|---|---|---|---|
| Agriculture—social science | 17 | 16 | 105 |
| Environmental science—social science | 24 | 26 | 91 |
| Computer science—undeclared | 42 | 32 | 132 |
| Engineering—undeclared | 137 | 27 | 512 |
| Biology/life sciences—undeclared | 173 | 18 | 981 |
| Math—social science | 28 | 27 | 104 |
| Physical sciences—social science | 25 | 22 | 116 |

[a]Should be read as "16% of all switchers in Agriculture switched to social science"

**Table 2.16** Specific destination majors in discipline-to-discipline pathways

| Discipline pathway | Specific majors | *N* | Percentage of discipline in pathway to major[a] |
|---|---|---|---|
| Life sciences—social sciences | Economics | 29 | 30 |
| Life sciences—education | Community health education | 20 | 52 |
| Engineering—social science | Economics | 50 | 50 |
| Engineering—business | Finance | 25 | 33 |
| Computer science—social science | Economics | 13 | 41 |
| Math—social science | Economics | 21 | 75 |
| Math—business | Finance | 9 | 45 |
| Life science—recreation/leisure | Health promotion | 26 | 19 |
| Engineering—recreation/leisure | Sports medicine | 17 | 34 |

[a]Should be read as "30% of all those in the life science to social science path chose the Economics major"

poorer academic records. In engineering and computer science, lower GPA students switched in higher numbers into an undeclared major. In life sciences, this was not the case: only slightly greater numbers of lower GPA students moved into undeclared majors.

**Table 2.17** Expected frequency ratios and number of students by gender for students switching out of STEM for each pathway

| Pathway | Women | Men |
|---|---|---|
| Engineering—business | O/E = 0.35<br>($N$ = 20) | O/E = 1.02<br>($N$ = 56) |
| Engineering—undeclared | 1.48<br>(31) | 5.27<br>(106) |
| Life science—psychology | 7.27<br>(127) | 2.37<br>(40) |
| Life science—recreation/leisure | 1.48<br>(31) | 5.27<br>(106) |
| Life science—undeclared | 6.5<br>(99) | 5.0<br>(74) |
| Life science—education | 2.8<br>(35) | – |
| Math—social science | – | 9.5<br>(21) |

*Note:* All differences by gender significant at $\alpha < 0.01$. Some pathways did not have enough cases to make reliable estimates

**Table 2.18** Expected frequency ratios and number of students switching for STEM switchers with high and low academic performance scores

| Pathway | Low | High |
|---|---|---|
| Engineering—business | O/E = 0.51<br>($N$ = 29) | O/E = 0.87<br>($N$ = 52) |
| Engineering—undeclared | 3.7<br>(82) | 2.4<br>(55) |
| Engineering—social science | 1.85<br>(70) | 0.87<br>(33) |
| Life science—psychology | 5.1<br>(96) | 3.8<br>(71) |
| Life science—recreation/leisure | 5.9<br>(77) | 4.6<br>(60) |
| Life science—social science | 2.8<br>(77) | 1.96<br>(54) |
| Life science—business | 0.6<br>(26) | 1.06<br>(46) |
| Computer science—social science | 6.7<br>(25) | 1.8<br>(7) |
| Computer science—undeclared | 12.3<br>(27) | 6.8<br>(15) |

*Note:* All differences by GPA significant at $\alpha < 0.01$

## Summary and Implications

We examined different national data sets in the United States to establish what proportion of STEM students leave their majors. In our appraisal, the best national estimate is provided by Chen's (2013) analysis of the 2004–2009 NCES survey: of those students who start out in a STEM major approximately 28% switch to another non-STEM major and another 20% leave school altogether. Longitudinal estimates provided to us from the CIRP survey are complicated by differences in sampling methods between years. However, these data do point to considerable improvements in switching rates since the original survey, although the extent of this trend is most likely overstated in the CIRP data due to differences in sampling methods between years.

Our analysis presented in this chapter uses a large aggregated dataset of student transcripts from six institutions. While the data set is large, it should be emphasized that it is by no means a random sample of students in the United States, although our institutions are representative in many ways of national averages, especially from research one institutions. Many of our patterns and findings also generalize across schools, suggesting that some factors affecting switching may be either independent of institutional setting and would be found at many other colleges. Our data shows a complex picture of differing rates of switching by particular student groups, including differences between students with higher and lower academic scores in their patterns of switching. They also show when students switch during their academic careers, patterns in the majors students choose after switching out of STEM.

## *Highlights from the Records Data*

Almost 14% of students at our six institutions started out in a STEM major and then switched to a non-STEM major sometime during their academic career. Rates of switching varied substantially between institutions (7–28%), and among STEM disciplines (8–24%). Biology, math and computer science showed the highest rates of switching, engineering the lowest.

Women switched at a 7% higher rate than men. Women switched at greater rates at all institutions in our sample, and across the range of standardized math scores, disciplines and GPA's. This suggests that whatever factors contribute to higher rates of switching for women seem to be present at all institutions in our sample and for all STEM majors. This finding also conforms with both the NCES and HERI estimates that show women switching more than men.

The picture for under-represented minority students was more complex. While URM students as a group switched at higher rates than non-URM students, this was not true at all institutions. One striking fact was that underrepresented minority women who entered the university with lower standardized math scores switched at much higher rates than any other group (41%). More than a quarter (26%) of URM women switched out of STEM majors versus 13.8% of all students. However, when demographic variables of math scores, GPA and URM status are assessed for their contribution to switching, students' math scores and GPA had a greater association to switching than URM status. In our logistic regression model, URM status by itself did not predict switching. This was also true in the NCES study where academic performance better explained the higher rates of switching than race/ethnicity in and of itself. The larger implication of this finding is that, raising the quality of math and science preparation in high schools that serve large numbers of students of color has the potential to significantly increase their STEM persistence rates.

Combining all demographic and academic factors, student GPA, standardized math scores, incomplete grades and the average difficulty of courses experienced by students all significantly predicted student switching. 'Being a woman' was also a significant predictor of switching with all other variables held constant. Switching also varied substantially by discipline and institution. The logistic model shows that academic duress and low-academic preparedness predicted switching. However it should be noted that the model is under-identified in that we may be missing variables about attitudes, behaviors or other institutional variables that contribute to switching. Thus, we turn to findings from the student interviews and SALG survey, and other sources that are discussed in this book to weigh the contributions of a wider array of factors contributing to switching and relocation and to explain the patterns that we have found in these analyses.

One striking finding of the analysis is that the majority of students who switch majors do so early in their academic careers. Fifty percent of students who switched did so by the end of the first year, and fully 80% had switched by the end of the second year. Students with higher standardized math scores, URM students and Pell grant recipients all switched later than their comparison groups. This finding suggests that efforts to intervene to support student persistence are best directed toward students during their first 2 years when they are taking introductory courses.

We also examined differences between students with the highest and lowest grades. Much higher proportions of women in the higher than the lower performing group (59% vs. 41%) switched out of STEM, as did much higher proportions of lower than higher performing URM students (19% vs. 6%). Some of the patterns evident in this comparison shed light on possible "push versus pull" reasons for switching. Perhaps most indicative of the role of academic duress in switching is the larger than expected number of lower-performing than higher-performing students (15% vs. 6%) who moved into undecided majors and remained in these for longer than one term. In these cases, it seems likely that these students did not choose their undecided majors in the same way that higher achieving students chose their individual majors.

We also examined the destination majors of all students leaving STEM and found that switchers followed some pathways more than others. Frequent pathways beyond STEM were, from biology to psychology and other social sciences, engineering to social sciences (mostly economics), and from all disciplines to undeclared majors. The most frequent pathways also attracted greater than expected numbers given the representation of STEM and non-STEM disciplines on each campus. Undeclared majors also attracted more switchers than would be expected. Pathways from engineering and computer science to undeclared majors were higher than expected for students with lower academic performances. Some pathways are based on affinity or similarity between subject matter, such as math majors switching to economics, or biology majors moving to health or sports medicine. Other pathways, such as undeclared majors, may simply be ways to escape from untenable STEM majors.

Mapping student switching patterns allows us some improved insights into who is most at risk for switching and when. For example, both academic duress and incoming level of preparation seemed to play a large role in switching. As will be discussed in Chap. 7, students who receive poor or incomplete grades in gateway courses in their first and second years are particularly prone to switching. Women switch more than men and are over-represented in the group switchers with high academic performance levels, as we detail in Chap. 10. Under-represented minority students switch more than non-URM students, although an important part of this greater risk is due to poor high school preparation (see Chap. 5). Some risks may be open to remediation in the form of reformed curriculum, better teaching and the help of counselors and others who support at-risk students. However, a major improvement of switching rates would be achieved by raising the level of science and math preparation in the K-12 system for all students.

## References

Chen, X. (2013). *STEM attrition: College students' paths into and out of STEM fields* (Statistical Analysis Report NCES 2014-001). National Center for Education Statistics.

Eagan, K., Hurtado, S., Figueroa, T., & Hughes, B. E. (2014). *Examining STEM pathways among students who begin college at four-year institutions*. National Academy of Sciences. Retrieved April 10, 2019, from http://sites.nationalacademies.org/cs/groups/dbassesite/documents/webpage/dbasse_088834.pdf

# Chapter 3
# Why Undergraduates Leave STEM Majors: Changes Over the Last Two Decades

**Anne-Barrie Hunter**

## Introduction

As discussed in Chap. 1, a significant finding of *Talking about Leaving* was that switchers and persisters were not distinctive types of students. They were not found to differ by individual attributes that could explain why one group left and the other group stayed. Both switchers and persisters identified the same set of 23 issues but had resolved them in different ways. From within this set of issues, those that prompted switching by some students were also an additional source of stress for those who switched for different reasons and were also troublesome to many who persisted in their STEM majors. In almost half of the identified problems, there was little difference in the proportions of switchers and persisters who reported them. White women and all students of color also reported a higher number of concerns influencing switching, and at higher percentages, than did their white male peers. The loss of high-performing students, notably women, was also a significant discovery. Although there was some variation in the ranking of problems by institutional type, all of the 23 problems were found at every participating institution and there was little institutional variation in the most highly ranked concerns. Seymour and Hewitt (1997) represented the structure of their main findings using the metaphor

A.-B. Hunter (✉)
Ethnography and Evaluation Research (E&ER), University of Colorado Boulder, Boulder, CO, USA

Center for STEM Learning, University of Colorado Boulder, Boulder, CO, USA
e-mail: abhunter@colorado.edu



E. Seymour, A.-B. Hunter (eds.), *Talking about Leaving Revisited*,
https://doi.org/10.1007/978-3-030-25304-2_3

| *Book* | *Average N of Concerns Contributing to Student Switching Decisions* | *Average N of Concerns Affecting All Switchers* | *Average N of Concerns Affecting All Persisters* |
|---|---|---|---|
| TAL (1997) | 4.2 | 8.6 | 5.4 |
| TALR (2019) | 12 | 64 | 23 |

**Fig. 3.1** Average number of concerns that contributed to students' decisions to switch from STEM to non-STEM majors and that also created problems for all switchers and for all persisters

of an iceberg because the same concerns that contributed to field-switching were reported—to some degree—by all the students that they interviewed, whether they left or stayed.

What distinguished switchers from persisters included their entry level of math and science preparation, whether they found and used appropriate and timely help with academic difficulties, developed ways to cope with issues in the design, pedagogy, and grading practices of STEM courses, and how many difficulties they encountered. Both the overall study findings and the detail in which participants explored each of them called into question the commonly referenced "rational choice" explanation for switching whereby students are thought to opt out of STEM because they discover non-STEM disciplines that are better suited to their intellectual interests and talents.

Results from the current study both replicate and validate findings from the original TAL study. All concerns identified by students in the first study were identified by students in the present study. While no new concerns emerged, there were changes in their relative ranking, and some were refined during analysis to further clarify their significance for switching decisions. As with TAL, the concerns described by TALR students as contributing to switching decisions were also found to affect other switchers and, to a lesser degree, persisters. However, the average number of problems reported by all students has increased. In the TAL study, the average number of concerns contributing to each switching decision was 4.2, with an average of 8.6 concerns reported by switchers, overall. Persisters were distinguished from switchers by a lower average of 5.4 concerns mentioned. Seymour and Hewitt concluded that one way to see switchers was "as people who have rather more problems with their original majors than do [persisters]" (p. 32, 1997). In the current study, we found that the number of concerns reported as contributing to switching decisions has increased by two-thirds to an average of 12 concerns per switcher (see Fig. 3.1). Discussions of all problems experienced yielded an average of 64 identified concerns for each switcher, and an average of 23 concerns for each persister. The higher averages recorded suggest that students are simultaneously handling more concerns than did students 20 years ago.

## *Changes to the "Problem Iceberg"*

The set of concerns that were identified in the current research as contributing to switching decisions did not differ significantly from those described in the original study. Table 3.1 shows results from the current study side-by-side with the original "problem iceberg" table published in *Talking about Leaving* (c.f. Seymour & Hewitt, 1997, p. 33). In Table 3.1 the first columns for each set of study findings (current and original) show how each concern ranked among all factors contributing to switching decisions; the second columns show the percentages of students who cited these concerns as prompting their decision to leave their STEM majors; the third columns contain the percentages of all switchers mentioning them; the fourth columns show the percentages of persisters also reporting these as concerns; the final columns show the average percentage of all students mentioning each concern.

Small differences from the original study in how students described particular problem areas led us to change how their concerns were coded and categorized. Following ethnographic practice, the content of categories and how they are labeled were derived from the way that speakers described their experiences. Accordingly, we made slight changes to two of the original study categories: "Difficulties in seeking and getting appropriate timely help" now includes two related concerns that were categorized separately in TAL (i.e., "Inadequate advising or help with academic problems" and "Lack of peer study group support"). Also, two distinct types of concerns that were not separated in TAL were broken out in this analysis; namely, "Difficult transition to college" and "Negative effects of weed-out courses." In the original analysis, these problems were well documented and discussed, but were not separately counted. In the current study, students specifically identify these concerns as contributing to decisions to leave their STEM majors. So having collapsed two previously independent issues into a single more inclusive category, we also added the two new discrete concerns. The net result was the final list of 24 factors identified as directly affecting students' decisions to switch out of STEM majors.

## *Types of Concern Comprising the Problem Iceberg*

The types of concerns contributing to switching decisions fall into the six broad groups shown in Fig. 3.2. Nine of 24 concerns prompting switching reference negative aspects of their STEM classroom experiences that, when combined, made learning new complex material more challenging. About one-fifth of switching-related concerns relate to projected careers and lifestyles in STEM fields and the competing appeal of non-STEM alternatives. They also reference ways to achieve original career goals by alternative means. These considerations exemplify the push-pull nature of the decision process which we also reported in the original study: while experiences in STEM majors were prompting students to rethink their

**Table 3.1** "The problem iceberg": Concerns that contributed to STEM switching decisions, to the concerns of all STEM switchers, all STEM persisters, and all STEM students, in the current and original studies

| *Significant Concerns Identified by Interviewees* | *Ranking of Concerns in TALR* | *% for Whom This Concern Contributed to Switching (N=96)* | *% All Switcher Concerns (N=96)* | *% All Persister Concerns (N=250)* | *% All Student Concerns (N=346)* | *Ranking of Concerns in TAL* | *% for Whom This Concern Contributed to Switching (N=184)* | *% All Switcher Concerns (N= 184)* | *% All Persister Concerns (N= 151)* | *% All Student Concerns (N=335)* |
|---|---|---|---|---|---|---|---|---|---|---|
| Discovery of an aptitude for a non-STEM subject | *1* | 76% * | 76% | 6% | 25% | *16* | 10% | 12% | 5% | 8% |
| Discouraged/lost confidence due to low grades in early years | *2* | 61% | 79% | 44% | 53% | *9* | 23% | 34% | 12% | 24% |
| Loss of incoming interest and motivation | *3* | 61% | 63% | 12% | 26% | *1* | 43% | 60% | 36% | 49% |
| Rejection of STEM careers and associated lifestyles | *4* | 58% | 58% | 5% | 20% | *6* | 29% | 43% | 21% | 33% |
| Shift to a more appealing career option | *5* | 54% | 54% | 4% | 18% | *7* | 27% | 33% | 16%** | 25% |
| Competitive, unsupportive STEM culture makes it hard to belong | *6* | 52% | 81% | 42% | 53% | *12* | 14% | 28% | 9% | 20% |
| Poor quality of STEM teaching | *7* | 48% | 96% | 72% | 78% | *3* | 36% | 90% | 74% | 83% |
| Reasons for choice of STEM major prove inappropriate | *8* | 48% | 68% | 22% | 34% | *13* | 14% | 82% | 40% | 63% |
| Difficult transition to college | *9* | 43% | 89% | 57% | 66% | - | - | - | - | - |
| Negative effects of weed-out classes | *10* | 35% | 43% | 18% | 25% | - | - | - | | - |
| STEM curricular design problems: pace, overload, labs, alignment | *11* | 31% | 86% | 56% | 65% | *4* | 35% | 45% | 41% | 44% |

| | | | | | | | | | | |
|---|---|---|---|---|---|---|---|---|---|---|
| Prefer teaching approach in non-STEM courses | *12* | 29% | 58% | 8% | 22% | *17* | 9% | 24% | 15% | 20% |
| Non-STEM major offers better education, holds more interest | *13* | 26% | 26% | 2% | 9% | *2* | 40% | 58% | 32% | 46% |
| System-playing as means to career goals | *14* | 26% | 26% | 1% | 8% | *19* | 7% | 9% | 3% | 6% |
| Inadequate high school preparation in subject and study skills | *15* | 19% | 64% | 34% | 42% | *11* | 15% | 40% | 38% | 39% |
| STEM career options and rewards felt not worth the effort | *16* | 17% | 17% | 1% | 5% | *5* | 31% | 43% | 20% | 33% |
| Conceptual difficulties with one or more STEM subject(s) | *17* | 16% | 80% | 42% | 53% | *14* | 13% | 27% | 25% | 26% |
| Problems related to class size | *18* | 15% | 56% | 29% | 36% | *21* | 0% | 20% | 11% | 16% |
| Difficulties in seeking and getting appropriate timely help | *19* | 13% | 80% | 31% | 45% | *8 & 15* | 24% | 75% | 52% | 65% |
| Time to degree influenced switching decision | *20* | 13% | 13% | 1% | 4% | *18* | 9% | 20% | 28% | 24% |
| Financial problems with completing STEM degree | *21* | 10% | 70% | 48% | 54% | *10* | 17% | 30% | 23% | 27% |
| Language difficulties with foreign faculty or TAs | *22* | 1% | 54% | 19% | 29% | *22* | 3% | 30% | 20% | 25% |
| Poor teaching, lab or recitation support by TAs | *22* | 3% | 3% | 2% | 3% | *20* | 0% | 20% | 11% | 16% |
| Poor lab/computer lab facilities | *24* | 0% | 0% | 0% | 0% | *23* | 0% | 4% | 4% | 4% |

* 0%-25% | 26%-50% | 51%-75% | 76%-100%

***Issue raised by persisters intending to move into non - STEM field following graduation.*

[a]Issue raised by persisters intending to move into non-STEM field following graduation

***Significant Concerns Identified by Interviewees***

| |
|---|
| **Issues of poor teaching, poor curricular design and the negative climate of STEM** |
| Competitive, unsupportive STEM culture makes it hard to belong |
| Poor quality of STEM teaching |
| Negative effects of weed-out classes |
| STEM curricular design problems: pace, overload, labs, alignment |
| Conceptual difficulties with one or more STEM subject(s) |
| Problems related to class size |
| Difficulties in seeking and getting appropriate timely help |
| Poor teaching, lab or recitation support by TAs |
| Language difficulties with foreign instructors or TAs |
| **Career-related issues** |
| Rejection of STEM careers and associated lifestyles |
| Shift to a more appealing career option |
| System-playing as means to career goals |
| STEM career options and rewards felt not worth the effort |
| **Pull factors** |
| Discovery of an aptitude for a non-STEM subject |
| Prefer teaching approach in non-STEM courses |
| Non-STEM major offers better education, holds more interest |
| **Issues Arising Outside of College** |
| Reasons for choice of STEM major prove inappropriate |
| Difficult transition to college |
| Inadequate high school preparation in subject and study skills |
| **Attitudinal factors** |
| Discouraged/lost confidence due to low grades in early years |
| Loss of incoming interest and motivation |
| **Financial issues** |
| Time to degree influenced switching decision |
| Financial problems with completing STEM degree |

**Fig. 3.2** Types of concern identified by students as contributing to decisions to switch out of a stem major

career plans, their decisions were simultaneously influenced by options offered by non-STEM majors. In another group of switching factors that exemplify the push-pull nature of switching decisions, while students were becoming discouraged or disenchanted with their STEM majors, they were also experiencing courses in non-STEM fields. In this process, they discovered majors that seemed better suited to their career or educational goals, interests and talents, and/or where they found more engaging teaching.

Students bring some issues with them into college that intersect with and exacerbate poor STEM learning experiences. Notably, these include ill-founded choices of a major; and under-preparation in mathematics and the sciences and in how to study for college-level work and navigate college as a system. Each of these deficiencies contributes to difficult transitions into college, including loss of confidence that is widespread among first-year STEM students. Financial difficulties in paying for college also influenced many decisions to leave STEM majors and having to work too many paid hours created problems for switchers and persisters alike.

In the balance of this chapter we compare and contrast differences in the switching factors cited in the original study over two decades ago with those cited by students in the current study. We also discuss these same issues as wider concerns for all switchers and for students who persist.

## The Problem Iceberg: Concerns Contributing to STEM Students' Decisions to Leave Their Major that Also Affect All Switchers and STEM Persisters: Then and Now

We noted above increases in the sheer number of issues that influenced STEM switching decisions and that all participants mentioned as concerns. As shown in Table 3.1, the top six concerns prompting switching were mentioned by at least half of all TALR switchers, as were 17 of the 24 factors (70%) that affected switchers more broadly; three of the concerns that prompted switching were also discussed by at least half of persisters. By comparison, concerns mentioned by more than 50% of participants in the original study were found only in the top seven contributors to switching. In the current study, we not only found a significantly greater number of concerns prompting both STEM switching decisions and affecting STEM students overall, we also found amplification and increased complexity in the array of concerns reported by STEM undergraduates.

## What Has Not Changed Since the Original Study

We now look at issues that contribute to switching and cause problems for persisters that are comparable across the two studies. Most important among these are students' assessments of their learning experiences in STEM classrooms and their effects on persistence.

- For students in the TAL study, *problems with poor teaching in STEM courses* ranked third (36%) of all reasons for switching. They were of concern to almost all (90%) switchers and were cited by three-quarters (74%) of persisters. In the current study, problems with STEM instructor pedagogy ranked seventh and echo, but are slightly higher than, those in the TAL study: almost half (48%) of switchers mentioned poor teaching in their STEM courses as prompting their decisions. Poor quality teaching continued to be a concern for nearly all switchers (96%), and for nearly three-quarters (72%) of persisting seniors.

- Related to issues with STEM teaching quality, *problems with STEM curricular design* include content overload, pace of delivery, and poor alignment between course elements. They were ranked fourth (35%) in the original study for all reasons contributing to switching; was of concern to 45% of TAL switchers and 41% of persisters. In the current study, this concern contributed to leaving decisions for a comparable proportion (31%) of switchers, but, in TALR, it was found to affect a large majority of switchers (86%) and over half of STEM persisters (56%).
- *Conceptual difficulties with one or more STEM subjects* ranked similarly as a concern in the two studies (17th in TALR and 14th in TAL). While this factor played little role in students' decisions to leave their STEM majors, then or now, it was of concern to 80% of TALR switchers, overall.

These three linked categories of concern about aspects of STEM learning experiences are discussed in Chaps. 6, 7, 8, and 9, both for STEM foundation courses and across all 4 years of STEM courses represented by our student sample.

The incidence of the following three contributors to switching decisions were either maintained or increased:

- *Loss of incoming interest and motivation to pursue a STEM major* ranked first (43%) in TAL among reasons for switching, was of concern to 60% of all switchers, and was cited by one-third (36%) of persisters. In the current study, loss of interest still ranked highly (third) in its contribution to switching decisions and (similar to TAL) was 61% of all switchers' concerns. However, it figured less than before in persisters' difficulties (viz., 12%).
- *Difficulties in seeking and getting appropriate timely help* was a problem for a large number of switchers and persisters, both in the present and the original study. Though it was a smaller factor in current decisions to leave STEM, finding and accessing resources—which could also prove critical in persistence—continues to be as serious a problem as it was 20 years ago. Over three-quarters (80%) of STEM switchers overall (75% in TAL) and 31% of TALR persisters (52% in TAL) continue to struggle to find the academic resources and support they need to survive.

We discuss the significance of finding and using appropriate and timely help in Chap. 12 which summarizes our findings on what enables persistence.

Issues of under-preparation were also found in similar proportions in both studies.

- *Inadequate high school preparation* was a causal factor for similar numbers of switchers in both studies. However, a slightly higher number of TALR switchers cite this as an important aspect of their difficulties, if not among their primary reasons for switching (64% compared with 40% in TAL). As before, under-preparation created issues to be overcome in order to survive for about one-third of persisters (34% and 38%).

## *What Has Changed Since the Original Study?*

Comparison of the two studies also reveals shifts over time in the weight of students' concerns. The most notable of these changes were evident in the following issues:

- In the current study, *discovery of an aptitude for a non-STEM subject* ranks first among all factors prompting switching. It was cited by three-quarters of switchers as directly influencing their decision to leave and as a consideration by 76% of all switchers. Its mention by 6% of persisters included students who relocated to other STEM majors. By comparison, in the original study, this reason for switching was cited by only 12% of switchers and ranked 16th as a contributor to their decisions. The large jump in the ranking of this concern, may reflect the large percentage of our switcher sample who were "high-performing switchers" that is, students who left their STEM major with a GPA of 3.5 (B+) or better. We have information about the academic records of our student samples in the present that was not available to us in the original study, so we know more than before about the caliber of students who are lost from STEM majors. High-academic performers accounted for roughly one-quarter (26%) of all STEM switchers across the six participating institutions. As we discuss in Chap. 10, high-achieving students often moved to non-STEM majors for reasons that reflected their multiple STEM and non-STEM interests and options. These are also reflected in their pursuit of multiple majors and minors in both STEM and non-STEM disciplines.
- For nearly two-thirds (61%) of TALR switchers, *loss of confidence* was a factor in their decision, and was also a concern for 79% of switchers overall. Losing confidence was also a problem for 44% of persisters. A much lower percentage of students in the first study described this issue; namely 23% as a switching factor, 34% for switchers overall, and only 12% of persisters. The increased ranking of losses of confidence from ninth to second place may, again, reflect the high proportion of high-performing switchers, two-thirds of whom were women and half of whom were also women of color. As already touched upon in findings presented in Chap. 2 on the institutional records analysis, we discuss this gendered question further in the following sections in this chapter and also in Chaps. 9 and 10.
- There was a large upward shift in students' *negative reactions to the competitive climate that they experienced in STEM classes*. In the original study, 14% of switchers cited the hostile, isolating atmosphere created by class peers as prompting their decisions to switch. This was also a concern for 28% of TAL switchers overall, and for 9% of TAL persisters. However, the percentages we report for TALR are much higher: 52% of TALR switchers cited negative class climate as a reason for switching. This experience also created problems for the majority (81%) of all switchers, and was an issue for 42% of persisting seniors. Thus,

class climate issues not only continue but appear to be growing as major deterrents to persistence. As we will discuss in Chaps. 6, 7, and 9, the intense status competitions among peers encouraged by steeply curved grading practices encourage isolation and failure to develop a sense of belonging that we found to be greatest among women of all races and ethnicities, and men of color.

- While *financial problems in completing a STEM degree* did not figure so prominently as other issues contributing to switching in either the TAL or TALR studies (10% and 17%, respectively), problems in financing college emerged as a far more widespread concern in the present study. While 30% of TAL switchers cited financial problems as a factor in switching, this proportion rose to 70% for TALR switchers. The same pattern was evident among persisters: although finding sufficient funding for college was a concern for one-quarter (23%) of the TAL persisters in the 1990's, in this study, half of persisting seniors (48%) registered financial problems as a serious concern. In Chap. 11, we discuss how students in this study were paying for college and note especially the increase in student working hours and how worry about large loans affects the career-related decisions of both switchers and persisters.
- *Choosing STEM majors for reasons that prove inappropriate* was a concern mentioned by almost half (48%) of the TALR participants as contributory cause for switching, compared with 14% of students in the first study. In both studies, it was also a concern for switchers overall (viz., 68% in TALR and 82%, in TAL) and also for persisters (viz., 22% in TALR and 40% in TAL). We discuss this and other issues of choice of majors in Chap. 4.
- *Problems related to class size* were not a major factor in students' decision to switch out of STEM in either study. However, far more (56%) of the TALR switchers than those in the TAL study (20%) raised this as concern. Two-thirds more TALR persisters also defined large classes as problematic compared to students in the original study (29% vs 11%).

Career-related concerns were found to be a more pressing influence on students' decisions in the current than in the original study. This pattern is evident in three out of the four career-related influences on switching decisions that students described:

- About half of switchers overall, both now (58%) and then (43%), *rejected the future careers and lifestyles* to which they projected STEM majors would lead. However, in the current study, twice the number of switchers (58%) than in the first study (29%) identified this a reason for their decision to switch.
- Similarly, nearly twice the number of switchers in this study (54%) than the prior one (27%) explained that they changed to a non-STEM major partly because it *offered more appealing career opportunities.*
- Making instrumental, *system-playing moves into other majors as a means to further career goals* was a far more prominent strategy among switchers in the current than the original study. One-quarter of all switchers (26%) either sought or considered non-STEM majors by which they could achieve their (unchanged) career goals while graduating with higher GPAs that would give them a competitive edge. By comparison, only 7% of students in the original study described

switching as a good way to achieve specific career objectives. In Chap. 12, we also discuss system-playing by STEM persisters.

- There was a decline, however, in the present study in the proportion of both switchers and persisters who expressed the view that the pursuit of *STEM career options and rewards were not worth the effort*. In TAL, this issue ranked fifth as a cause for switching (31%) and was a concern for half (48%) of switchers and one-fifth (20%) of persisters. In the current study, students weighing the benefits and costs of a STEM degree assessed its career value more positively with only 17% of all switchers deciding that the rewards were not worth the costs of securing them. We discuss students' thinking and decision-making about their careers in Chap. 11.

In the next section we discuss our overall findings on what did and did not distinguish the contributory influences on the switching and persistence decisions of women and men in STEM majors.

## Concerns Contributing to Men's and Women's Difficulties in STEM Majors, Then and Now

A number of differences and similarities between men's and women's concerns over time are evident, whether they contributed to switching directly, or were of wider concern for switchers and persisters alike.

As intrinsic interest in the subject matter is one of the single best predictors of persistence in STEM, students' loss of incoming interest and motivation to pursue a STEM degree is concerning. Indeed, a surprising finding from the original study was that students were leaving their STEM majors because they were being "turned off science"—largely by poor faculty pedagogy. Similar percentages of men and women in TAL described how their original passion for the subject matter had fallen away since embarking on their STEM coursework (44% men and 43% women). Today, higher rates of men and women mentioned *loss of interest and motivation to continue in their major* as a reason for switching, women in particular (54% and 66%, respectively) (see Table 3.2 below and Table F.1 in Appendix F).

Then, as now, issues of poor quality teaching and course design, and other problems related to STEM teaching and learning experiences continue to dominate as factors influencing students' switching decisions, and are of wider concern for switchers overall, as well as for the students who stay. In TAL, bad teaching caused roughly one-third of men's and women's switch out of STEM (39% and 33%), and was a problem for almost all male and female switchers (92% and 89%); it remained an issue for 66% of male seniors and for 80% of female seniors. Today, poor teaching is cited by over half of men and women as a reason for switching (46% and 49%) and as a problem by nearly all male and female switchers (94% and 97%) and nearly three-quarters of male and female persisters (71% and 72%).

**Table 3.2** "The problem iceberg" by gender: Concerns currently contributing to men's and women's STEM switching decisions, and to the concerns of all men and women STEM switchers, of all men and women STEM persisters, and of all STEM students

| *Ranking of Concerns in TAL* | *Significant Concerns Identified by Interviewees* | *% for Whom This Factor Contributed to Switching (N=96)* | *% for Whom This Factor Contributed to Switching MEN (N=35)* | *% for Whom This Factor Contributed to Switching WOMEN (N=61)* | *% All Switcher Concerns MEN (N=35)* | *% All Switcher Concerns WOMEN (N=61)* | *% All Persister Concerns MEN (N=102)* | *% All Persister Concerns WOMEN (N=148)* | *% Concerns of all MEN (N=137)* | *% Concerns of all WOMEN (N=209)* | *% All Student Concerns (N=346)* |
|---|---|---|---|---|---|---|---|---|---|---|---|
| *16* | Discovery of an aptitude for a non-STEM subject | 76% | 83% | 72% | 83% | 72% | 6% | 7% | 26% | 26% | 26% |
| *9* | Discouraged/lost confidence due to low grades in early years | 61% | 51% | 67% | 71% | 84% | 37% | 49% | 46% | 59% | 54% |
| *1* | Loss of incoming interest and motivation | 61% | 54% | 66% | 54% | 67% | 17% | 9% | 26% | 26% | 26% |
| *6* | Rejection of STEM careers and associated lifestyles | 58% | 54% | 61% | 54% | 61% | 6% | 5% | 18% | 21% | 20% |
| *7* | Shift to a more appealing career option | 54% | 46% | 59% | 46% | 59% | 5% | 4% | 15% | 20% | 18% |
| *12* | Competitive, unsupportive STEM culture makes it hard to belong | 52% | 46% | 56% | 74% | 85% | 30% | 52% | 42% | 62% | 54% |
| *3* | Poor quality of STEM teaching | 48% | 46% | 49% | 94% | 97% | 71% | 72% | 77% | 79% | 78% |
| *13* | Reasons for choice of STEM major prove inappropriate | 48% | 54% | 44% | 74% | 64% | 22% | 22% | 35% | 34% | 35% |

| | | | | | | | | | | | |
|---|---|---|---|---|---|---|---|---|---|---|---|
| | Difficult transition to college | 43% | 34% | 48% | 89% | 89% | 53% | 61% | 62% | 69% | 66% |
| | Negative effects of weed-out classes | 35% | 34% | 36% | 37% | 46% | 21% | 16% | 25% | 24% | 25% |
| *4* | STEM curricular design problems: pace, overload, labs, alignment | 31% | 34% | 30% | 94% | 82% | 55% | 59% | 65% | 66% | 65% |
| *17* | Prefer teaching approach in non-STEM courses | 29% | 31% | 28% | 63% | 56% | 8% | 8% | 22% | 22% | 22% |
| *2* | Non-STEM major offers better education, holds more interest | 26% | 31% | 23% | 31% | 23% | 2% | 3% | 9% | 9% | 9% |
| *19* | System-playing as means to career goals | 26% | 29% | 25% | 29% | 25% | 3% | 0% | 9% | 7% | 8% |
| *11* | Inadequate high school preparation in subject and study skills | 19% | 14% | 21% | 66% | 62% | 37% | 33% | 45% | 42% | 43% |
| *5* | STEM career options and rewards felt not worth the effort | 17% | 11% | 20% | 11% | 20% | 1% | 1% | 4% | 6% | 5% |

(continued)

**Table 3.2** (continued)

| *Ranking of Concerns in TAL* | *Significant Concerns Identified by Interviewees* | *% for Whom This Factor Contributed to Switching (N=96)* | *% for Whom This Factor Contributed to Switching MEN (N=35)* | *% for Whom This Factor Contributed to Switching WOMEN (N=61)* | *% All Switcher Concerns MEN (N=35)* | *% All Switcher Concerns WOMEN (N=61)* | *% All Persister Concerns MEN (N=102)* | *% All Persister Concerns WOMEN (N=148)* | *% Concerns of all MEN (N=137)* | *% Concerns of all WOMEN (N=209)* | *% All Student Concerns (N=346)* |
|---|---|---|---|---|---|---|---|---|---|---|---|
| *14* | Conceptual difficulties with one or more STEM subject(s) | 16% | 20% | 13% | 83% | 79% | 50% | 38% | 58% | 50% | 53% |
| *21* | Problems related to class size | 15% | 14% | 15% | 51% | 59% | 31% | 28% | 36% | 37% | 37% |
| *8 &15* | Difficulties in seeking and getting appropriate timely help | 13% | 14% | 11% | 80% | 80% | 24% | 37% | 38% | 50% | 45% |
| *18* | Time to degree influenced switching decision | 13% | 14% | 11% | 14% | 11% | 0% | 1% | 4% | 4% | 4% |
| *10* | Financial problems with completing STEM degree | 10% | 9% | 11% | 60% | 75% | 45% | 51% | 49% | 58% | 54% |
| *22* | Poor teaching, lab or recitation support by TAs | 3% | 6% | 2% | 6% | 2% | 1% | 3% | 2% | 3% | 3% |
| *20* | Language difficulties with foreign faculty or TAs | 1% | 3% | 0% | 57% | 52% | 17% | 22% | 27% | 31% | 29% |
| *23* | Poor lab/computer lab facilities | 0% | 0% | 0% | 0% | 0% | 0% | 0% | 0% | 0% | 0% |

| * | | | | |
|---|---|---|---|---|
| | 0%-25% | 26%-50% | 51%-75% | 76%-100% |

Prominent in student decisions to switch from STEM 20 years ago were problems of STEM curricular design, notably, too much subject material covered at too rapid pace and poor fit between course elements. These issues contributed to switching decisions for 42% men vs 29% of women. In the present study, the incidence of students' issues with flaws in course design were slightly lower than hitherto for men (34%) but the same in their contribution to women's switching decisions (30%). Problems with course design, however, affected nearly all switchers (95% of men; 82% of women) and over half of all persisters (55% of men; 59% of women).

Finding and securing appropriate, timely help with understanding subject material or other academic difficulties was an issue for students 20 years ago remains problematic. While this concern did not figure prominently in student decisions to switch, it was a larger, ongoing issue for switchers and persisters alike—both then and now. In TAL, 41% of men and 42% of women discussed problems of inadequate help, as did 32% of male seniors and 41% of female persisters. Today, 80% of male and female switchers mentioned difficulties accessing necessary help as a wider concern, and this proved an issue for one-quarter (24%) of male persisters and over one-third of female seniors (39%).

In the original study, men much more so than women described how the competitive ethos encountered in STEM classrooms had factored into their decision to switch (26% vs 4%). In this study, almost half of men (46%) and over half of women (56%) described the ways in which the competitive, unsupportive atmosphere of STEM classrooms had pushed them out of STEM majors. Problems with class climate, which were largely generated by peer status competitions that were encouraged by curve grading practices, were reported by a large majority of both male and female switchers (74% and 85%) by one-third of male seniors and over half (52%) of female seniors.

As found in TAL, overly large classes ranked low in students' reasons for switching in the current study. We note, however, that much larger numbers of today's switchers see it as a wider problem (51% of men and 59% of women), as do about one-third of today's persisters (31% of men and 28% of women).

As for the overall student sample, men's and women's discouragement and loss of confidence due to low grades in early years has shifted significantly. In the previous study 27% of men and 20% of women talked about the negative impact low grades in their introductory STEM courses had on their morale. Today 51% of men and 67% of women cited this as a reason for switching, and this was of broader concern to a large majority of men and women switchers overall (54% and 67%) and remains a problem for over one-third of male seniors (37%) and nearly half of female seniors (49%).

Thus, we find that the factors contributing to students' decisions to switch out of STEM majors two decades ago are predominately the same today. However, in our current study, we found that multiple issues were reported by larger percentages of students. That more students are experiencing more of the same issues would seem to suggest that students' experiences have been converging over time.

## *Changes in Men's and Women's Career-Related Concerns Over Time*

As discussed earlier in this chapter, career-related issues figure more prominently in students' concerns today than previously.

The original study found that men were more willing to switch majors as a means to improve their overall chances of achieving their career goals: "system playing" accounted for 10% of men's, compared to 4% women's, switching decisions. Mentioned much more frequently in this study, one-quarter of both men and women (29% and women 25%) discussed using this tactic as a means towards securing a desired career objective.

Many men and women switchers two decades ago came to *doubt the benefits of a STEM degree were worth the effort* (36% vs 27%). This strategy was mentioned less in the present study as a reason for switching, but more women (20%) than men (11%) chose to switch their major where they assessed the expected outcome would not be worth the effort necessary to achieve it.

As discussed earlier in this chapter, the original study found that few students switched from their STEM major due to the *discovery of an aptitude for a non-STEM major* (10% of men; 11% of women). We found a significantly different pattern in the current study: 83% of men and 72% of women described being "pulled" into a non-STEM field in which they found a good fit for their interests and talents. This experience both prompted switching and was also widely considered by other switchers and seniors.

In the TAL study, nearly half of women (46%) compared to one-third of men (35%) *switched to a major that offered more intrinsic interest or better education.* In TALR, this concern was reported less often by women (23%) and reports by men remained at the same level (viz., 31%).

More women previously *rejected the STEM career and lifestyles they associated with them* than male students (38% women vs 20% men). Today this is even truer: 61% of women and 54% of men said they no longer aspired to the STEM careers they intended upon college entry.

Comparison of the career-related concerns prompting men and women to switch from STEM shows that, then as now, women tend to take a holistic, longer-term view about prospective job fit and satisfaction.

As in the original study, we found that STEM losses included many high-achieving students (those with a 3.5 or better GPA). As reported from the institutional records analyses described in Chap. 2, 26% of STEM switchers were high performers. High risks of switching for high-performing women were also reported and women in under-represented minority groups had greater risks of switching than both white women and all men. Also reported in TAL, women switched from STEM majors with higher average performance scores than men who stayed. As described in Chap. 1, we found that 69% (42 of 61) of the interview sample of women switchers had left their STEM major with GPAs of 3.5 (B+) and half of these ($N = 21$) were women of color. (Three white men and five men of color were also high-performing switchers.)

Given that losses among students of this caliber are likely to be of concern, we wished to understand what distinguished their reasons for switching from those of other students. In Chaps. 9 and 10, we explore the experiences of women in STEM majors and their consequences, including those of high-performing women of all races and ethnicities.

In interviews, we found that choice of a STEM major (as discussed in Chap. 4) was often influenced by parents' views and experiences. By students' accounts, parental attitudes towards a daughter's choice of a STEM major have changed dramatically over the last two decades. Twenty years ago, parents commonly supported their daughters' moves to more traditional, non-STEM majors. By contrast, in the current study, we found little difference in parents' attitudes towards sons and daughters: male and female students, and their parents alike, were largely concerned to secure careers that would provide financial security. These findings are explored in Chap. 11.

## *Changes in Gendered Issues Arising Outside of College*

We found similar patterns to those reported in TAL for issues that students bring with them into college and that affect their performance in early coursework. While these concerns rarely had a direct effect on decisions to switch, they were of broader concern to much larger percentages of switchers and persisters in the current study:

- *Inadequate high school preparation*: while poor high school preparation ranked low for men and women as a reason for leaving their major two decades ago, in this study struggling to overcome deficiencies in subject matter and/or poor study habits and skills was mentioned by 21% of women vs 14% of men as a reason for switching. However, 66% of male, and 62% of female, switchers, as well as for one-third of all seniors (37% of males and 33% of females) discussed problems related to their high school education.

In addition, in the current study, 48% of women, compared to 34% of men talked about the *difficulties they faced adjusting to college*, though comparable numbers of male and female switchers (both 89%) and persisters (53% and 61%) also mentioned issues with transitioning to college.

- *Financial problems completing a STEM degree:* In TAL, men were more vocal about financial concerns than women, particularly as reason for switching (24% vs 11%); in TALR, while similar small percentages of men and women worried about the cost of the degree as a reason for switching (9% and 11%), financial concerns were much higher for women than for men who switched (75% vs 60%) and for female than for male persisters (51% vs 45%).

One important change in women's reports from two decades ago concerns their interactions with male peers and instructors in STEM classrooms. In TAL, women routinely experienced hostile, sexually inappropriate behavior by their male classmates and, on occasion, by male STEM faculty that made them feel uncomfortable and unwelcome. Although we specifically asked all women about this kind of behavior in the TALR study, we did not hear these same stories. However, as discussed in Chaps. 6, 7, and 9, women's struggle to belong in STEM was still affected by the competitive ethos that was widely reported as dominating these fields.

In sum, our overall findings on differences between concerns affecting women's and men's decisions to switch out of STEM point to a marked convergence between the sexes in accounts of their persistence difficulties: more than half of all men and women in the current study mentioned the same 17 of 24 factors (70%) as concerns. Where there were differences between men and women in the nature of their concerns, these were consistent with results from the original study.

As with the original study, the "problem iceberg" findings, in and of themselves, do not offer explanations for the greater loss of women (or of students of color) from STEM majors. As before, we turn to the ethnographic analysis of the interview data for explanation of differential losses. Throughout the following chapters, we address particular issues that disproportionately affect women, along with their consequences.

## Differences in Concerns Contributing to Decisions to Leave STEM Majors by Race/Ethnicity, Then and Now

Students of color were 36% of our student interviews sample (26 switchers and 100 persisters), comprised of Latinos/as, African-Americans, Native Americans, and students who identified as multi-racial. Though Asian-Americans (or Pacific Islanders) are often excluded from discussions on under-represented minorities, they were included both in the original and the current study. Of all students of color, 12 were the first in their family to attend college and seven students were immigrants.

The original *Talking about Leaving* study did not provide a "problem iceberg" broken out by race/ethnicity. However, from their ethnographic analysis of the text data, the authors distilled differences in students' experiences in STEM majors based on their racial and ethnic minority status (see pp. 27–28 for discussion in Chap. 1 and Table 1.1, and Chap. 6 of TAL). As with other results we have reported for TALR, we did not find major differences in switchers' or persisters' accounts among students of color from those found 20 years ago. Table 3.3 shows the "problem iceberg" broken out by concerns affecting students of color and those affecting white students. As with issues of gender, a number of important differences between students of color and white students are evident.

The sharpest distinctions between the concerns of students of color and those of white students are found in four areas: *inadequate high school preparation, difficult transition to college,* the *competitive, unsupportive STEM culture makes it difficult to belong*, and *discouragement/loss of confidence due to low grades in early years.* Students of color, more than did their white peers, described the following concerns

**Table 3.3** "The problem iceberg" by race/ethnicity: Current concerns contributing to STEM switching decisions, to the concerns of STEM switchers and STEM persisters, by race/ethnicity, and to the concerns of all STEM students

| *Ranking of Concerns in TAL* | *Significant Concerns Identified by Interviewees* | *% for Whom This Factor Contributed to Switching (N=96)* | *% for Whom This Factor Contributed to Switching ALL SOC (N=26)* | *% for Whom This Factor Contributed to Switching ALL WHITE STUDENTS (N=70)* | *% All Switcher Concerns ALL SOC (N=26)* | *% All Switcher Concerns ALL WHITE STUDENTS (N=70)* | *% All Persister Concerns ALL SOC (N=100)* | *% All Persister Concerns ALL WHITE STUDENTS (N=150)* | *% All Students (N=346)* |
|---|---|---|---|---|---|---|---|---|---|
| *16* | Discovery of an aptitude for a non-STEM subject | 76% | 77% | 76% | 77% | 76% | 9% | 5% | 26% |
| *9* | Discouraged/lost confidence due to low grades in early years | 61% | 69% | 59% | 92% | 74% | 59% | 35% | 54% |
| *1* | Loss of incoming interest and motivation | 61% | 62% | 61% | 62% | 63% | 15% | 10% | 26% |
| *6* | Rejection of STEM careers and associated lifestyles | 58% | 62% | 57% | 62% | 57% | 8% | 3% | 20% |
| *7* | Shift to a more appealing career option | 54% | 46% | 57% | 46% | 57% | 7% | 3% | 18% |
| *12* | Competitive, unsupportive STEM culture makes it hard to belong | 52% | 62% | 49% | 88% | 79% | 60% | 32% | 54% |
| *3* | Poor quality of STEM teaching | 48% | 35% | 53% | 96% | 96% | 92% | 58% | 78% |
| *13* | Reasons for choice of STEM major prove inappropriate | 48% | 38% | 51% | 58% | 71% | 29% | 17% | 35% |
| - | Difficult transition to college | 43% | 73% | 31% | 96% | 86% | 78% | 45% | 66% |
| - | Negative effects of weed-out classes | 35% | 27% | 39% | 42% | 43% | 14% | 20% | 25% |
| *4* | STEM curricular design problems: pace, overload, labs, alignment | 31% | 19% | 36% | 92% | 84% | 74% | 46% | 65% |
| *17* | Prefer teaching approach in non-STEM courses | 29% | 27% | 30% | 54% | 60% | 13% | 5% | 22% |

(continued)

**Table 3.3** (continued)

| *Ranking of Concerns in TAL* | *Significant Concerns Identified by Interviewees* | *% for Whom This Factor Contributed to Switching (N=96)* | *% for Whom This Factor Contributed to Switching ALL SOC (N=26)* | *% for Whom This Factor Contributed to Switching ALL WHITE STUDENTS (N=70)* | *% All Switcher Concerns ALL SOC (N=26)* | *% All Switcher Concerns ALL WHITE STUDENTS (N=70)* | *% All Persister Concerns ALL SOC (N=100)* | *% All Persister Concerns ALL WHITE STUDENTS (N=150)* | *% All Students (N=346)* |
|---|---|---|---|---|---|---|---|---|---|
| *2* | Non-STEM major offers better education, holds more interest | 26% | 31% | 24% | 31% | 24% | 3% | 2% | 9% |
| *19* | System-playing as means to career goals | 26% | 27% | 26% | 27% | 26% | 1% | 1% | 8% |
| *11* | Inadequate high school preparation in subject and study skills | 19% | 35% | 13% | 73% | 60% | 41% | 31% | 43% |
| *5* | STEM career options and rewards felt not worth the effort | 17% | 12% | 19% | 12% | 19% | 0% | 1% | 5% |
| *14* | Conceptual difficulties with one or more SME subject(s) | 16% | 12% | 17% | 92% | 76% | 54% | 35% | 53% |
| *21* | Problems related to class size | 15% | 15% | 14% | 77% | 49% | 42% | 21% | 37% |
| *8 and 15* | Difficulties in seeking and getting appropriate timely help | 13% | 15% | 11% | 92% | 76% | 36% | 29% | 45% |
| *18* | Time to degree influenced switching decision | 13% | 15% | 11% | 15% | 11% | 0% | 1% | 4% |
| *10* | Financial problems with completing STEM degree | 10% | 8% | 11% | 69% | 70% | 61% | 40% | 54% |
| *22* | Poor teaching , lab or recitation support by TAs | 3% | 0% | 4% | 0% | 4% | 1% | 3% | 3% |
| *20* | Language difficulties with foreign faculty or TAs | 1% | 0% | 1% | 58% | 53% | 19% | 20% | 29% |
| *23* | Poor lab/computer lab facilities | 0% | 0% | 0% | 0% | 0% | 0% | 0% | 0% |

* | 0%-25% | 26%-50% | 51%-75% | 76%-100% |

as contributing to switching decisions, and as general concerns for all switchers and all persisters.

- *Inadequate high school preparation* disproportionately affected students of color. One-third of students of color referenced the poor preparation provided by their high schools as a reason for switching (35% vs 13%). For switchers more broadly, the ongoing problems arising from insufficient content knowledge, study habits, and cultural capital necessary to negotiate college affected nearly three-quarters of students of color (73%) and nearly two-thirds of white students (60%); it was also a problem for 41% of students of color and 31% of white students persisting in their STEM major.

Highly related to students' accounts of the ways in which their high school education failed to adequately prepare them for college-level coursework, much greater percentages of students of color also reported problems adjusting to college.

- Strikingly, students of color said that a *difficult transition to college* contributed to their decisions to switch to a non-STEM major at a much higher rate than did white students (73% vs 31%). Yet this problem affected similar proportions of students of color and white students and more generally (86% vs 96%). As persisting seniors, three-quarters (78%) of students of color also described difficulties adjusting to college; this was an issue for less than half (46%) of white peristers. Issues of inadequate high school preparation and students difficulties transitioning to college are explored in depth in Chap. 5.
- *Competitive, unsupportive STEM culture makes it difficult to belong:* Sixty-two percent of switchers of color, compared to 49% of white switchers, said that their decision to leave their major had been prompted by the unwelcoming, competitive ethos that they experienced in STEM courses. These experiences were also widely reported by all switchers (viz., 88% students of color vs 79% of white students), and were an ongoing problem for over half of seniors of color (60%) and one-third of white seniors (32%).
- *Discouraged/lost confidence due to low grades in early years:* This concern was reported as a reason for switching by a slightly higher percentage of students of color than white students (69% compared with 59%). However, loss of confidence was a concern for nearly all students of color who switched (92%) and for three-quarters (74%) of white students. Among persisters, over half of students of color (59%) described this problem compared with one-third (35%) of white students. This topic and its effects on students of color are discussed in Chaps. 2, 6, 7, and 9.

As reported in Chap. 2, students of color were more likely to feel that they were poorly prepared in math, particularly women of color. Women also were more likely to attribute their decision to switch to a lack of STEM preparation.

Mentioned by students of color as contributing to switching decisions *less* often than their white peers were concerns about:

- *poor quality teaching* (35% vs 53%)
- *problems with curricular design* (19% vs 36%)
- *conceptual difficulties with one or more STEM subject(s)* (12% vs 17%)

However, these problems were reported as overall concerns by nearly all switchers—whether students of color or white students.

While difficulties in seeking and getting appropriate timely help was a minor factor prompting switching (15% for students of color compared to 11% for white students), this was a problem for almost all students of color (92%) and for three-quarters (76%) of white students. It was a problem for one-third of persisters (36% of students of color and 29% for white students).

Concerns that were found to be of similar concern for switchers and persisters as causes for switching, and as overall concerns for switchers and persisters, regardless of race and ethnicity also include:

- *Discovery of an aptitude for a non-STEM major.* While this was a concern for approximately three-quarters of switchers (77% students of color and 76% white students), it was negligible for persisters of all races and ethnicities.
- *Loss of incoming interest and motivation* affected nearly two-thirds of all switchers (viz., 62% students of color and 63% of white students).

Students' views were also similar for two career-related concerns: both students of color and white students mentioned *dissatisfaction with the prospects and lifestyles associated with future STEM careers* (62% vs 57%, respectively). About one-quarter of both groups also discussed *system-playing to achieve their career goals* (27% of students of color vs 26% of whites). White students somewhat more often than students of color discussed weighing the pros and cons of completing their STEM degrees, and based on a 'hedonistic calculus', concluded that the *benefits weren't worth the effort* (19% vs 12%). Students of color were also slightly less likely to report that their switch to a non-STEM field was due to the *pull of a more appealing career option* (46% vs 57%).

We did not find in this analysis that the overall number and type of concerns that students of color and white students described as prompting switching or broader concerns for all switchers, differed much. As with our findings for men and women, over half of both students of color and white students mentioned the same 16 of 24 concerns (67%) as wider problems. Where there was divergence was among persisters: over half of seniors of color mentioned seven persistence-related concerns compared to only one concern mentioned by over half of white students. As with our earlier observation on the convergence of women's and men's problematic experiences, it also appears that, for some concerns, the experiences of students of color and of white students show strong similarities.

This apparent trend does not, however, detract from the larger finding that students of color continue to confront multiple, intersecting concerns which they report at higher rates than white males in STEM fields. Findings from the institutional data analysis presented in Chap. 1 show the dampening effects on the persistence of students of color who enter with lower mathematics preparation. We also see in the "problem iceberg" table on race/ethnicity that greater percentages of students of color—switchers and persisters alike—than of white students describe ongoing issues arising from poor high school preparation and subsequent difficulties in transitioning to college. The significance of these issues for switching rates, including those of students of color, is addressed in Chaps. 5, 7, and 12.

## *Differences in the "Problem Iceberg" by Institutions Participating in TALR*

Similar to findings presented in TALR, we did not find that student concerns varied greatly by institution, regardless of the different sizes, types, and student populations of the schools. Table 3.4 shows the top five ranking of concerns contributing to students' switching decisions, and concerns affecting switchers and persisters more broadly. Overall, these tables demonstrate that student problems ranked similarly across the six institutions.

The top three concerns contributing to switching were also the same concerns for all students—switchers and persisters alike:

- *Poor quality teaching in STEM courses*
- *Difficult transition to college,* and
- *STEM curricular design problems*

The only significant difference in switchers' and persisters' concerns is discernible in career-related issues. Not surprisingly, these concerns factored far lower in rank for persisters than for switchers overall. Concerns ranking lowest as issues prompting switching, and for switchers and persisters overall, are factors gleaned from the literature which hypothesized reasons for switching, and, upon checking with students were, as in the original study, shown to have little to no influence on students' switching decisions.

## Conclusions

The results from this study remain congruent with those presented in *Talking about Leaving*, not only in validating the type and range of concerns contributing to students' switching decisions, but also in identifying a set of linked problems with their STEM learning experiences as the most commonly-cited persistence problems in STEM majors. There were many other similarities in the findings of both studies: switchers and persisters in both studies experienced the same types of problems; decisions to switch were again found to be the result of complex "push-pull" processes in struggles with the same set of problems that also affected other switchers and, to a lesser extent, persisters; and persisters were, again, more successful than switchers in locating appropriate and timely sources of help to surmount these shared problems. Our current findings also replicated many of the earlier findings about women and students of color, especially, the loss of high-performers. As with TAL, we did not find that our results differed much by institution, regardless of type, size, or student population.

The greatest difference in our findings from the original study is that "the problem iceberg" has grown larger: switchers and persisters in the present study report a higher average number of the same problems than were identified hitherto, and

**Table 3.4** Comparative ranking by students at six institutions of the top five concerns contributing to switching; Concerns raised by switchers overall; Concerns raised by persisters; and Concerns of students overall

| *Rank of Concern in TAL* | *Significant Concerns Identified by Interviewees* | *Rank of Concerns Contributing to Switching Decisions TALR* | | | | | | *Rank of Switchers' Concerns, Overall, TALR* | | | | | |
|---|---|---|---|---|---|---|---|---|---|---|---|---|---|
| | | *PV1R3* | *PB1R1* | *PB4R1* | *PV2R1* | *PB2R1* | *PB3R1* | *PV1R3* | *PB1R1* | *PB4R1* | *PV2R1* | *PB2R1* | *PB3R1* |
| *1* | Loss of incoming interest and motivation | 2 | 3 | 2 | 3 | 3 | 2 | 3 | | 5 | | | |
| *2* | Non-STEM major offers better education, holds more interest | 5 | | | | | 4 | 4 | | | | | |
| *3* | Poor quality of STEM teaching | 4 | 3 | 3 | 5 | | 3 | 1 | 1 | 1 | 1 | 3 | 1 |
| *4* | STEM curricular design problems: pace, overload, labs, alignment | 5 | | | | | 5 | 1 | 3 | 3 | 2 | 1 | 5 |
| *5* | STEM career options and rewards felt not worth the effort | 5 | | | | | | 3 | | | | | |
| *6* | Rejection of STEM careers and associated lifestyles | 2 | 3 | 3 | | 1 | 3 | 1 | | | | 5 | |
| *7* | Shift to a more appealing career option | 2 | 5 | | 3 | 2 | 3 | 1 | | | | | |
| *9* | Discouraged/lost confidence due to low grades in early years | 1 | 1 | 4 | 4 | 3 | 3 | 2 | 1 | 5 | 3 | 5 | 5 |
| *10* | Financial problems with completing STEM degree | | | | | | | 1 | 4 | 3 | | | |
| *11* | Inadequate high school preparation in subject and study skills | 5 | | | | | | 2 | | | 5 | | 5 |
| *12* | Competitive, unsupportive STEM culture makes it hard to belong | 3 | 4 | 4 | 2 | 5 | 3 | 2 | 3 | 4 | 4 | 4 | 3 |

| | | | | | | | | | | | | | |
|---|---|---|---|---|---|---|---|---|---|---|---|---|---|
| *13* | Reasons for choice of STEM major prove inappropriate | 4 | 4 | 5 | 5 | | 3 | 2 | | | 4 | 5 | 5 |
| *14* | Conceptual difficulties with one or more SME subject(s) | 5 | | | | | | 1 | | 3 | 2 | 4 | 3 |
| *16* | Discovery of an aptitude for a non-STEM subject | 1 | 2 | 1 | 1 | 4 | 1 | 2 | 4 | 2 | 4 | - | 2 |
| *17* | Prefer teaching approach in non-STEM courses | 5 | | | | | 5 | 3 | | 5 | 4 | | |
| *18* | Time to degree influenced switching decision | | | | | | | | | | | | |
| *19* | System-playing as means to career goals | 5 | | | | | | 3 | | | | | |
| *20* | Language difficulties with foreign faculty or TAs | | | | | | | | | | | | 5 |
| *21* | Problems related to class size | | | | | | | 4 | | | | | |
| *22* | Poor teaching, lab or recitation support by TAs | | | | | | | | | | | | |
| *23* | Poor lab/computer lab facilities | | | | | | | | | | | | |
| *8 & 1* | Difficulties in seeking and getting appropriate timely help | | | | | | | 4 | 5 | 2 | 4 | 3 | 1 |
| - | Difficult transition to college | 4 | | | 5 | 3 | 3 | 2 | 2 | 2 | 2 | 2 | 4 |
| - | Negative effects of weed-out classes | 4 | | 3 | | | 5 | | | 4 | | | |

(continued)

**Table 3.4** (continued)

| *Rank of Concern in TAL* | *Significant Concerns Identified by Interviewees* | *Persisters' Concerns TALR* | | | | | | *All Students' Concerns TALR* | | | | | |
|---|---|---|---|---|---|---|---|---|---|---|---|---|---|
| | | *PV1R3* | *PB1R1* | *PB4R1* | *PV2R1* | *PB2R1* | *PB3R1* | *PV1R3* | *PB1R1* | *PB4R1* | *PV2R1* | *PB2R1* | *PB3R1* |
| *1* | Loss of incoming interest and motivation | | | | | | | | | | | | |
| *2* | Non-STEM major offers better education, holds more interest | | | | | | | | | | | | |
| *3* | Poor quality of STEM teaching | 1 | 1 | 1 | 1 | 1 | 1 | 1 | 1 | 1 | 1 | 2 | 1 |
| *4* | STEM curricular design problems: pace, overload, labs, alignment | 2 | 2 | 4 | 2 | 2 | 3 | 2 | 3 | 4 | 2 | 3 | 3 |
| *5* | STEM career options and rewards felt not worth the effort | | | | | | | | | | | | |
| *6* | Rejection of STEM careers and associated lifestyles | | | | | | | | | | | | |
| *7* | Shift to a more appealing career option | | | | | | | | | | | | |
| *9* | Discouraged/lost confidence due to low grades in early years | 5 | | | 4 | 3 | 4 | 5 | | | 4 | 5 | 4 |
| *10* | Financial problems with completing STEM degree | | 2 | 3 | 5 | 4 | | | 4 | 3 | | | |
| *11* | Inadequate high school preparation in subject and study skills | 4 | | | | | | 4 | | | | | |
| *12* | Competitive, unsupportive STEM culture makes it hard to belong | | 5 | 4 | | | 5 | | 5 | 5 | | | 5 |

| | | | | | | | | | | | | | |
|---|---|---|---|---|---|---|---|---|---|---|---|---|---|
| *13* | Reasons for choice of STEM major prove inappropriate | | | | | | | | | | | | |
| *14* | Conceptual difficulties with one or more SME subject(s) | | 4 | 5 | | 5 | 5 | | | | 5 | 4 | |
| *16* | Discovery of an aptitude for a non-STEM subject | | | | | | | | | | | | |
| *17* | Prefer teaching approach in non-STEM courses | | | | | | | | | | | | |
| *18* | Time to degree influenced switching decision | | | | | | | | | | | | |
| *19* | System-playing as means to career goals | | | | | | | | | | | | |
| *20* | Language difficulties with foreign faculty or TAs | | | | | | | | | | | | |
| *21* | Problems related to class size | | | 5 | | | 5 | | | | | | |
| *22* | Poor teaching, lab or recitation support by TAs | | | | | | | | | | | | |
| *23* | Poor lab/computer lab facilities | | | | | | | | | | | | |
| *8 & 15* | Difficulties in seeking and getting appropriate timely help | | | | | | | | | | | | |
| - | Difficult transition to college | 3 | 3 | 2 | 3 | 1 | 2 | 3 | 2 | 2 | 3 | 1 | 2 |
| - | Negative effects of weed-out classes | | | | | | | | | | | | |

persistence to graduation in STEM degrees appears to be a more complex achievement than it was for students two decades ago. Compared to students' accounts in TAL, the accounts of both switchers and persisters in this study contain higher average numbers of interacting concerns, being negotiated simultaneously. There were also differences in the frequency with students cited switching factors that also created problems for other students. There were also some important shifts in the ranking of students' concerns. Notably, TALR switchers described discovery of an aptitude for a non-STEM subject above all other factors prompting their switching. There were also increases in accounts of lost confidence and in difficulties with the competitive, unsupportive climate of STEM classrooms. We also found financial difficulties both more prominent in switching decisions and widespread among all students. Our findings also indicate that, over time, the experiences of students of both sexes and across races and ethnicities have somewhat converged. However, as we describe in the following chapters, there are still problems that that are distinctive to particular student groups in driving switching decisions and informing persistence problems.

In the balance of this book each chapter draws on findings from the contributory studies that comprise the TALR project in order to focus on particular persistence issues.

## Reference

Seymour, E., & Hewitt, N. M. (1997). *Talking about leaving: Why undergraduates leave the sciences*. Boulder, CO: Westview Press.

# Chapter 4
# Choosing STEM Majors

**Heather Thiry and Timothy J. Weston**

## Introduction

Many aspects of students' experiences and backgrounds influence whether they will choose to enroll in a STEM major. Most of the prior research examining students' entrance into STEM majors has focused on the demographic and educational factors associated with declaring or completing a STEM major (Bonous-Hammarth, 2000; Chang, Cerna, Han, & Saenz, 2008; Crisp, Nora, & Taggart, 2009; Eagan, Hurtado, & Chang, 2010; Gayles & Ampaw, 2014; Rask, 2010; Riegle-Crumb & King, 2012). Yet less is known about the reasons or decision-making processes underlying students' choices to enter a STEM major and whether these reasons do or do not hold up over time, questions we sought to answer in this study.

Still, a growing body of research has begun to focus on students' decision-making processes in weighing and selecting a STEM college major. Parents and teachers play a large role in shaping students' choices to enter a STEM major with parents often exerting more influence than classroom teachers (Sjaastad, 2012; VanMeter Adams et al., 2014). Fathers can be more influential than mothers in

H. Thiry (✉)
Ethnography and Evaluation Research (E&ER), University of Colorado Boulder, Boulder, CO, USA
e-mail: Heather.Thiry@Colorado.edu

T. J. Weston
Ethnography and Evaluation Research (E&ER), University of Colorado Boulder, Boulder, CO, USA

National Center for Women and Information Technology (NCWIT), University of Colorado Boulder, Boulder, CO, USA
e-mail: westont@colorado.edu



E. Seymour, A.-B. Hunter (eds.), *Talking about Leaving Revisited*,
https://doi.org/10.1007/978-3-030-25304-2_4

students' choice of STEM majors (Sjaastad, 2012; Simpson, 2003). While parental support may be beneficial, parental interference in students' choices can lead to less career exploration for students (Dietrich, Kracki, & Nurmi, 2011), and at least one study has found that parents do not influence engineering students' choice of major (Sheppard et al., 2010). Teachers may be more influential on students' major choices in some fields than others, such as disciplines that are represented in high school curricula (e.g., biology or chemistry) rather than fields less frequently offered in high school, such as geosciences or computer science (Sjaastad, 2012). Pre-college experiences and early science experiences also shape students' interest and their decisions to major in STEM (VanMeter Adams et al., 2014; McGill, Decker, & Settle, 2015). Finally, outreach or after-school programs introduce underrepresented minority students to STEM fields and careers who may not have had such exposure otherwise and, thus, influence their enrollment in STEM majors (Sheppard et al., 2010).

Interest also appears to be a key factor influencing students' choice of STEM major (Lapan & Shaughenessy, 1996; Shehab et al., 2015). Early interest in STEM, particularly during the middle school years, is related to enrollment in and completion of STEM degrees (Maltese, Melki, & Wiebke, 2014; Tai, Liu, Maltese, & Fan, 2006). Yet others have documented that identity, or students' sense of self in the discipline, is a more important determinant than interest for engineers' selection of major (Matusovich, 2010). Math self-concept is an important predictor of STEM aspirations, yet it has become a weaker predictor of entrance into STEM majors for women than it used to be, indicating that lower math self-concept does not currently account for gender differences in STEM as much as it has in the past (Sax, Kanny, Riggers-Piehl, Whang, & Paulson, 2015). Finally, STEM fields hold the promise of high-status, stable, and lucrative careers and this has also been an important factor in students' selection of STEM majors (Hall et al., 2011; Stinebrickner, 2013; Sheppard et al., 2010), yet altruism and interest in the common good also draw students into STEM (Sheppard et al., 2010), especially women of color (Carlone & Johnson, 2007).

## *Findings from TAL1*

Similarly, in TAL1, students discussed multiple influences on their choice of STEM major and, occasionally, their reasons for choosing their major directly contributed to their subsequent choice to switch to a non-STEM major. Indeed, 14% of switchers in TAL1 chose STEM for the wrong reasons which contributed to their decision to move to a non-STEM major. Moreover, 18% of all students in TAL1 chose a STEM major because of the influence of others, including parents or teachers, although the influence of others on students' initial choice did not subsequently influence the decision of some students to leave STEM. Women were more likely than men to cite the influence of others, especially their fathers. In TAL1, students also chose STEM majors because they were good at math or science in high school, yet they often confused aptitude or a good teacher for true interest in the field. Students also sought stable, financially secure careers, yet altruism motivated some students, most notably women and students of color. Finally, a small minority of students (13%) switched because they felt that they had made an uninformed choice of STEM major.

## Reasons for STEM Students' Initial Choice of Major: Findings from the Interview Study and SALG Survey

There is a time difference in the collection of data between the SALG survey and the interview study. The respondents to the SALG survey were taking foundational STEM courses within their first year of entering a STEM major. Their answers to questions about why they had chosen their STEM major reflect their high school thinking about what they enjoy and can do well, their career aspirations which sometimes express idealism, and how to realize these. A few (1% of written responses) were already reconsidering those choices in light of their early college experiences. However, most have not yet acted upon any doubts about whether to remain in their major. The interviewees are juniors who have already chosen to leave STEM, or seniors who have chosen to remain in their STEM major or to relocate to a different one. Thus, the interview sample has reflected on the viability of their original choice to a far greater degree than the survey respondents.

### *What Reasons Did the SALG Respondents Give for Choosing Their Major?*

Over 1400 students completed the TALR-SALG survey which, as described in Chap. 3, asked questions about learning gains made in STEM foundational courses and student perceptions of majors and institutions. We asked the 1167 students in this sample who had declared STEM majors to rate (on a four-point agree/disagree scale) a set of reasons for choosing their major that were grounded in prior research. Their responses are summarized in Table 4.1. We also asked students to offer written comments in answer to the question: "What was your primary reason for choosing your major?" This time, students were not constrained by pre-coded options but could freely explain the most important determinant of their choices. Their responses are grouped into seven broad categories in Table 4.2. It is important to note that students could register multiple possible reasons for choosing their major.

The highest rated reason, "I wanted a career in this field" (average = 3.52, on a 4 point scale) was one of the four reasons that focused on ultimate careers rather than intrinsic interest in the discipline. However, it implies interest in the work options to which the degree may lead. The other three career-related ratings express gaining a

**Table 4.1** SALG respondents' rating of reasons for choosing a SALG major (4 point scale)

| I chose (4 point scale) this major because: (Strongly agree—Strongly disagree) (*n* = 1167) | Av. Rating | SD |
|---|---|---|
| I wanted a career in this field | 3.52 | 0.7 |
| A career in this major allows me to help others | 3.39 | 0.72 |
| I am good at this subject | 3.34 | 0.63 |
| I enjoy studying for courses in my major | 3.26 | 0.62 |
| My major will allow me to make a good income after I graduate | 3.16 | 0.77 |
| My major is necessary to go on to graduate or other professional school | 3.01 | 0.9 |
| I can get a job easily with my chosen major | 2.95 | 0.78 |
| I have friends who have entered this major | 2.45 | 0.94 |
| My family wants me to pursue this major | 2.4 | 0.9 |
| This major is easier than others | 1.78 | 0.87 |

**Table 4.2** Primary reasons for choosing a STEM major in SALG respondents' written answers (4 point scale)

| Broad category | % in category | Reason for choosing major | *N* | % |
|---|---|---|---|---|
| Affective reasons | 53 | Interested in major | 213 | 20 |
| | | Loves, likes, is passionate about major | 202 | 19 |
| | | Enjoys area of study | 90 | 9 |
| | | Skilled in area of study (efficacy) | 30 | 3 |
| | | Challenged by area of study | 14 | 1 |
| Degree is means to an end | 34 | Major needed for future career | 182 | 17 |
| | | Major needed for future education such as medical school, veterinary school, or graduate school | 111 | 11 |
| | | Major needed to achieve lifestyle goal ancillary to career such as "being outdoors" or "working with horses" | 38 | 4 |
| | | Major will allow student to earn money | 20 | 2 |
| | | Major allows student to get a job easily | 11 | 1 |
| Altruistic reasons | 9 | Want to help others, solve social problems, help animals, or help environment | 92 | 9 |
| External reasons | 1 | Family wanted student to pursue major | 14 | 1 |
| | | Student received scholarship for major | 2 | <1 |
| | | Advisor told student to pursue major | 1 | <1 |
| Characteristic of degree program | <1 | Classes in major looked easy | 2 | <1 |
| | | Liked reputation of degree program | 1 | <1 |
| Wrong major | <1 | Student said they chose wrong major, were leaving major | 9 | <1 |
| Not applicable | 2 | Answer did not relate to question | 20 | 2 |
| Total | (100) | | 1052 | (100) |

Note: <1% = "less than one percent"

particular STEM education as a means to an end—a good income, job security, or as a stepping stone to a higher degree. A comparable set of written comments (accounting for 34% of answers) also related to the choice of STEM majors as a means towards a desired career. They included, choosing a major because it was required for particular careers (17%), or to secure graduate education in medical, veterinary, dental, or graduate schools (11%). Such apparently instrumental choices were not necessarily materialistic or lacking in intrinsic interest. Indeed, only small numbers of students wrote that they chose the major to earn a lot of money (2%), because it was easier to get a job (1%) or a scholarship (less than 1%) with their chosen major. And the rating for "This major is easier than others" was only 1.78.

When students are asked to describe their primary reason for choosing their major, the picture (reflected in Table 4.2) is slightly different than when they are presented with a set of pre-coded alternatives. For example, in Table 4.1, altruistic reasons were rated second highest, but only 9% wrote that it was their primary reason. This would seem to suggest that, although altruistic motives are not the largest single reason for students' choices, when given the opportunity to name other contributory reasons, altruism emerges as an important secondary motive for many students.

Also, choosing a major as means to a career-focused goal was mentioned by 34% of students, but received only the sixth highest rating. However, when students are asked to state only their primary reason for their choice, the results still reflect that most students choose what they enjoy doing, what they are good at, and what they want to pursue as a career.

Some writers described STEM majors as the entrée to careers that offered an appealing lifestyle such as forestry that would enable them to work outdoors and pre-veterinary majors who wanted to work with animals (4%). Thus, for many of these respondents, it was the career field that was chosen more than the discipline leading to it:

> I would like to attend medical school and become a doctor. (White woman, biology major)
>
> It will help me in the career I would like to pursue. (Asian woman, neurology and physiology major)
>
> It's most applicable to my chosen career path, wildland fire management. (Multi-racial man, agricultural major,)

The risk implicit in such choices is that of discovering that the nature of the major is less appealing than the aspired career to which it may lead.

It is notable that the second-highest rated reason in Table 4.1 is "A career in this major allows me to help others" (3.39). Choices prompted by altruism were reported in the original study (TAL1), and nine percent of students also wrote comments in which they described altruistic reasons as their primary motivation for choosing their major. Many of the students wrote that they wanted to help others through careers in medicine; others wanted to help solve social or environmental problems through a wide range of majors including engineering and biological research, for example:

> I am and always have been an environmentalist and I want do all I can to help and conserve the environment, and encourage myself and others to live a simpler more sustainable life (White man, ecosystem science major)

To help people better their lives through proper nutrition and an active lifestyle. (White man, food science major)

Some students gave altruistic reasons in conjunction with career specific aspirations:

One day, I would like to enter the medical field and become a doctor because it is a career that truly serves others and helps improve the lives of many people. This major, integrated sciences, provided me with a path to medical school, which will help me ultimately achieve my goal of being a doctor. (Multi-racial woman, integrated sciences major)

Again, there is some risk that a major chosen for its capacity to enable a desired goal rather than out of intrinsic interest in the discipline leaves students open to disillusionment. In the original study, we found that altruistic aspirations that were developed in childhood and not re-examined as students approached college entry were apt to give way as students grappled with the academic realities of realizing them. However, we also found in TAL1 that long-term altruistic motivations could sustain students through difficult times in their STEM education.

Of all 1052 written responses, 53% fit the broad category we have labeled "affective reasons" in that they express interest in, enjoyment of, attraction to, and efficacy in the content and skills embodied by the major. These were the most frequently described set of reasons for a choice of major. They also include the third most-rated survey reason for choice, "I am good at this subject" (3.34). Examples include:

I am interested in how the human body works and how it relates to exercise. (Asian-American woman, physiology major)

I enjoy making things work and building things in a hands-on environment. (White man, mechanical engineering major)

I am really interested in space and subjects such as black holes and quasars. They are fascinating! I also want to go to graduate school for this. (White male, astrophysics major)

The original study reported that this set of reasons for choosing a STEM major were the least risky and the most likely to sustain students through to achievement of their aspirations. We shall consider in our discussion of interview study findings (and also in Chap. 12 on what enables persistence) whether this remains the case.

Only small numbers of students (1.6%) gave external reasons for choosing their majors, with just 14 students reporting that they chose their major because of their family's wishes or to follow a parent into the same career. Indeed, more students disagreed than agreed with the statements, "My family wants me to pursue this major" (2.4), and I before have friends who have entered this major after (2.45), and less than 1% had chosen a major on the recommendation of an advisor. However, as we shall discover when discussing findings from the interviews, low survey ratings of external reasons for the choice of a major do not deny the importance of parents and other advisors in helping students reach their decisions. What we learn from the survey are the primary and secondary reasons that they weigh in coming to a choice. Families, advisors, and friends may not provide the reasons for their choices, but as we shall hear from the interview findings, they are often critical in helping students process their alternatives and reach a decision.

We saw no obvious (or statistically significant) differences by gender or underrepresented racial minority (URM) status for most of the reasons for choosing a major. However, the items "My major is necessary to go on to graduate or other professional school" and "A career in this major allows me to help others" did show significant statistical differences by gender, with women giving higher ratings than men for both of these items (3.1 women to 2.9 men, 3.5–3.1). Some differences by gender were also evident at particular institutions where pre-medical and pre-veterinary programs enrolled more women than men. URM students gave lower ratings than non-URM students to the questions "I have friends who have entered this major" (2.5–2.3) and "I am good at this subject" (3.4–3.2).

Overall, although both the SALG survey pre-coded questions and written comments point to the importance of students' career aspirations in guiding the choice of a STEM major, when students are invited to state the primary reason for their choice, intrinsic interest in the field and a sense of their own ability to pursue it are dominant. Altruistic motivations and lifestyle goals are also well-represented and shape particular career aspirations.

We now turn to a discussion of findings from the interview data to see what more we can learn about what shapes students' choice of a STEM major.

## *What Reasons Did STEM Switchers and Persisters Give for Choosing Their Major?*

In contrast to the original study, in the current study, students' incoming level of knowledge and awareness about their major was the primary choice-related factor that most influenced whether they would stay in the STEM major or not. More than half of switchers (56.3%) moved to a non-STEM major, at least in part, because they were underinformed about their chosen field and the career options within that field, more than four times the proportion of TAL1 students with the same problem. Otherwise, switchers and persisters in TALR did not differ markedly in their reasons for choosing a STEM major. Many students chose a STEM major based on a long-standing interest or because they aspired to a STEM career. A small number of students had altruistic reasons for choosing STEM that were grounded in the possibility of helping people or making a difference through science or engineering. Others chose STEM because they were talented at STEM in their prior education or they had enjoyed STEM in their K-12 schooling. Students were also influenced to pursue a STEM degree by their network of family, teachers, or role models (Table 4.3).

As can be seen, although the language of the SALG survey and that of students talking about their choices differs, the reasons given in interviews map fairly closely to those reflected in the survey findings shown in Table 4.1.

**Table 4.3** Reasons for STEM students' initial choice of major

| Reason for initial choice of STEM major | Switcher | | Persister | |
|---|---|---|---|---|
| Long standing interest | 43 | 44.8% | 105 | 42.0% |
| Enjoyed math, science in high school | 38 | 39.6% | 59 | 23.6% |
| Did well in high school science, math | 26 | 27.1% | 53 | 21.2% |
| Aspired to a STEM-based career | 25 | 26.0% | 85 | 34.0% |
| Economic/financial conditions, ability to find well-paid STEM job | 12 | 12.5% | 35 | 14.0% |
| Altruism | 11 | 11.5% | 14 | 5.6% |
| Underinformed about initial choice of STEM major | 54 | 56.3% | 47 | 18.8% |

## The Role of Careers in Students' Initial Choice of STEM Major

Persisters were slightly more likely than switchers to note that their choice of STEM major emanated from a specific career aspiration or goal. Medical professions were a common career goal upon college entry for both switchers and persisters, although many students in both groups also aspired to be an engineer. Career goals in other fields, such as the physical sciences or math, were less common. Although there were a few differences between switchers and persisters in initial career aspirations, there were notable differences in the nature of their decision-making processes. Switchers were either strongly committed to a very specific career goal or chose STEM after a lengthy process of weighing competing interests, often between a STEM and non-STEM major. On the other hand, persisters more often chose their major because of an interest in broadly pursuing a career in a certain discipline and then began a process of narrowing down and honing their specific career choice within that discipline as they learned more about disciplinary sub-fields and career options during their undergraduate studies.

Generally, persisters entered the major with career interests that developed earlier in life than those of switchers, although not always. For example, an astronomy major discussed her long-term interest in the field which was sparked by an early interest in becoming an astronaut. Her expression of a "career" interest was typical of many persisters in that she expressed a strong interest in simply wanting to "do" or "be" in a certain field, yet did not necessarily express interest in a precise career goal. Many persisters refined their career ambitions throughout college as they learned more about the options available, although they were initially drawn into the major by a strong interest in simply wanting to "do" or "be" in a certain STEM field.

> I knew that this was what I wanted to do since forever, really. I mean, I remember being three and watching NOVA and thinking that's what I'm gonna do. Originally, I wanted to be an astronaut because I knew that Saturn's rings were solid, and I knew they were made of ice, so I just imagined this giant ice slide, and so I wanted to be an astronaut so I could go to Saturn and slide on the rings. And then I figured out that's not really how it works. But when I was really little, I wanted to be an astronaut, and then, I kinda refined that as I got older to, 'Oh, really what I wanna do is study astronomy and how the universe works.'... I figured out the name for this thing I wanna do is astrophysics. From then on, I knew that that was what I wanted to do, but really the whole time. (Native American woman, astronomy persister)

Many switchers and persisters alike changed majors as a result of the process of exploring and honing their career interests during their undergraduate studies—the main difference was that switchers moved to a non-STEM major and persisters refined their focus within STEM, hence the term "relocator" for this type of within-STEM switching. Either way, students' initial career aspirations evolved, shifting to more suitable and desirable careers. In fact, 17% of persisters who initially chose their major for career-related reasons ended up switching to another STEM major at some point during their undergraduate career. This suggests that both switchers and persisters, especially those who were more career-focused in their initial selection of major, explored and refined career options as their understanding and awareness of STEM careers matured. Unlike switchers who were more narrow in their specific career goals, persisters and relocators were more likely to choose their STEM majors based on the breadth of available careers and opportunities in that discipline. This relocator changed her major from chemical engineering to materials science in the process of honing her future interests and learning more about the range of career opportunities in materials science.

> So I came in thinking I wanted to do chemical engineering as a freshman and then I changed right away. I found out what chemical engineers do and that just didn't interest me, and then materials science engineering was in the same department and so I just talked to a whole bunch of professors in material sciences and what they seemed to be doing is something that I could see myself doing.... So materials science you work with all different types of materials and what I liked about it the most was that I could work in a whole variety of industries. I could go into bio-med or aero-space or packaging is what I did this summer or a polymers company. So I could go into almost anything, any engineering company needs a materials engineer, so that's why I really like it. (Hispanic woman, relocator, chemical engineering to materials science)

Relocators occasionally discovered new fields or careers within STEM that they were not aware of when they entered college. For instance, this relocator changed his major from chemical engineering to food science as he came to better understand the nature of both of those fields and the type of work that professionals in those fields do. He acknowledged that Food Science is a major that many students are not familiar with when they are in high school and first choosing their college major. Similarly, he learned about the field after starting his studies in the university and came to realize that it was a more suitable major for his interests and his recently selected career goal of "making beer and salsa." He described his decision-making process:

> I went through the sciences in my prior education and I'm like, 'Alright, the next step is, I guess, engineering.' Because that's, when you first go into college they give you like five main majors that everyone knows about and you just choose that. And then once you get to the college or the university, you can find other colleges with other majors catered towards your interest. And that was what happened to me. So I came in with the idea of chemical engineering like, 'Oh, I get to learn how to process many units, mass produce stuff, but food science actually had the same thing but catered more towards the specific individual. A little bit more of the fine details about how to make products.' (Multi-racial man, relocator, chemical engineering to food science)

Likewise, 58% of switchers moved from a STEM major into a non-STEM major, at least in part, because they became dissatisfied with their career-related choice of major and, ultimately, they found the career paths in a non-STEM field to be more appealing. However, the career exploration process of relocators and switchers was markedly different. Switchers tended to enter the major with more specific, and often less realistic, career goals than persisters or relocators, who often simply wanted to "do" or "be" in the discipline. Switchers who entered their STEM major with a career focus tended to have a narrow interest deeply rooted in a personal experience and not always grounded in a broader interest in the discipline as a whole, as shown in the following typical quote from a life sciences switcher:

> It's interesting because in high school I didn't particularly like biology either. I took AP Bio and it wasn't like it was difficult for me or anything but when I was younger, my older cousin, he was maybe two years older than me, he had diabetes. And from the time that I was like five, I had said in my head, you know, I'm going to grow and become a doctor so I can fix my cousin. (White man, switcher, biological science to film studies)

While persisters and relocators had broad interests within STEM, switchers were more inclined to have broad interests both within and outside of STEM that shaped their initial career aspirations, and ultimately, their choice to move to a non-STEM major. For instance, the switcher in the quote below described the influence of her varied interests, including non-STEM fields, on her career aspirations and choice of major. Certainly, some persisters and relocators also had interest and talent in non-STEM fields, but switchers were more likely to note an aptitude and passion for a non-STEM field and to consider these competing interests before originally selecting a major in STEM.

> I have always had a very, very, very diverse set of interests. Even though I did sort of focus a little bit more in high school on the physics and chemistry and biology and mathematics, I still also had probably about 40 to 50% of my classes were still English and psychology and history, and I took Spanish for 4 years, and I loved everything about all of my classes. So, at one point I wanted to be a teacher, at one point I wanted to be a meteorologist. I mean I was all over the place and I still to this day could be very happy doing a lot of different things but I just kind of went for engineering 'cause it matched my skillset and interests, so I was like 'Sure.' (White woman, switcher, biological engineering to management)

A small group of switchers and persisters had career ambitions that were altruistic in nature and chose their major based on its potential for helping people or addressing global problems. Switchers were more likely than persisters to hold altruistic motives for choosing their major, yet only a small proportion of students in either group chose STEM for altruistic reasons. Still, women were much more likely than men to choose their major for altruistic reasons (e.g., 73% of switchers with altruistic intentions were women). Likewise, students of color were also more likely to base their choice of STEM major on altruistic rationale (e.g., 60% of persisters with altruistic intentions were students of color). Similar to students' general career aspirations upon their entry into college, persisters expressed more general, vague altruistic ambitions than switchers. Thus, persisters were broader in their interest in "making a difference" or helping and therefore were more open to a variety of possibilities in which they could use their STEM degrees for the common good. For some persisters who also felt the pull between STEM and the arts or humanities, the opportunity to make a tangible difference in the world tipped the decision-making process towards STEM, as shown in the quote below.

> I'd always been interested in chemistry. When I was a high schooler my strengths were actually in English and humanities and languages and that was what I was really convinced I was going to do for a long time. And then I got the opportunity to talk to one of my brother's friends who was in college at the time studying chemistry and biochemistry and he really inspired me to put my mental abilities to the test and try to make a difference in the world 'cause he was studying alternative fuel sources and that's what I wanted to do at the time too. (Hispanic man, chemistry persister)

In contrast, switchers often had very concrete and specific, yet potentially unrealistic, altruistic ambitions, such as a desire to cure a certain type of cancer or to clean up a specific lake or river, in contrast to a general desire to help the environment. For instance, the student in the quote below was inspired to enter cancer research because of a personal relationship but was not necessarily interested in other careers in the life sciences. Additionally, some switchers, such as the student highlighted below, only connected their STEM major to altruistic intentions later in the decision-making process, unlike persisters who were more often inspired by altruistic motives at a younger age.

> Well biology was my favorite class in high school. And I took AP Bio as well, and I love biology. So, molecular biology seemed like a good choice. And the summer before I came into college … I really got inspired, 'cause my, my boss, her son developed a brain tumor. And that was really like emotionally stirring, and part of my motivation was going into cancer research. And doing stuff with the cells and finding that out. But I've sort of figured out that I'm not gonna be the person that cures cancer. (LAUGHS) It just wasn't in my playing cards, I guess. (White woman, switcher, biological science to psychology)

Many students selected their STEM major based on career interests or aspirations, yet the nature of those choices and the decision-making processes differed between switchers and persisters. Switchers were more likely to choose STEM after a process of reconciling multiple, competing STEM and non-STEM interests, only to return to a long-standing interest in a non-STEM field. On the other hand, switchers were also more likely to hold very narrow, specific career aspirations that may have been unrealistic or not fully informed. In contrast, persisters expressed a more general interest in "being" in the field and then used their undergraduate studies to refine their interests and select a suitable career option within STEM.

## Social Influences on Students' Initial Choice of STEM Major

Many students, switchers and persisters alike, noted the influence of a parent, teacher, or other significant person on their decision to major in a STEM field. Parents exerted the most influence on students' choice of STEM major, and there were a few differences between switchers and persisters in this regard. Switchers were slightly more likely to cite the influence of a parent on their choice, although many persisters also noted parental influence. The real difference between the groups lay more in the type of influence exerted by parents as switchers were more likely to have a parent who pressured or pushed them into a STEM field. Nonetheless, the majority of students in both groups mentioned general parental encouragement and support (though not pressure) to pursue STEM. Many students, switchers and persisters alike, also had a parent who was a scientific or technical professional and whose career path they followed (Table 4.4).

We think it is, again, important to distinguish between the SALG survey which asked for the reasons why students chose particular majors, and interview conversations about who or what influenced their thinking in coming to a decision. Parents can have great influence on choices (whether as pressure or support) but that may or may not constitute a reason (or the primary reason) for a student's choice. What, then, is the role of parents in the process of making the choice of STEM major?

## *The Role of Parents in Students' Initial Choice of STEM Major*

Many students benefited from supportive parents who encouraged them to pursue their passions and interests. These parents especially encouraged their children to pursue STEM fields, yet without pressuring or forcing. For instance, in the quote below, a persister described how her parents supported her inclination towards the sciences even though they were artistic types.

> Both of my parents are very artistic, and I don't feel like I inherited any of that. They're both very creative beings. But my dad reads a lot about science and I think he realized early on that I was not going to play music and I was not going to paint like he did, but that I was actually pretty skilled in science. And so he I think he tried to stoke my interest more. (Multi-racial woman, biological sciences persister)

Other students mentioned that their parents influenced their choice of STEM major by introducing them to their field of study or by suggesting a certain career path. In this way, parents helped their children identify fields of study that suited their talents, temperament, or interests.

> I had no idea what I was doing until maybe near the end of junior year [in high school], but that's when you kinda have to start thinking about it. And even then, I was just like 'I don't know.' I think it was just presented to me by my mom to just look into. And actually at the time, we were thinking about either biomedical or aerospace [engineering], just 'cause I

**Table 4.4** Social influences on students' initial choice of STEM major

| Influence | Switchers | | Persisters | |
|---|---|---|---|---|
| Parent or other family member | 52 | 54.2% | 101 | 40.4% |
| Teacher, mentor, role model | 32 | 33.3% | 69 | 27.6% |
| Friends or other than family | 8 | 8.3% | 13 | 5.2% |
| High school counselor; high school or university advisors | 4 | 4.2% | 11 | 4.4% |
| Media influence | 4 | 4.2% | 4 | 1.6% |

> have had a lot of experience of medical things coming up and I was fascinated with those things going on, so I'm actually going to grad school for biomedical, so I'm doing both. But yeah, it kinda came together, like, 'Oh, you love working with your hands, you love building things, and you love space. That's what they do.' I was like, 'Ah, I'll give it a try.' (African-American man, aerospace engineering persister)

Many students, including both switchers and persisters, followed a parent into a STEM career. These students had a parent who was a STEM professional and often chose the same, or at least a very similar, field as their initial major in college. Although whether a student had a parent in a STEM field did not necessarily influence their ultimate persistence in their initial field. Students' persistence was influenced more by whether the decision was an informed choice, whether the student had a prior interest in their initial choice of STEM major, and whether the parent had pressured the student, rather than simply encouraging them, to pursue the major. In the quote below, a student described how his parents' technical career paths had inspired his own interest in becoming a mechanical engineer.

> So my dad is an electrical engineer and my mom is an architect so I grew up in a household with a lot of design around and I love, I ended up loving to build things and so the major which could give me a really broad, technical education and allow me to build stuff would be mechanical engineering. So I applied as a mechanical engineer, got in as a mechanical engineer and I'm still a mechanical engineer. That's my major. (Asian/Pacific Islander man, mechanical engineering persister)

While engineering was a common path that students learned about from their parents, many students also followed their parents or other family members into health professions.

> My dad's a doctor and so I knew that I was probably gonna go into some kind of health care field. But I didn't know necessarily what. And so I kinda entered- cause they're just like, 'Oh, declare a major!' So I just declared biology because that's what I thought the main science pre-med major was. (Asian/Pacific Islander woman, biological sciences persister)

Other students blended their parent's STEM career path with their own unique talents and interests to find an appropriate STEM major. For instance, a biomedical engineering major described how she blended her father's medical career with her own interest in math to choose a biomedical engineering major.

> Ever since I was like, at least since fourth grade, I've always wanted to be a doctor. My father was a doctor so I wanted to follow his footsteps, but at the same time, when I entered high school, I really liked biology, but I also really, really liked math. And I also was very good at it. And I thought it was very interesting in terms of the applications that could be used because biology seems to not have as much quantitative analysis as things like chemistry and physics would. So, I was trying to look for a major and I was like, 'Okay, biomedical sciences is part of the classic pre-med major.' But then I also had an upper-classman talking to me about her engineering aspect. And she was doing biomedical and chemical engineering. And I was like, 'That was really interesting.' So I decided to join. (Asian/Pacific Islander woman, biomedical engineering persister)

On the other hand, some students tried to resist their parent's STEM career path, in the hopes of defining their own identity independently, only to find that their interest did indeed lie in their parent's field.

> So my dad is an electrical engineer. I think that might have played a little bit of a role because on take your kid to workday or just whenever he would pick me up and have to drop by work, I'd get to go there and kind of see the fun toys that he might be working on or, some of those companies put out fun little demo boards or something. And I actually didn't go into electrical engineering right away because I wanted to do my own thing. I didn't want to be just going into it because one of my parents was in it and I was…I got to do my own thing. But then I found out that it's what I like. (Hispanic woman, relocator. chemical engineering to electrical engineering)

Thus, the majority of students who cited some sort of parental influence on their choice of a STEM major noted that the influence was positive. Parents encouraged and supported STEM career paths, introduced or suggested appropriate career paths, or inspired their children to pursue a similar STEM career path to their own. In contrast, some parents pushed or pressured their children into STEM majors that the parents viewed as the only acceptable majors. A few parents even asserted that their financial support for college was contingent upon the student pursuing a STEM degree. Parental pressure was more common in immigrant communities which highly valued certain STEM professions, as mentioned in this remark from an immigrant student.

> From my ethnic background anyway, in India, there are only three professions people go to, computer science, engineering, and medical school. That's like, almost exclusively everyone will pick one of those three, so I can't really speak to switching out of that, but as a girl, I guess the biggest thing was that I was never deterred from going into engineering or I guess the thing is that my parents always expected me to go into a technical major. It's like if I told my parents that I wanted to be an actor, they would laugh me out of the house. They really would. (Asian/Pacific Islander woman, biological engineering persister)

Occasionally, students chose the wrong major because their parents had pressured them to pursue a certain STEM field. Consequently, these students switched fields within STEM or switched out of STEM entirely.

> Student: Both my parents are in STEM fields. And they're both from India so they both had college over there and then they came over here after that. And so I've kind of grown up around people who knew a lot about science and that always interested me as a kid, too, science stuff. So then I got into more of the learning about it in high school, like actual chemistry classes and stuff, and physics, and that still like really interests me, so I decided to stick with that.
>
> Interviewer: And why engineering per se to begin with?
>
> Student: That was basically my parents. My parents were really pushing me to get into engineering because it's more opportunities, job wise. Which I've noticed, but I couldn't really stick with it because I wasn't really devoted to it. And my GPA kind of suffered because of that. (Asian/Pacific Islander woman, relocator, chemical engineering to chemistry)

Parental pressure to major in a STEM field was particularly troublesome for switchers who later decided to enter a non-STEM field. These students often entered their STEM major with less interest than their persister counterparts and selected the major, in large part, due to parental pressure. Many of these switchers' parents strongly disapproved of their child's choice of a non-STEM field, often causing conflict and stress for the student.

> My parents have science backgrounds and so they just kind of ingrained in me, like, 'Oh you're smart. You're our daughter. You'll probably do something science-y.' And I guess I figured since I like everything, like yeah I am interested in science, I am interested in physiology and anatomy and like whatever, that it would be cool and I'd be fine with it until I realized I wasn't fine with it and did not want to do that. And to my parents, getting an English degree or a history degree or whatever is a huge joke to them. It was just kind of a constant battle with them 'cause they had in their head like, 'This is what she's gonna do.' (White woman, switcher, physiology to film studies)

## *The Role of Teachers in Students' Choice of Major*

Teachers also were influential in students' decisions to enter a STEM major, although less so than parents. Nevertheless, teachers sparked students' interest in a STEM field, encouraged students to pursue a STEM career, and offered career and educational advice and mentoring. Although switchers and persisters varied little in the extent to which teachers influenced their pursuit of a STEM major, they did differ in the type of influence exerted by teachers. For instance, persisters were more likely to be actively encouraged or mentored by a high school STEM teacher, as a biological sciences persister described.

> I think one of the unique things in my case was that I had high school science teachers that were so willing to push me towards this field. I had an AP Biology teacher that helped me get a research position at the University. And I also had an AP Chemistry teacher that was very active in trying to encourage students to really learn about science and chemistry. I think that was probably one of the biggest things that has helped me I guess, helped me find a home faster. I suppose it was the encouragement I received in high school, and being taught to be flexible with what I really wanted to learn because I know there are a lot of people who go into college with very explicit ideas of what they want to do. Like, 'I want to go to medical school!' or 'I want to do this, and I want to do this.' And I think part of society encourages people to do that, to kind of have a heading before they even start. And I think that's kind of difficult. It results in an effect that I think is kind of opposite of what we want, because while it's nice that they know they want a heading, that they know they want to go to something, when something turns out, it's not what they expected. They begin to falter and begin to not want to continue because it's not what they wanted. I suppose the idea that you need to start college with an explicit plan of action for your future. That's what limits a lot of students. (Asian/Pacific Islander man, biological sciences persister)

On the other hand, switchers were more likely to say that the STEM class in their high school was fun or enjoyable and that they liked the teacher. Less often, they described a teacher who actively mentored or encouraged them. Switchers' choice of major was inspired by their enjoyment of the teacher or the particular course, more so than because of a deep interest in the discipline overall, as described by a biology switcher.

> Student: I think I had a really great teacher in high school. He really engaged all the students and I loved what we were learning, I think 'cause I loved his personality and how he taught everything. So it it was all interesting to me, it wasn't like hard work.
>
> Interviewer: So, did that teacher greatly influence your choice of a major, then?
>
> Student: I think so. Yeah.

Interviewer: Was it him, or the biology, or both?
Student: I think it was probably more the teacher than maybe the discipline. (Hispanic woman, switcher, biological sciences to sociology)

In sum, parents and teachers influenced students' decisions to enroll in a STEM major, although parents exerted greater influence than teachers on students' choices. Nonetheless, the type of influence exerted by both parents and teachers differed between switchers and persisters. Switchers were more likely to have parents who pressured them to declare a STEM major even if their interests and temperaments did not support that choice. Switchers were also more likely to choose STEM because they liked their high school STEM teacher, while persisters were more likely to be actively encouraged or mentored by their high school STEM teachers.

## The Influence of Underinformed Choice of Major on Persistence

The primary way that students' choice of major ultimately influenced their persistence was in whether their choice was an informed one. Students with little knowledge of their discipline were much more likely to switch to a non-STEM field. Students' incoming knowledge about their intended major was based on a variety of factors, such as whether a parent worked in that field, whether the subject was offered at their high school, or whether they knew anyone in their circle of family and friends who had completed that major. Still, knowledge about some majors, especially engineering, remained elusive even for those who knew engineers in their immediate or extended family. Nonetheless, persisters generally entered their STEM major with a higher level of incoming knowledge about what the major may be like and had more awareness of the range of career options available within that field.

Many persisters relied on multiple sources of information to research the major and potential career options, using internet searches, conversations with STEM professionals, and other sources of information besides family or word of mouth. Many students who were underinformed about their STEM major had declared their initial major because they had been told that they would be good at it or because they had done well in STEM in high school, yet they did not have a full understanding of what the major entailed nor were they aware of available career options in that field. Switchers were much more likely to feel that they were underinformed about their major when they entered college (56.3% of switchers vs. 18.4% of persisters). Nonetheless, persisters were still affected by a lack of incoming knowledge about their selected major; indeed, about one-third of persisters who were underinformed about their major were relocators who ended up switching to another major within STEM.

So I entered as a computer engineering major and I did one quarter of that. I enjoyed it. I was always a strong math and science student in high school and I had interviewed a couple

> alumni from the engineering department and I wasn't as interested in that career path as I thought I would be. Both the guys I met with loved their jobs but they spent a lot of time inside, I mean most of their time inside and I didn't see as much of a passion in it. I was taking a geography class at the time and I changed my major at the end of the quarter to environmental science. And so I'm studying environmental science and sociology. (Multi-racial man, relocator, computer engineering to environmental science/sociology double major,)

Students often knew little about the academic content of disciplines that are not typically offered in high school, such as biochemistry or any of the engineering fields. For instance, a persister described her lack of knowledge about biochemistry when she initially declared her major.

> I really didn't learn anything until like I was already in it, so when I signed up for it I, they gave me like a basically like an estimated timeline of when you should take certain classes and stuff, so first semester you should take like general chemistry, etcetera. So I kind of knew like what classes I had to take but I didn't really figure out like what those classes were and what they entailed until I was actually in them. So like for me choosing biochemistry like in my mind I was like it's just like taking half chemistry courses and half biology courses cause that's what I thought it would be, but then when I got here I was like, no there are actually biochemistry courses that you have to take. So I think the most I learned was from talking to upper classmen and like people that were in that major and like going into advising and talking about like what I need to take for next semesters. So I feel like that was nice getting to know that, but if I had known that coming in I probably would have not declared biochemistry major. I just didn't know what I was doing really, but it wasn't, like I had no like concrete examples of what being, like what that major was and what it entailed. So it was more like an abstract idea before I got here. (Asian/Pacific Islander woman, biochemistry persister)

Persisters who were underinformed about their initial choice of major gradually learned more about the major itself and potential career options through various sources during their undergraduate career. Consequently, some of them learned that their initial field may not suit their interests or that it did not have the range of career opportunities that they would like. Some of these relocators moved to a suitable STEM major that offered the options or the "fit" that they did not find in their initial major. For instance, in the quote below, a relocator described learning about her new major through student academic groups and a campus career fair. She had chosen her initial major based on the tragic loss of her father which had sparked an interest that did not hold up over time.

> I chose the sciences because I was always like better at math and science when I was in, like elementary school, middle school and high school. I really enjoyed those courses whereas history and English or like the arts were courses that I really did not enjoy. So I applied to the Engineering School but I was just going to decide later which engineering I wanted, but then during my senior year when I was applying for college my dad actually died of colon cancer. So, that's why I chose biomedical. So I chose biomedical because like I could work in the medical devices and like do something about it. But then the reason I switched was because I learned that as a biomedical engineering like I knew a couple girls, because I'm in the Society of Women Engineers, and so I knew a couple girls that were graduating with that degree and a lot of them like have commented on like the lack of options they have after graduation, so a lot of them feel as if, you know, they either enter the medical devices industry or they go to grad school. Like it's not a degree that's very versatile. So I went to the

career fair as a biomedical engineering major and found out there were only like five companies that would hire biomedical, but then I learned about material science 'cause one of my friends was a material science major and like I learned that Medtronic and Boston Scientific, like medical device industry leaders still hire material science but then a ton of other companies do also, so I was like, well if I don't end up going down this medical device industry road, at least I'm going to have other options as a material science major. So that's why I decided to switch to material science. (Asian-American woman, relocator, biomedical engineering to materials science)

In contrast, switchers were noticeably less informed than persisters about the nature of their chosen majors, the required classes or sequencing of courses within the major, and the potential career options within the field, even for disciplines that they had been exposed to in high school. Switchers were often unaware of how the curriculum of courses in these familiar fields would differ from high school to college. Moreover, some switchers had not put extensive time or thought into selecting their initial major, as evidenced in the following comment from a biology switcher:

So I think a lot of it was I clicked molecular biology when I registered and I just went into it. A lot of people go in undeclared, and I didn't wanna do that, 'cause I was like, 'No! I have to have some direction.' But I feel like if I had known what I was getting in to, I probably wouldn't have gone into it all. 'Cause like you just see molecular biology, and you're like, 'Oh, what does that mean? It's okay. I'll go for it.' And then you look at the class list after you've enrolled and you've already bought all the books and everything, and you're like, 'Oh my god. I have the most rigorous four years ahead of me, and I hate everything.' (White woman, switcher, biological sciences to psychology)

Other switchers struggled because of a lack of awareness of the course sequencing and required courses within STEM majors. They were discouraged by the challenge of taking non-STEM classes within the stringent requirements of many STEM majors. Given that many switchers had broader incoming interests than some of the non-switchers, it was a disappointing realization that they could not take as many non-STEM classes as they would like within their STEM major.

I understood what it was going to be but I also didn't realize…I think I knew what was coming but I didn't understand what that really meant. Like, I knew the classes that I was going to have to take but then I didn't realize that it was gonna exclude a lot of the like other… it was so science-based and I wanted to take other things too and didn't want to just be focused on the sciences. And I didn't realize that wasn't going to be possible, really. I just wanted a more holistic approach. 'Cause I wanna go into medicine so at first I wanted to do premed and now I'm pre-nursing. But I felt like the public health classes gave you a more global view of health. And biology was just too specific and too focused right from the get-go and so I was missing out on some of the broader aspects of it. (White woman, switcher, biological sciences to community health)

While there were a few disciplinary differences among persisters in relation to which majors were most likely to entail a lack of information, there were marked differences among switchers as computer science and engineering majors struggled the most with a lack of knowledge about their majors. In fact, 76% of engineering and 75% of computer science switchers felt that their lack of incoming information about their major contributed at least partially to their decision to switch. In contrast, only a third of physical science switchers left their major for the same reason.

Not only did engineering switchers have the least amount of incoming information about what is required for their major or what the major will actually be like, they also seemed to select the major using a less reflective process than many of their persister counterparts. Some engineering switchers felt that engineering would be a good fit or they had been encouraged to pursue engineering because they are good at math, but the field of engineering did not turn out to be what they expected.

> I guess it would mostly be because I didn't really know what else to do sitting there in high school filling out college applications you have check your major, they list everything and I thought about, my brother did psychology so I thought about that for a little while. My dad was a geologist so I also considered that but I landed on engineering mostly because it interested me the most I think. It sounded fun and exciting and I like figuring out how things work and stuff like that and ended up being different than I expected. I'd like to think that I knew before I started college but I think I learned mostly first and second semester freshmen year, what it actually entailed and what I was going to be doing. (White man, mechanical engineering to construction management switcher)

## *Conclusion*

Overall, both the SALG survey pre-coded questions and written comments point to the importance of students' career aspirations in guiding the choice of a STEM major. However, when students are invited to state the primary reason for their choice, intrinsic interest in the field and a sense of their own ability to pursue it are dominant. Altruistic motivations and lifestyle goals are also well-represented and shape particular career aspirations.

From the interview study we found that the reasons why students in TALR chose STEM majors were similar to those cited by students in the original study. However, the factors weighed by students have become more complex, and many students, including both switchers and persisters, chose between multiple, often competing, interests. A lack of incoming knowledge about the nature of STEM majors and career options within STEM fields was one of the leading contributors to relocating within STEM or switching to a non-STEM field. Yet, the decision-making processes that prompted these moves differed for switchers and relocators. Switchers, in general, had broader interests than relocators and a greater number of them held strong incoming interest in non-STEM fields. Ultimately, they decided a non-STEM major better suited their talents and interests. On the other hand, many switchers also held a more limited conception of the nature of STEM disciplines and the career options available to them with STEM degrees, often aspiring to narrow and unrealistic career goals. In this "Goldilocks" conundrum, incoming interests that were too narrow or too broad did not necessarily hold up over time and students were more likely to switch as they came to realize that their initial STEM major did not suit their interests, goals, or temperament. Persisters, on the other hand, entered STEM majors with a more general desire to "do" or "be" in a certain field and then gradually honed and refined their interests as they progressed in their studies.

## References

Bonous-Hammarth. (2000). Pathways to success: Affirming opportunities for science, mathematics, and engineering majors. *The Journal of Negro Education, 69*(1/2), 92–111.
Carlone, H. B., & Johnson, A. (2007). Understanding the science experiences of successful women of color: Science identity as an analytic lens. *Journal of Research in Science Teaching., 44*(8), 1187–1218.
Chang, M. J., Cerna, O., Han, J., & Saenz, V. (2008). The contradictory roles of institutional status in retaining underrepresented minorities in biomedical and Behavioral science majors. *The Review of Higher Education, 31*(4), 433–464.
Crisp, G., Nora, A., & Taggart, A. (2009). Student characteristics, pre-college, college, and environmental factors as predictors of majoring in and earning a STEM degree: An analysis of students attending a Hispanic serving institution. *American Education Research Journal., 46*(4), 924–942.
Dietrich, J., Kracki, B., & Nurmi, J.-E. (2011). Parents' role in adolescents' decision on a college major: A weekly diary study. *Journal of Vocational Behavior, 79*, 134–144.
Eagan, M.K., Hurtado, S. & Chang, M. J. (2010, November). *What matters in STEM: Institutional contexts that influence STEM bachelor's degree completion rates*. Paper presented at the 2010 Association for the Study of Higher Education (ASHE) annual meeting, Indianapolis, IN.
Gayles, J. G., & Ampaw, F. (2014). The impact of college experiences on degree completion in STEM fields at four-year institutions: Does gender matter? *The Journal of Higher Education, 85*(4), 439–468.
Lapan, R. T., & Shaughenessy, P. (1996). Efficacy expectations and vocational interests as mediators between sex and choice of math/science college majors: A longitudinal study. *Journal of Vocational Behavior, 49*, 277–291.
Maltese, A. V., Melki, C. S., & Wiebke, H. L. (2014). The nature of experiences responsible for the generation and maintenance of interest in STEM. *Science Education, 98*(6), 937–962.
Matusovich, H.M., Streveler, R.A. & Miller, R.L. (2010). Why Do Students Choose Engineering? A Qualitative, Longitudinal Investigation of Students' Motivational Values. *Journal of Engineering Education, 99*(4), 289–303.
McGill, M. M., Decker, A., & Settle, A. (2015, August 9–13). *Does outreach impact choices of major for underrepresented undergraduate students?* Paper presented at the International Computing Education Research Conference, Omaha, NE.
Rask, K. (2010). Attrition in STEM fields at Liberal Arts College: The importance of grades and pre-collegiate factors. *Economics of Education Review, 29*, 892–900.
Riegle-Crumb, C., & King, B. (2012). Questioning a white male advantage in STEM: Examining disparities in college major by gender and race/ethnicity. *Educational Researcher., 39*(9), 656–664.
Sax, L. J., Kanny, M. A., Riggers-Piehl, T. A., Whang, H., & Paulson, L. N. (2015). "But I'm not good at math": The changing salience of mathematical self-concept in shaping Women's and Men's STEM aspirations. *Research in Higher Education, 56*, 813–842.
Shehab, R. L., Walden, S. E., & Wellborn, E. E. (2015, June 14–17). *Motivating factors for choosing engineering as reported by racial and ethnic minority students*. Paper presented at the 122nd American Society for Engineering Education (ASEE) Conference and Exposition, Seattle, WA.
Sheppard, S., Gilmartin, S., Chen, H. L., Donaldson, K., Lichenstein, G., Eris, O., … Toye, G. (2010). *Exploring the engineering student experience: Findings from the academic pathways of people learning engineering survey (APPLES)*. Seattle, WA: Center for the Advancement of Engineering Education Technical Report CAEE-TR-10-01. http://www.engr.washington.edu/caee/
Simpson, J.C. (2003). Mom Matters: Maternal Influence on the Choice of Academic Major. *Sex Roles, 48*(9/10), 447–460.

Sjaastad, J. (2012). Sources of inspiration: The role of significant persons in young people's choice of science in higher education. *International Journal of Science Education, 34*(10), 1615–1636.
Stinebrickner, T. (2013). A major in science? Initial beliefs and final outcomes for college major and dropout. *Review of Economic Studies., 81*(1), 426–472.
Tai, R. H., Liu, C. Q., Maltese, A. V., & Fan, X. (2006). Planning early for careers in science. *Science, 312*, 1143–1144.
VanMeter-Adams, A., Frankenfeld, C.L., Bases, J., Espina, V. & Liotta, L.A. (2014). Students who Demonstrate Strong Talent and Interest in STEM are Initially Attracted to STEM through Extracurricular Experiences. *CBE-Life Sciences Education, 13*, 687–697.

# Chapter 5
# Issues with High School Preparation and Transition to College

**Heather Thiry**

## Introduction

Students' academic backgrounds are important because their prior learning experiences will inherently shape their knowledge, understanding, attitudes, and identities in subsequent learning environments. It is widely believed that high math achievement in high school or on standardized tests, such as the SAT or ACT, indicates that a student is well-prepared for college-level STEM coursework. This folk wisdom is substantiated by research on the relationship between students' academic backgrounds and their subsequent college outcomes. Students' high school preparation, including enrollment in AP courses, high school GPA, and SAT or ACT scores, often predicts their success and persistence in STEM majors (Bonous-Hammarth, 2000; Chang, Cerna, Han, & Saenz, 2008; Crisp, Nora, & Taggart, 2009; Eagan, Hurtado, & Chang, 2010; Gayles & Ampaw, 2014; Rask, 2010). Differences in academic backgrounds may also account for racial and ethnic disparities in STEM majors as the persistent racial gap is largely explained by unequal academic preparation and access to STEM opportunities prior to college entrance (Chang et al., 2008; Price, 2010; Riegle-Crumb & King, 2010). Nevertheless, nearly all of the switchers in our study had excelled at these metrics and many had taken advanced coursework in high school. Thus, they should not have encountered many difficulties in their early college coursework, yet their accomplishments on these measures belied the struggles of some students in transitioning from high school to college-level coursework and expectations.

H. Thiry (✉)
Ethnography and Evaluation Research (E&ER), University of Colorado Boulder, Boulder, CO, USA
e-mail: Heather.Thiry@Colorado.edu



E. Seymour, A.-B. Hunter (eds.), *Talking about Leaving Revisited*,
https://doi.org/10.1007/978-3-030-25304-2_5

## *Findings from TAL1*

In TAL1, inadequate academic preparation played a small role in students' switching decisions; nearly 15% of students cited poor high school preparation as part of their reason for shifting to a non-STEM major in the original study (Seymour & Hewitt, 1997). Minority students in TALI were most affected by inadequate preparation as they were more likely to attend under-resourced high schools and, given the time period when the study was conducted (the early 1990s), were also more likely to lack prior exposure to computers or STEM laboratory equipment. High school preparation remains an impactful issue for students in this study at nearly equivalent rates as TAL1. Switchers in TALR were more likely to report both good preparation and poor preparation in key subject areas, indicating that academic background alone was not the sole cause of switching. Nevertheless, nearly 20% of switchers in TALR attributed at least part of their decision to switch to their lack of academic preparation; however, we purposefully selected an interview sample of switchers with high incoming math scores so that we could rule out prior preparation as a factor in their switching decisions to better understand the other underlying causes of switching. So it is somewhat significant that nearly one in five switchers still based at least part of their decision to switch on their inadequate background in core subjects or academic skills.

## Switchers' and Persisters' College Readiness

Preparation issues created great difficulty for some students in their transition to college, in relation to STEM coursework and in their adjustment to the general college atmosphere, which also impacted some students' decisions to leave STEM. A lack of college readiness and inadequate preparation in core subjects also caused difficulties for a few persisters who addressed these challenges by moving to a different STEM major in which they were better able to overcome their lack of preparation for college-level STEM coursework. Table 5.1 compares switchers' and

**Table 5.1** Switchers' and persisters' beliefs about their high school preparation

| Preparation | Switcher (%) | Persister (%) |
|---|---|---|
| Good math preparation | 34.4 | 28.4 |
| Good science preparation | 42.7 | 27.6 |
| Good general preparation | 34.4 | 17.2 |
| Poor math preparation | 28.1 | 18.0 |
| Poor science preparation | 33.3 | 8.8 |
| Poor general preparation | 19.8 | 7.2 |
| Preparation contributed to switching from original major | 18.8 | 3.2 |

persisters' perceptions of their college readiness in different areas, highlighting that a fair number of switchers were well-prepared for college, although they were also more likely than persisters to lack preparation in fundamental areas.

## *Who Are the Underprepared Students?*

Students' perceptions of their college readiness slightly differed by race, gender, and first-generation college status, especially for switchers. One-third of Hispanic switchers and 30% of White switchers cited poor high school preparation as contributing to their switching decision, although no African-American students cited this reason for switching. Women of color were the most likely of all switchers to report that preparation issues contributed to their decision to leave their STEM major. Although Asian-American students are not underrepresented in STEM, four out of the 18 students (22%) who cited preparation issues as their reason for leaving STEM were Asian-American women, representing nearly half of the Asian-American women switchers in the overall interview sample. These women were all first-generation college students from immigrant families. Likewise, 7 of the 18 switchers most negatively impacted by preparation issues were first-generation college students, reflecting 39% of those who switched because of preparation issues, though only 13% of the larger sample of switcher students. There were no gender differences in persister' reports of poor high school preparation. However, students of color who stayed in their STEM majors were more likely to report poor high school preparation in certain areas; for instance, 22% of Hispanic persisters and 36% of African-American persisters reported poor math preparation, while 16% of White persisters reported the same.

There were disciplinary differences as well among students who switched majors because of high school preparation issues: 66% of them were life sciences majors, 22% were physical sciences majors, 11% were engineering, and 0% were mathematics or computer science majors. In contrast, only 45% of the total sample of switchers were life science majors. Thus, life and physical science majors were over-represented among the most underprepared switchers, while engineering was less represented. Indeed, underprepared students most often discussed difficulties in chemistry coursework and, to a lesser extent, in calculus. Therefore, prior preparation impacted science majors more so than other disciplines where chemistry is not required for graduation. Students' struggles in chemistry were also reflected in their high school STEM background. Most switchers (68%) had taken AP or IB Calculus in high school, yet only 38% of switchers had taken AP or IB Chemistry. Switchers were more likely than persisters to report that they were inadequately prepared for college coursework, yet there was a small group of persisters who reported similar issues and yet persisted in the major. Despite the challenges that poorly prepared students faced in adjusting to college, persisters were not as negatively impacted by these issues as switchers.

## The Causes of Poor Preparation

### *Poor Math and Science Preparation*

A multitude of factors in students' previous schooling contributed to their lack of readiness for college math and science coursework. Many of the most unprepared switchers noted that they had not experienced good teaching in high school or had taken unchallenging math and science classes that emphasized worksheets and rote memorization. Students found that they were unprepared for the higher-level, conceptual, and abstract thinking required in college-level STEM courses. The focus on application of knowledge in college, rather than strict memorization, was a challenging adjustment for many students.

> In chemistry there was just so much more material in the introductory classes here that high school didn't really prepare me for. And the way it's taught also 'cause here it's very concept heavy and making sure you can apply complex concepts to whatever you're trying to learn, whereas I felt like in high school everything was memorization based and you're just trying to learn everything you can and then after I took the IB exam it kind of just went away. (Asian/Pacific Islander woman, switcher, molecular biology to international business)

> The physics class at my high school was nowhere close to the first physics class I had to take in college. The content, just the concepts were not nearly as heavily stressed, you know, it's just sort of equations on paper. 'And here's m and here's a, what's f?' You know? And stuff like that. Not too much actual insight into the real physics. And that's why, before coming into college, I watched all these lectures and read all these books because I even felt unprepared before I got to college. I guess I would explain the gap between high school and college there as much less emphasis on concepts. The concepts didn't seem to be emphasized at all. (White man, astronomy persister)

The high school that students had attended greatly impacted their preparedness for college. Many of the students who felt they were underprepared had attended small or poorly resourced high schools. These high schools typically did not offer calculus or other advanced math classes, often housed inadequately stocked laboratory facilities, and did not offer advanced science coursework. Many of these schools were in rural or low-income urban areas. For example, the two quotes below are both from students who attended small, rural high schools which lacked advanced STEM course offerings.

> [My preparation was] not great. I took two years of chemistry in high school and that got me through the first week of Chem I. Or not even, actually. And I took, the most rigorous classes that I could at my high school—that was offered at a really tiny rural high school, and so they didn't have much. And we didn't have pre-calc, we just had trig, and so calc was kind of tough when I entered it [because] they did all the pre-calculus stuff in two weeks. (White man, switcher, physiology to architecture)

> And I would say the reason I ended up changing [majors] was because my high school didn't offer a wide variety of classes. It was a small public rural high school, and our lab situation was, we basically used marbles and rulers to do physics experiments. And we actually didn't offer a physics class until my senior year. And that was just because the

teacher who was there decided to. We had one teacher for every subject. So we had one math teacher for four years. One science teacher for four years. (White woman, computer science computer science to information technology persister)

Other students had access to advanced science and math curricula in high school, yet were unable to enroll in those courses because they had not completed necessary prerequisites. While this may appear to be a lack of planning or inadequate advising, several switchers traced their difficulties back to elementary or middle school where they were placed in lower science or math tracks, or were not encouraged in STEM. Tracking practices have been shown to have negative impacts on students of color who are disproportionately placed into lower academic groups early in their schooling (Oakes, 2005; Rubin, 2003). Once on these lower academic tracks, students are often unable to move ahead to more advanced, college preparatory curricula. Similarly, women, students of color, and first-generation college students at our study sites were most likely to report that they had been placed into a low-ability math group or experienced a general lack of encouragement in STEM in their early schooling. A biology switcher connected her struggles in college coursework, especially chemistry, to her lack of mastery of algebra in middle school and the general lack of encouragement in math she experienced throughout her schooling.

It started in middle school, and all through high school, no one really pushed me to learn the math, and I'm really good at memorizing, so I would just memorize the equations, regurgitate it, and never learn it. I still struggle with basic algebra, so whenever I have to do that in classes, I freak out (LAUGHS). So the math was definitely lacking, and that's part of the reason why I couldn't stand the major I was in, 'cause the math is so important for the chemistry. And even though people call it a science class, I feel like it's more of a math class…'cause you're dealing with chemicals, but the equations and the balancing and the whatnot, that's more math. (White woman, switcher, biology to psychology)

Despite the challenges experienced by some students in accessing rigorous high school coursework, most switchers reported that they had indeed taken AP/IB or other advanced STEM coursework. For instance, slightly more than 2/3 of switchers had taken calculus in high school. Additionally, 61% of switchers had taken at least one AP or IB science course. Yet advanced coursework did not necessarily ensure adequate preparation as some students described poor teaching, lack of challenge, superficial coverage of important concepts, or a focus on memorization without conceptual understanding in their advanced high school STEM classes. Many students were surprised that these rigorous high school courses had not prepared them better as they struggled to learn the same material again in a college environment.

For me personally, it was a pretty big shock coming from high school level. I went to a high school which was, for my town, pretty science-oriented. And they claimed to be preparing us. I took both Calc 1 and Calc 2 at the college level, but I took them in my high school. And it was just nothing compared to what we had to step into coming here. I ended up having to take it here as well, which I don't actually regret. I learned a lot more taking it the second time. But it was still a very, very different class. (White male, mechanical engineering persister)

Some students, notably students of color, attended high schools with advanced course offerings but found that these courses had not prepared them as well as their peers were prepared in their introductory STEM classes. In the comment below, a biology persister acknowledged the structural disadvantages she faced compared to her peers because she attended an under-resourced high school.

> In high school even though I took AP Biology, I took the more rigorous courses in high school and those didn't quite help me even when I went to college compared to other people who were like, 'Oh this is what we did' or 'This is what I did in high school.' And they said, 'Oh that's what I did in my freshman year of high school.' And I was like, 'Oh that is what I did my senior year.' So it seemed like a lot of other people had a lot more advantages than where I came from. My high school was not that good. (Hispanic woman, biological sciences persister)

## Poor General Preparation

Some students were well-prepared in math or science, yet were still unprepared in other ways for college, especially in terms of their readiness for the demands and expectations of college classes. One of the most frequently mentioned academic issues by both switchers and persisters was a lack of study skills. A lack of knowledge about how to study properly for college tests and how to organize and manage their work was widespread among students. A few students, switchers in particular, did not realize upon entering college that it was even necessary to study or to prepare at all for college classes because they had not needed to do so in high school. Students found that the study practices and behaviors that had earned them As in high school did not always work in college. Some of these students adjusted their study practices and were able to "recover" relatively quickly, as noted by a computer engineering persister in the first quote, while others took longer to adjust. Many switchers were not able to develop effective study habits and strategies until sophomore year or beyond.

> In high school, they did what they could, or maybe I didn't challenge myself enough in high school or maybe the high school system in general wasn't preparation enough. When I came in it was a wakeup call, needless to say, and I had to get my stuff together pretty quickly if I wanted to stay in the major. My first semester I took calculus, it was college level math, it was Calc 2, and I was used to in high school kind of just going through, as long as you just showed up, it was really good enough to do well. Show up, pay attention in class, and do the homework, you could come in and get an A on the test. But it doesn't work like that in college. So I found that out pretty quickly. I came in, went to lecture, did the homework, and that wasn't enough. I had to study, I had to spend a lot of extra time studying. So I learned that I wasn't as prepared as I needed to be but I also learned that I just needed to spend a lot more time preparing for exams and quizzes and whatnot outside of class. (White man, computer engineering persister)

> I don't know what changed between high school and college but high school I didn't feel like I needed to study and I always got good grades. So I don't know if it was just the level of difficulty increased or what, but I got here and I don't know. Studying just didn't pay off for me here. Studying in the way that I did, didn't pay off for me. (White man, switcher, mechanical engineering to sports medicine)

While many students, switchers and persisters alike, struggled with the necessary study skills and time management to succeed in college, other students struggled with navigating the college environment. First-generation college students and students of color were more likely to enter college with limited knowledge about the college environment, including the differences in expectations between high school and college and the sequencing of courses that students need to take in college to graduate on time. An African-American man who is a biology major discussed the difficulty of entering college with little knowledge of how to navigate the college environment and to succeed in college, a consequence of attending an under-resourced high school. Like the Hispanic woman in a previous quote, he acknowledged the structural disadvantage that he experienced by attending the high school that he did which he compared to more affluent, high-achieving high schools that strongly emphasize college attendance.

> The high school I went to I felt didn't prepare me as well as if I'd gone to a different high school. So I realized the importance of high school after the fact. I took all these classes but it wasn't stressed to me just how important it was, the transition between college and high school. So I took these classes without the proper, I felt like I didn't know enough about college in my high school. Like my advisors didn't give me enough information as to even the amount of credits or what I should try to do. So I felt like I could've been better prepared if I had gone to a better high school. It was a public school and actually I think in the county, its grades were one of the lowest. So I felt that it definitely hindered my ability to understand what college was about. So even in just preparation and what I should plan on, like how I should make my schedules or just what classes I would be probably taking in college. (African-American man, biological sciences persister)

## The Impact of Poor Preparation on Students' Transition to College

Although a fair number of students struggled with a lack of study skills and most cited some sort of difficulty during college adjustment, negative outcomes resulting from college transition were often more pronounced for first-generation college students and students of color, especially for switchers who were often demoralized by their transition to college. This group of demoralized switchers included all but three of the sub-set of 18 seriously underprepared students. However, the overall number of demoralized switchers was larger than the sub-set of most underprepared students, so even some students who felt well-prepared were still discouraged by the college transition process. In fact, 11 out of 12 first-generation college students and 69% of all students of color switchers felt discouraged about some aspect of their early college experience, leading many to question their choice of major and even whether they belonged in college. Some students' struggles with transition issues lingered beyond their first year. Overall, one-third of switchers described feeling demoralized during their initial transition to college in a way that affected their confidence or academic identity or caused them to doubt their academic trajectory.

Many of these struggles and the consequent loss of confidence and academic identity were rooted in difficulties in gateway STEM courses.

> Student: I just felt like I wasn't being successful and I just felt like, even though I love biology, being a biology major was making me hate biology. And I didn't want to hate biology so I knew I had to switch.
> Interviewer: What did you hate about biology?
> Student: All the chemistry. Like in all my actual biology classes, I was doing pretty well. But then my chemistry classes were making it difficult… I just felt like I went from being a really like good student in high school to basically a terrible student in college. And it just made me feel like I didn't belong in that major. Or like in this field. (African-American woman, switcher, biological sciences to community health)

While some students struggled academically leading to difficult transitions, others struggled with the alienation and isolation of large classes and other impersonal aspects of college life compared to their high school experiences.

> I didn't personally feel prepared, when I first came here, 'cause you kinda go from a classroom where there's like… thirty people and you know 'em all because you've been together for years. And then you kinda go here and you're like. 'Oh, there's three hundred people in my classroom. Okay this is, it's not even a classroom.' It was a little crazy. And my first semester I really struggled. I felt really out of place. (Hispanic woman, switcher, physiology to psychology)

## Good Preparation

In contrast to poorly prepared students, students who felt they were well-prepared in terms of background content knowledge or academic skills described much different educational trajectories than their less well-prepared peers. Most of the well-prepared students had taken AP/IB Calculus or even more advanced mathematics classes, such as Calculus II or III during high school, and they were more likely to have taken AP or IB science classes. Many of them described high school STEM curriculum that was challenging and taught by good, highly qualified teachers; several mentioned that their high school teachers had PhDs or had been practicing scientists or engineers. Students described curriculum that focused on inquiry or creative exploration as well as exposure to scientific research or scholarly articles. Some students also took rigorous high school STEM electives that provided college-level content, such as Physiology, Zoology, C++, Java, Organic Chemistry, or Genetics. In contrast to the underprepared students, some of the well-prepared students had been placed on accelerated STEM tracks prior to high school which allowed them to advance in STEM beyond their peers while still in high school. For example, an engineering switcher described her advanced math trajectory in her K-12 education.

> I was put on an accelerated math course in seventh grade. So in my school district, you could, if you passed a certain algebra test, you could start taking algebra one. You had to take a zero hour, so it meant getting up at five in the morning and going to the junior high to go take it with the other little nerdy seventh graders. And then you could be two years

> accelerated. And then I was lucky enough to live in a town where we had a college, and then I could enroll in the college for Calc 3 my senior year [in high school]. (White woman, switcher, civil engineering to finance)

In contrast to underprepared students who struggled with the workload, pace, and amount of studying required in introductory STEM courses, many well-prepared students attended high schools with rigorous, advanced STEM curricula and high expectations. Subsequently, these students adjusted easily to the demands of college, with some even feeling that college required less effort than their high school.

> My first year of high school I was pulling all-nighters because I had a lot of math homework- I was in pre-calculus in ninth grade and- and, taking various classes. So by the time I got here to college, college was actually easier for me, (Hispanic woman, switcher, biological sciences to journalism)

While underprepared students often described under-resourced high schools with limited STEM offerings, their well-prepared counterparts attended high schools with abundant resources and high-quality teachers with strong backgrounds in math and science.

> I felt very well prepared. I went to a relatively new high school so I think there was a better focus on math and science. I would say both my junior and senior year my best teachers were math teachers. So I was really lucky to have, and both of them were ex-industry engineers who had went and made their million and came back to teach. So they had a really good passion about how they taught and they were very good at communicating the relevance of what we were learning. (Multi-racial man, computer engineering persister)

Besides being on accelerated, advanced math and science course sequences, many well-prepared students took challenging course loads in high school that allowed them to complete multiple science and math courses that provided college-level curricula and, in some cases, transferable credit.

> Biology and chemistry wise, I felt completely set. I actually came in with the credit to skip over the first year, and part of the second years in both of those areas. And I think my high school did an amazing job of preparing students who wanted to pursue careers in science. (Asian/Pacific Islander man, biochemistry persister)

## *Conclusion*

The experience of some students, most notably switchers, illustrates the complexity of students' transition to college and questions the notion that one either "is" or "is not" ready for college, especially based on traditionally used criteria such as SAT or ACT scores or high school GPA. Although these metrics, especially high school GPA, have been shown to predict first year GPA in college (Belfield & Crosta, 2012), they may mask more subtle ways in which students are not prepared for college-level coursework and, in the most worrisome cases, these preparation issues can affect students' academic identity, their sense of belonging to the institution or their major, and even their decision to stay in STEM. General Chemistry and, to a lesser extent, Calculus I and II were the most troublesome courses for underprepared students. When students' problems spanned multiple areas, their difficulties seemed to compound and they were more likely to cite preparation issues as a major influence in their decision to leave STEM. Still, many switchers were quite well-prepared for STEM coursework yet switched to non-STEM majors anyway, as preparation issues only influenced the switching decisions of about 20% of all of the switchers. This proportion is slightly larger than the 15% of students in the original *Talking about Leaving* study who cited preparation issues as influential in their switching decision. Thus, preparation issues and subsequent difficulties in college transition continue to play an important, though not primary, role in the reasons why high-achieving students in selective institutions may switch from STEM to non-STEM majors.

These findings parallel findings from the gateway course study of instructors' beliefs about teaching, learning, and persistence that complemented the persistence interview analysis of the TALR study. Although instructors' linkage between persistence and high school preparation was not widespread (only 10%), some instructors at the six institutions identified students' prior preparation as an obstacle to their persistence in STEM majors (Ferrare & Miller, 2019). In contrast to students' nuanced perspective on their own preparation which highlighted the role of K-12 teaching practices, access to resources and curricula, and conceptual difficulty of prior coursework, postsecondary instructors generally pointed toward lack of role models or poor teaching as the significant factors in students' lack of preparation. Therefore, students, for the most part, identified multiple, structural variables within their backgrounds that contributed to their challenging transitions to college, while instructors were more likely to focus on a single variable, K-12 teachers. Students' analysis of their poor preparation highlights more leverage points for possible action, including instructional practices, teacher preparation, resources, curricula, and access to STEM opportunities.

## References


Belfield, C. R., & Crosta, P. M. (2012). *Predicting success in college: The importance of placement tests and high school transcripts* (CCRC Working Paper No. 42). Community College Research Center, Teacher's College, Columbia University.

Bonous-Hammarth, M. (2000). Pathways to success: Affirming opportunities for science, mathematics, and engineering majors. *Journal of Negro Education, 69*(1/2), 92–111.

Chang, M. J., Cerna, O., Han, J., & Saenz, V. (2008). The contradictory roles of institutional status in retaining underrepresented minorities in biomedical and Behavioral science majors. *The Review of Higher Education, 31*(4), 433–464.

Crisp, G., Nora, A., & Taggart, A. (2009). Student characteristics, pre-college, college, and environmental factors as predictors of majoring in and earning a STEM degree: An analysis of students attending a Hispanic-serving institution. *American Educational Research Journal, 46*(4), 924–942.

Eagan, M. K., Hurtado, S., & Chang, M. J. (2010, November). *What matters in STEM: Institutional contexts that influence bachelor's degree completion rates.* Paper presented at the annual meeting of the Association for the Study of Higher Education, Indianapolis, IN.

Ferrare, J. J., & Miller, J. M. (2019). Making sense of persistence in scientific purgatory: A multi-institutional analysis of instructors in introductory science, technology, engineering, and mathematics (STEM) courses. *Journal of Higher Education*. https://doi.org/10.1080/00221546.2019.1602392

Gayles, J. G., & Ampaw, F. (2014). The impact of college experiences on degree completion in STEM fields at four-year institutions: Does gender matter? *The Journal of Higher Education, 85*(4), 439–468.

Oakes, J. (2005). *Keeping track: How schools structure inequality* (2nd ed.). New Haven, CT: Yale University Press.

Price, J. (2010). The effect of instructor race and gender on student persistence in STEM fields. *Economics of Education Review, 29*(6), 901–910.

Rask, K. (2010). Attrition in STEM fields at a liberal arts college: The importance of grades and pre-collegiate preferences. *Economics of Education Review, 29*(6), 892–900. https://doi.org/10.1016/j.econedurev.2010.06.013

Riegle-Crumb, C., & King, B. (2010). Questioning a white male advantage in STEM: Examining disparities in college major by gender and race/ethnicity. *Educational Researcher., 39*(9), 656–664.

Rubin, B. C. (2003). Unpacking detracking: When progressive pedagogy meets students' social worlds. *American Educational Research Journal, 40*(2), 539–573.

Seymour, E., & Hewitt, N. M. (1997). *Talking about leaving: Why undergraduates leave the sciences*. Boulder, CO: Westview Press.

# Chapter 6
# Student Responses to Problematic STEM Teaching Methods

**Raquel P. Harper, Timothy J. Weston, and Elaine Seymour**

## The Consequences of Poor Quality STEM Teaching

As indicated in the main iceberg table (cf., Chap. 3), 78% of all students expressed frustrations with poor quality teaching, curriculum design, and assessment practices. Three-quarters (72%) of persisters and 96% of switchers described how aspects of poor STEM teaching had negatively affected them, with almost half of switchers (48%) offering these experiences as a reason for deciding to leave their STEM major. Problems with STEM teaching also played a significant role in switching decisions in the original study where they ranked third (at 36%) among all reasons for leaving STEM majors and were described as creating problems for 90% of all switchers and 74% of persisters. Sadly, two decades later, poor quality in

R. P. Harper (✉)
Ethnography and Evaluation Research (E&ER), University of Colorado Boulder, Boulder, CO, USA

Center for STEM Learning, University of Colorado Boulder, Boulder, CO, USA
e-mail: Raquel.Harper@Colorado.EDU

T. J. Weston
Ethnography and Evaluation Research (E&ER), University of Colorado Boulder, Boulder, CO, USA

National Center for Women and Information Technology (NCWIT), University of Colorado Boulder, Boulder, CO, USA
e-mail: westont@colorado.edu

E. Seymour
Ethnography and Evaluation Research (E&ER), University of Colorado Boulder, Boulder, CO, USA
e-mail: Elaine.Seymour@colorado.edu



E. Seymour, A.-B. Hunter (eds.), *Talking about Leaving Revisited*,
https://doi.org/10.1007/978-3-030-25304-2_6

**Table 6.1** Switchers and persisters reporting negative effects of poor teaching

| Switchers | Negatively affected (%) | Contributed to switching (%) | Persisters | Negatively affected (%) |
|---|---|---|---|---|
| Men of color | 100 | 20 | Men of color | 90 |
| Women of color | 95 | 38 | Women of color | 93 |
| White men | 93 | 50 | White men | 60 |
| White women | 98 | 55 | White women | 56 |

**Table 6.2** Contributions to switching by poor quality teaching by institution

| College/university | Contributed to switching (%) | Negatively affected (%) |
|---|---|---|
| PB4R1 | 60 | 95 |
| PB1R1 | 53 | 95 |
| PB3R1 | 53 | 100 |
| PB2R1 | 41 | 91 |
| PV1R3 | 36 | 100 |
| PV2R1 | 20 | 100 |

STEM teaching—and its negative consequences—persists as the leading contributor both to the loss of STEM majors and to negative learning experiences for the majority of STEM students in this study.

As also found in the original study, some student groups are more negatively affected by poor quality teaching than others. Table 6.1 compares the effects of poor teaching by switching status, racial background, and sex, and Table 6.2 by type of institution attended.

White male and female STEM persisters were the least negatively affected, nevertheless over half (60% and 56%, respectively) registered concerns. Men and women of color, both switchers and persisters, reported more problems with STEM teaching than their white peers. Indeed, persisters of color reported problems with STEM pedagogy almost as often as switchers overall. We observe that students of color are often more reluctant to blame teachers for their learning problems than are their white peers, internalizing these issues by faulting themselves for poor performance on an exam or low grades at the end of a class. More white women than any other group reported poor STEM teaching as a contributor to their decisions to switch (viz., 55% of white women, compared with 20% of men of color, 38% of women of color, and 50% of white men).

We also checked for variations in students' reports of poor quality learning experiences and their consequences in our six sample institutions. As shown in Table 6.2, students at every institution reported concerns with poor quality STEM teaching, but over half of the switchers at the three of the four large universities in the sample gave this as a major reason for their decision to leave.

We also offer negative critiques of teaching offered by 252 students as written comments in the 1432 SALG surveys completed by (primarily) first- and second-year students in the 80 introductory STEM courses surveyed. The SALG survey respondents represent a wider sample than the interviewees in that they include not only STEM majors but students in other majors that require introductory STEM courses, students fulfilling general education requirements, and students taking science and math courses out of interest. Problems with learning related to poor quality teaching were offered in answer to two open-ended questions: "Please comment on how (and if) this class helped your learning" (212 comments), and "Please comment on how your understanding of the subject has changed as a result of this class" (further 71 comments). As many of the written comments reflect categories of response from the student interviews, we present comments on the same issue together.

## What Kinds of Teaching Methods Did Students Experience?

To answer this question, we compare findings from three different studies that were undertaken concurrently at the six study sites—the interview study with STEM switchers and persisters; results from a sample of 28 (of 80) STEM foundation courses surveyed via the SALG instrument; and a classroom observation study in the same 80 foundation courses (matched across the six sites) in which the SALG survey was administered. The (largely first-year) students who responded to the SALG survey were at the end of the same courses that were observed by the researchers. Thus, these sets of results are in "present time." Switchers and persisters in the interview study reflected back on their experiences, both in foundation courses and those in other years. Notwithstanding some variations between researchers in the ways in which they categorized student experiences, we found a high degree of concurrence across study results in their portrayals of the kinds of teaching methods used and the nature and extent of student participation in their own learning.

***The Interview Study*** As in the original study, students report that non-interactive lectures are the dominant mode of STEM teaching, especially in introductory courses. All but one of the 95 switchers in our study (99%) reported that STEM instructors used "straight lectures" as their main teaching method, and 57% of the 143 persisters reported the same. Table 6.3 shows the breakdown (distilled from the coded interview data) of switchers' and persisters' experiences of all modes of teaching reported.

***The SALG Surveys*** We drew on the results from 28 of the 80 STEM foundational courses whose students took the SALG survey to determine what teaching methods were used and what student learning activities were experienced in each class. As is reflected in Fig. 6.1, the most frequently reported teaching methods and student activities in class were the most conventional. As in the interview study, in all or

**Table 6.3** Teaching methods experienced by switchers and persisters

| Reported teaching methods | Switchers | Persisters |
|---|---|---|
| Straight lectures | 99% (95) | 57% (143) |
| Lectures with PowerPoint | 15% (14) | 14% (35) |
| Lectures with handwritten materials | 21% (20) | 10% (25) |
| Interactive lectures | 26% (25) | 33% (83) |
| Clickers | 8% (8) | 9% (22) |
| Multimedia | 1% (1) | Less than 1% (2) |
| Small group work | 35% (34) | 32% (79) |
| Whole class discussions | 2% (2) | 7% (17) |
| Student presentations | 1% (1) | 2% (5) |
| Classroom demos | 7% (7) | 4% (9) |
| Online instruction | 11% (11) | 4% (10) |
| Flipped classroom | 1% (1) | Less than 1% (2) |

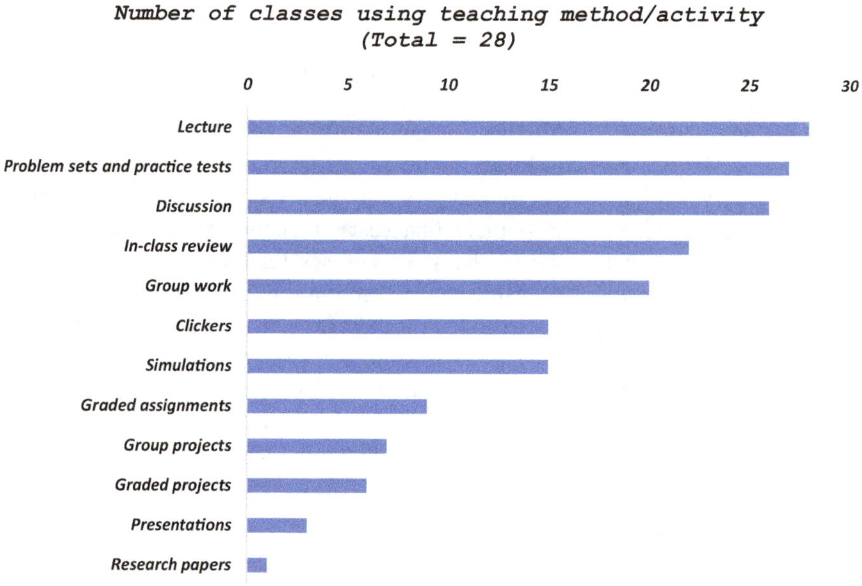


**Fig. 6.1** Reports of teaching methods and student activities in 28 STEM foundation classes by students answering the Student Assessment of their Learning Gains (SALG) survey

almost all classes, students were taught by lecturing. Interaction was predominantly via problem sets and practice tests, in-class discussion and review. Students in 20 of the 28 classes participated in group work, 15 classes used clickers and seven incorporated group projects.

Other commonly used teaching methods were identified in answers to the survey's open-ended questions. Students' written descriptions included a range of methods whose common characteristic was their incorporation of learning technologies. They included online homework systems (e.g., ALEKs and Sapling), the use of MATLAB software, short YouTube videos, notes, and PowerPoint slides (all posted online by the instructor), and other non-course-affiliated internet resources. We saw evidence of the use of online homework systems in four of the courses. Although group work was reported, other active or interactive teaching methods involving student participation were evident in very few classes (e.g., group projects occurred in seven courses).

***The Observation Study*** The findings from both the interview study and the SALG survey are highly comparable to those reported in the classroom observation study of 71 STEM foundation courses (Ferrare, 2019), which was undertaken using the Teaching Dimensions Observation Protocol (TDOP) developed by Hora, Oleson, and Ferarre (2013). In two-minute intervals throughout each class, Ferrare and his team of class observers identified two different forms of the lecture method that, taken together, accounted for 75% of teaching styles in foundation courses. Lecturing while writing on a board ("chalk talks") was observed in 41% of courses, and lecturing aided by pre-made slides ("slide shows") was observed in 34% (Ferrare & Miller, 2017). "Question and answer" sessions were the most interactive aspects of the chalk-talk lecture form (found in 9%); small group work was observed in only 4% of classes; and clicker questions and digital tablets were rarely used. Chalk talks were over-represented in math courses (38%) and were rarest in biology. Slide show lectures made greater use of clickers and small group work (both 11%) and were most commonly found in biology and physics but not found at all in math courses.

Ferrare also conducted semi-structured interviews with the instructors in the courses sampled and, following his own earlier work (Ferrare & Hora, 2014), reports that instructors' beliefs about how students come to understand the foundational concepts and skills of science and mathematics, and the teaching strategies that they deploy to achieve this, are entirely related. What instructors who used each form of lecturing shared was a distinct and coherent set of tacitly understood beliefs about how students learn. The learning beliefs of instructors who used chalk talks emphasized what students should be doing to facilitate their own learning. An important focus in their teaching, therefore, was to pose problems that would facilitate practice by "perseverance." Students should "grind away" at conceptual problems until mastery was achieved. Their use of Q&A also reflects the importance that they accorded to intellectual risk-taking where students posed and responded to questions through open dialog with the instructor. Slide show lecturers believed it was important to introduce students to theory and mathematics of new concepts, then model applications through repetition and variability until students could solve the same types of problems—a process enabled by clicker questions.

In 17% of introductory STEM classes, Ferrare's team also observed "multimodal" teaching that made extensive use of digital tablets, a variety of visual materials, clickers to answer questions, and small group work. A final teaching approach (found in 9% of courses) made little use of lectures and worked 70% of the time with small group work in which the instructor worked directly with each student group. In other teaching forms observed in foundation courses, the boundary between instructors and students was (as students in the interview study also recount) sharply preserved. Only in these classes did students experience the kind of interactive dialog with their instructors that they strongly preferred. These two more interactive teaching forms accounted for 26% of all course observations and were also (as noted above) reported by 26% of switchers and 33% of persisters in the interview study. Small group work was the most common interactive method recorded in all three studies: group work was reported in 71% and the use of clickers in 53% of the foundation courses surveyed by the SALG instrument. The modest moves into research-based instructional strategies (RBIS) evident in these three studies will be encouraging to the STEM education improvement effort that has been ongoing between the original and present studies. Ferrare's findings, however, underscore conclusions from educational change research (e.g., Lotter, Harwood, & Bonner, 2007; Lund & Stains, 2015; Wieman, Perkins, & Gilbert's, 2010; Woodbury & Gess-Newsome, 2002) that wider uptake of RBIS has to begin with acknowledging how instructors conceptualize the student learning process and persuading them to consider the research-grounded learning theories that underpin research-grounded teaching.

Other researchers (Aronson, Fried, & Good, 2002; Blackwell, Trzesniewski, & Dweck, 2007; Canning, Muenks, Green, & Murphy, 2019; Dweck, 2008, 2015) have identified a belief about intelligence that is common among STEM instructors, namely, that the ability to "do science" is innate and fixed rather than something that grows with interest and effort. Through their modes of teaching, assessment, and contact with students, instructors who believe in fixed intelligence convey the message that only "innately gifted" students are likely to succeed—a message that many of our interviewees had encountered and that some had internalized as "true" (cf., Dweck, 2008, 2015). STEM instructors who believe that it is part of their job to identify students with natural ability and to encourage others to do something more suited to their presumed abilities were clearly doing that. Those instructors who believed that intelligence grows were more likely to show students how to become better learners and motivate them do their best. In controlled experiments, Canning et al. (2019) found that the racial achievement gaps in courses taught by more "fixed mindset" instructors were twice as large as the achievement gaps in courses taught by more "growth mindset" instructors. Indeed, their teachers' mindset beliefs predicted student motivation and achievement far better than any other instructor characteristics (gender, race/ethnicity, age, teaching experience, or tenure status).

Taken together, beliefs about the nature of students' intellectual capacity and about how science must be learned are powerful influences on how students are taught and on the student outcomes of teaching than their instructors' beliefs.

As Ferrare makes clear, the teaching methods that instructors in foundation courses deployed were entirely consistent with their beliefs about how students learn science. However, from student accounts of how they learn best, it is also clear that the learning theories of students and their instructors were sharply divergent. In students' descriptions of "chalk talks" (which, in both studies are reported to be most common in math-heavy courses), instructors face the board with their back to students and write while they talk (often problems for students to solve). Students were disappointed and unmotivated by this form of instruction. They wished for more explanation, interaction, and discussion with their teachers:

> The very first day, I will never forget this, we were all sitting in this room, and he walks in, sets his stuff down, doesn't even say anything to us, turns to the board, and just starts writing stuff. And we were all like, 'Is this the class? Is this, like, what's going on?' And then, he just turned around at one point, and said, 'Okay?' Then he grabbed his stuff and sat down and we were like, 'Is class over?' (White woman, switcher, mathematics to economics)

> I'm in Fluid Dynamics, for Chemical Engineering specifically, and it's a very terrible class.... Like the professor writes on the board ... but he sorta jumps all over the board so it's really hard to follow along in lecture. (White woman, chemical engineering persister)

> It's hard to stay awake while he just does math on the board for an entire hour. Like you see the board get covered up in just numbers. It was like, numbers are cool, and I see how you did that, but what does it mean? (African-American man, physics persister)

Many students felt that STEM faculty also relied too heavily on lecturing with PowerPoint slides which many teachers (though not all) simply read aloud:

> I think most teachers in the sciences tend to teach the same way. You know, PowerPoint lecture sort of deal. (Asian woman, chemistry persister)

> He just came in and put on the PowerPoint, left, came in the next day, PowerPoint, and left. And that's the class I did not do good in. (White woman, chemistry persister)

Sometimes, the slides did not seem to enable the instructor to explain the main conceptual ideas to their students:

> I'm not a fan of the method where he has a PowerPoint presentation and will just put a whole paragraph of text up on the PowerPoint and then he'll just talk about what the paragraph says. He's not really saying anything that's not on the PowerPoint. (White woman, geology persister)

> She wasn't really good at explaining concepts, like she would have all these PowerPoint slides. And she would just point to things. She's like, 'This is this and this and this.' But she wouldn't really break it down, so I really had to teach myself. (African-American woman, microbiology persister)

The slides themselves were not the problem. When made available to students before class, they could free up the instructor to expand, pose questions, hold discussions, and give examples. However, based on students' reports, instructors rarely seem to grasp these opportunities. Students in engineering, life sciences, and physical sciences experience traditional lecturing as the main mode of teaching at the highest rates, as shown in Table 6.4.

Our three sets of results may be compared with those of a large observational study (Stains et al., 2018) that deployed an adaptation of the TDOP in 2000 STEM classes across 25 North American institutions. Lower-level classes accounted for 71.4% of the sample with the balance from upper-level, cross-listed, level-unspecified, and graduate courses. Lecturing with little or no student involvement was observed in 55% of all classes; lectures supplemented with

**Table 6.4** Teaching methods experienced by STEM majors by disciplinary group

| Discipline | Straight lecture | Interactive |
|---|---|---|
| Engineering | 80% (66) | 9% (7) |
| Life sciences | 75% (105) | 11% (15) |
| Physical sciences | 74% (25) | 9% (3) |
| Geosciences | 58% (7) | 17% (2) |
| Computer sciences | 41% (7) | 12% (20) |
| Mathematical sciences | 38% (9) | 13% (3) |

interactive strategies including clicker questions and group work were observed in 27% of classes; and student-centered teaching methods were observed in 18% of classes. The lower proportion of "didactic" teaching methods across classes of all types may suggest that "chalk talks" and "slide shows"—or, in the students' words, "straight lecturing"—are less dominant in later courses than in the foundation courses that comprised our study samples.

As in the original (1997) study, lecturing apparently remains the customary form of teaching in introductory STEM courses. However, what were not found in 1997 were any of the interactive and multi-methods teaching methods that were reported in the current study by 26% of switchers and 33% of persisters and was observed in 25% of introductory STEM classes in Ferrare's observational study. Small group work was the most common type of interactive methods recorded in all three of our studies. Students answering the SALG survey reported the use of clickers in 53% of their classes. Although SALG comments suggested peer–peer and teacher–student interactions in lectures were still limited, some experience of group work was reported in 71% of the courses surveyed. This modest, but demonstrable move towards active, research-based forms of teaching will be encouraging to the STEM education reform effort that has been ongoing between the original and present studies.

## What Do Students Define as "Poor Quality Teaching"?

The six most commonly described characteristics of ineffectual teaching offered by both switchers and persisters in the interview study are outlined in Table 6.5. In what follows, we explain and illustrate what students meant by each of these dimensions of "poor teaching."

### *Disorganized Teaching*

The majority of switchers (64%) and nearly half of the persisters (43%) included disorganization in their descriptions of poor quality instruction. Descriptions of teachers' disorganization frequently included: failure to present topics in a logical

**Table 6.5** How students define poor quality teaching

| Definition of poor teaching | Switchers ($N = 96$) | Persisters ($N = 250$) |
|---|---|---|
| Disorganized teaching | 64 | 43 |
| Delivering course material at inappropriate level | 49 | 27 |
| Unapproachable demeanor | 34 | 33 |
| Inadequacies in presentation | 30 | 38 |
| Disengaged teaching | 30 | 22 |
| Indications that teaching is under-valued | 21 | 22 |

sequence, presentations of material that were incoherent or that missed out foundational information, inappropriate pacing, and poor management of class time:

> The instructor was non-directive. I remember my physics notes were just so ugly because he would go back and alter things. It was very nonlinear. It was just all over the board. (White man, switcher, microbiology to political science)

> I felt like the content—the lecture was just all over the place to me so I didn't understand … And it was hard to focus…He just kinda talked without a set agenda. So I didn't know what was necessary and what wasn't. (Asian woman, switcher, biology to international studies)

Descriptions of disorganized teaching included comments about the lack of structure in course design or organization in its implementation and were also cited as problems for learning in 25% of the SALG written comments describing poor teaching—as in this example:

> The professor was late every day. He gave scattered explanations, and didn't give homework or anything to reinforce the content. He didn't post lectures either. (White woman, electrical engineering major)

As 4% of critical SALG comments described, some disorganization was rooted in the structural misalignment of course content, activities (including labs), and assessments. Disjunction between class and lab attention to the same content was often a consequence of departmental class sequencing:

> Class activities and labs were often completely disjointed from the labs, or occurred at completely different times than the labs. (White man, aerospace engineering major)

However, other misalignments, such as the focus of class and test content, were usually within the instructor's control.

Finding a course "difficult" is often explained in terms of students' inability to grasp complex ideas. However, 23% of SALG commentators explained that the difficulty of their courses was closely related to the pace of the class, the amount of content presented, and the speed with which it was delivered:

> Too much content was presented too fast. Cell Biology is a *huge* subject and there are so many little parts. Cramming them into a semester is too much. (White woman, microbiology major)

> There was an absurd amount of content covered, of which about one-third was actually mentioned in lecture. Most of the class came from screencasts, which were just as useless as her actual lecturing. (White man, integrative physiology major)

> The pace of this class was way too fast. The professor often did not have time to complete all the material and had to make screencasts, which were not helpful. (White woman, integrative physiology major)

Conceptual disorganization and incoherence was frustrating for many students, but for some students, it was debilitating. When teachers covered topics out of order, and skipped steps or content that linked topics, they were unable to follow the logic of the presentation:

> The lectures weren't hierarchically organized and it's a lot easier to learn in hierarchies…. They would jump off to different points and be kinda scatter-brained and you wouldn't be able to follow what they were saying. (White woman, switcher, biological sciences to international studies)

> There would be dramatic jumps … and there'd be nothing in between. I'd be like, 'What happened to everything else?' and I couldn't follow along. (White woman, switcher, microbiology to dietetics and nutrition management)

Many of the SALG survey comments also focused on instructors' failure to explain concepts clearly. This linguistics major identifies an instructor's difficulty in communicating effectively as partly a lack of pedagogical understanding and technique:

> The instructor was bad at explaining concepts, explained examples too fast, and asked students about what they thought far too often—almost relying on them to answer the question for her because she was too incoherent to do it herself. (White man, linguistics major)

Non-sequential presentation of new material created learning problems for many students, and was a grumbling irritation for others. Students who were accustomed to learning in a linear progression—especially switchers— found jumps in the ordering of content and dipping in and out of topics to be chaotic and confusing, impeded learning, and undermined students' ability to progress, and were a contributor to two-thirds of decisions to withdraw or switch to other majors.

Uneven treatment in time and emphasis given to essential material also created problems for students. They described teachers who took longer than necessary to cover some topics but who rushed through others:

> He would just kind of expand on certain details that weren't worth mentioning. But he would constantly cover one topic for an entire week and then go over another topic in like a day. He couldn't regulate his own pace. (Asian woman, biological sciences persister)

> I think a lot of teachers waste time, or spend all of class time explaining in-depth why things are a certain way, but they don't have time to give you an example of how to actually do the problems they give you. (White woman, engineering persister)

Some teachers were apt to stray from the main focus of the syllabus and spend too long in off-topic detours:

> He didn't explain things very well, he'd get caught up in specific cases that you don't need to know…and he just sort of trails off and keeps going and going…He gets side-tracked, and everyone in class is like, 'What're you doing?' He spent too much time explaining things that no one in the class really can understand yet. (White man, switcher, aerospace to studio art)

> I had some bad professors who were not able to cover the material in that amount of time and then just rushed through important points that they should have covered. (Asian woman, biological sciences persister)

The SALG commentators also complained about instructors who did not cover conceptual material in sufficient depth for a solid understanding, skimmed the surface of topics, or focused on one subject for too long at the expense of other important areas:

> The lecture did not help my understanding of so many concepts, there is not enough time to explore a subject in depth in only 50 minutes. (White woman, integrated physiology major)

Moving too fast through some topics and too slow for others suggests that these teachers did not know how best to pace through material, had not prepared for the class, or approached all their lessons in a disorganized way. Pacing is particularly problematic where knowledge, understanding, and skills must be built up sequentially:

> There would be certain topics, certain algorithms that the professors would blow through, and your head is spinning. It's just like, 'Oh, my gosh! How did he jump from that to that?' (White woman switcher, microbiology to dietetics and nutrition management)

> I've found the pace of the class has been challenging, because when he goes through problems on the board, he's assuming that we can go from point A to point B with these 10 steps that are in between and see exactly what's happened. And so he'll rush through that and I feel totally lost. (White woman, horticulture persister)

Where STEM faculty moved too quickly through material that required progressive build up, students felt lost and unable to catch up, especially where important pieces of information in the sequence were missing.

A related problem, especially for switchers, arose when teachers moved straight to advanced concepts without first reviewing foundational information, or offering a brief overview of what students should have already learned before launching into a new topic:

> I feel like some professors, because they've been teaching for so long, don't really start from square one. Sometimes professors will be like, 'Oh so you learned all about this in that class so I'm not going to go over it.' It's like, 'Well, please go over it 'cause I don't get it yet.' (Asian woman, biomedical engineering persister)

SALG commentators described this structural flaw in the presentation of content as another cause of what made some courses "difficult":

> I don't feel I was given enough resources to fully grasp the background foundational material before I was expected to do more complex problems. (White woman, biology major)

This problem also arose when critical material was insufficiently explained at the outset, instructors did not check that everyone understood the conceptual foundations on which the effectiveness of the class depended:

> For chemistry and physics…they expected you to know exactly what they were talking about the first time they said it. And they didn't bother to explain it. They just wrote out the math problem, solved it on the board, and said, 'This is how you do this.' (White woman, switcher, engineering to psychology)

Students felt they deserved at least a short review of essential material to enable them to make the transition to more advanced content:

> They expected you to know the basics already … it was too fast for me … they wouldn't explain it enough … I'd rather they spent more time in depth describing it. (White woman, switcher, biology to psychology)

> My stats class really sucked. The teacher just gave us an equation and then she'd move on to the next thing…. She'd give you the quick version in class and then expect you to fully understand it on the test. Unless she explains it in depth to you then you're not going to understand it in depth. (White man, switcher, mathematics to psychology)

Failure to review foundational material was one of the indicators to students that some teachers simply did not care about enabling students' learning. This was also evident to students when instructors did not pause to check whether students were understanding the material by making time for questions and discussion. Students described this teaching style as "being talked at":

> The professor would just keep working through it really fast, almost like there was no one else in the room. That was one thing that really bugged me in trying to learn. (White man, switcher, life sciences to architecture)

> The teachers are just throwing up slides that they got from books and just going through it and not really pausing for questions, or they're just talking at you. (White woman, switcher, microbiology to dietetics and nutrition management)

> They don't make time to ensure you understand it, or ask 'Does everybody understand that?' I really don't like that. (White man, mechanical engineering persister)

As the SALG commentators also pointed out, one reason why the content of some classes did not engage them was the one-way transmission method by which it was delivered. Failure to stimulate dialogue, participation, or questions was a sure way to make lectures dull and uninteresting:

> Lectures were almost uniformly silent on the part of the students and he did not often ask for a response, aside from inquiring if we had any questions. (White woman, chemistry major)

Even instructors who did pause to answer questions did not necessarily manage their time well, either in the length and detail of their responses, or how much time they spent addressing some questions at the expense of others:

> It would take 10-15 minutes for him to answer someone else's question before getting to mine. And sometimes we had the same question and he's answering it all over again. (Hispanic woman, switcher, life sciences to psychology)

Failing to provide opportunities for students to ask questions and engage in discussion, and poor management of question breaks, suggest the need for professional development in classroom management techniques. Also needed according to SALG complainants was advice on how to prevent or manage disruptive classrooms:

> Class sessions were not well managed. When students got too loud or left early there was no repercussion for such actions. (African-American man, engineering major)

Seymour argues from data (2005, 2016) that dis-attentive and disruptive student behaviors are forms of student protest or resistance in lecture classes where the teaching lacks stimulation, interaction, or coherence. Instructors have long normalized such behavior as a characteristic of students rather than as an outcome of poor quality teaching.

Students also found some team-taught courses were disorderly and confusing. Lack of coordination between the participating instructors could make the learning experience disjointed and chaotic. Additional problems were created by multiple teaching styles, instructors who contradicted one another, and insufficient communication among the instruction team:

> There are four professors and so they would rotate. They would each come in to speak about the different things they specialized in. This was difficult because they all had different teaching styles. Some would concentrate on a small portion of the chapter; others would try to expand on it. Some would relate it back to what was being taught, and others would just talk about their own personal experiences (Hispanic woman, switcher, mathematics to business)

> So, there's three or four different professors, but there's one exam, so they all write questions, but it's like you'll have a different professor that emphasizes, do this or do that. But then the questions on the test aren't like that at all, because some other professor wrote them… So, it was just really frustrating. (White woman, persister, life sciences)

> One professor would say, 'Oh, I really like your project, it's great.' And the other one would come by and say, 'I don't think you should do this project. I don't think it's going to be sustainable for you.' (White man, switcher, life sciences to visual and performing arts)

We have described the different facets of what the interview and SALG studies (and also that of Ferrare, 2019) found to be the commonest student complaint about STEM teaching, namely incoherence in the organization and presentation of conceptual material. As we have illustrated, the main dimensions of incoherent teaching as described by both switchers and persisters were:

- Poor organization of content and poor management of class time. These were evident in disjointed or illogical ordering of content, non-sequential handling of important new material, and missing out or glossing over essential concepts;
- Uneven treatment in the time and emphasis given to more and less essential content; rushed explanation of steps involved in understanding hierarchical information, pacing that was not appropriate to the importance of particular topics—too fast for essential material and too slow for less significant content; and straying from central themes into over-long illustrative detours.

Students regarded these aspects of disorganized teaching as evidence of insufficient preparation of the content and lack of planning in the mode and pace of delivery.

### *Delivering Course Material at an Inappropriate Level*

Teaching content at a level that was inappropriate for the course designation was the second-most commonly described attribute of poor teaching cited by switchers (49%). A quarter of persisters (27%) also criticized STEM instructors for pitching materials at unsuitable levels:

> I was expecting them to teach me the material. And they expected me to already know it. (White woman, switcher, biological sciences to psychology)

> They just run through the material without relatability—without taking into account that we don't know the information already. (White man, switcher, mechanical engineering to finance)

> I feel like most professors in college forgot how it was to be an undergrad—trying to understand things that you have never seen before. (Asian woman, mathematics persister)

Students described getting off to a bad start where instructors launched straight into new, complex subject matter without providing a review or a link to previously learned material:

> I think they really were pushing for us to think for ourselves and solve the problems … but without having a very strong baseline, it's hard to do the critical thinking … The [foundational] material that was required to be successful in that class just wasn't being taught. (White woman, switcher, biological engineering to business)

Some of the earliest switchers were first-semester freshmen who were unable to connect what they knew of the subject from high school with material pitched at too high a level at the start of introductory classes. Failure to review essential material at the outset before moving to more advanced concepts left many students unable to grasp new complex concepts:

> Some teachers say, 'Oh you guys know this stuff.' But high school did not prepare me in a lot of ways. That first couple of years it was like, 'No, we don't really know this stuff.' (White man, civil engineering persister)

> Because they've been teaching something for so long, they don't really start from square one. They sort of have a train of thought and they just start somewhere in the middle versus going back to the beginning and realizing that, you know, we're 18 and we don't know what's going on. (Asian woman, relocator from biomedical engineering to materials science)

Many persisters reported that failure to connect new material to what they had previously learned continued to be a problem in STEM courses:

> It was assumed that you had a strong background in organic chemistry and in biochemistry. You also needed a strong background in genetics. … So he hit the ground running very quickly. You get a lot of the students get left behind and that's the problem is because the concepts are really abstract and if you miss and trip and fall, the train just keeps going. (Hispanic man, biological sciences persister)

For freshmen, this problem was exacerbated where high school preparation had been inadequate (as discussed in Chap. 5). Where instructors expected them to have a solid grasp of prerequisite course material that many students simply did not have, underprepared STEM majors were commonly overwhelmed and took an early decision to leave. However, lack of reviews at the start of courses and help with transition to new material was frustrating to switchers who persisted longer and to over one-quarter of all those who stayed in their STEM major.

A related question was whether instructors pitched their content at a level that was appropriate for the course designation. Indeed, some students assessed the level at which particular classes were taught would be appropriate for a graduate-level course:

> He teaches the class as though it's a grad-level course, but we haven't learned the undergrad stuff yet. He just does complicated algebra on the board all class. (Native American woman, astronomy persister)

> I think he was used to teaching upper grads ...because he would go into more detail than we needed to know...Most of the class just stopped taking notes when he was talking, just 'cause he went over everyone's head. (White woman, switcher, biological sciences to psychology)

> It was so hard because he is a graduate-level teacher who is trying to teach to freshmen. And he was just so scattered and all over the place and it was really hard to follow. And his homework was really difficult because he didn't know how to gear it towards freshmen ... He was just teaching above our knowledge level. ... This is like my third or fourth class where they're just teaching at like almost a graduate level and we're all undergrads. (White woman, engineering persister)

Some of these observations call into question whether the instructors described knew how to assess the level of depth and complexity that was appropriate for the level indicated by the formal course designation. Some students thought that it would be possible to approach the same topics at a level appropriate to their stage in the major. Others concluded that STEM faculty were deliberately making it harder for students to understand perhaps to weed them out. Persisters were more apt to make sense of teachers' apparent inability to explain fundamental concepts at a level appropriate to their students in terms of their instructors' advanced subject knowledge or their sophisticated and complex research:

> A lot of professors that I've had that are absolutely brilliant people, and their research is fascinating, but they're not good at conveying information to students. They might be the first-rated person in the lab, making ground-breaking discoveries, and that's what the university wants to see. But when they get into the classroom and they have to relate this information to people who've never heard it before, people who are not familiar with the field, they're just useless. They can't really do that. (White man, mechanical engineering persister)

> They're just so smart, have such high levels of education and research, and teach a lot of the upper-level classes. So teaching the lower-level classes, they kind of forget that we're down here. (White woman, biological sciences persister)

> The worst professors that I've experienced have actually been some of the smartest. I realize that they're experts in their field; they're really excellent at what they do... but they're unable to simplify things in ways that they can express to students ... They will look at me and not be sure why I'm confused. And I'm confused why they don't understand that I'm confused. And it's just an ongoing cycle of that. I think it's a lack of knowing how to express really detailed concepts in a baseline way. (White woman, neuroscience persister)

Both switchers and persisters struggled with STEM faculty who seemed unable or unwilling to simplify complex material, break it down into manageable pieces, and pitch it at a level appropriate to the course. As illustrated, persisters often interpreted this incapacity in terms of the sophisticated levels at which research faculty customarily worked. However, they also saw the need for STEM instructors to gain professional development in teaching methods, syllabus design, and course preparation. Adding these skills to their advanced disciplinary knowledge would address many learning problems that they described and illustrated. Primary among these would be to pitch material at appropriate levels, and build conceptual scaffolds that can take young learners with them into increasing levels of complexity.

### *Intimidation and Distancing Behavior*

One-third of both switchers and persisters (34% and 33%, respectively) reported rejecting, hostile or other negative interactions with STEM instructors. There were two main types of experience: being (often publicly) intimidated, belittled, or made to feel too stupid to belong in a STEM major; and instructors who were consistently unavailable for individual help with problems, avoided answering students' questions or answered them begrudgingly. Switchers contrasted these experiences with interactions with non-STEM teachers whom they found to be more approachable and open towards them, both in class and as individuals.

A very common complaint by both switchers and persisters was being intimidated, belittled, or mocked by instructors when they asked questions in class:

> If someone asked a 'dumb' question, he would call it out, and be like, 'That's not a very good question. Did you do the reading?' And he would belittle that student in front of everyone. (White woman, switcher, biological sciences to psychology)

> I asked her a question and she laughed at me 'cause I didn't know the answer. (Hispanic woman, switcher, biological sciences to journalism)

> He was kind of blunt—a little bit abrasive…He would make you feel stupid if you didn't know what was going on. A lot of people were just intimidated to ask him questions. (Hispanic woman, switcher, biological sciences to sociology)

SALG commentators also cited instructors whose disdainful demeanor towards students was a deterrent to taking further courses in that discipline:

> The professor was extremely condescending and arrogant, which did not facilitate student learning whatsoever. (White man, biochemistry major)

> This class was unnecessarily antagonistic. The attitude of the instructor was not one of helpfulness, but was hostile to the students. I was interested in chemistry at the beginning of the semester, now I am completely turned off by the subject. (White woman, biology major)

Students reported experiences where their teacher had expressed incredulity that they did not understand the material being presented, or conveyed the impression that the student was alone in his or her confusion while the rest of the class understood the content:

> If you tried to ask him a question he was really condescending and he would be, 'I can't believe you don't know this. Why don't you know this?' (Multi-racial woman, relocator, materials sciences, to chemistry)

> There was a student that asked the professor a question because they didn't understand the way the professor explained a particular reaction on the board. And he literally erased what he did and drew the exact same explanation, and said, 'I don't understand why you don't get this.' (White man, biochemistry, relocator, to biological sciences persister)

Descriptions of faculty who did not appear to understand why students were unable to comprehend the material were not limited to in-class interactions, but were also experienced during office hours:

> In those science classes… where there's 300 students… I've met with professors during office hours and they're always short with you and kind of frustrated. Like, 'Why don't you get it? You should get it.' (White woman, physiology persister)

> When we go talk to them during their office hours, we'd be belittled, like, 'Why don't you know this? Why don't you understand this?' And I was thinking, 'I don't know. That's why I'm here.' It was always hard to go back to them, because we knew as soon we asked our question we'd be belittled again. So why bother going to back when we already know we're struggling. (Hispanic man, biomedical sciences persister)

> One professor in biomedical engineering would always tell us, 'Don't come to office hours and ask me stupid questions,' or if we're emailing him about a certain question he would rudely email back, 'Go ask the TA. You shouldn't be asking me this question.' (Asian, woman, relocator, biomedical engineering to materials science)

There were frequent reports of instructors who expressed incredulity at a student's confusion or failure to understand, along with the impression that they were annoying the instructor by asking questions:

> If you went in to ask for help during office hours—it was more of a bother towards the end. It was more of like, 'You're here again?' So, they weren't willing to help. (Hispanic woman, switcher, biological sciences to communications)

> It felt like any time I went and talked to a professor I was wasting their time. They were more intent on what they were doing outside of class than teaching. There was just a lot of frustration in that. (White man, switcher, biomedical sciences to marketing)

> After a couple times of trying to explain a problem, if a student can't get it, their solution is just like, 'Oh, okay, go read.' Just brush this student away. (White man, switcher, computer science to economics)

> He would come into class, expecting you to have read a certain portion of the material, which is fair. But if a student asked a question about the material or any kind of question pertaining to it, he would say, 'Didn't you read the material? You should have read this. Don't you know?' Like, 'Don't you understand this?' (Asian woman, biological sciences persister)

Some students described this instructor behavior in adversarial terms in which students were viewed by the instructor as seeking to gain some kind of advantage:

> It felt like a power conflict rather than a classroom because any time a student asked him, 'What should we study? What's on the test?' He would just say, 'Everything. Because

everything is important and I don't want to tell you that one thing is more important than another.' (White woman, switcher, biological sciences to European studies)

Sensing that teachers are bothered by questions made students feel less confident in approaching them and left them confused about the material without any clear way to gain an understanding of it. These instructor attitudes and behaviors contributed to students' feeling that they did not belong in these classes and programs and thereby to their decisions to switch. Persisters also reported the same instructor attitudes and behaviors and described their consequences—both in changes in direction, and how students felt about themselves:

I was literally scraping by with help from all the other people because when I would talk to him he would look down on me, like, 'How do you not understand what this is doing?' (White man, mathematics persister)

In P-chem, which I recently dropped, because the teacher talked in this tone that was so condescending that I couldn't manage to go to class and listen to her talk to me like that. Her tone was implying that I should still be in high school. … It was insulting. And that made me just drop the class. (White man, physics persister).

I'm remembering a class, the class that I actually failed, the professor was condescending and I didn't want to ask him questions because I felt that I would have gotten a snooty response. I felt that by asking him a question, he would have given me some sort of answer that made me feel a lot worse about myself while answering the question. So I didn't really want to ask him questions. (White man, computer engineering persister)

I took two math classes in the spring, which is when I did terribly in math… Neither of my professors made me feel I like I belonged. They made me feel really stupid all the time. (Asian woman, mathematics persister)

Although all of the last four speakers persisted in STEM majors, one of them dropped a course in his major, another questioned that she belonged in STEM, and all felt humiliated or insulted.

Students greatly value teacher openness to interaction with students. Indeed, for some students, the discounting and distancing behavior that they experienced from STEM instructors contributed to their decision to leave STEM majors—especially where non-STEM faculty were experienced as accessible and welcoming by comparison. We describe this as part of the push–pull processes that contribute to many moves out of original STEM majors:

My biology teachers and my chemistry teachers here are more separated from the students, and not very friendly or helpful. Whereas the psychology teachers are more friendly and joyful. They will work with you. (White woman, switcher, biological sciences to psychology)

I tried some religious classes … And I really enjoyed my professors, it was much easier to go up and talk to them about things in class, about my own ideas, about my interpretation of a particular set of texts that they had given us … I felt like there was a lot more potential for me to interact with the professors, and to show myself to the professors in a way that I couldn't in the engineering department. (Hispanic man, switcher, mechanical engineering to religious studies)

[The Biology teachers] just don't seem to want to get to know you on an individual level or care about knowing your name. … Some of my health professors, even in a 200-person

> lecture, they get the roster at the beginning of the year with pictures of everyone ... one teacher for sure knows everybody's name because I have her for two classes and I asked her a question in the first class and she didn't know the answer to it, and then she came up and found me a couple days later in the second class and was like, 'The answer to your question is this...' I was kind of surprised she knew who I was because it's a big lecture, but I appreciated it. (White woman, switcher, biological sciences to community health)

As we have illustrated, discounting and distancing experiences with STEM instructors affect both switchers and persisters. As we discuss in Chap. 12, persisters found coping strategies to deal with intimidating interactions, but were nevertheless troubled by them. Some switchers found the rejecting behavior of STEM faculty sufficiently damaging to their confidence that it contributed to pushing them out of STEM majors. As these observations also illustrate, this kind of behavior by instructors was particularly upsetting to women, especially women of color, and contributed to loss of confidence and failure to develop a sense of belonging among women that we point to throughout this book.

## *Inadequacies in Presentation*

Inadequacies in delivery of course content was the second-most discussed characteristic of poor teaching among persisters (38%) and the fourth by switchers (30%). It was also the subject of 25% of SALG written comments. Students described the delivery style of many STEM instructors as "boring," "unenthusiastic," "monotonous," "dull," and "uninteresting." Persisters offered more examples of boring presentations than switchers, possibly because they had endured them more often and for longer than the switchers:

> Just straight lecture for two hours without any break or any questions or anything -- that's like the dullest way to learn I think, especially when it's done in a monotone... It was just an awful class [referring to Ecology and Evolution]. (Asian woman, molecular biology persister)

Classes with more than 100 students were often described as the most boring and least interactive:

> Right now, I'm in a stats class where there's over 200 students and if I didn't like stats I shouldn't even be taking it ... It's boring and he just explains things horribly and I see these kids around me struggling ... He stands up there and just talks to us for fifty minutes ... I'm so glad I [already] understand stats ... I feel bad for people who just don't because he's horrible. (Multi-racial woman, switcher, microbiology to psychology)

> It's just a guy standing at a microphone with either a PowerPoint or a white board. Just very monotone voice; just kinda droning on and on and on ... All pretty stagnant—just the professor at the microphone speaking. (White man, switcher, biomedical engineering to Asian languages and literature)

Finding ways to interact with students in large classes is a challenge for all teachers. This may help to explain why many STEM instructors opt for a traditional lecture method in which they talk to students in one-way transmission with the aid of a microphone. Students, however, were extremely dissatisfied with one-way teaching experiences in so many of their STEM courses. They disliked the feeling of being "lectured at" and were especially annoyed when teachers read directly from PowerPoint slides without adding any additional information or incorporating some discussion. Four switchers described their reactions to this method of teaching thus:

> In general, I'm not a tired person, so when I feel exhausted in class it's because I'm not feeling engaged… In a lot of the science classes you're just being lectured at, and you're kind of a sponge to soak in the material and turn it around. (White woman, switcher, neuroscience to sociology)

> I really cannot pay attention to the teachers who literally read the PowerPoint—like the entire PowerPoint. I don't need to come to your class, obviously, because I can read your PowerPoint on my own. So it's like you're wasting your time. I can't even stay awake during stuff like that because you're staring at someone read something and that's just not interesting. (White woman, switcher, biological sciences to human development)

> They're just throwing up slides that they got from books and just going through it and not really pausing to ask for questions. They're just talking at you. (White woman, switcher, microbiology to dietetics and nutrition management)

> It was not up to the level that I had been expecting. It was just reading directly from PowerPoints that were then posted online. I didn't see a real motivation to come to class. (White woman, switcher, biological sciences to literature and languages)

Overall, 10% of the SALG commentators wrote that the various inadequacies of the straight lecture format forced them to learn much of the course content on their own:

> Lecture materials were completely irrelevant. All knowledge gained was self-taught via outside resources (Khan Academy, etc) (Multi-racial man, computer science major)

SALG commentators also described instructors who used poorly designed visuals, overly busy PowerPoint slides, or spent much of class time writing notes on the board:

> The teaching method in this class was horrible. Reading off of a slide and writing in illegible writing is absolutely horrible. (White man, biomedical engineering major)

> I didn't like how the professor would take so much time to write the notes. I would rather have her explain the examples much better than spend time writing everything out. (White woman, physics major)

And the sheer size of some introductory science classes exacerbated student learning difficulties:

> Very large courses such as this often do not have an atmosphere conducive for everyone's learning. (White man, physics major)

Two STEM relocators also expressed their frustration with the straight lecture format and its consequences:

> I just can't sit there and listen for 50 minutes if you're not gonna do something. … It really helps if the professor is talking and drawing something on the board and explaining it through what he's doing. And then I just do it myself and then I get it. That's how I learn. (African-American woman, relocator, chemical engineering to civil engineering)

> Professors just reading straight off a PowerPoint ... It's become so bad that I skip a lot of classes because there's no point. I can just look at the PowerPoints online and figure out the same little amount of material in a quarter of the time. And my time would be better used sleeping or something else. (White woman, relocator, mathematics to computer science)

A persisting senior suggests that the instructor's choice not to take questions, or engage in any two-way exchanges, is probably driven by trying to cover too much material in the available time. However, the student is concerned that this choice threatens student comprehension:

> Some professors have a certain amount of material that they want to get through in that lecture and they don't want to sacrifice any time answering questions. Some professors will say, 'Just ask me after class.' But that question might be important in order for us all to understand what's going on in that specific lecture. (African-American man, biological sciences persister)

Students were also disappointed when instructors provided too few examples, did not work through sample problems, offer applications, or make connections between concepts. These omissions made it harder for students to apply what they were learning for themselves:

> They wouldn't do any problems in class. It was more conceptual, and they would just write the concepts and a bunch of variables on the board, but they wouldn't actually go through any examples ... So, I wasn't sure if I would actually be able to translate the information into what I needed to do. (White woman, switcher, mathematics to hearing and speech science)

> It was just rambling—not many examples and all theory. And I realize it's math, but we're engineers and we need the examples of how to apply it. (Asian woman, electrical engineering persister)

> The worst is when they're just giving you theory with no connection. I think the more examples the better. (African-American woman, biological engineering persister)

Eight percent of the SALG negative comments specifically addressed students' need for real-world applications and examples to enable conceptual understanding:

> A lot of the concepts did not have real world application, which, at times, made it very hard to follow or care. It would have helped if the professor could come up with his own examples in class instead of taking examples straight out of the book. (White man, mechanical engineering major)

Persisters also lamented the shortage of illustrations and applications in their upper-division STEM courses:

> Genetics is a very content-heavy course. But the professor just laid information out there and didn't give you any background on how certain discoveries were made. I just memorized it and didn't really retain it later on. (Asian woman, biological sciences persister)

> The class is just theory. 'Here's the proof of how to do this.' But that doesn't help me at all. That means nothing to me. Sometimes it's so abstract that I don't have anything concrete to grab onto. That's a big issue for me. ... I do a lot better with examples. (White woman, mathematics persister)

Persisting seniors who had experienced a wide range of STEM courses over time had very clear thoughts about what distinguished good from poor delivery and what were its root causes. The prevalence of dull lectures and limited application of concepts were commonly explained in terms of STEM faculty simply not caring enough about their teaching role to do it well:

> I feel the difference is having a professor who likes teaching, and you can very clearly tell who does and who does not. Some people that like teaching understand that some material is boring and they try to help you through it. The people that don't like teaching, know that it's boring, and don't care that it's boring. They're just gonna throw it up on the blackboard and make you understand it and then that's it. The worst classes are the ones where they throw all this information at you, and that's all you get. (White woman, chemistry persister)

> The worst teachers are those who don't really seem to care what they're doing. I think the biggest thing that makes a teacher bad is a teacher who doesn't care about the students or the material. (White woman, relocator, geosciences to biology education)

As our persister informants pointed out, dull lectures have counterproductive consequences:

> The teacher reads straight from a PowerPoint … it's just super monotone … those classes don't make me think at all and don't challenge me; I actually end up doing worse in them just because I'm so bored. (White woman, biological sciences persister)

> It's a dark room and a PowerPoint presentation for the entire three hours. They use a monotone voice, don't go back over things, won't take questions, and just keep going on a very linear path. That's why I'm zoning out, don't understand, and then I'm asking myself, 'What did I just learn?' (White woman, biological sciences persister)

> In freshman year chemistry, the lectures were so boring … It got to the point where everyone just stopped going. (African-American man, aerospace engineering persister)

As noted in the original study and discussed throughout this book, dissipating students' incoming interest by presenting material in a dull, uninspiring way risks losing students that departments might prefer to retain, including high-performing and multi-talented students who are disappointed by under-stimulating teaching and choose alternative majors—whether within or beyond the sciences—that better capture and sustain their interest.

Some persisters also blamed their departments and universities for not caring about teaching skills enough to require, or support, education in teaching methods for instructors or graduate student teachers. One mathematics senior explained his view of the system and why there is so much poor delivery in his math courses in these terms:

> They don't hire math *teachers* … they don't hire people with teaching degrees or teaching interest at all to actually teach those courses. … Calculus 1, 2 and 3 are taught by grad students. … They will not pay people to teach those courses here, and that causes the product to be very poor and even if you have somebody who understands the topic well, it doesn't make them teach it well. (Multi-racial man, relocator, engineering to hydrogeology)

There were indications scattered throughout the interview and SALG comments data that some instructors were aware of more active and interactive teaching methods and were trying them out. However, 4% of SALG negative comments on teaching methods described innovations that were counterproductive because instructors had not learned how to deploy them properly:

> The class was a lot of discussion with other students (who also may not know the answer) so it felt like a lot of pointless back and forth. The excess amount of discussion took up valuable time. (Asian woman, biology major)

> This class was the most confusing class I have ever taken. I despised the group learning. I hated it when, instead of explaining concepts succinctly and clearly, the professor assigned us the task of arguing among ourselves. This was counter-productive. Not only were we not given the correct answer, we convinced ourselves of the wrong answer. I would often, if not always, leave class frustrated and confused. (Asian woman, biology major)

In summary, both switchers and persisting seniors described and illustrated what they meant by "poor" delivery of course material. Traditional lecturing with one-way transmission was reported to be common in foundational courses but was still evident in some senior-level courses. The method was faulted for:

- dull, disengaged, spiritless presentation;
- over-dependence on PowerPoint slides whose content was often read without amplification;
- little opportunity for questions or discussion,
- lack of applications, examples, and sample problems;
- failure to provide context for conceptual and theoretical material by connecting it to other bodies of knowledge or real-world phenomena.

Above all, students lamented that many instructors showed little or no responsibility for stimulating and sustaining student interest and lacked professional skills to deliver their content to best effect. As students in the original study explained, there is no inherently dull material—only dull teaching (Seymour and Hewitt 1997 Chap. 3, pp. 151–156).

## *Disengaged Teaching*

Often mentioned alongside poor delivery, 30% of switchers and 22% of persisters expressed disappointment with STEM instructors who showed little engagement with either the course material or the learners. Sadly, there were substantially more descriptions of unenthusiastic and disengaged STEM teaching than there were of passionate and engaged teaching. Students described teachers who seemed "checked out" during their delivery and indifferent to whether or not students learned. This apparent disengagement demotivated students who wanted to be stimulated by their teachers' passion for their discipline and encouraged by their teachers' interest in getting them to share it. As noted above, it was apparent both to switchers and persisters that many STEM instructors were uninterested in teaching or in motivating students to work hard and learn. Switchers described the dampening effects thus:

> I was less encouraged to do well when they didn't even care and weren't interested in their work or what they taught. That was a big thing. It was almost as if, because they were teaching just basic stuff, that they were bored with it. They didn't really have an interest. (Asian woman, switcher, biology sciences to strategic communication).

> None of the teachers seemed like they cared if you succeeded or failed. They had this mindset that if you drop out, you drop out—one less doctor in the world. And they taught their class in such a manner that it was like almost a bare minimum. 'This is what you need to know.' (Male, white, switcher, microbiology to political science).

Persisters also expressed annoyance with apathetic STEM instructors:

> Of all the classes that I really disliked, the one thing that links them all is that you can tell that the professors didn't really enjoy what they were doing. (Hispanic man, biological sciences persister)

> The teacher may know the subject, but if they're not motivated to teach, that's a big determining factor for my motivation in the class. When they're more motivated, it sparks [students'] interest. (White man, zoology persister).

When teachers portray indifference, students feel less motivated to attend class and work hard. Students reported feeling deprived in their learning, frustrated at wasted time, and resentful for having to pay for courses where the teachers were disengaged. Their indications of half-heated teaching included some of the criticisms already discussed—one-way lecturing without discussion or interaction, delivery in a monotone, frequently getting off-topic, and coming to class unprepared:

> He would just talk for the whole hour and 50 minutes without engaging with the class. And I've taken very little from the course because of it. (White man, switcher, environmental science to marketing).

> I feel that if I'm paying a certain amount of money to go to school, I should be able to take a class with a professor who is willing to teach and not just read off of Power Points … I want them to be more dynamic … It's pretty important to me, I haven't met too many professors here who have inspired me to work hard. (Asian woman, switcher, biological sciences to finance).

> [I thought] he should be fired because of his lack of interest and inability to teach. At one point he left our tests in his refrigerator. So he left class for two hours to go get them. We were sitting in class waiting for him to return and when he came back he said, 'Well you guys are gonna have to stay for lunch 'cause I just got them back.' (White woman, biological sciences persister)

As we discuss below, students commonly explained STEM faculty's lack of engagement in their teaching or with their students in terms of their research priorities.

## What's the Problem with "Straight Lecturing"?

Several attributes that students identified as poor quality teaching were characteristic of lectures that were devoid of interaction. As established at the outset of this chapter, "straight lecturing" remained the predominant method of teaching evident in this study, especially in STEM courses for under-classmen. Switchers were more negatively affected than persisters by lecturing that was unrelieved by interaction (i.e., 41% in contrast with 13% for persisters). However, positive assessments of traditional lectures were offered by only 6% of switchers and 4% of persisters. Some switchers expressed disappointment with the lack of interaction that was common in university science and math courses. They contrasted this with high school teaching that they described as interactive, engaging, and as having sparked their interest in the sciences. For these students, lecture-only STEM courses in their early semesters were shocking and could significantly undermine their incoming interest and deplete their motivation. Lack of interaction was one of several indicators reported here from which students inferred that their instructors took little interest in whether or how they learned:

> There wasn't any interaction. The professor just walks in, comes to the front, tells us all these things. We could not come to class and they would still keep going. And then they leave, and we leave. (White man, switcher, biochemistry to visual and performing arts)

> The [science] teachers just weren't interactive. They didn't ask, 'Do you understand this?' Based on that, I didn't take the material as seriously, because I don't really care about your class if you don't take any interest in what I'm trying to learn. (Pacific Islander woman, switcher, biology to strategic communication)

As the last speaker notes, another indicator to students of instructor indifference to their learning was the common failure of lecturers to pause and check (for example, with a clicker question or a Q&A break) the degree to which students were following their line of thought and understanding the concepts being laid down.

Some switchers migrated to non-STEM majors, in part, because they offered more interactive and engaging learning experiences. As reported in the iceberg table (Chap. 3) 58% of switchers preferred the more interactive teaching approaches used in non-STEM courses, and 29% cited this as a major factor in their switching decision. As one African-American man who switched from physics to cultural studies and literature explained, "Part of the reason why I left STEM is that I like discussion way more and that's just something you don't get." His rationale is also articulated by these two women switchers:

> I enjoy the ability to discuss things with my fellow students and my professors and I feel like that was something that was missing from my science classes. I wished there was more discussion in the classroom. (White woman, switcher, biological sciences to European studies)

> I took 'Thinking Sociologically' and absolutely loved it. I found it to be much more interesting and to be conversation-based instead of just being lectured at. I ended up taking more sociology classes, and I haven't gone back to the sciences since. (White woman, switcher, neuroscience to sociology)

Persisters also expressed frustration about lack of interaction in lower-division STEM courses and reported that they did not experience interactive teaching until upper-division courses. As one survivor of gateway course lecturing explained, the challenge was to make an essentially passive experience active and personal:

> I don't learn through straight lectures. It just washes over my head. I have to read it myself and learn it alone. Lecture just doesn't work for me, which is bad because all of the science

classes here are lecture. I always make sure that, when the teacher is lecturing, I am reading the text at the same time. That's the way I learn. (Hispanic woman, biological sciences persister)

This student's problem and her solution reflect Akiki's (2014) findings that the straight lecture method induces passive learning that does not allow students to engage with the subject material in depth. A comprehensive review of studies by the National Research Council (2012) concludes, from evidence, that traditional lectures are less effective than the available array of research-based instructional strategies (RBIS) in building conceptual understanding. The report cites interactive lecturing, the incorporation of active, hands-on activities, including small group work, the use of authentic problems and "appropriately sequenced" experiences such as research, as all out-performing "straight lecturing" in enabling a solid grasp of core concepts and their appropriate application. Reports by the National Science Foundation, American Association for the Advancement of Science, the National Academy of Sciences, and many private foundations stress the value of evidence-based interactive classroom strategies in engaging students in their own in-depth learning (cf., Ejiwale, 2012; Mulnix & Vandergrift, 2014; Seymour, 2007). To which Beichner et al., 2007, and others, add the positive effects of active and interactive learning methods in reducing or eliminating achievement gaps between black and white students and first-generation and other students.

## Compensatory Learning

We examined students' numeric SALG ratings of how much they gained from aspects of the class alongside their written comments and found that students appear to compensate for what they define as poor class instruction by turning to other resources in the same course. Students reported, for example, that clickers, labs, recitations, textbooks, study groups, and other teaching resources, "helped them learn." However, these positive ratings were often made in contrast to poor overall ratings of their learning gains from the lecture along with critical written comments about the instructor.

This pattern of compensatory learning can be clearly seen in the ratings of class activities by students who gave the lowest rating (*"no gain"*) to the question: *How much did the following parts of the class help your learning: the lecture?* Many of these students who also wrote strongly negative comments about their instructor's teaching methods, gave positive ratings to other class resources such as the teaching assistants and study groups (Fig. 6.2).

Students commented that they learned from resources such as clicker questions, homework, and online resources, but added that they turned to these resources because they did not learn from the lecture or from another part of the course:

I did not like how the class was taught. Physics is already a hard topic, there is no need to make it boring too. I would have preferred more enthusiasm from my professor about the subject. The class was very fast-paced and boring. There were lots of example problems but not a lot of depth went into explaining why the subjects mattered in a larger context.

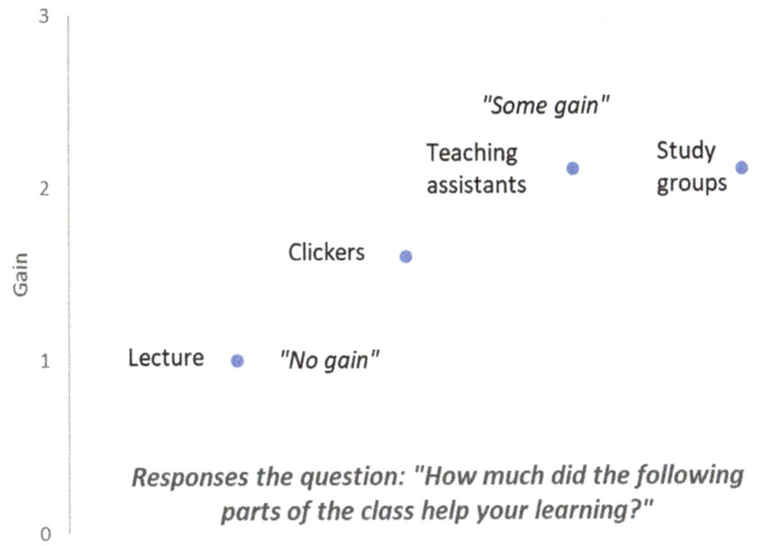


**Fig. 6.2** Compensatory learning patterns for students who recorded "no learning gain" from their lecture

> However, clicker questions and in-class worksheets definitely helped my learning of the material. They helped us interact as a class and learn by peer help. (White man, physics major)

> The lecture slides were very frustrating to follow in class because of the amount of information on them. It was difficult to both listen to [the teacher] and read the slides. However, since I ignored the slides in class, I looked over them when studying for exams and they were very helpful. I think it was great to have unlimited tries on the Smartwork [online system]. It helped me to actually work through the problem instead of getting frustrated and giving up if I lost all the credit. (Asian-American woman, cell biology major)

> The professor was late every day, gave scattered explanations, and didn't give homework or anything to reinforce the content. He didn't post lectures either. But I did learn from the labs and studying with peers. (White man, electrical engineering major)

This pattern was often evident among students who wrote that they learned independently from the textbook:

> This class did not help my learning. I taught myself all semester from the book. There were no class activities that I found helpful. When I did the simulations on my own that was helpful, just not in class. Reading the book helped a lot. (White woman, integrative physiology)

Some students wrote that they learned more from online homework systems (e.g., ALEKs) than from the teachers:

> This class was unnecessarily antagonistic. The attitude of the instructor was not one of helpfulness but was hostile to the students. I was interested in chemistry at the beginning of the semester, now I am completely turned off by the subject. I purchased ALEKS online learning software for chemistry at the same level as this class and used it to help me drill with memorization: names of ions, naming conventions, etc. (White woman, biology major)

And some students felt they received the most help from the teaching assistant or at a learning center:

> This professor is seriously the worst that I have ever had in my life. Clickers are expensive and useless I couldn't think of a way to waste more money except maybe the useless textbook. The TA was the best part of the class. I loved discussion and working through problems with classmates and having my TA help. The TA was great! (White male, biology major)

Finally, some students gave up on their instructor altogether and hired their own. They explained that, as they did not benefit from any part of the course, they were learning effectively with the help of a private tutor:

> The professor was always sarcastic and was not patient with people's questions or would simply say she couldn't answer it. I don't think this class helped my learning at all. I had a tutor that helped a lot to talk about things and work through things that I wasn't getting taught in lecture. (White woman, psychology major)

Later in this chapter, we discuss students' negative experiences with graduate students in teaching and learning support roles, and in Chap. 8 we report their positive experiences with TAs. Also, in Chap. 8, we expand upon the various ways in which students resolve their learning difficulties by seeking available sources of help. Finally, in Chap. 12, we report on the significance for persistence of seeking and finding appropriate and timely academic help and support.

## *Students Conclude that Faculty Value Research More than Teaching*

Twenty-one percent of switchers and 22% of persisters expressed frustrations about STEM teaching being treated as a secondary concern with faculty giving most of their time and energy to their research:

> Most of the tenured professors in the sciences are here for research. They don't care about teaching. My final chemistry class was taught by a guy who I just could not stand. He couldn't teach and he was just excited to get back to his laboratory. (White woman, switcher, biochemistry to film studies).

> I feel like a lot of the [science] professors tend to concentrate on research, and they treat teaching as a secondary obligation. I felt they didn't try to make any attachment to students. (White man, switcher, computer science to economics)

> I feel like a big issue is that some of the [STEM] professors aren't interested in teaching, they just teach so they can keep their research positions. The professors I had that weren't that great. (Female, white, switcher, engineering to psychology)

> They just wanna go and do their research. They couldn't care less whether a student's having trouble understanding a problem. (White man, relocator, chemistry to chemical engineering)

> Who knows if he cares about educating or whether he's just in it for the research. I find a lot of professors [in STEM] don't really want to be a part of the education side of the university. (White man, physical sciences persister)

Some STEM faculty were quite open with students about their research priorities:

> One of my friends interviewed our calculus teacher, 'So why did you decide you want to teach here at the university?' And he said, 'Well, I had to teach if I was going to be able to do research here.' (White man, switcher, biological sciences to dance)

> I've definitely had teachers [in life sciences] say things like, 'Well, I just really wanna do my research. This isn't my main priority. I'm just here 'cause I have to be here.' (White woman, switcher, life sciences to sociology.)

Some students were aware of the importance of research funding to their departments and university, and also that certain faculty brought in more grant money than others. They surmised that universities rewarded research excellence more than good teaching, which helped them make sense of why so many faculty lacked good teaching skills:

> The physics class is taught by a guy who brings in all the research money for the physics department here and he is very highly regarded, but he is the worst professor I've ever had. (White man, switcher, engineering to construction management)

> Most of the professors make the majority of their salary through the research that they do. So a lot of them don't know how to interact with students. (Multi-racial man, zoology persister)

> Clearly, they were just being paid to do research and had to teach a class… Their teaching wasn't the greatest … They would be like: 'Here's the PowerPoint. This is the stuff. Don't talk to me after class. I'm busy.' (Multi-racial man, biomedical sciences persister)

Students did not necessarily understand how research funding is deployed by their university. However, they recognized its high value, and inferred from their learning experiences that investment in research must be greater than investment in improving the quality of teaching. They were also convinced that STEM departments select new faculty largely by their research and fund-raising pedigree rather than their teaching ability. And they suspected that STEM departments were not invested in helping current or future faculty learn the skills of effective teaching:

> I think they're looking for people that do well in research. Whether they can communicate the information is kind of a secondary thing to the universities. (White man, mechanical engineering persister)

> I think that the chem program here … It's not student-oriented. It's very research-oriented. … The chem program as a whole just isn't interested in the students. (White man, switcher, chemistry to theater).

> I wasn't expecting colleges to hire professors based on their research abilities and not necessarily because they're good teachers … That was something I was unprepared for, I guess. (Female, white, persister, chemical engineering)

> The math department in this school is a bit shaky; the chemistry department is a bit shaky as well. And I think it's because they don't hire math *teachers* … especially they don't hire people with teaching degrees or teaching interest at all to actually put on those courses. (Male, multi-racial, persister, hydro-geology)

Many students were convinced that STEM faculty simply do not enjoy teaching and are not interested in learning to teach effectively. They surmised that STEM faculty enter professorships to pursue their research, and never intend to put much effort into their teaching:

> Professors are here to do research but the only way they can do it is to teach. They seem almost annoyed, like, 'I have better things to do,' when you go and talk to them.... When they're there, you feel like they're not giving it everything they have. That they're thinking about other stuff. (White woman, biological sciences persister)

> I would love for [my STEM teachers] to care. It just feels like they don't. I've heard it from other people, and that's how I feel myself... A lot of them are here because they are teaching to do research, and so, 'I'll teach you on the side, but I really don't know how to teach, and I don't really enjoy teaching.' (Native American woman, physiology persister)

> The professors love their research, which is great, but then they have to teach the lower level classes and some like teaching some don't. Some just do it 'cuz they have to. The professor doesn't value what they're doing in the classroom because they value what they're doing for fun. (Asian-Pacific Islander man, mechanical engineering persister)

> I've had instructors that didn't take an interest in their students...They're more interested in their research and wanna work on that rather than teach the one class they got stuck with that semester. (White woman, chemical engineering persister)

Although both switchers and persisters were frustrated with STEM faculty's preoccupation with research and resulting lack of time and skill dedicated to helping them learn, the consequences were especially strong for students for whom this was a major reason to switch or relocate:

> I came in with biomedical sciences and I think that the big reason I shifted out of it was I had a lot of frustration with the tendency of professors to be more focused on their research than what they were teaching. Specifically, in introductory chemistry, I went and talked to a professor and felt I was wasting his time. He was more intent on what he was doing outside of class. I was very frustrated because, with the amount of money that I'm paying for college, I wanted a professor who's going to be present. (White man, switcher, biomedical sciences to marketing)

> The physics department here sucks. There are a few teachers who are very good, but there are some major duds. Like this one lady I took it with, she does very well at getting grants and she's the best researcher in the department, which is why she has tenure, but she puts zero energy into her classes. I remember I had missed a day 'cause I was sick and I go in the day before our final to ask her for help on something and I was, like, 'Can you go over this with me?' And she said, 'No.' And she said I had to do the test. I was, like, 'Are you kidding me?' I think if I had felt like I was supported by the teachers then I would have stayed. (White man, relocator, physics to computer science)

As we have illustrated, students widely interpret faculty's obligations to their research as a primary, sometimes only, commitment to the department, with nothing to offer to student learners. This conclusion is enough to prompt some students to switch or relocate.

In summary, almost all switchers (96%) said that poor quality teaching had negatively affected them, and almost half (48%) identified poor teaching as a leading factor in their decision to leave STEM. Most persisters (72%) also reported that poor STEM teaching had caused them difficulties. Some had persisted by finding ways to learn notwithstanding their instructors' limitations or had side-stepped these deficiencies by relocating to other STEM majors where they found the teaching to be better. However, we offer as a concern that, for 90% of students of color who persisted in STEM majors, poor quality teaching was cited as a constant problem throughout their academic career. It also remained a concern for 58% of white persisters.

## Students' Experiences with Graduate Students in Teaching and Learning Support Roles

Graduate teaching assistants play a significant role in undergraduate students' learning, especially in introductory courses. At research universities, undergraduate teaching in STEM disciplines often rests largely on the backs of graduate teaching assistants (TAs) who teach laboratory courses, recitations, and sometimes components of larger lecture sections (Dotger, 2010; Gardner & Jones, 2011; Kurdziel & Libarkin, 2003). Indeed, a recent survey of the biological sciences reports that graduate teaching assistants were responsible for teaching 91% of undergraduate laboratory sections in research institutions in the USA (Sundberg et al., 2005). They also tend to have more contact with STEM undergraduates, especially in first-year courses than do faculty (Gardner & Jones, op.cit., Kendall & Schussler, 2012; Zehnder, 2016). Given the importance of first-year coursework on student retention in STEM disciplines, teaching assistants must be considered key contributors to the quality and outcomes of STEM undergraduate learning (Benjamin, 2002; Dotger, 2010; Kendall & Schussler, 2012).

In the original study, 15.5% of the students interviewed complained about their graduate teaching assistants' instruction, which included comments about poor pedagogical skills, indifference towards teaching, and insufficient familiarity with the course material (Seymour & Hewitt, 1997). However, none of the students attributed the poor teaching they received from their TAs as having contributed to their decision to leave STEM. Dissatisfaction with language-related problems of foreign TAs was a bigger concern for 29.5% of switchers and 20.4% of persisters. But again, the effect of both foreign teaching assistants and faculty on students' decisions to switch out of STEM was negligible (3.3%). Overall, students did not place primary responsibility for their learning difficulties on their TAs. Indeed, many students credited their TAs with a higher level of interest in teaching than their instructors as evidenced by their TAs being readily available for questions, providing helpful alternative ways to approach problems, explaining material in ways that they could understand, and strengthening their confidence and enthusiasm in the discipline.

In the current study, nearly half of all switchers (47%) reported negative learning experiences with their teaching assistants while only 10% of persisters complained of difficulties. However, similar to the original study results, the effect of negative TA teaching experiences on students' reasons for leaving STEM was negligible (viz., 6%). By contrast, 53% of switchers and 33% of persisters cited good quality instruction and learning support from their teaching assistants. That said, students had much to say about their experiences with teaching assistants—both positive and negative—and it was clear that TAs played a significant role in their learning. Tables 6.6, 6.7, and 6.8 provide an overview of students' negative and positive experiences with TAs by gender, institution, and low- and high-math status. We found no significant differences in race or ethnicity.

Considerably more female than male switchers reported positive experiences with their teaching assistants (66% compared with only 31%, respectively). Several explanations for this are suggested by our analyses. Women more readily than men turn to TAs for learning help and support. They also feel less intimidated by teaching assistants than by faculty, and can more easily relate to their TAs, especially where they are also women. Women of color especially find it easier to approach a TA than a male instructor. One Asian woman explains how having a female TA in a course with a male professor was both comforting and encouraging:

> What really helped me in one of my classes—optical neurology—was that our TA was a woman. I really liked her and it comforted me that she was doing this. So I would go and

**Table 6.6** Students' experiences with teaching assistants by gender

| Gender | Negative TA experiences (%) | Positive TA experiences (%) |
|---|---|---|
| Switchers male | 49 | 31 |
| Switchers female | 46 | 66 |
| Persisters male | 8 | 37 |
| Persisters female | 12 | 30 |

**Table 6.7** Students' experiences with teaching assistants based on math readiness

| Type of student | Negative TA experiences (%) | Positive TA experiences (%) |
|---|---|---|
| Persisters, HIGH math | 11 | 35 |
| Persisters, LOW math | 17 | 55 |
| Switchers, HIGH math | 47 | 49 |
| Switchers, LOW math | 50 | 68 |

**Table 6.8** Students' experiences with teaching assistants by institution

| Institution | Negative TA experiences (%) | Positive TA experiences |
|---|---|---|
| PV2R1 persisters | 0 | 7 |
| PV2R1 switchers | 0 | 60 |
| PB1R1 persisters | 13 | 38 |
| PB1R1 switchers | 16 | 42 |
| PB4R1 persisters | 4 | 32 |
| PB4R1 switchers | 70 | 60 |
| PV1R3 persisters | 8 | 35 |
| PV1R3 switchers | 18 | 55 |
| PB2R1 persisters | 13 | 49 |
| PB2R1 switchers | 64 | 55 |
| PB3R1 persisters | 20 | 34 |
| PB3R1 switchers | 63 | 53 |

talk to her sometimes. TAs that are women—that's encouraging, you know? (Asian woman, switcher, geosciences to business)

Given the role of TAs in helping students who enter college with poorer preparation in math or science to catch up with their better-prepared peers, we checked to see how students with higher and lower math scores on college entry assessed their learning experiences with TAs. As is evident in Table 6.7, the significance of TA assistance to students who enter with low-math SAT/or ACT scores is confirmed. Low-math switchers reported more positive experiences with teaching assistants than their high-math peers (68% to 49%, respectively). Similarly, low-math persisters also reported more positive experiences with teaching assistants than their high-math peers (55% and 35%, respectively). Low-math students used the services of their TAs more; they also made more comments on how much more comfortable they felt in asking questions of their TAs versus their professors, and that TAs were generally more approachable than faculty.

We were also interested to learn whether STEM students' assessments of their TAs varied between the six institutions in our sample, and indeed, they did. As indicated in Table 6.7, negative TA experiences ranged from no complaints about TA teaching and support by either switchers or persisters at one institution to 70% of switchers registering concerns at another. Over half of STEM switchers expressed frustrations with teaching assistants at three of the universities in our sample. However, less than 20% of students—regardless of switching status—expressed concerns with TAs at the other three universities in our sample. Positive TA experiences ranged from 7% of persisters at one university to 49% at another; and from 42% of switchers at one university to 60% at two other universities. Notwithstanding institutional variations, at all six institutions, switchers had more to say about their experiences with teaching assistants—both positive and negative—than did persisters. This, in itself, is one indicator of how important TAs are to students who are struggling to survive and thrive in STEM majors. Details about their negative experiences are described further in this chapter and in Chap. 8 for positive accounts.

For this study, we did not collect data that would enable us to explain why institutions varied in the balance of positive to negative appraisals of TA teaching and support. Drawing on our findings in an earlier study of TAs (Seymour, Melton, Pedersen-Gallegos, & Wiese, 2005), we posit that institutional and departmental variations in the quality of TA education for their teaching roles is a key explanatory factor. Where there is limited or no professional development in teaching methods, graduate students take their cues about how to teach and what attitudes to take towards both teaching and undergraduates from those faculty whose courses they serve and from their research advisers. The extent to which instructors do or do not model research-based instructional strategies for their TAs thereby shapes the quality of TA pedagogical practices and attitudes.

## *Negative Experiences with Teaching Assistants*

### Language Problems with Teaching Assistants

The most common complaint about teaching assistants was that students had a difficult time understanding foreign TAs whose first language was not English. One switcher reported that being unable to understand the TAs teaching the calculus sequence played a key role in his decision to shift from microbiology to political science:

> This is going to sound really, really horrible, but my TAs in my calculus classes—English was not their first language—and the language barrier was one of the top three things that turned me off [of pursuing microbiology]. It got worse as I went up [in calculus courses] and I hated it because they couldn't explain things to me in the way I wanted them to, and I didn't need someone stumbling over their words when they were trying to explain. (White man, switcher, high math, microbiology to political science)

This student entered university with high-math scores and had taken AP Calculus in high school, but his difficulties in getting clear explanations from his TA calculus teachers proved a significant barrier to persistence. He was, however, the only student who attributed part of his decision to switch majors to this issue.

Many other students, however, expressed frustrations about their difficulties in understanding their non-native-English speaking graduate teaching assistants. Although these TAs spoke English, they might not speak it fluently enough to explain complex and technical material:

> I'm not trying to be racist, but it was an Asian man who was the TA, and I could not understand a single word he said … he was very smart, but when he tried to explain it to us in his own words … because of the language barrier, it did not translate at all. (White man, switcher, physics to theater)

> My chemistry TA isn't very helpful. He's not very good at English and so it's harder for him to say his ideas to us and explain what he wants to tell us, and it's hard for us to understand what he's talking about. (African-American man, switcher, physiology to psychology)

> My TA was from Russia and didn't speak very much English … He would explain it to you in one way and if you didn't understand it, then he'd just shrug and walk away. (White woman, switcher, civil engineering to geography)

> One TA I had was from China. Sometimes he would say words, but we wouldn't quite understand what he was meaning. And that made it hard to learn. But also, I think he had a language barrier between him and the professor. The professor would tell him when things are due or how to do things, but he didn't quite understand, so then when he relayed it to us, we didn't understand either. (White woman, physiology persister)

Foreign accents (as opposed to English deficiencies) also frustrated some students. Several complained that it was difficult to understand English spoken with an accent and this made them disinclined to attend class:

> My TA for Calculus 2 … She had a really strong accent and was kind of monotone, so I would either not be able to understand her or I would tune out or would just not show up. (White woman, geosciences persister)

> A few of the TAs that I had—they had extremely heavy [foreign] accents and they didn't bother repeating stuff when we told them we couldn't understand them. (White woman, switcher, engineering to psychology)

> In my physics lab, I had a TA that I just couldn't understand. He was Persian, I think. And he had a thick accent and wasn't incredibly helpful. (White man, switcher, engineering to construction management)

These students struggled to understand TAs from other countries because, as well as strong accents and unfamiliar pronunciation, translations of complex material into English from another language could be confusing or incomplete.

How widespread is this problem? About two-thirds of foreign students who study at universities in the USA major in STEM disciplines (Ruiz, 2014), and over half of the graduate students who major in engineering, biology, and physics are foreign-born (CGS, 2006). A tradition of dependence on graduate students to teach many introductory STEM courses means that many graduate teaching assistants are non-native English speakers. However, many others are native English speakers from countries such as India, Pakistan, and African and Caribbean nations where the problem raised by US students is not one of command or fluency of English but of English spoken with unfamiliar inflections. This sensitive subject has promoted support for measures to prevent foreign-born instructors from teaching unless they can do so in clear English. For example, a 2005 Chronicle of Higher Education article cites a bill by a North Dakota Representative to prevent instructors who do not speak English clearly from being assigned instructional tasks (Gravois, 2005). Several other reports also document problems with non-native graduate students who speak poor English (Chiang, 2009; Pickering, 2004; Williams, 1992). However, other studies report that students' attitudes toward teachers who are non-native English speakers influence the success of communication between the two (Kang, Rubin, and Lindemann, 2015; Travers, 1989). They also cite the success of English language programs for international teaching assistants (Chiang, 2009; Travers, 1989). In the original study, as here, we reported that, although difficulties in understanding English spoken with a foreign accent was a fairly common complaint, it was not a contributor to STEM switching. We also cited students' testimony that a TA's willingness to spend time helping them through conceptual difficulties was more important than any initial problems with communication. As in the present study, students' dominant concerns focused not on language communication issues but on the experience of ineffective teaching whether the instructor was a TA or a member of the faculty.

## *Poor Quality Teaching from Teaching Assistants*

Another common student complaint was that some of their graduate teaching assistants had insufficient disciplinary knowledge to be effective as instructors. Undergraduates expected TAs to be knowledgeable in the subjects for which they were instructors, recitation leaders, or lab assistants. They expressed surprise and irritation about teaching assistants who had insufficient knowledge of the discipline to answer their questions or provide the specific kinds of support that they needed:

> I'd show the TAs the standard problem … and the TAs really didn't have an idea of what the professor was thinking. These problems were difficult for the TAs … and they can't do them on their own. (Hispanic man, switcher, mechanical engineering to religious studies)

> The TAs were very unhelpful. In math, the TAs would sometimes not fully understand the concepts we were doing. (White man, switcher, physics to theater)

> I'll ask the TA if my answer is right, and he'll be like, 'I don't know, I have to check on that.' And I kinda find that unacceptable, like, why are you the TA for this course if you're not sure? (Asian woman, biological engineering persister)

> The teaching assistants are not as well prepared as they need to be, and so a lot of times people just don't go [to class] because they don't feel like it's helpful because the teaching assistants don't understand the material well enough. It would be a lot better if the teaching assistants were held to a higher standard. (Asian man, materials science and engineering persister)

> You would ask them a question, and they wouldn't know the answer … I mean they took the class like a year ago or something, so it's like they probably just don't remember the material. (Hispanic woman, switcher, physiology to psychology)

Poor teaching skills were also a frequent complaint. Students often felt that they had to teach themselves because the graduate students struggled to instruct effectively:

> My statistics class was taught by a grad student … I found it very hard to learn with her teaching because she wasn't able to explain anything really … I had to teach myself … (White woman, equine science persister)

> He was a grad student … He just didn't put much effort into the class and would halfway go through things and not really describe them, and then he didn't do examples. (White woman, relocator, mechanical engineering to actuarial science)

> I was taking mechanics of solids. And the TA would just face the board and talk, make his diagrams too small, so no one beyond the first three rows could see anything. (White man, mechanical engineering persister)

As the last two examples illustrate, the TAs' poor teaching methods mirrored those of instructors described earlier. Perhaps more disturbing was that many of the students reported their TAs barely did anything at all. Students were frustrated with the lack of instruction and support, and found a few TAs to be simply lazy:

> We get to lab and he'd pretty much just says, 'Alright, you guys can start your labs. Next week, we have this due. Make sure you bring that to class.' And that's it. … Not even kidding. Doesn't explain anything that we're doing. … He starts out by walking around and checking our notebooks to see if we did the warm-up problems, and then he'll check off our name for attendance. (White man, switcher, physiology to architecture)

> I remember some of the TAs would not teach the students. They would say, 'Okay, well here's a quiz, and here are the answers to the quiz before the quiz.' And then the students would all get tens out of tens, and then they would go into the test and they would wonder, 'Why don't I understand this material?' … How irresponsible as a teaching assistant, where you are being paid to ensure the proficiency of learning for other students. – (Hispanic woman, switcher, biological sciences to journalism)

> My Chem TA—she was so lazy. … She would just plop herself on the counter at the front of the room. … One time I was looking at something under a UV light and I had to ask her a question and she was like, 'I'm not going to move. You're going to need to bring the UV light over here.' I had to unplug the stuff to bring it to her. And I was like, 'You're a horrible TA.' (White woman, chemical engineering persister)

Again, students found the same indifferent attitude toward teaching among graduate teaching assistants that they found with some faculty. They described TAs who were not invested in their students' learning, were not easy to approach, did not care about teaching, were largely focused on their own studies and research, and seemed deliberately unavailable:

> I did not like my TA at all in calculus. I felt like he did not care at all what he was doing and he didn't show up two weeks in a row to our 8:00am recitation. (White woman, switcher, chemical engineering to accounting)

> I get a new TA every semester and they weren't really that invested in us. 'Cause they also have their own schooling to do. (White woman, switcher, physics to social sciences)

> Most of my TAs—they don't wanna teach. They're not very personable. They're angry. (White woman, switcher, life sciences to film studies)

> It felt like they were doing the TA job because they were working on their PhD in engineering. They weren't there because they enjoyed doing it. It felt a little closed off, it felt like they were too busy and it would just be a hassle [to talk to them]. (White man, switcher, engineering to business)

> I felt like most of the TAs in engineering were just there because they have to do it, but they're not really inclined to teach. (Asian woman, biomedical engineering persister)

> The TAs were non-existent. I couldn't get an email response back from them. … I struggled to make a connection with the material because I struggled to make a connection with the people I think. (White woman, switcher, biochemistry to film studies).

A few graduate assistants had especially cold attitudes in that they actively discouraged questions and dismissed students' requests for extra help:

> My sassy TA for [molecular biology] he would definitely be classified as rude. But that was sort of how he taught. … I mean, if someone asked a 'dumb' question, he would call it out, and be like, 'That's not a very good question. Did you do the reading?' And would belittle that student in front of everyone. (White woman, switcher, biological sciences to psychology)

> I was very frustrated with my physics TA. I presented him with the letter from Disability Services and said I'm sorry, I've been having a lot of difficulty in my personal life and I will try to get things in on time, but I may not always be able to, so let me know what the options are. And he was kind of dismissive of the problem. He mentioned that he felt like medication for depression wasn't really a valid option. He didn't feel like it should be interfering with my school work and he kind of wasn't going to cut me any slack. (Multiracial woman, switcher, microbiology to psychology)

In an earlier study (Seymour & Hunter, 1998), we found a tendency by instructors to conduct a "lay diagnosis" of students who presented them with a formal request from the Disability Services Office for a "reasonable accommodation" to their condition, as provided in the Americans with Disabilities Act, 1990.

A different form of discrimination arose where male TAs undermined female students' confidence by treating women unfairly:

> I was not comfortable talking to my physics TA. I think that giving equal treatment to men and women – well, I didn't feel that my physics TA did that, and I felt very discouraged. And I was afraid to talk to him. And I was afraid to talk to my teacher even more. (White woman, switcher, chemistry to social science)

In summary, the undergraduate learners in our study were dissatisfied with the teaching they received from graduate teaching assistants when they lacked knowledge in the discipline, had poor pedagogical skills, seemed disinterested in teaching or in their students, or were unavailable, dismissive, or intimidating. We again observe that the teaching attitudes, behaviors, and methods of graduate student teaching assistants described here reflect those reported among some instructors. As also found in our earlier study of TAs, where no program of professional teaching for graduate students in place, the tendency is for TAs simply to replicate what it modeled and legitimated for them by STEM faculty who have significant influence on their careers (cf., Seymour et al., 2005). This informal, but very powerful, form of professional socialization ensures the perpetuation of poor quality teaching into the next generation of young faculty.

## *Negative Interactions with Teaching Assistants Outside of the Classroom*

Some students, especially switchers, were frustrated by the inadequate help they received from teaching assistants in "help rooms" and TA's office hours. By contrast, as will be discussed in Chap. 8, there were an almost equal number of positive comments offered by switchers about the help from TAs that they received outside of class. Persisters also offered more favorable assessments of TA help both inside and outside of the classroom.

The under-staffed and over-crowded conditions under which TAs operated were, however, a source of concern. Switchers lamented that help rooms were often very crowded and the ratio of teaching assistants to students needing support was inadequate:

> When I went to the chemistry help room hours it was always super packed in there. And there would be like one TA in there, and we're all trying to get help from him. It was just frustrating to try and get help. (White woman, switcher, biology to psychology)

> The physics help room—there would be like one TA – so even if there were three or four people in there, it wasn't helpful. You'd wait like 20, 30 minutes to get one question answered, and you'd get a minute of the TA's time. So that was really frustrating. (White woman, switcher, civil engineering to geography)

> I went to the physics help room every week, but there would be ten other people there at the same time and two TAs. So it was frustrating. (White woman, switcher, chemistry to social science)

As illustrated, students were irritated with having to wait a long time to get their questions answered in help rooms. One student who switched out of physics was especially exasperated with the crowded help rooms and gave up:

> They didn't have enough TAs—they just didn't budget for it or something. So I'd get there and there would be like 15 people there and I'd be like, 'Well, my question is never going to get answered.' So I'd just leave. (African-American man, switcher, physics to cultural studies)

A few switchers also complained that the assistance they received from TAs in the help rooms often created more confusion rather than clarity:

> There's a room where the tutors sit all day ... The tutors can only do so much: you're sitting there, with other people that are confused, and then you're all talking it out, and then you're confusing each other more, and then the tutors do the same thing where they don't wanna spoon-feed you ... And you're like, 'No, seriously. We don't get it. Can you teach it to us?' But they don't and so we're all confusing each other more. (White woman, switcher, biology to Spanish studies)

> It was frustrating because I just absolutely could not get certain aspects and they wouldn't teach it to me. They were trying to do the 'lead to the answer' sort of thing, and I just needed them to re-teach it to me. They didn't have time to do that. (White woman, switcher, chemistry to social science)

In conclusion, students' most common complaints about their graduate teaching assistants were language problems with foreign TAs, insufficient knowledge of the subject, under-preparation for sessions, lack of effective teaching skills, laziness, and uncaring or, indifferent attitudes towards students and their learning. Switchers were especially affected by the poor quality of support and long wait times provided in TA-lead help rooms, some of which was created by over-crowded facilities and poor TA-to-student ratios.

## Placing Student Appraisals of STEM Teaching in Context

As explained at the outset, the student sample from which the interview data analysis was drawn was structured in the same manner as for the original study 20 years earlier and was located at six of the seven original institutions. Both then and now, switchers reported on teaching in largely foundational STEM courses during their first two years in college, while persisting seniors reflected on all four (or more) years of courses in their STEM majors. Thus, the text analysis addresses STEM learning experiences and their consequences for students across all four years—some described in real time, and some in retrospect.

Findings from the SALG survey—an addition to the original study—reflect the views of students who were taking one of the same set of 80 foundational STEM courses at each participating institution. As explained, most, though not all, foundational courses were offered in the first two years of STEM majors. Thus, the SALG survey results include students who will subsequently switch out of STEM, some

**Table 6.9** Student problems with STEM teachers and teaching methods as factors contributing to STEM switching and as concerns of the switchers, persisters and all interviewees

| | Switchers | | Persisters | |
|---|---|---|---|---|
| Book | contributed to decision to switch (%) | Described as a concern (%) | Described as a concern (%) | All students described as a concern (%) |
| TAL (1997) | 36 | 90 | 74 | 83 |
| TALR (2019) | 48 | 96 | 72 | 78 |

who will relocate to other STEM majors, and some who will persist in their original choice of a STEM major.

Table 6.9 compares the extent of students' problems with STEM instructors and teaching methods from analysis of interview data in the present and original studies.

The similarity of the present findings to those of the original study is striking, and, many of the types of problems with STEM instructors and their teaching methods described here echo those reported in the original study (cf., Chap. 3). As before, most of the concerns that lead to switching for some students, and to ongoing learning problems for others, relate to particular aspects of teaching and course design. In the current study, over half of the persisters (57%) and 99% of the switchers reported that they mainly experienced traditional lectures without interactivity. In describing unsatisfactory experiences, students now, as then, often made invidious comparisons between the quality of STEM course teaching and that offered by former high school science teachers, or faculty in non-STEM disciplines. Students also drew similar conclusions about what explained their predominant experience of poor quality in STEM teaching, particularly in foundational courses. They identified lack of professional education in teaching methods in an institutional and professional culture that rewards excellence in research over effectiveness in teaching. Without benefit of the insights provided by Ferrare's interviews with their foundation courses instructors, what students could not also deduce was that the methods their teachers deployed were grounded in widely shared beliefs about how students learn science and, thus, how concepts, skills, and ways of thinking had to be taught. Nor could they track how these related learning theories and classroom practices were laid down as graduate students in the course of professional socialization (Seymour et al., 2005).

Where the original and present studies diverge, however, are that, in this study, 26% of switchers and 33% of persisters also described STEM courses (across all years) in which they experienced interactive, active, or multi-methods teaching. In Chap. 8, we present these positive students' accounts of teacher and teaching and what made them effective in enabling learning and motivation. In the class observation study of the same 71 foundational courses surveyed by the SALG survey, Ferrare and Miller (2017) also documented methods used in 25% of these courses. They included: Frequent use of small group work that included instructor engagement with student groups, and multiple forms of lecturing mixed with group work. Ferrare also observed numerous illustrations of course content, questions posed by the instructor in different forms. As we noted at the outset, none of these methods was reported in any of the STEM foundational courses represented in the original study.

Since the publication of TAL in 1997, researchers have continued to explore the role of instructional practice in "gateway" STEM courses in undermining persistence. One large group of studies (reviewed in Chap. 1) confirms TAL's finding that inadequacies in students' learning experiences in freshman and sophomore STEM courses play a major role in promoting losses from original STEM majors. As with TAL, their findings include: loss of students' incoming interest in the discipline as a result of poor teaching, rigid curricula, and negative departmental climates (Biggers, Brauer, & Yilmaz, 2008; DeAngelo, Franke, Hurtado, Pryor, & Tran, 2011; Suresh, 2007). Researchers have focused more on problems with the quality of teaching than on issues in curriculum design. However, as in TAL, a number of subsequent studies also cite the combination of poor course design, teaching methods, and assessment practices experienced in gateway courses as contributing to loss of confidence and interest particularly among men of color and women of all races and ethnicities (Barr, Gonzales, & Wanat, 2008; Crisp, Nora, & Taggart, 2009; Ellis, Fosdick, & Rasmussen, 2016; Hurtado, Eagan, & Chang, 2010a; Johnson, 2011; Strayhorn, Long, Kitchen, Williams, & Stenz, 2013).

Another large body of literature has explored and demonstrated the efficacy in significantly improving student performance and retention of a variety of active and interactive classroom strategies that actively engage students in their own learning (cf., Crouch & Mazur, 2001; Drane, Smith, Light, Pinto, & Swarat, 2005; Felder & Brent, 2016; Freeman et al., 2014; Henderson, Beach, & Finkelstein, 2011; Hurtado et al., 2010a; Mazur, 1997; National Research Council, 2012; Thiry, Hug, & Weston, 2011; Yoder & Cook, 2014). Accumulating research also finds that one-way transmission lectures without interaction are substantially less effective for promoting learning of more complex concepts (McCray, DeHaan,, & Shuck, 2003). Curricula choices and pedagogical methods that enable students to see relevance and conceptual connections also encourage the engagement of students of color and are effective in securing their academic and social adjustment (Hurtado, Newman, Tran, & Chang, 2010b). The collective result is a body of what are widely referred to as "science-"or "evidence-based" teaching methods, or "research-based" instructional strategies (RBIS) that may be deployed with increasing confidence to improve student performance and retention, and foster a more diverse supply of STEM graduates (Olson & Riordan, 2012; Nielsen, 2010; Wieman et al., 2010).

Well before the availability of solid evidence that these alternatives to "straight lecturing" did indeed produce significant improvements in student achievement and retention, many STEM faculty had already begun their own experiments with such methods. Some early STEM "reformers" coalesced into within- and across-institution coalitions to share their methods and findings, and support each other in promoting them through peer-to-peer workshops and presentations at professional meetings (notably, Project Kaleidoscope, ChemLinks and ModularChem, and the

EXCEL engineering coalitions). By the time the research evidence that underwrote their work was being published, a considerable nationwide STEM education improvement movement was already underway, with encouragement and financial support from public and private funding agencies, and with the National Science Foundation in the lead. As a historical note, we add that, the SALG online classroom evaluation survey was developed by Seymour in 1997 (with support from the Exxon Educational Foundation and the NSF) to provide feedback on specific areas of course design, teaching, and assessment to instructors who are seeking ways to improve their pedagogy and monitor their progress. It was also conceived as a course evaluation tool for departments and institutions that focuses exclusively on what students are gaining from their STEM learning experiences. Though progressively enhanced to meet specific user needs (e.g., undergraduate research programs, workshop evaluation) the present version, SALG 3.0, continues to serve these purposes for a wide range of disciplines, educational settings, and evaluation purposes (www.salgsite.org).

Increasingly, these nationwide efforts have also carried the active support of academic, disciplinary and professional associations. Their engagement was stimulated and endorsed by the 2012 Report of the President's Council of Advisors on Science and Technology (PCAST), *Engage to Excel: Producing One Million Additional College Graduates with Degrees in Science, Technology, Engineering, and Mathematics* (PCAST), and by a series of reports from the National Academies of Sciences and of Engineering. Among these, the National Research Council's (2012) synthesis of empirical research on teaching and learning in the sciences, recognized the emergence of "discipline-based education research" (DBER). The national associations of higher education institutions have also become active in promoting the uptake of instructional methods grounded in education theory and research to improve the quality of STEM education and create greater access for hitherto underrepresented student groups. For example, in 2012, the Association of American Universities launched a five-year initiative to influence the culture of STEM departments in member institutions such that faculty are encouraged and supported in the use of research-based instructional strategies (RBIS). Their progress reports grapple with the difficulties encountered across all STEM reform efforts, notably how to get whole departments to adopt methods shown to be effective in enabling students' engagement and learning (AAU, 2013, 2017b), and how to secure institutional recognition and rewards for good teaching (AAU, 2017a).

It is also critical for STEM faculty that their own disciplinary societies legitimate an increasing focus on quality teaching. By 2008, several disciplinary societies had begun to set educational standards for their disciplines, make endorsed educational resources available to their members, advocate for professional development in teaching methods, and press for institutional recognition of teaching excellence. Disciplinary and professional societies that were early to act in these ways include: The American Association for the Advancement of Science, the Mathematical Association of America, the American Society for Engineering Education, the American Association of Physics Teachers, and the American Association of Colleges and Universities. Disciplinary societies across all STEM disciplines now provide platforms for discussion of results from a growing scholarship of teaching and learning among instructors by establishing educational sections within their annual meetings, publishing reports and newsletters, maintaining informative websites, mounting conferences focused on pedagogical improvement, and devoting funding and staff to these initiatives (Matyas, Ruedi, Engan, & Chang, 2017).

Also since publication of the original study, there is clear evidence of an increasing desire to see improvement in access, quality, and student outcomes in STEM education. However, it is a matter of ongoing debate whether evidence of the efficacy of alternative modes of instruction, their endorsement by professional and disciplinary bodies, and financial support from myriad educational improvement grants by public and private foundations have secured progress towards these aims (discussed in Seymour & Fry, 2016; Seymour & DeWelde, 2016). As noted in Chap. 1, scholars report that faculty are increasingly aware of interactive teaching methods and have some inclination to try them, but also report a considerable gap between knowledge and uptake, plus high discontinuation rates after trial use (Dancy & Henderson, 2008; Hora, 2013; Hurtado, Eagan, Pryor, Whang, & Tran, 2012; McCray et al., 2003). Indeed, via a survey of 23,824 faculty in higher education, Hurtado et al. (2012) found that traditional lecturing was the primary form of instruction in 63% of courses across the USA. When asked, faculty report that the teaching strategies they mainly use are heavily influenced by the way they were taught as undergraduates (Balschweid, Knowbloch, & Hains, 2014). Furthermore, the role models that STEM faculty tend to follow in deciding how to teach are colleagues with high research prestige rather than those honored for their teaching or educational scholarship (Foertsch, Millar, Squire, & Gunter, 1997). Our student informants are correct in their assessment that most STEM faculty have no formal training in teaching. However, their theories of how students learn science and mathematics, and the methods that reflect these, are indeed learned, in the process of professional socialization in their graduate and post-graduate education (Seymour et al., 2005). Awareness of good alternatives to "straight lecturing" may be a necessary condition for uptake of research-based instructional strategies, but it is clearly not sufficient. Ferrare's findings underscore conclusions from educational change research (c.f., Woodbury & Gess-Newsome, 2002; Lotter et al., 2007; Wieman et al.'s, 2010; Lund & Stains, 2015) that wider uptake of RBIS has to begin with acknowledging how instructors conceptualize the student learning process, then persuading them to consider the research-grounded learning theories that underwrite research-grounded teaching.

Active engagement as researchers and project evaluators in many STEM improvement initiatives from publication of the original study to the present, and awareness of many others, prompts us to consider our findings in light of the considerable effort that has been, and is being, made to move from successful experiments in improved learning to their widespread implementation by STEM departments. Thus, throughout this study, we have looked for signs of change towards evidence-based teaching practices in our six sample schools. We discuss some of these indicators in Chap. 8.

## References

Akiki, T. K. (2014). A review on effective teaching and learning in higher education. *European Scientific Journal, 1*, 159–164.

Aronson, J., Fried, C., & Good, C. (2002). Reducing the effects of stereotype threat on African-American college students by shaping theories of intelligence. *Journal of Experimental Social Psychology., 38*, 113–125.

Association of American Universities. (2013). *Framework for systemic change in undergraduate STEM teaching and learning* (p. 2013). Washington, DC: Author.

Association of American Universities. (2017a). *Essential questions and data sources for continuous improvement of undergraduate STEM teaching and learning*. Washington, DC: Author.

Association of American Universities. (2017b). *Aligning practice to policies: Changing the culture to recognize and reward teaching at research universities*. Washington, DC: Author.

Balschweid, M., Knowbloch, N. A., & Hains, B. J. (2014). Teaching introductory life sciences courses in colleges of agriculture: Faculty experiences. *Journal of Agricultural Education, 55*(4), 162–175.

Barr, D. A., Gonzales, M. E., & Wanat, S. F. (2008). The leaky pipeline: Factors associated with early decline in interest in premedical studies among underrepresented minority undergraduate students. *Academic Medicine, 83*(5), 503–511.

Beichner, R. J., Saul, J. M., Abbott, D. S., Morse, J. J., Deardorff, D. L., Allain, R. J., … Risley, J. S. (2007). The student-centered activities for large enrolment undergraduate programs (SACALE-UP) project. In E. Redish & P. J. Cooney (Eds.), *Research-based reform of university introductory physics*. College Park, MD: American Association of Physics Teachers. Retrieved April, 2015, from: http://www.per-central.org/document/ServeFile.cfm?ID=4517

Benjamin, E. (2002). How over reliance upon contingent appointments diminishes faculty involvement in student learning. *Peer Review, 5*, 4–10.

Biggers, M., Brauer, A., & Yilmaz, T. (2008, March). *Student perceptions of computer science: A retention study comparing graduating seniors vs. CS leavers*. Paper presented at the 39th SIGCSE technical symposium on computer science education, Portland, OR. https://doi.org/10.1145/1352135.1352274

Blackwell, L., Trzesniewski, K., & Dweck, C. S. (2007). Implicit theories of intelligence predict achievement across an adolescent transition: A longitudinal study and an intervention. *Child Development, 78*, 246–263.

Canning, E. A., Muenks, K., Green, D. J., & Murphy, M. C. (2019). STEM faculty who believe ability is fixed have larger racial achievement gaps and inspire less student motivation in their classes. *Science Advances, 5*(2), eaau4734. https://doi.org/10.1126/sciadv.aau473

Chiang, S.-Y. (2009). Dealing with communication problems in the instructional interactions between international teaching assistants and American college students. *Language and Education, 23*(5), 461–478.

Crisp, G., Nora, A., & Taggart, A. (2009). Student characteristics, pre-college, college, and environmental factors as predictors of majoring in and earning a STEM degree: An analysis of students attending a Hispanic-serving institution. *American Educational Research Journal, 46*(4), 924–942.

Crouch, C. H., & Mazur, E. (2001). Peer instruction: Ten years of experience and results. *American Journal of Physics, 69*, 970. https://doi.org/10.1119/1.1374249

Dancy, M. H., & Henderson, C. (2008). *Barriers and promises in STEM reform*. Commissioned Paper for the National Academies Workshop: Evidence on Promising Practices in Undergraduate Science, Technology, Engineering, and Mathematics (STEM) Education. Retrieved November 2, 2010, from http://www7.nationalacademies.org/bose/PP_Commissioned_Papers.html

DeAngelo, L., Franke, R., Hurtado, S., Pryor, J. H., & Tran, S. (2011). *The American freshman: National norms for fall 2009*. Los Angeles, CA: Higher Education Research Institute, UCLA.

Dotger, S. (2010). Offering more than "here is the textbook": Teaching assistants' perspectives on introductory science courses. *Journal of College Science Teaching, 39*(3), 71–76.

Drane, D., Smith, H. D., Light, G., Pinto, L., & Swarat, S. (2005). The gateway science workshop program: Enhancing performance and retention in the sciences through peer-facilitated discussion. *Journal of Science Education & Technology., 14*(3), 337–352.

Dweck, C. (2015). Carol Dweck revisits the growth mindset. *Education Week, 35*(5), 20–24.

Dweck, C. S. (2008). *Mindset: The new psychology of success*. New York, NY: Random House Digital, Inc.

Ejiwale, J. A. (2012). Facilitating teaching and learning across STEM fields. *Journal of STEM Education, 13*(3), 87–94.

Ellis, J., Fosdick, B. K., & Rasmussen, C. (2016). Women 1.5 times more likely to leave STEM pipeline after calculus compared to men: Lack of mathematical confidence a potential culprit. *PLoS One, 11*(7), e0157447. https://doi.org/10.1371/journal.pone.0157447

Felder, R. M., & Brent, R. (2016). *Teaching and learning STEM: A practical guide* (1st ed.). San Francisco, CA: Jossey-Bass, a Wiley Brand.

Ferrare, J. J. (2019). A multi-institutional analysis of instructional beliefs and practices in gateway courses to the sciences. *CBE-Life Sciences Education*. https://doi.org/10.1187/cbe.17-12-0257

Ferrare, J. J., & Miller J. (2017). *Cultural models of persistence and inequality in introductory STEM courses*. Annual Meeting of the Sociology of Education Association, Monterey, CA.

Ferrare, J. J., & Hora, M. T. (2014). Cultural models of teaching and learning in math and science: Exploring the intersections of culture, cognition, and pedagogical situations. *The Journal of Higher Education, 85*(6), 792–825.

Freeman, S., Eddy, S. L., McDonough, M., Smith, M. K., Okoroafor, N., Jordt, H., & Wenderoth, M. P. (2014). Active learning increases student performance in science, engineering, and mathematics. *Proceedings of the National Academy of Sciences, 111*(23), 8410–8415.

Foertsch, J. M., Millar, S. B., Squire, L., & Gunter, R. (1997). *Persuading professors: A study in the dissemination of educational reform in research institutions*. Report to the NSF education and human resources directorate, Division of Research, Evaluation, and Communication. Washington DC: University of Wisconsin-Madison, LEAD Center.

Gardner, G. E., & Jones, M. G. (2011). Pedagogical preparation of the science graduate teaching assistant: Challenges and implications. *Science Educator, 20*(2), 31–41.

Gravois, J. (2005). Teach impediment. *The Chronical of Higher Education, 51*(31), A10.

Henderson, C., Beach, A., & Finkelstein, N. (2011). Facilitating change in undergraduate STEM instructional practices: An analytic review of the literature. *Journal of Research in Science Teaching, 48*(8), 952–984.

Hora, M. T., Oleson, A., & Ferarre, J. J. (2013) *Teaching dimensions observation protocol (TDOP) user's manual" Wisconsin Center for Education Research*, University of Wisconsin.

Hora, M. T. (2013). *Exploring the use of Teaching Dimensions Observation Protocol to develop fine-grained measures of interactive teaching in undergraduate science classrooms* (WCER Working Paper No. 2013-6 October).

Hurtado, S., Eagan, K., Pryor, J. H., Whang, H., & Tran, S. (2012). *Undergraduate teachign faculty: The 2010-2011 HERI Faculty Survey*. Los Angeles, CA: Higher Education Research Institute, UCLA.

Hurtado, S., Eagan, K., & Chang, M. (2010a). *Degrees of success: Bachelor's degree completion rates among initial STEM majors*. Los Angeles, CA: Higher Education Research Institute, University of California.

Hurtado, S., Newman, C. B., Tran, M. C., & Chang, M. J. (2010b, Winter). Improving the rate of success for underrepresented racial minorities in STEM fields: Insights from a national project. *New Directions for Institutional Research*, (148), 5–15. https://doi.org/10.1002/ir.357

Johnson, A. C. (2011). Accomplishments and challenges for women in STEM: Implications for future research and programs. *Journal of Women and Minorities in Science and Engineering, 71*(1), 5–10.

Kang, O., Rubin, D., & Lindemann, S. (2015). Mitigating U.S. undergraduates' attitudes toward international teaching assistants. *Tesol Quarterly, 49*(4), 681–706.

Kendall, K. D., & Schussler, E. E. (2012). Does instructor type matter? Undergraduate student perception of graduate teaching assistants and professors. *CBE – Life Sciences Education, 11*, 187–199.

Kurdziel, J. P., & Libarkin, J. C. (2003). Research methodologies in science education: Training graduate teaching assistants to teach. *Journal of Geoscience Education, 51*, 347–351.

Lotter, C., Harwood, W. W., & Bonner, J. J. (2007). The influence of core teaching conceptions on teachers' use of inquiry teaching practices. *Journal of Research in Science Teaching, 44*(9), 1318–1347.

Lund, T. J., & Stains, M. (2015). The importance of context: An exploration of factors influencing the adoption of student-centered teaching among chemistry, biology and physics faculty. *International Journal of STEM Education, 2*(1), 13. https://doi.org/10.1186/s40594-015-0026-8

Matyas, M. L., Ruedi, E. A., Engan, K., & Chang, A. L. (2017, Spring). Life science professional societies expand undergraduate education efforts. *CBE Life Sciences Education*, (16)1. https://doi.org/10.1187/cbe.16-01-0019

Mazur, E. (1997). *Peer instruction: A user's manual*. Upper Saddle River, NJ: Prentice Hall.

McCray, R. A., DeHaan, R. L., Schuck, J. A., & National Research Council (U.S.). (2003). *Improving undergraduate instruction in science, technology, engineering, and mathematics: Report of a workshop*. Washington, DC: National Academies Press.

Mulnix, A. B., & Vandergrift, E. V. H. (2014). A tipping point in STEM education reform. *Journal of College Science Teaching, 43*(3), 14–16.

National Research Council. (2012). Discipline-based education research: Understanding and improving learning in undergraduate science and engineering. In S. Singer, N. R. Nielsen, & H. A. Schweingruber (Eds.), *Committee on the staus, contributions, and future directions of discipline-based education research. Board of Education, Division of Behavioral and Social Sciences and Education*. Washington, DC: National Academies Press.

Nielsen, N., National Academies Press, & National Research Council. (2010). Planning Committee on Evidence on Selected Innovations in Undergraduate STEM Education. *Promising practices in undergraduate science, technology, engineering, and mathematics education: Summary of two workshops*. Washington, DC: National Academies Press.

Olson, S., & Riordan, D. G. (2012). *Engage to excel: Producing one million additional college graduates with degrees in science, technology, engineering, and mathematics*. Report to the President. Executive Office of the President1600 Pennsylvania Avenue NW, Washington, DC. Web site: http://www.whitehousegov

Pickering, L. (2004). The structure and function of intonational paragraphs in native and nonnative speaker instructional discourse. *English for Specific Purposes, 23*, 19–43.

Ruiz, N. G. (2014). The geography of foreign students in U.S. higher education: Origins and destinations. Accessed February 5, 2018 from https://www.brookings.edu/interactives/the-geography-of-foreign-students-in-u-s-higher-education-origins-and-destinations/

Seymour, E., & Hewitt, N. M. (1997). *Talking about leaving: Why undergraduates leave the sciences*. Boulder, CO: Westview Press.

Seymour, E., & Fry, C. F. (2016). The reformers' tale: Determining progress in improving undergraduate STEM education. In G. C. Weaver, W. D. Burgess, A. L. Childress, & L. Slakey (Eds.), *Transforming institutions: Undergraduate STEM education for the 21st Century*. Purdue, IN: Purdue University Press.

Seymour, E., & DeWelde, C. (2016). Why doesn't knowing change anything? Constraints and resistance, leverage and sustainability. In G. C. Weaver, W. D. Burgess, A. L. Childress, & L. Slakey (Eds.), *Transforming institutions: Undergraduate STEM education for the 21st century*. Purdue, IN: Purdue University Press.

Seymour, E. (2007). *The US experience of reform in science, technology, engineering, and mathematics (STEM) undergraduate education*. Paper presented at the Policies and Practices for Academic Enquiry: An International Colloquium, Marwell Conference Centre, Winchester, UK.

Seymour, E., Melton, G., Pedersen-Gallegos, L., & Wiese, D. J. (2005). *Partners in Innovation: Teaching assistants in college science courses*. Boulder, CO: Rowman and Littlefield.

Seymour, E., & Hunter, A.-B. (1998). *Talking about disability: The education and work experiences of graduates and undergraduates with disabilities, in science, mathematics and engineering*. Washington, DC: AAAS.

Stains, M., Harshman, J., Barker, M. K., Chasteen, S. V., Cole, R., DeChenne-Peters, S. E., … Young, A. M. (2018). Anatomy of STEM teaching n North American universities. *Science, 359*(6382), 1468–1470. https://doi.org/10.1126/science.aap8892

Strayhorn, T. L., Long, L. III, Kitchen, J. A., Williams, M. S., & Stenz, M. E. (2013). Academic and social barriers to black and Latino male Collegians' success in engineering and related STEM fields. In *Proceedings, 2013 ASEE annual conference and exposition*, Atlanta, GA. Retrieved from https://commons.erau.edu/publication/295

Sundberg, M. D., Armstrong, J. E., & Wischusen, E. W. (2005). A reappraisal of the status of introductory biology laboratory education in U.S. colleges and universities. *The American Biology Teacher, 67*, 526–529.

Suresh, R. (2007). The relationship between barrier courses and persistence in engineering. *Journal of College Student Retention: Research, Theory, and Practice, 8*(2), 215–239.

Thiry, H., Hug, S., & Weston, T. J. (2011). The Computing Alliance of Hispanic-Serving Institutions: Enhancing the success of Hispanic undergraduates in computing disciplines. *Journal of Enrollment Management, 5*(1), 32–56.

Travers, P. L. (1989). Better training for teaching assistants. *College Teaching, 37*, 147–149.

Wieman, C., Perkins, K., & Gilbert, S. (2010). Transforming science education at large research universities: A case study in progress. *Change: The Magazine of Higher Learning, 42*(2), 7–14.

Williams, J. (1992). Planning, discourse marking, and the comprehensibility of international teaching assistants. *TESOL Quarterly, 26*, 693–711.

Woodbury, S., & Gess-Newsome, J. (2002). Overcoming the paradox of change without difference: A model of change in the arena of fundamental school reform. *Educational Policy, 16*(5), 763–782.

Yoder, G., & Cook, J. (2014). Rapid conversion of traditional introductory physics sequences to an activity-based format. *Journal of STEM Education: Innovations and Research, 15*(2), 16.

Zehnder, C. (2016). Assessment of graduate teaching assistants enrolled in a teaching techniques course. *Journal of College Science Teaching, 46*(1), 76–83.

# Chapter 7
# Weed-Out Classes and Their Consequences

**Timothy J. Weston, Elaine Seymour, Andrew K. Koch, and Brent M. Drake**

## Defining and Identifying "Weed-Out" Classes

Weed-out courses are defined in a limited research literature as a subset of introductory, foundational courses that students normally take in the first or second years of college. Also described as "gateway" or "barrier" courses (Flanders, 2017; Koch, 2017; Koch & Drake, 2018; Suresh, 2006), they are (typically) required courses, often in a series. Classes are often large, and instruction is generally lecture-based with a recitation or laboratory section. Successfully completing these courses during the first 2 years of college is a strong predictor of persistence in STEM majors (Flanders, 2017; Rask, 2010). However, passing these courses is difficult because grading is much more severe than in other university courses. Instructors may grade on a curve or give D and F grades to a predetermined quota of students. High proportions of students withdraw from these courses and many repeat the same

T. J. Weston (✉)
Ethnography and Evaluation Research (E&ER), University of Colorado Boulder, Boulder, CO, USA

National Center for Women and Information Technology (NCWIT), University of Colorado Boulder, Boulder, CO, USA
e-mail: westont@colorado.edu

E. Seymour
Ethnography and Evaluation Research (E&ER), University of Colorado Boulder, Boulder, CO, USA
e-mail: Elaine.Seymour@colorado.edu

A. K. Koch · B. M. Drake
John N. Gardner Institute for Excellence in Undergraduate Education, Brevard, NC, USA
e-mail: koch@jngi.org; drake@jngi.org



E. Seymour, A.-B. Hunter (eds.), *Talking about Leaving Revisited*,
https://doi.org/10.1007/978-3-030-25304-2_7

courses—some more than twice; some without ever passing. Rask's (2010) analysis of transcript data reveals that these courses depress STEM departments' grade averages, and students who receive low grades or repeat these foundational courses are at higher risk of leaving their STEM majors. As Koch and Drake also report in this chapter, students who are otherwise in good academic standing who earn a D, F, W, or I grade in a STEM gateway course are at high risk of not returning to their institution for the subsequent year.

Notwithstanding the high visibility of such courses and their effects, the term "weed-out" is reputational, traditional, and pervasive, but entirely unofficial. Both from direct and shared experience, students apply the term "weed-out" to particular courses (e.g., Organic Chemistry), to course sequences (e.g., the calculus series in engineering), and also to particular course sections or classes. Instructors and advisors also use this term in a familiar and even colloquial manner, and (with the exceptions cited above) authors are inclined to assume a tacit understanding that we all know what this term means. Thus, one limitation of the small body of studies that examine weed-out courses is lack of precision about what this term denotes and what features distinguish these courses from other foundational courses. We draw upon on our available data sources to address the questions: What are weed-out courses, and what are their distinctive features?

*The interview study* defines, identifies, and characterizes weed-out courses in terms of student accounts of problems with their learning experiences in particular courses and their consequences. Thus, in student terms, weed-out courses are those that manifest, in a highly concentrated way, several of the problems reported in the "iceberg" tables. In weed-out courses, these issues form an intensive cluster of encompassing problems with curriculum design and teaching methods, and with assessment and grading practices. These issues combine to dissipate confidence and incoming interest, expose deficiencies in high school preparation, and exacerbate college transition difficulties. Ultimately, the combined negative outcomes as both our student informants and other researchers attest are high failure rates, course repetition, STEM switching and relocation, and college exit rates all of which far exceed those in other foundational courses.

*From our analysis of institutional records*, we identified 293 STEM foundational courses with the following characteristics at our six sample sites. They were:

- Large (i.e., individual class sizes are greater than 100 students);
- Lower Division: First or second year courses with primarily first and second year students;
- Established (i.e., class records exceeded 4 years);
- Required and/or prerequisite for a STEM major (often serving multiple STEM majors).

For the sake of brevity, we called these courses Large Foundational Courses (LFC).

To identify "weed-out" courses from among all LF courses, we added a further distinction based on the "severity" of course grading. We define course severity by the percentage of students who received a D or F grade, withdrew, or received an incomplete course grade. This statistic is called "DFWI" and is a commonly used

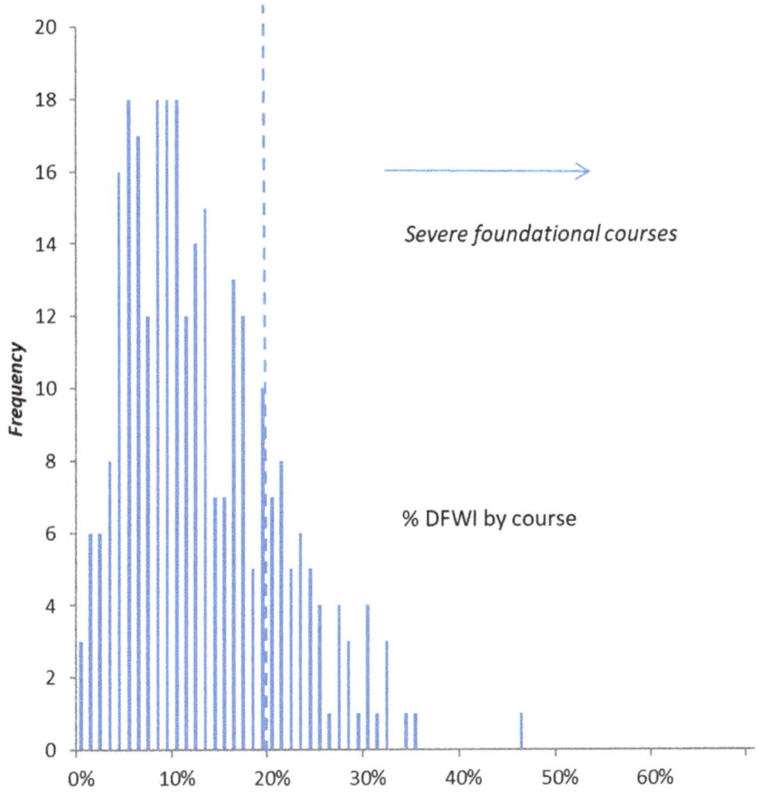


**Fig. 7.1** Distribution of DFWI percentages in Large Foundational Courses (LFCs)

metric for higher education researchers (cf. Freeman et al., 2014). We thereby created a subset of courses labeled Severe Foundational Courses (SFCs) comprising only those LF courses with higher than 20% DFWI rates. We propose that courses with these characteristics provide the best approximation of weed-out courses. As Fig. 7.1 shows, the distribution of DFWI rates for LF courses at our six institutions ranged from 0% to 45% with an average DFWI rate of 13% and a median rate of 10%. We considered courses with above 20% DFWI to be Severe Foundational Courses (SFCs).

## *Which Courses Are Identified as "Weed-Out" or Severe Foundational Courses?*

*In the institutional records study analysis*, the first and second year courses with the highest numbers of students receiving a D, F, or incomplete/withdrawal grades were calculus and chemistry courses, both of which averaged 20% DFWI.

Calculus, chemistry, and both first and second year computer science courses, which together represent 25% of all LF courses, are greatly over-represented as SFCs: 60% of these courses are in the high severity group. Individual courses in all three disciplines had much higher DFWI rates than their disciplinary averages. For instance, in *Calculus for Biological Scientists* at WPUB3, 29% of the students received a poor grade or an incomplete, while 45% of students in *Calculus for Physical Sciences* received a D, F, or incomplete/withdrawal, making it the course with the highest rate of students leaving a LF course at all six institutions. Both biology and the physical sciences had lower DFWI rates—around 10% each.

Calculus and chemistry courses also "service" students in many STEM and non-STEM majors as well as their own majors. For example, a typical calculus course with high severity at WPUB3 (viz., 21% DFWI) enrolled students primarily from the physical sciences (28%) and biology (28%). A chemistry course at ECPUB4 (26% DFW) serviced a mixture of 42% biology and 14% engineering majors. By contrast, most other LF courses (i.e., not SFCs) primarily served their own students. This is true in foundational engineering courses where 80% of students are engineering majors, physical sciences courses where 62% are physical science majors, and computer science (CS) courses where 60% are CS majors. Only calculus includes less than half, (48%) of its own math majors. We saw no obvious relationship between the percentage of STEM majors in foundational courses and their DFWI rates, that is, the rate of DFWI did not go up or down depending on the proportion of STEM students in a course. It may be important to note that STEM majors had only slightly lower DFWI rates than non-STEM majors (14% vs. 16%) in LF courses overall.

*The interview data and SALG survey responses* identify weed-out courses largely by direct experience, but somewhat by reputation. In interviews, 147 students (both switchers and persisters) identified 172 courses as weed-out in character. Respondents to the SALG survey in foundation courses identified 12 out of 53 courses surveyed as weed-out courses. The combined student list of courses identified as "weed-out" from these two sources is similar to the list of courses with high DFWI rates shown in Fig. 7.2. Chemistry courses were the most often cited (i.e., by 50% of students), notably General Chemistry and Organic Chemistry 1 and 2.

## What Happens in Weed-Out Classes that Prompts Students to Apply the "Weed-Out" Label to Particular Courses?

To answer this question, we asked interviewees to explain by what characteristics they identified particular classes as intended to weed students out. The seven resulting categories of weed-out class identifiers are summarized in Table 7.1 for all students, and for switchers and persisters, proportionate to their sample size.

First among these (and congruent with Greenwood & North's, 1999 finding) are assessments that are misaligned with course content and curved grading that distorts the extent to which students have gained mastery of the material presented. Cited by

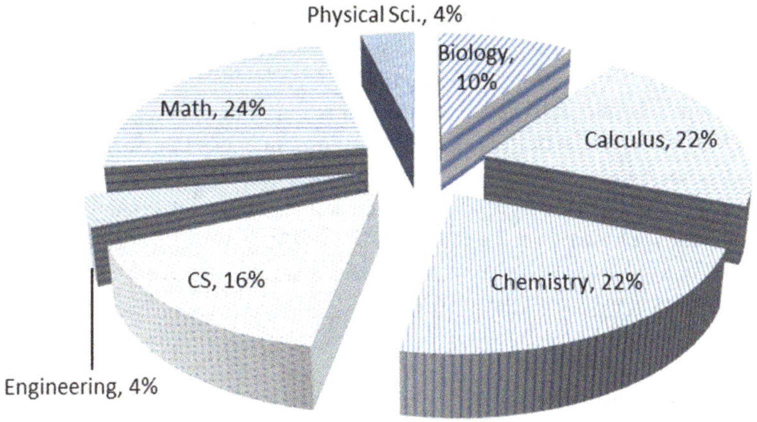


**Fig. 7.2** Distribution of course types for severe foundational courses (DFWI >20%)

**Table 7.1** Characteristics of weed-out classes as identified by all interviewees, and by switchers and persisters, as a percentage of their group

| | All students *N* = 346 | | Switchers *N* = 96 | | Persisters *N* = 250 | |
|---|---|---|---|---|---|---|
| Characteristics | *N* | % | *N* | % | *N* | % |
| Assessments misaligned w/content and understanding (including curved grading) | 116 | 29 | 30 | 31 | 86 | 34 |
| Heavy volume and pace | 82 | 20 | 15 | 16 | 67 | 27 |
| Level too high/abstract for intro class | 60 | 15 | 18 | 19 | 42 | 17 |
| Rote learning/dull content in lecture mode | 54 | 13 | 13 | 14 | 41 | 16 |
| Teacher indifferent whether learn: Learning alone w/out help | 44 | 11 | 15 | 16 | 29 | 12 |
| Incoherent organization, missed steps or explanations | 30 | 7 | 8 | 8 | 22 | 9 |
| Competitive class culture | 19 | 5 | 9 | 9 | 10 | 4 |
| All observations | 405 | 100 | 100 | 113[a] | 297 | 118[a] |

[a]Group percentages total more than 100% because students cited multiple characteristics

approximately one-third of both switchers and persisters, these assessment practices were described as promoting the hostile, competitive classroom cultures that were also cited as characterizing weed-out courses. Echoing a similar finding by Kardash and Wallace (2001), 20% of interviewees cited apparent indifference about whether students are following or understanding the class presentation as an indicator of an instructor's weed-out intent. This intent was also inferred where courses cover too much content for the time available, resulting in a pace that is too rapid for comprehension, assimilation, and application. Twenty-seven percent of our sample of persisting seniors could, with hindsight, see no reason, other than intimidation, for over-cramming courses with content that they encountered later in a more measured sequence. The effects of content overload are exacerbated when the level is pitched too high, or is inappropriately abstract, for an introductory course. Flaws in teaching methods said to typify weed-out courses include incoherent organization of material, incomplete explanation, and presentations whose dullness is exacerbated by rote learning in large classes without interaction.

Both switchers and persisters described all seven of the weed-out class attributes reflected in Table 7.1 as promoting negative consequences that included, but were not limited to, STEM switching and relocation. They are also the same weed-out course characteristics that were identified by students in the original "Talking about Leaving" study. Drawing on the interview data, SALG survey responses, and findings from relevant research, we examine the characteristics and consequences of weed-out courses in greater detail below.

## Testing and Grading Practices

Prominent among the testing and grading practices described as key characteristics of weed-out course experiences are lack of alignment between test questions and emphases in class content and misaligned levels of difficulty between homework and exams:

> You're just chasing your tail because the homework problems don't reflect the difficulty of the exams, and the exams are not based on your understanding of the topic—just whether you can pump the right answer to the bottom…And the test is on something that has nothing to do with what we've worked on. (African-American man, low-math, biology persister)

Assessments also trip students up where they do not address core class concepts:

> It's not about content and knowledge, it's all about details…It's grading by irrelevancies. (White man, high math chemical engineering persister)

However, students described the most serious cause of disconnection between understanding and assessment as the practice of "curve grading" (also called "norm referencing"). As discussed in Chap. 9, its use is also common in other STEM (and some non-STEM) courses. However, it is the norm in weed-out classes. As also reported in the original study, curve grading makes it difficult for students to make sense of their grades, that is, to align them with their understanding of course material and gauge their own progress:

> I am happy that I got a B+ but what does this mean since I failed every exam. And what does it mean for me next year when I am taking the next level of class. How can I judge how I am doing and whether it's fine to continue? (African American man, low math, life sciences persister)

> Mathematically, I was well prepared, but I wasn't prepared for classes with a 75% failure rate. In some of the weed-out classes, 30% was an A. (White man, high math mechanical engineering persister)

> You know it's a weed-out class when half the class fails. And when they have to curve the test so much that you didn't actually learn anything but you got a good enough grade to just keep going. (White woman, high math switcher, life sciences to psychology)

> I was prepped beforehand—'This is what you should expect from organic chemistry. Just be happy with the average and you'll do alright.' But why is knowing only 50% good? It's not conducive to learning. And people who didn't adjust to that started falling behind, then dropping out. (Hispanic man, high math bio-medical engineering persister)

The idea that survival requires adjustment to grading practices that seem irrational and to grades that are inconsistent with self-assessment of comprehension and ability was unacceptable to many students. As we discuss in subsequent chapters, these included high-performing students accustomed to high grades for whom such adjustments were a struggle that might not succeed.

Not only does curve grading limit students' ability to assess how they are progressing, they often see it as linked to poor quality teaching:

> If everyone's not getting it, is everyone stupid? Or is it something about the way you're teaching it? Maybe there's a disconnect between how it's being taught and what people are actually learning. We needed some way to evaluate that, but with curve grading, that's impossible. So you just chug along and hope to do better on the next test. It obviously didn't work for me. (White woman, high math switcher from life sciences to psychology)

> For the most part, I think it is irresponsible. You're not actually teaching your students if your average is 40. You are just inflating the test scores to make it look like someone has a higher understanding than they actually do--my A versus his B could be a 60 and a 55. Does that truly mean that I know more than he does, or was I just lucky with one of the equations? (Multi-race man, high math relocator, mechanical to bio-medical engineering)

Grades may be so low, or damage to GPA so serious (especially for aspirants to medical and other professional schools) that otherwise interested and competent students feel they must leave. Although most weed-out courses are in freshman year, in some majors competitive grading continues to weed students out beyond that:

> That organic chem class where I told you I went drinking after one of the exams, that's probably a pretty good weed-out story. There were a people in there whose scores were so low, they just had to switch out. That was a sophomore level class but then there's a few junior and even senior level classes that weed you out—like in neuroscience which has a massively disgusting weed-out system. If you don't maintain 81%, even seniors get weeded out. (White man, high math chemical engineering persister)

> Aesthetics is another weed-out course that comes in your sophomore year, which I think is very unfair. It's far too late in the process. You should focus on the material and making sure that students understand what they're studying. (White man, high math switcher, engineering to construction management)

Adjusting to curve grading also carries the risk that habitual focus on peer comparison can over time create ill-founded confidence in one's own level of understanding:

> With really inflated curves, over time, people come to care less about what they do individually. The test score becomes 'How did I do compared with other students?' The average for our last test ranged from 40-60, so I don't think anyone knows what they are doing, but they are content because they did well compared to others. (Asian woman, high math biomedical engineering persister)

> At the time, I didn't really mind because it was nice to my grade. But, looking back, it's not conducive to learning. The message is, 'Hey, you can only learn half of it and you will probably do well.' (African-American man, low math physics persister)

One SALG respondent who commented on grading practices in a course that he identified as "weed-out" wrote, "This class completely destroyed my confidence, made me decide to drop my computer science minor, and eliminated any interest I had in this area." An older returning student was disappointed to find the same curve grading practices in a chemistry course that she had taken before. She wrote:

> After my second year [at previous school] I left science because the atmosphere in the chemistry and physics classes was one in which every student was struggling to keep their head above water in a similar environment in which a student with a 50% would pass with a C. I had regretted not going for a science degree which is why I have gone back to school, but I am disappointed to see the same type of unrealistic standards in these chemistry classes set that require such a drastic curve in order to allow half the students to pass. I really feel that this is a significant cause of students leaving science majors because it is very discouraging. Furthermore, the students who pass with a C at 50% find themselves incredibly unprepared for their future classes because the subsequent classes assume at least a 75% understanding of the prerequisite material.

Curve grading practices and their effects are reported both within and beyond introductory STEM courses by 29% of switchers and 24% of persisters. Their wider use and consequences are further discussed in Chap. 9.

## Curriculum Design and Instruction in Weed-Out Courses

Four of the seven student-identified characteristics of weed-out courses reflected in Table 7.1 point to problems with design of the curriculum and the quality of instructional methods. Too much content for the time allotted, often at too high a level or too abstract for introductory courses, forces a pace that one senior in the original study described as, "like drinking from a fire hose" (see also Breussoud, 2015). In line with other studies that highlight shortcomings in curricula and modes of instruction (PCAST, 2012; Tinto, 2000), Prosser and Trigwell (2013) describe a typical first year lecture course instruction thus, "Teacher-focused, teacher activity with the intention of transferring information about the discipline. It is presumed that students do not need to be active in the teaching/learning process." (p. 786).

Many of the same concerns about instruction are found in the interview study where students describe instruction that requires them to memorize content rather than understand concepts, actively construct knowledge, or engage in active learning. Students complain of incoherent organization and delivery of course materials—sometimes pitched above introductory level, with little scaffolding, insufficient or unclear explanation, missing steps, confusing expectations for student work, and dull presentations. Problems with course design and teaching methods are exacerbated by an emphasis on rote learning for tests. Teachers are portrayed as discouraging questions in class and as inaccessible out of class. Students are not introduced to authentic science nor do they learn the skills needed to master content in later courses.

As Breussoud (2015) also reports, to meet the challenge of teaching large numbers of students in Calculus 1, mathematics departments either use their best instructors in very large classes or they opt for smaller classes taught by part-time instructors—often graduate students—whose preparation for teaching is often minimal. At one of the six institutions in this study, Calculus courses 1–3 were entirely taught by graduate students. However, it is not only graduate student teachers that have little or no preparation for their teaching work. As documented by Seymour, Melton, Pedersen-Gallegos, and Wiese (2005), rarely do faculty and instructors receive professional education in the principles and practice of effective teaching—a situation that, compared with other professions, is anomalous.

Both in this study and the original, it is not class size per se that students report as problematic so much as the limitations that large classes place on interactive teaching methods, including opportunities for questions, discussion, and instructor-student interaction—issues that are addressed in Chaps. 6 and 9. Although student accounts characterize weed-out classes as "hard," all seven student-identified weed-out characteristics are described as instances of "constructed" rather than intrinsic hardness, that is, where the level of difficulty becomes irrelevant to the understanding and skills needed by a student in the sciences. (We discuss this conceptual distinction further in Chap. 9.) As with curve grading and content-test alignment issues, students distinguish between the problems that they experience with curriculum design, teaching, and assessment methods chosen by the instructor and those that are determined by the nature of the material itself. As persisters attest, many aspects of STEM disciplines are "just plain hard." However, in weed-out courses, it is predominantly instructor-created difficulties rather than conceptual complexity that students see as making course content "hard":

> I won't say that weed-out classes are the most conceptually difficult. They aren't. And they may not be the hardest in terms of workload. But they are designed to deliver a culture shock of what your major might be like. So they are the most frustrating and stressful experience—like they go further in depth than is needed to frighten and overwhelm you. (White man, low math relocator, chemistry to chemical engineering)

Assessments are often the vehicle to carry this message:

> It's not the content, it's the assessments—like assigning something that even the professor doesn't solve…Five whiteboards—that's the average for this particular professor. It has little to do with weeding out people with less intelligence, and more to do with who just doesn't have the perseverance to wade through that much material every single time. Even when you're doing it, and you're on page three of one problem, you are thinking, 'There's no way I am right with all this crazy algebra.' But routinely that was the case for a correct answer. (Asian man high math electrical engineering persister)

Hardness is also constructed by setting the conceptual level too high for the course, over-loading the syllabus with content, and thereby forcing a killing pace. It is, as both seniors in interviews and commentators in the SALG surveys explained, less a test of interest and competence than of physical and mental endurance:

> For neuroscience, it was the most information any of us had ever tried to take in. We could be tested on any detail in the reading or the lecture. It was just a test of dedication, motivation and endurance. Can you spend ten hours a day studying? If you can't, you are clearly

> not dedicated enough and you won't get the grade. For the final, I studied 12 hours a day for four days straight. Didn't stop. Didn't go outside. That's what it took—just an absurd amount of work, an absurd grade curve; absurd standards. Looking back--did I have to be like that? Not at all. And it discourages a lot of student who would otherwise be interested. Neuroscience is an incredible field that holds your interest. I don't think you need to weed people out. (White woman, high math bio-sciences persister)

> There was an absurd amount of content covered, of which about one-third was actually mentioned in lecture. Most of the class came from screencasts. (White male, biology major)

> I felt that the class was very fast paced. This made it hard to keep up and learn the material well. I also think class time should be used to teach the material, not just review it, which is what the lectures feel like. It felt like we had to teach ourselves. (Multi-racial woman, psychology major, SALG survey respondent)

Switchers and persisters alike tried to make sense of constructed forms of hardness evident in weed-out classes. Having never experienced anything like this from high school teachers, freshmen were shocked to find college instructors who teach as if they do not want students to understand or succeed:

> I don't know why, but it seems like they don't want you to pass—it's like they don't want you to succeed at all…The teachers purposively make the tests more difficult than makes any sense. But why? I don't understand the purpose. I feel like learning should be for the sake of learning, but I see now that it's just about scores. And it's like all the teachers want their class to be the hardest—like a professional reputational thing. (White woman, high math mathematics persister)

> The professor spent more time trying to trip up students instead of teaching us what we needed to know. I understand the science department wants to weed kids out because there are so many of us, but I feel like science is either something people get or don't get; no need to try to trick the kids who have the potential to be good at it. (White woman, bio-chemistry major)

> The teacher tried to be creative in incorporating simulations that made you hate the class, hate the concepts, and hate chemistry. (Asian American man, chemistry major)

Students somewhat endorse the idea that instructors and their departments want to be sure that only students with the requisite interest, capability, and determination proceed to the next (intrinsically harder) stages of the major, but making the introductory learning experiences artificially hard appears counterproductive to this end:

> I sort of understand wanting to weed-out people who can't handle it or are no good, but making the classes so hard to the point where it's discouraging to people who can do the work and who want to stay makes no sense. (Hispanic woman, low math life sciences perister)

> It's made more difficult than it needs to be. There are subjects that are just flat difficult but there are ways to help students understand. You don't have to make it so easy that everyone is passing without trying hard, but you shouldn't be breaking your students when they are working hard to understand something important. (African American man, low math physics persister)

By "making learning hard" weed-out classes may successfully discourage less interested and capable students from continuing, but, even if they achieve this, they risk undermining, or failing to stimulate, interest among students that they might prefer to retain:

> I feel if they wanted to get more students, they'd get us more interested in engineering. By focusing on culling out the ones who are on the fence, they also lose a lot more of us who are actually interested. Getting us more interested in what you can be doing with this major would, right off the bat, make more sense. (Multi-race man, high math, relocator from chemical engineering to food science)

## Making Sense of Weed-Out Courses

The conclusion that students reach in their efforts to make sense of what they view as counterproductive instructor behavior is that these classes are not intended to foster learning and interest, but are deliberate, structured attempts to reduce student numbers:

> All the applied classes you had to get through in the math and engineering curriculum. Every single class it was, 'Let's try to cut off another row of students.' So you come in being used to high school teachers that want to figure out how to get you to understand something. Then that changes to 'Well; last one standing will make it into the major.' It was all about 'How can we get our numbers down?' We have 1,000 freshmen who want to be engineers and we can only graduate 250. So, we are going to cut three-quarters of them over the next two years before they ever get to the core engineering classes. (White man, high math switcher from engineering to economics)

> There's no concern for whether or not you understand the process. It's just whether or not you got the right answers and they just check the end. That's not having a concern or an interest in fostering learning and critical thinking.…And it's rampant. Everybody on campus knows this. (Multi-race man, low math relocator from engineering to hydro-geology)

As the last speaker implies, students come to realize that weeding out is a widespread, deliberate, and normalized process of structured wastage. As some students also noted, these courses weed-out the same proportion of students, regardless of the caliber of class entrants:

> It's like there's a requirement that there's an equal balance in the grades. It doesn't matter how good the students are that enter. They are going to come up with the requisite number of Cs and Ds. There' no sense of trying to help students to understand things and do well. The mentality is, we mark down and fail 30%. And my GPA was a victim to that mentality. (White man, high math chemical engineering persister)

The suspicion that the wastage rate is preconceived and deliberately achieved is reflected in the caution delivered in freshmen orientation sessions by senior faculty:

> During welcome week, the associate dean came in and told us, 'The people sitting next to you are probably not going to be here by the time that you graduate.' And faculty would look at us with that mentality rather than being encouraging. They were deliberately trying to get rid of the weakest link and, for a while, that's what I felt like. (White woman, low math switcher from math to economics)

This story, also widely reported by students in the original study 20 years earlier, predicts a self-fulfilling student loss rate of one-third. Students also come to realize that institutionalized, legitimated weed-out practices are an open secret: their instructors, advisors, TAs are well aware of them and treat them as normal. More experienced students pass along reputational folklore about which instructors and classes it is best to avoid:

> I get it that you have to learn fast in college, but the pace at which we were forced to go was no way to understand anything. It's stupid. But none of the TAs seemed to think it was an issue: 'Oh it's just the way it works round here.' And you can see what they mean. In Chem 140, the room is full, but in upper level chemistry there's just seven in the class. But, along the way you are extinguishing people's dreams. Of all the kids I met in that major, only one stayed, and he changed into another science major this semester. So, none of them stayed in chemistry. They all changed their majors. (White man, high-math, switcher, life sciences to communications)

> Everybody know that advisors tell you that cell biology is the hardest class in the program, and that most people don't pass organic chemistry. So, you're screwed before you even start. (Hispanic man, low math biology persister)

> Not passing my first weed-out classes was a hit to self-esteem, and then I had to retake it. The teacher just joked—which was very rude—that the first semester in the series he had 200 students and next semester they only needed a classroom for 30 people…This is a weed-out class and we were meant to be cut. (White woman, low math relocator, molecular biology to evolutionary biology)

As students come to understand that weed-out instructors are primarily focused on getting rid of a high proportion of incoming students, the instructors' behavior (although hard to bear) makes more sense. It provides a way to explain the poor quality of teaching that students report as characterizing weed-out classes:

> I felt like the content—the lecture—was just all over the place. It was hard to stay focused. He talked without any really set agenda to it. (Asian woman, high math switcher, life sciences to international relations)

> If you were below that curve there were professors who wouldn't even talk to you. And the people who are below the curve are the ones that need most help. There was a host of students who were asking questions and we were all ignored. (White woman, low math switcher, life sciences to sociology)

> I'd work my ass off and I'm proud of it…And I'm thinking every way I can to understand the concepts. But I would routinely get a failing grade. There's no way to show understanding and to get it corrected along the way. (Multi-race woman, low math relocator, engineering to hydro-geology)

This interpretation of weed-out's departmental purpose also provides students with an explanation for why less able teachers or graduate students may be assigned to teach courses that are critical to students' progress in chosen degrees and careers:

> I had an absolutely abysmal teacher—a TA—because there's no instructors to do this course. He's the worst person I've ever seen in front of a class. He didn't try to teach anybody anything. He never did a lesson plan of any kind. I felt very upset about this because, in effect, he was deciding whether or not people got to do certain careers. And I took that very personally. I wanted to do environmental engineering and all I needed was for someone to lay out the math in a reasonable manner so I could take a step up in understanding, and do well enough to proceed. That wasn't the case….Does it disadvantage you? Absolutely. I think I could have been an incredible environmental engineer… They don't

> hire people with teaching degrees or teaching interests at all… Calc 1, 2, and 3 all are taught by grad students. They will not pay people to teach those courses here. I am now in geology and the instructors in that department are absolutely phenomenal. (Multi-race woman, low math chemical engineering persister)

Even where instructors are experienced, carelessness about how they teach introductory classes damages those who persist as well as those who switch. Seniors reported that failure to establish a solid understanding of particular concepts in these courses had undermined their grasp of more complex ideas later:

> The whole first year felt like weed-out classes. What made them hard classes was a lot of the teachers taught upper level classes and they really didn't care about the lower level ones--which is what helped weed-out so many students. But it also wasn't helpful with learning the stuff we needed to know for those of us who stayed. (Multi-race man, high math relocator, mechanical engineering to bio-medical engineering)

As in their observations on curve grading effects, persisters explain the need to adjust in order to survive. However, adjustment comes at a cost:

> It's not teaching kids how to think or how to synthesize. It teaches them how to float--just get by. The classes and the content are not structured well, and you often don't get anything from them…It's a system that doesn't foster intelligence and creative thinking… It fosters submission…You can get through all of these classes if you learn most of these stuff on your own—but at the cost of a degree that doesn't mean anything. (Asian woman, high math relocator from biology to physiology)

Persisters were also affronted where "good" teachers appear to deliberately teach particular courses badly:

> There's a block of absolute weed-out classes in sophomore year that discourage a lot of people I had one teacher who was completely arrogant—he made it clear that he didn't wanna be there. He told us we were stupid, and then made it seem true. The average on our first test was 25% and 40% of the class failed. I have him now for a grad-level class and his attitude and his teaching are completely different. So you can't tell me that wasn't a weed-out class. It's the same guy. (White man, high math food science persister)

In Chap. 6, we reported on teaching methods that were observed in use in 80 introductory STEM courses, and in Chap. 8, we offer a more nuanced analysis of what student interviewees defined as "good" and "bad" teaching. Here we have focused on how students make sense of their learning experiences in foundational courses that they defined as "weed-out" in nature and intention.

## *Comparisons of Weed-Out and Non-Weed Courses in the Student Assessment of Their Learning Gains (SALG) Survey*

As explained in Chap. 1, the SALG online survey is designed to solicit students' assessments of how much they gained from specific aspects of any course. We deployed a version of the SALG instrument tailored to the research questions of our study to 1431 students in STEM introductory courses at our six sample institutions. Students rated their learning gains in 53 courses and also answered open-ended questions about course elements that did or did not enable their learning. Students in each class also rated the item "This course was a "weed-out" course for the major" on a 1–4 agree/disagree scale. More than half of students in 12 classes agreed or strongly agreed with this statement, and these same courses also showed greater than 20% DFWI. Six of the 12 classes were in chemistry, three in calculus, two in computer science, plus one physics course for biology majors. Nine of these 12 classes had more than 100 students.

We compared ratings between students in the student-identified weed-out classes (SFCs) with those of students in the non-SF courses in our survey sample. We saw large gaps between the two groups that consistently favored the non-SF courses in average ratings of learning gains and the effectiveness of particular course elements. The largest differences between SFC and non-SFC student ratings of how much particular activities had (or had not) enabled their learning were for assignments, the number and spacing of tests, and in-class reviews. Students in the SFCs also gave much lower overall ratings for their courses. Smaller yet still sizable differences were evident also for class pace, the amount of content covered, and class atmosphere—all of which were identified in the interviews as prevalent in weed-out courses. The contrast between weed-out and non-weed-out foundational STEM courses in terms of how much 14 aspects of the courses listed in the survey had enabled their learning is laid out in Fig. 7.3a–c.

We also coded responses to the open-ended survey questions, "Please comment on how your understanding of the subject has changed as a result of this class," and "Please comment on how (and if) this class helped your learning." Students answered these questions by describing both the characteristics of courses that helped their learning, and what, and how much, they had learned. Most comments were unambiguously dichotomous—either praising or critical of the teacher and the teaching methods used. In the SF courses identified by students, the proportion of negative comments about their learning experiences was far higher for the SF classes than it was for the non-SF classes (viz., 54% to 30%, respectively). Written comments mirrored many of the concerns expressed by interviewees, although survey respondents focused more than the interviewees on the poor teaching they had encountered in SF courses than on criticisms of assessment practices. However, the following samples of what students wrote about particular experiences endorse the assessments offered by interviewees.

### Poor Teaching

A higher proportion of students' comments in SFCs reported that they learned little or no new content from the class (67% vs. 29%). This was also the most commonly offered comment overall. Students in weed-out (SFCs) also blamed their teachers for their learning difficulties more than did students in non-weed-out-courses, and 47% of comments about specific teaching methods in SF courses were negative compared with 31% in non-SF courses:

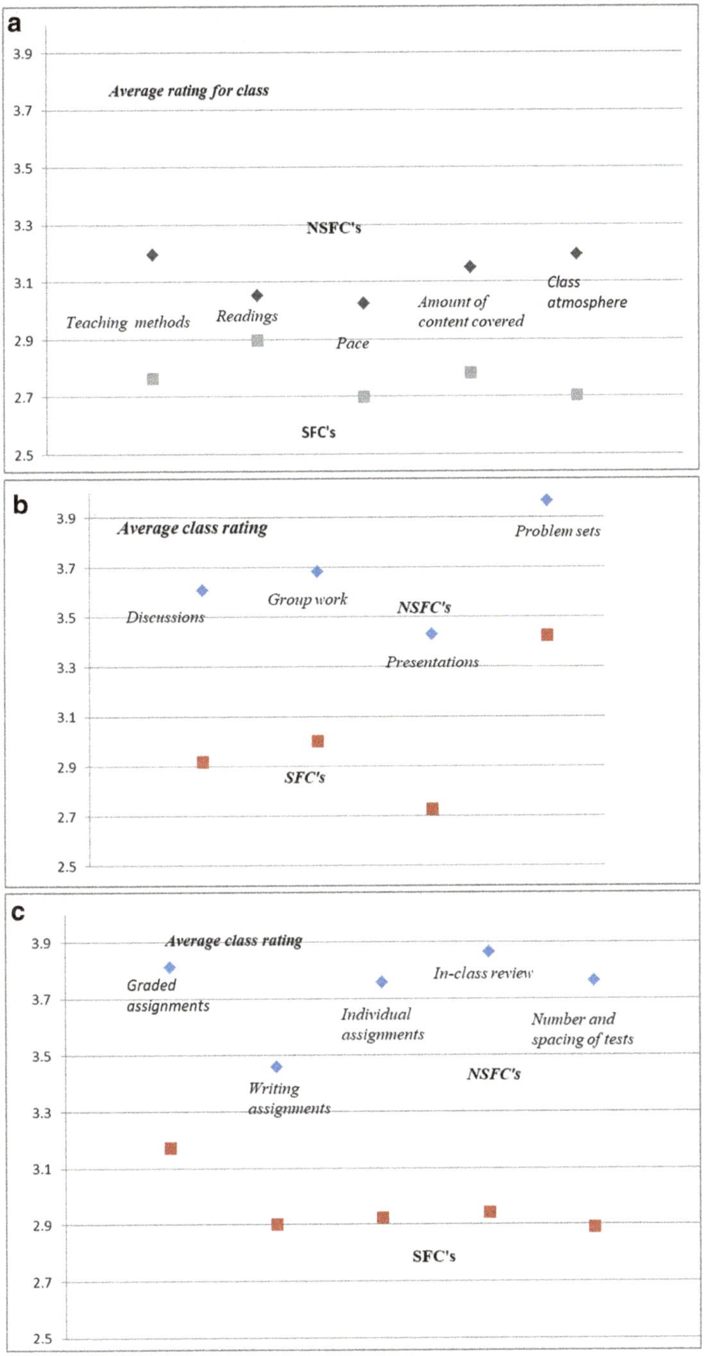


**Fig. 7.3** (**a**–**c**) Comparison of SFCs and non-SFCs by student answers to the SALG. *Question, "How much did the following parts of this class help your learning?"*

> This class didn't really help my learning that much. I want to point out that I am genuinely curious about chemistry, that it is something that interests me; I'm not just taking the course to get past it. That said, I feel more defeated than anything else at this point, and I think that has much to do with how the material was presented. (White man, physiology major)

> I never really learned chemistry. I felt that the tests were supposed to make you think about how it related to the real world, however, we were never taught how it related to the real world, so I did not understand. (White woman, neuroscience major)

In non-SFC classes 20% of students praised their teacher in at least one comment but only 4% wrote positive comments about their SF class teacher.

In another category of comments, students discussed the personal characteristics of teachers, such as their demonstrated interest in the subject matter, their willingness to answer questions, and the interest they took in their students. Again, students in the SF classes made fewer positive comments about the teacher's personal style. Most wrote general negative comments about "poor teaching" as blanket, "thumbs down" judgments that did not include details about specific teaching methods—as in these examples:

> This class was very poorly taught in my opinion. (Hispanic man, Evolutionary Biology major)

> This is the worst class I have ever taken. (White man, engineering major).

> This class was my least favorite because the professor questioned himself and didn't teach well at all. He shouldn't have his job. (Native American woman, Aerospace Engineering major)

## Rote Learning and Dull Content Presented in Lecture Mode

Those students in both types of classes commented on specific teaching methods that had helped them learn the course content described a wide variety of helpful teaching methods and learning activities. These included, effective styles of lecturing, the use of clickers (interactive question and answer devices), group work, the use of "flipped" classrooms, and how well course content learned in labs and homework fitted with course assessments.

However, many students took the opportunity offered by the survey to criticize the teaching methods that they experienced in weed-out (i.e., SF) classes and negative assessments of instructors' lecturing styles greatly outnumbered positive comments. Far fewer innovative or participatory instructional techniques were reported for weed-out courses, and by far the most common types of complaint about weed-out teaching were that lectures were disorganized and dull. In addition, complaints about the over-use of PowerPoint slides and other visual aids were reflected in, or compounded by, an apparent inability of their teachers to explain concepts clearly:

> I thought the lectures were poor. They lacked clarity and topics seemed disconnected. A lot of information was skimmed, and core concepts weren't elaborated upon. (White man, Astrophysics major).

> This class was taught purely with Power Point slides... it was very boring and made it hard to pay attention to the concepts presented. (White woman, engineering major)

Other comments stressed the use of lecture to the exclusion of other teaching methods or learning activities, and, as (in interviewee complaints), pointed to poor intellectual articulation between lecture and lab content:

> I would have gained much more had the lecture been divided up into smaller sections. I personally learn much better in an environment where there is little distraction and I can feel free to ask questions. Being in a lecture with 300-400 students was extremely distracting and made it difficult to focus and ask the questions I needed too. (White woman, physiology major)

> The lecture did not help my understanding of many concepts...Also, lecture material did not correspond well with the lab, this made it difficult to understand the labs thoroughly and often led to confusion. (White man, Integrated Physiology major)

Many students in the SF courses commented they could have learned class content without actually attending classes. Indeed, many respondents wrote that they could (or did) learn the same content from reading the text book or from online information:

> This class did not help my learning. I taught myself all semester from the book. (Hispanic woman, physiology major)

> It seemed like going to lecture was helpful to a minor degree but not completely necessary. At some points I found myself thinking that I might as well read the book and supplement the parts that I don't understand with other material found online. I rarely found that I had learned something in a completely novel way from sitting in lecture. (White man, Evolutionary Biology major)

## Heavy Volume and Pace

As in the interviews, many students reported that weed-out classes presented too much information in too little time. In the SALG survey responses, critical comments about the pace of courses were much more common in SF than in non-SF courses (10% vs. 5% of all written comments):

> [The professor] stated that the course moved at a fast pace, but I had no idea that we would be attempting to cover so much material over the semester. When a professor has to send out multiple screencasts of material each week just to keep up with the course, you know it's too much material, even for them. I personally have felt like I never had time to completely grasp any concepts in the course before we moved on to new topics, causing me to feel like I was drowning in the course by the end. (White male, environmental studies major)

## Large Classes

Finally, large class sizes encouraged disruptive student behavior which also made learning difficult:

> The atmosphere of the class - This was my first large lecture class, but having a class where everyone around you is talking is very distracting. Especially in a class like chemistry. I understand that many people have to take the class, but it would be much better for the students of the class size was smaller. (White female, evolutionary biology major).

> I wish it were smaller; a lecture hall of 200+ students is just too large. (White male, physics major)

As we have argued elsewhere (Seymour & DeWelde, 2016), large classes offer opportunities for various forms of resistance to disgruntled students, such as overt disattention to the lecturer, arriving late, leaving early, eating, drinking, sleeping, chatting to friends, and use of cell phones. Poorly taught large classes also encourage an adversarial relationship between teachers and students that can support hostile attitudes by instructors toward students who approach them for individual help, and a generalized suspicion that students care more about grades than learning and are apt to cheat. The roots of such dysfunctional relations lie, however, in what our interviewees and survey respondents portray as the structured intentionality of the weed-out system to get rid of a high proportion of them rather than teach them.

Overall, the class ratings and written comments from the SALG survey present a negative picture of teaching and curriculum in STEM weed-out (i.e., SF) classes. Many students make general criticisms of the class overall. However, by the 14 attributes shown in Fig. 7.3, students in weed-out classes were distinctly more negative about their learning experiences than were students in other foundation courses in our sample. We have highlighted some of the facets of teaching methods, including instructors' poor lecturing styles and course structure that make learning difficult, the fast pace and the overwhelming amount of material covered, and problems arising from class sizes that were not conducive to learning. Many students reported that they had learned little from these classes, and many also wrote that they had gained more by studying course content from books and online sources than by attending class. In short, *the results of the SALG surveys that were gathered in a separate sample from the student interviews draw the same conclusions*.

## *How Were Weed-Out Courses Taught?*

In Chap. 6, we described some results from the classroom observation study of 71 STEM foundation courses in the same set of six institutions (Ferrare, 2019). Here we examine the results from structured observations that were made in the 12 (of 71) courses that we identified as weed-out (i.e., SF) courses. The structured observations in these courses supported the claim made by students in interviews and in SALG survey responses that the prevailing method of teaching was by lecture. As is illustrated in Fig. 7.4, in nine of 12 courses, instructors lectured more than 80% of the two-minute intervals observed, either using premade visual slides or writing on a chalkboard. In the three remaining courses, instructors lectured more than 60% of the time.

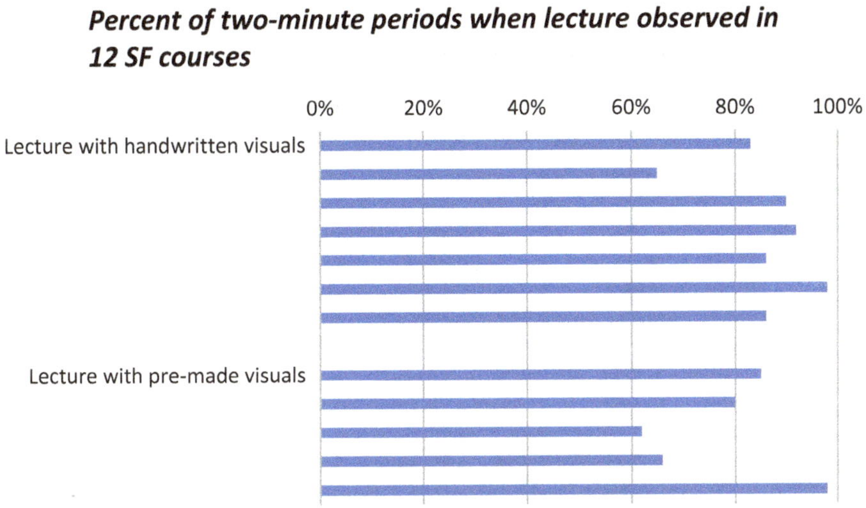


**Fig. 7.4** Percent of two-minute periods when lecturing was observed in 12 SF courses

Alternative teaching methods such as group work were rarely observed. Some work in groups was observed in two courses, and in three courses, observers recorded desk work and the use of clicker questions. Interactive lectures, where instructors asked students questions and students discussed answers, happened very infrequently. As is indicated in Fig. 7.5, in three courses, these alternatives to "straight lecturing" occurred at rates of only 6%, 4%, and 2% of recorded two-minute intervals. These were the only non-lecturing methods observed. In none of the 12 SF courses did observers report other active or interactive method. They recorded these, however, in regular use in 16% of other foundational courses and in occasional use in other lecture-dominant non-SF courses. Methods included peer interactions that required creating and problem-solving, teacher engagement with small groups, or multi-methods teaching that combined multiple forms of lecturing with demonstrations, small group work, and multimedia to illustrate content and make connections.

The findings from the three sources of data available in this overall study (viz., student interviews, SALG survey responses, and classroom observations), when compared, show the same lecture-dominant teaching methods in foundational STEM courses that we identified as "weed-out" (SF) by their teaching and assessment methods and DFWI rates. The interviews and survey responses also concurred on a high degree of student dissatisfaction and low estimates of their learning gains in weed-out courses. As many students testify, the difficulty level in many of these courses is artificial and fails to support development of authentic scientific understanding and skills. In the interview data especially, we noted that students regarded the poor quality of teaching practices that they experienced as deliberately designed and departmentally legitimated for the purpose of reducing the numbers of students who could proceed in STEM majors. Whether weed-out course instructors and their

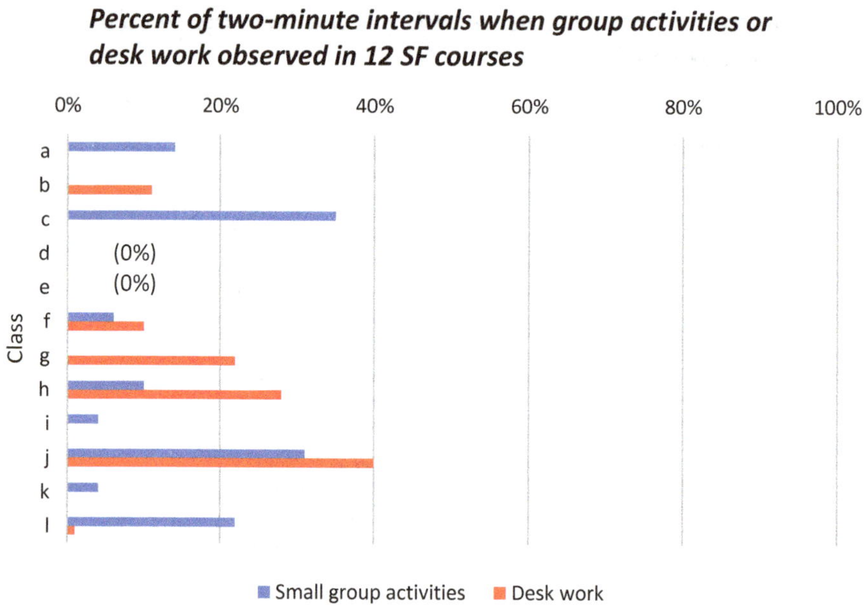


**Fig. 7.5** Percent of two-minute intervals when group activities or desk work were observed in 12 SF courses

departments check to see which students change their majors following weed-out courses, and whether these are the kinds of student that they aim to discourage from continuing, are open questions. We therefore turn our attention to the question of which groups of students switch or relocate as a direct consequence of their weed-out class experiences.

## The Consequences of Weed-Out Classes

Accounts of the consequences of weed-out course experiences were provided by 85 students in interviews. Of these, 50 students (59%) took a decision to change their majors that they explicitly ascribed to weed-out course experiences. The moves that these students recounted were of three kinds: switching out of STEM majors ($N = 34$); relocation from one STEM major to another ($N = 14$); and dropping one of two original majors ($N = 2$). Another 35 students who had persisted in their STEM major reported adverse effects from weed-out courses that mirrored those of those who changed their majors. We did not have access to students who had left college altogether following weed-out class experiences. However, later in this section, we present data on students in this group in the account of the (2019) study by our collaborators at the Gardner Institute, Andrew Koch and Brent Drake.

As all three studies reveal, students who switch following a severe foundational (i.e., weed-out) course are not a random group. As we shall illustrate, these students differ from persisters by a set of intersecting factors that include how they respond to poorer-than-accustomed grades, their gender, race/ethnicity, socio-economic status, and major. Weed-out course related switchers also include multi-talented and high ability students as well as those who enter college with poorer levels of high school preparation and lower SAT or ACT math scores.

## *What Is the Association Between Switching Rates and Receiving DFWI Grades in Severe Foundational (i.e., "Weed-Out") Courses?*

As discussed, we can identify courses that appear to push students out of their STEM majors by grading and assessment practices that discourage students from persisting. As we have also established by the institutional records analyses reported in Chap. 2, switchers get lower grades than persisters and show a significantly higher pattern of DFWIs. We wanted to know whether switchers were more likely than non-switchers to receive poor grades or withdraw from SFCs, and how switchers and non-switchers compare on the same indicators in other foundational courses.

In their first year, STEM switchers take more SFCs than persisters—an average of 2.09 for switchers and 1.34 for persisters. The difference is not due to over-representation of switchers in disciplines requiring SF courses. Indeed, we found that switchers take more SF courses across all majors. A more likely explanation is that switchers are more prone to "front load" with difficult courses during their first year—a pattern that is also likely to contribute to their academic duress. That said, not all STEM students take severe foundational courses or take them early in their major. In our sample of six institutions, just over two-thirds of students (70%) take one or more SFCs during their first year and 76% do so during their second year. In the interview study, we noted that some students avoid or defer courses with a weed-out reputation and some take them concurrently at a community college.

We examined the association between receiving a poor grade (DF) or an incomplete/withdrawal (IW) in SFCs and switching rates. Our transcript data analyses revealed which students received poor grades or incompletes/withdrawals in SFCs, and how many students who switched had received zero, one, or two poor grades during their first year. We compared groups of students who received none, one, or two poor grades or incompletes in SFCs.

As represented in Fig. 7.6, increases in switching rates for these groups are somewhat striking. While 12% of students who received no DFWI switched, this number jumped to 23% (almost double) for those students who received one DFWI and to 33% for those receiving two DFWIs. We found that:

- *Switching rates double for students who receive one DFWI.*

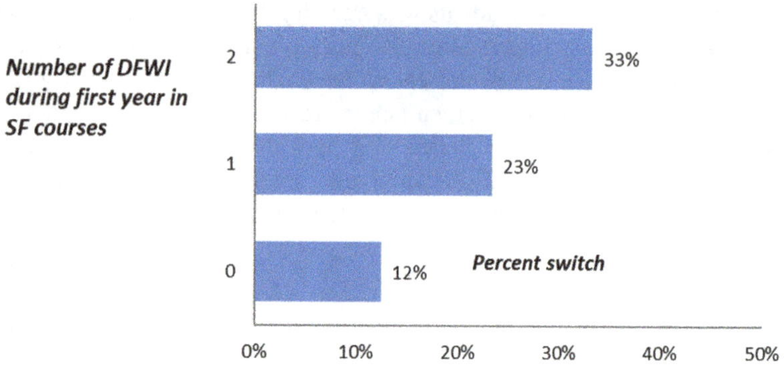


**Fig. 7.6** Percentage of students that switched during their first year for zero, one, or two DFWIs

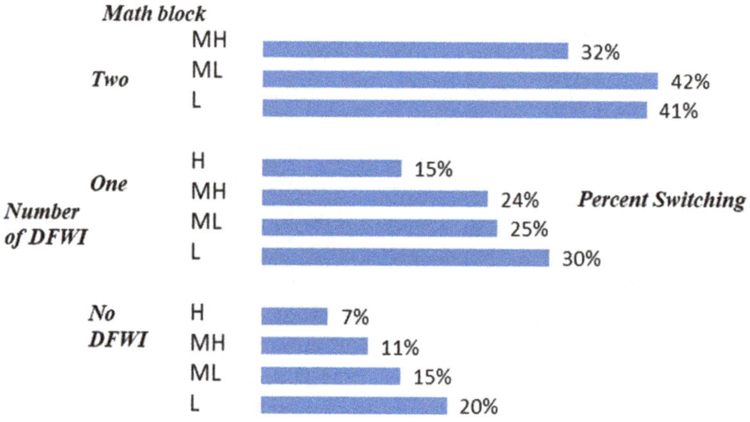


**Fig. 7.7** Percentage of students switching after receiving DFWI grades by math score block. *Note*: Math block, *L* low, *LM* medium low, *MH* medium high, *H* High. *Note:* Group size not large enough for confident estimate for two DFWI high math

We also looked at students with different levels of SAT and ACT standardized math scores and assessed the association between receiving a DFWI in an SF courses and the percentage of students switching. As described in Chap. 2, students with lower math scores switched out of STEM at higher rates. As Fig. 7.7 illustrates, the rates of switching rise proportionally across levels of math scores, with rates doubling between zero and one DFWI for students in the higher math groups. We concluded that:

- *While students with higher standardized math scores switch less than those with lower scores, more students switch at all ability levels when receiving one DFWI.*

In Chap. 2, we reported that women had higher rates of switching than men. As is shown in Fig. 7.8, we also found that women switched at higher rates than men across the math score distribution.

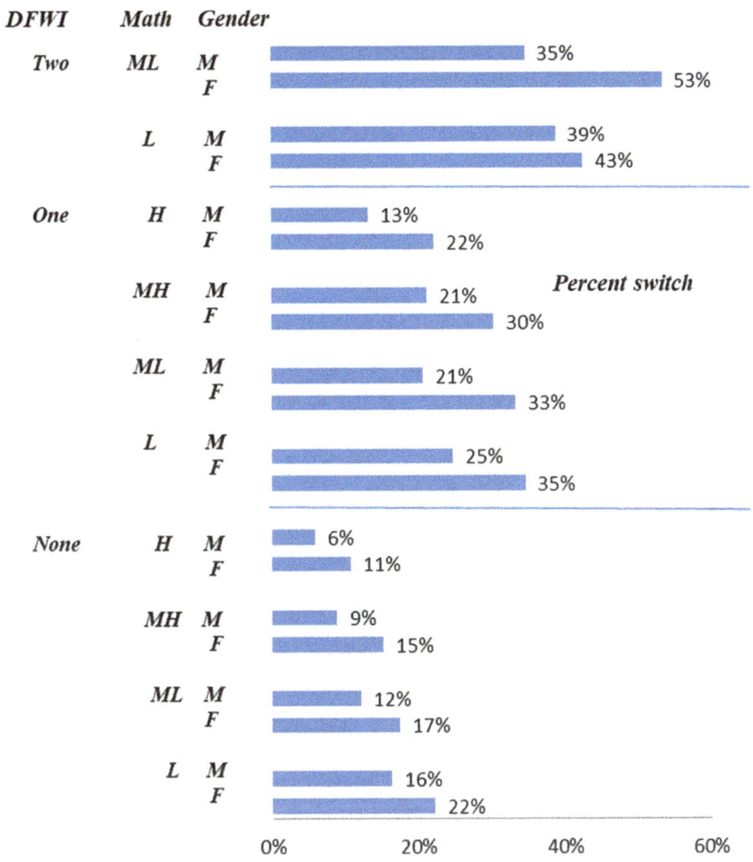


**Fig. 7.8** Percentage of male and female students who switch following receipt of a DFWI grade in an SF course

The difference in switching rates for both men and women double, or nearly double, for students who receive one DFWI—even for those students with higher math scores. Thus:

- *Rates of switching increase for both men and women proportionally when they receive a single DFWI in an SF course.*

We also examined how poor grades and incompletes in SF courses functioned in the two largest disciplinary groups—engineering and biology/life sciences. We were especially interested in the role of gender in this comparison, given large differences in gender representation by discipline such that proportionally more women majored in biology/life sciences (60%) than in engineering majors (21%). While more students switch out of biology/life sciences than engineering (20% vs. 9%), the rate of increase from zero to one DFWI in biology was proportionally less than in engineering. In engineering, switching rates increased 2.4 times (from 7% to 17%)

between students' receiving no DFWIs versus one DFWI. In biology, this increase was 1.8 (17% to 32%). In both engineering and biology this rate of increase was only slightly different for women than for men, suggesting that *gender representation in the major was unlikely to be a factor in the association between receiving poor grades and switching*. We therefore concluded that:

- *The gap between the switching rates of men and women remains fairly constant in different disciplines regardless of differences in the representation of women in these disciplines.*

We assessed the relative effects of receiving poor grades or incompletes in SFCs. Our logistic regression model tested whether receiving a poor grade, or an incomplete, in an SF course predicted switching; also, whether receiving one of these grades was a better predictor of switching than receiving a DFWI in another course. Holding other variables constant, we found that one poor grade in an SFC (i.e., a weed-out course) was associated with switching at a significantly higher rate than one poor grade in other courses. The risk of switching because of incomplete grades in SFCs was not only greater than in non-SFC courses but also showed a smaller difference. This finding suggests that:

- *Receiving either a poor grade or an incomplete in an SFC course is associated with students' decisions to leave a STEM major.*

In Fig. 7.9 we show the relative association of grades, student characteristics, institution, and degree on switching. The "odds ratio" (exponent-B) figure shows

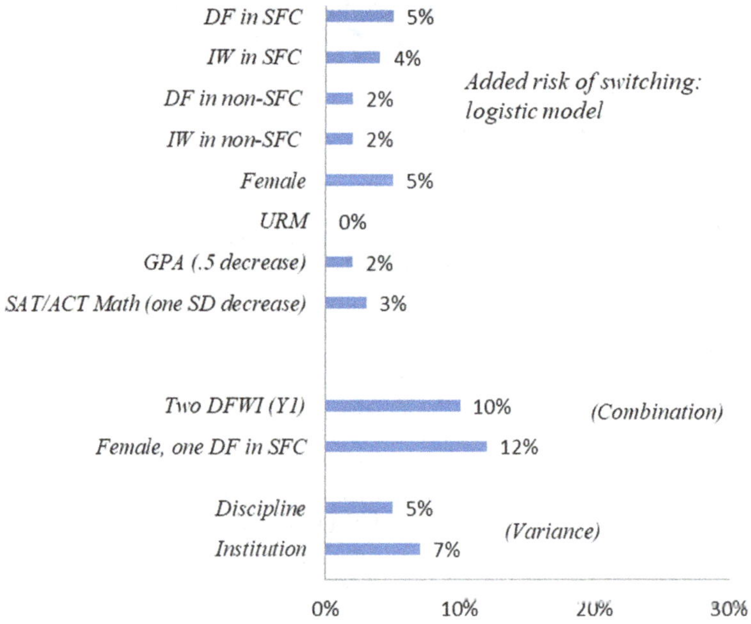


**Fig. 7.9** Added risk of switching out of STEM major for DFWI and other variables in logistic model

the magnitude of the predictive relationship between a variable and switching. Values greater than one predict the odds of switching; values less than one predict persistence. The full logistic regression model can be found in Appendix B.

From the logistic model, we calculated the increased risk of leaving a STEM major based upon student characteristics, grades, major discipline, and institution. Table 7.2 shows the increased probability of switching related to either an individual characteristic (e.g., female versus male) or a specific change in the value of a variable such as GPA.

We also showed the observed differences for each comparison. (The two numbers are usually different because the logistic model combines factors to calculate risk, and also takes into account the individual contribution of each factor toward predicting switching.) Combinations of student characteristics can increase the risk of switching substantially; for example, a woman with one D or F in an SF course has an 11% higher chance of switching than a man with no D's or F's. We therefore concluded that:

- *DFWI in SF courses is a good predictor of switching even when student characteristics, institution, discipline, and other variables are held constant.*

**Table 7.2** The increased risk of switching created by DF grades, IW, and other variables

| Variable/combination | Status/magnitude of difference | Increased risk of switching (%) | Observed difference (%) |
|---|---|---|---|
| DF in SFC | One DF in SFW course (Y1) | 5 | 12 |
| IW in SFC | One IW in SFW course | 4 | 10 |
| DF in non-SFC | One DF in non-SFW course | 2 | 7 |
| IW in non-SFC | One IW in non-SFW course | 2 | 5 |
| Gender | Female | 5 | 7 |
| Underrepresented minority | URM | 0 | 6 |
| Grade point average | Half point decrease (3.2 to 2.7) | 2 | 4 |
| SAT/ACT math | One standard deviation decrease (e.g., 600 to 500 SAT) | 3 | 6 |
| Average difficulty of courses encountered | One point decrease, equal to 0.25 of grade point average difference between courses | 3 | 0 |
| Both DF and IW in SFC | Receive both a DF and an IW in an SFC course | 10 | 21 |
| Female with one DF in SFC | Receive one DF in SFC v. male with no DF | 12 | 22 |
| Discipline | Average difference with median discipline | 5 | 5 |
| Institution | Average difference with median institution | 7 | 7 |

*Note*: Risk estimates based upon logistic regression model in table L Appendix
*Note*: Effects for Discipline and Institution are based on the standard deviation of b-coefficients for each factor

## *Summary of Risk Factors Derived from the Institutional Records Analyses*

As our analysis indicates, some students are more at risk of switching into non-STEM majors after receiving one poor grade in, or failing to complete, a severe foundational (i.e., weed-out) course. Combining our SF course findings with those reported in Chap. 2 for all STEM courses sampled, we found students have a higher risk of leaving a STEM major if they:

- Receive one or more poor grades in, or do not complete, an SF course during their first year;
- Have a lower GPA overall (even with no DFWIs);
- Enter college with a low SAT/ACT mathematics score;
- Are a woman;
- Are a woman of color with below average ACT/SAT math scores;
- Are a woman accustomed to getting good grades who receives a poor grade, an incomplete/withdrawal (or even a single C grade) in an SFC course;
- Come from a family with lower socio-economic status (indicated by a PELL award) or are a first-generation college student (The risk for all women is enhanced for women of color in this group.);
- Are a high-performing (including multi-talented) student whose interest is discouraged by SFC experiences. Their leaving is associated with receiving one poor grade;
- Choose mathematics or computer science as their major.

Receiving poor grades or failing to complete an SFC course, by itself, increases a student's chances of leaving a STEM major by 5%, a value that is similar to that of other variables such as gender or institution. It is important to note that *combinations of characteristics substantially increase risk*. Thus, women with one D or F in an SF (weed-out) course have an 11% greater chance of leaving than a man with a higher GPA and no poor grades. In Chap. 2 we also explained how some characteristics increase risk only in specific combinations. In these cases, groups such as women of color with low math scores show very different rates of switching than do men of color with higher math scores.

The patterned, non-random, nature of losses from STEM majors arising from foundation course experiences found in the institutional records data analyses are corroborated by findings from the Gardner Institute study of the patterned consequences of DFWI grades in STEM gateway courses by our research collaborators, Andrew Koch and Brent Drake. In the following sections we present these findings along with a discussion of explanations for the patterns documented by our three triangulating studies.

## Patterns of Structured Disadvantage and Their Consequences in STEM Gateway Courses in 36 Institutions

These findings form part of a seven-year study of gateway course redesign initiatives conducted by the non-profit Gardner Institute. The whole study sample includes both STEM and non-STEM disciplines in 36 institutions in a broad array of public and private institutional types. Institutional records were provided for academic years 2005–2006 through 2015–2016, with the earlier ranges allowing for analysis of 4-, 5-, and 6-year graduation rates. The full data file comprises over 1.2 million records and contains: (1) course data (course type, instructor type, mode of instruction, course number, section, initial grade, final grade); (2) student registration data (enrollment, retention, degree completion, student classification, degree type, academic program CIP, credits, credits source); (3) student demographic data (race/ethnicity, gender, first-generation status, entry term, birth date, high school location, high school GPA, prior degrees, GED status); and (4) financial aid data (Pell grant eligibility and recipient status, FASFA filing date, grant award status). The STEM gateway course analysis comprised four STEM subject areas (biology, chemistry, algebra, and calculus) and was limited to the academic year prior to implementation of any undergraduate education redesign efforts at each of the 36 participating institutions. These baseline years ranged between 2012–2013 and 2014–2015. Gateway courses were defined in a similar manner to our institutional records study, namely, as foundational, high-enrollment, and high risk by dint of their high levels of DFWI rates (cf.Koch, 2017 and Koch & Rodier, 2014).

Koch and Drake used the baseline data to calculate a DFWI rate for each institution by course subject, namely, the percentage of DFWI grades of the total grades awarded in the course, and the mean DFWI rate for each subject from the institutions' individual DFWI rates. This allowed focus on differences between rather than within course subject areas and avoided skewing the data towards institutions with larger student populations. As Table 7.3 shows, not all institutions had data available for all four STEM gateway course disciplines: 33 institutions had data for biology, 31 for chemistry, 34 for algebra, and 32 for calculus. Findings were disaggregated by race/ethnicity, gender, first-generation, and family income statuses, and by subsequent retention of the students. Students of color were defined as those from African American, Native American, or Latino/Hispanic groups. Eligibility for a Pell grant was used as a proxy for low income because, based on the Federal formula, all such students come from families with the nation's lowest incomes. Although there is no definition of "first-generation students" that is applied at all colleges and universities, this designation is usually given where a student's parents have not earned a baccalaureate-level degree. It is also correlated with possessing less of the social capital needed to navigate an undergraduate education.

What follows are both the aggregate and disaggregated findings for the four STEM gateway courses. The broader study of DFWI rates and correlated outcomes for all eight (STEM and non-STEM) courses will be published by the Gardner Institute in 2019.

## *Study Findings*

As supported by the data displayed in Table 7.3 (Columns D, E, F, G, and H), students of color almost always had DFWI rates that exceeded both the course average and the average for their white peers. In many instances, the differences are striking. For example, in general biology, African American students had a DFWI rate that is 12.1 percentage points (40.6%) higher than the course average and 16.0 percentage points (61.8%) higher than white students in their course. In general chemistry, Latino/Hispanic students had a DFWI rate that was 12.6 percentage points (42.9%) higher than the course average and 15.7 percentage points (59.7%) higher than white students in the course.

A similar pattern of disadvantage is evident in the data in Table 7.4: both first-generation and Pell grant-eligible students did worse, on average, in the four STEM courses when compared both to their peers from more educated and affluent

**Table 7.3** Rates of D, F, W, and I grades by STEM gateway course and race/ethnicity

| A - Course | B - Institutions (*N*) | C - Students (*N* across courses) | D - DFWI rate (%) | E - African -American DFWI rate (%) | F - Native -American DFWI rate (%) | G - Latino/ Hispanic DFWI rate (%) | H - White DFWI rate (%) |
|---|---|---|---|---|---|---|---|
| General biology | 33 | 24,636 | 29.8 | 41.9 | 37.4 | 35.0 | 25.9 |
| General chemistry | 31 | 20,987 | 29.4 | 47.2 | 54.5 | 42.0 | 26.3 |
| Math—algebra | 34 | 55,075 | 34.4 | 49.9 | 39.0 | 38.3 | 30.6 |
| Math—calculus | 32 | 13,253 | 34.3 | 47.8 | 32.3 | 47.9 | 31.5 |

**Table 7.4** Rates of D, F, W, and I grades for STEM gateway courses by first-generation and Pell grant eligibility statuses

| A - Course | B - Institutions (*N*) | C - Students (*N* across courses) | D - DFWI rate (%) | E - First-Gen. DFWI rate (%) | F - Non first-Gen. DFWI rate (%) | G - Pell DFWI rate (%) | H - Non-pell DFWI rate (%) |
|---|---|---|---|---|---|---|---|
| General biology | 33 | 24,636 | 29.8 | 34.0 | 29.0 | 34.0 | 28.8 |
| General chemistry | 31 | 20,987 | 29.4 | 32.8 | 26.9 | 32.4 | 28.3 |
| Math—Algebra | 34 | 55,075 | 34.4 | 38.7 | 31.5 | 38.0 | 31.7 |
| Math—Calculus | 32 | 13,253 | 34.3 | 36.8 | 33.4 | 37.5 | 29.9 |

families and to the overall course DFWI rate average. First-generation students had DFWI rates ranging between 3.4 percentage points higher than their non-first-generation counterparts in calculus (36.8% first-generation students compared with 33.4% for non-first-generation) to 7.2 percentage points higher in college algebra (38.7% compared with 31.5%). (See Table 7.4, Columns E and F.)

Similarly, Pell grant-eligible students had DFWI rates ranging between 4.1 percentage points higher in chemistry (32.4% Pell-eligible students compared with 28.3% for students who were not Pell-eligible) to 7.6 percentage points higher in calculus (37.5% compared with 29.9%). (See Table 7.4, Columns G and H.) While the differences are not so large as those found for groups by race/ethnicity (in Table 7.3), they suggest that students from families with greater financial (and likely social) capital have a clear advantage in STEM gateway courses over their peers from poorer families.

Table 7.5 shows the relationship between earning a DFWI in one of the four STEM courses and the three retention-related statuses available in the institutional record. These are, students who: (1) were retained in the institution; (2) chose to leave or were not retained in the institution; and (3) graduated. (Groups 1 and 2 are wider group than our switcher sample which did not include students who had left school). In all four STEM courses, those students who returned the following year to the institution at which they had taken the course (identified as "retained" in Table 7.5) had lower DFWI rates compared to their non-retained peers. (See Columns E and F.) The differences in DFWI rates for students who left the institution after they took a STEM gateway course (Column F) compared with their retained counterparts (Column E) range between 18.9 percentage points (59.4%) higher in calculus (31.8% DFWI rate in calculus for retained students compared to 50.7% DFWI rate in calculus for students who chose to leave) and 23.2 percentage points (89.2%) higher in general chemistry (26.0% compared with 49.2%).

The DFWI rate for students who chose to leave the institution (Column F) may be of particular concern to STEM departments and their institutions, and to STEM educators and policy makers at large. These students were eligible to return to the institution because they were otherwise "in good academic standing." This is defined as having a 2.0 grade point average (GPA) or better—the minimum GPA needed to

**Table 7.5** Rates of D, F, W, and I grades by STEM gateway course and graduation rates

| A - Course | B - Institutions (*N*) | C - Students (*N* across courses) | D - DFWI rate (%) | E - DFWI retained rate (%) | F - DFWI non-retained rate (%) | G - DFWI graduation rate (%) |
|---|---|---|---|---|---|---|
| General biology | 33 | 24,636 | 29.8 | 26.5 | 47.6 | 17.5 |
| General chemistry | 31 | 20,987 | 29.4 | 26.0 | 49.2 | 17.5 |
| Math—algebra | 34 | 55,075 | 34.4 | 31.0 | 52.5 | 18.6 |
| Math—calculus | 32 | 13,253 | 34.3 | 31.8 | 50.7 | 22.6 |

graduate from nearly all accredited postsecondary institutions in the USA. It is important to note that these students were not dismissed from their institution. Rather, having earned a D, F, W, or I grade in one of these four STEM gateway courses, they elected not to return to the institution the following year. Thus, an unsuccessful outcome in just one of these gateway courses is related to the decision to leave the institution, even when the student is otherwise in good academic standing.

Column G in Table 7.5 also merits examination. This is the DFWI rate for students who took one of the STEM courses in this study and subsequently graduated from the institution at which they took the course. These DFWI rates range between 11.7 percentage points (34.1%) lower for graduates compared with the course average rate in calculus and 15.8 percentage points (45.9%) lower for graduates compared to the course average rate in college algebra. (See Columns G and D in Table 7.5.) These data indicate that:

- *Not only does earning a D, F, W, or I grade serve as a predictor for attrition, these grades are also a predictor of who ultimately graduates.*

Koch and Drake also examined DFWI rates by gender and, although the same patterns of disadvantage apply both to women and men, nevertheless, the data in Table 7.6 indicate that women did better on average than men in the four STEM gateway courses studied. Women had DFWI rates ranging from 1.2 percentage points lower than their male counterparts in calculus (34.5% for women compared with 35.7% for men) to 8.1 percentage points lower than men in general chemistry (25.8% compared with 33.9%). (See Table 7.6, Columns D through F). In all but one case, women always had lower DFWI rates than the course average. (See Table 7.6, Columns D. through F.). The exception were calculus courses where students whose gender was not recorded had much lower DFWI rates that pulled down the overall mean.

Koch and Drake offer some speculations about how to interpret the patterns that they found among women in their sample of STEM gateway courses. We discuss these in the following section alongside findings from the interview study on this issue.

**Table 7.6** Rates of D, F, W and I grades by STEM gateway course and gender

| A - Course | B - Institutions (*N*) | C - Students (*N* across courses) | D - DFWI rate (%) | E - Women DFWI rate (%) | F - Men DFWI rate (%) | G - Unreported DFWI rate (%) |
|---|---|---|---|---|---|---|
| General biology | 33 | 24,636 | 29.8 | 28.2 | 32.5 | 20.4 |
| General chemistry | 31 | 20,987 | 29.4 | 25.8 | 33.9 | 23.3 |
| Math—algebra | 34 | 55,075 | 34.4 | 31.5 | 37.4 | 38.1 |
| Math—calculus | 32 | 13,253 | 34.3 | 34.5 | 35.7 | 7.9 |

## *Implications of Findings*

Koch and Drake's findings from 36 colleges and universities are in accord with our institutional records study from our six sample universities. Both studies clearly demonstrate that a student's chances of passing "severe" STEM gateway courses (i.e., those with high DFWI rates that are informally referred to as weed-out courses), and of remaining in college to successfully graduate, are greatly diminished by structured disadvantages of race/ethnicity, family income, and education level. These disadvantages present among entrants to STEM majors as under-preparation by poorly resourced schools and limited social and financial capital. As both studies show, these risks are evident when examined singly, but, they greatly increase when, as our institutional records analysis establishes, they occur in combination. Expressed from the perspective of STEM departments and their institutions, students who receive DFWI grades in STEM gateway courses are found by both studies not to be random individuals. Rather, they are disproportionately found among students who enter with one or more socio-economic disadvantage.

Koch and Drake also found this patterning to be evident also in the consequences for graduation of poor performance in these "severe" gateway courses: students with disadvantaged socio-economic profiles are at risk of leaving their institution following one DFWI grade even when their grades in other courses place them in good academic standing. They are, therefore, less likely to graduate than students who enter STEM gateway courses without these disadvantages. This finding makes a significant contribution to our understanding of why 20% of students who enter STEM majors leave college without a degree in any major—a major form of loss whose causes we described in Chap. 2 as under-studied. Koch and Drake identify STEM gateway courses as major contributors to the patterns of under-performance and loss in STEM majors by students from low-income and first-generation families who may also be African American, Native American, or Latino/Hispanic. These have long been reported as populations that do not flourish in STEM majors (e.g., Eagan, Hurtado, Figueroa, & Hughes, 2014; Hrabowski, 2003 & National Action Council for Minorities in Engineering, 2013). What this study achieves is to clearly identify STEM courses with high DFWI rates as major contributors to these patterned losses. In doing so, it confirms and reinforces the findings from both our institutional and interview studies.

As the Gardner Institute initiative focuses on improving the quality of teaching and assessment practices in STEM gateway courses, Koch (2018) is reluctant to use the term "weed-out" to describe foundational courses that produce these patterned outcomes. However, the study's findings align with our evidence that weeding out some proportion of students is evidently intended, although the students that are weeded out are not be the ones that are consciously (if hypothetically) targeted.

The most critical implication of their study in Koch and Drake's view is that, if the culture and practices in gateway courses remain un-remediated, they will continue to exacerbate the situation discussed throughout this book whereby only 40% of students who start in STEM fields complete a STEM degree within 6 years

(President's Council of Advisors on Science and Technology, 2012). Furthermore, as Grave points out, "over the next 15 years, persistent trends in immigration, migration, and differential birth rates, coupled with the recent acute birth dearth, will markedly alter the college-age population" (Grave, 2018, p. 18). The populations that, historically, have not done well in the STEM fields, and that did not do well in the STEM gateway courses included in these studies, will become the growing majority of the college-age population.

Koch and Drake argue that to ignore their findings in the context of these demographic shifts would be unwise and, ultimately, self-defeating. For decades a more affluent and privileged college-going majority has masked failure rates and enabled STEM educators to ignore structural racism and classism at work in gateway courses. They conclude that gateway course failure rates can no longer be ignored or viewed as a badge of distinction and rigor. To do this would be to ignore evidence about which student groups do and do not succeed in STEM fields and endorse outcomes that are clearly inequitable. It would also disregard the growing body of evidence about pedagogical and policy practices that can improve, even reverse, these outcomes. They express the hope that, for the betterment of the STEM disciplines, the institutions that house them, and the communities they serve, STEM educators will not ignore the weight of this evidence and will make gateway courses a primary place for action and agency.

## Explaining the Loss of Women from Weed-Out Courses

Consistent with the findings from both institutional records studies (and as further discussed in Chap. 10), women of all races and ethnicities often described themselves at greater risk than men in weed-out classes. Their explanations include their own socialized responses to courses where the instructor's overt aim appears to be to fail a high proportion of students, and, in some cases, discriminatory practices by instructors. We describe in Chaps. 3 and 9 (and also Chap. 5 of the original study) loss of confidence and extrinsically dependent assessments of their own abilities reported by many young women, both switchers and persisters. In the original study, women expressed more dependence on the good opinion and encouragement of their high school and college teachers than they do in the current study. However, 20 years later, the need for validation from significant others has not disappeared, and many young women are still less able than their male peers to diminish the significance of reversals, take them in stride, and not allow low grades or college teacher indifference to throw them off track.

The institutional study points to the elevated risk to young women of color of abandoning their aspirations for a STEM degree. In the interview study, we noted that fragility of self-confidence was particularly marked among women of color, especially the daughters of more recent immigrant families, as in this case:

> Yes, weed-out classes do affect women differently because men go in there with attitude, 'I am going to be in the top 10%.' Women are thinking, 'Oh shoot, I am going to have to try

> hard not to fail.' And when you have a guy saying, 'Oh well, I didn't study,' it just makes it worse. You feel you aren't good enough. And you leave. I had an African-American woman friend who struggled with the same thing—although she managed to stay in the field. (Asian woman, low math, switcher, life sciences to strategic communication)

This white woman stayed in her STEM major notwithstanding a similar struggle with damage to her confidence and a fear of failure:

> The weed-out courses are trying to get rid of students rather than bring everyone along—it's psychologically crushing if you are in a class where you know their one objective is to get rid of you. (White woman, high math persister in mathematics)

A woman who also describes her college transition problems linked to under-preparation also points to the damage that weed-out courses can do to confidence—especially when this is already shaky:

> People who are more sensitive, and people who don't have a lot of confidence in their abilities; it's just reaffirming what they already believe—that they can't do it—that they're not smart enough. Though these classes can be hard on a lot of different people, some people can push through a lot easier and keep going. (White woman, high math switcher, life sciences to psychology)

Fragile, other-dependent confidence is also undercut by a tendency to link grades to identity. We described this in the original study as the reaction that, "If I get a B, I am a B." This tendency may now be greater. Parents expect more of their daughters in terms of academic and career success and future financial independence than at an earlier time when marriage was still viewed as a source of long-term financial security for women. Disenchantment with marriage as a durable institution appears to include some rethinking (especially by fathers) about how best to secure a self-supporting future for their daughters. This is reflected in pressures to do well in high school and in a college major that is expected to lead to a financially secure career. These pressures (described in more detail in Chaps. 10 and 11) are internalized in young women as very high demands on themselves. This "perfectionist" tendency helps to explain why women abandon original majors with grades and GPAs that male peers view as good enough to keep going. As described in Chap. 2, women with the same SAT math scores as men get higher grades. And it is not a new discovery that women both enter and leave STEM majors with higher average performance scores than both male switchers and persisters. Indeed, we reported this in the 1997 study. What may have increased is the intensity of self-demands under parental and societal pressures that women internalize to such a high degree that many seniors described their ongoing difficulties in letting go of them:

> Interviewer: Is it difficult to accept anything other than the perfect score? Why are you smiling?
>
> Student: Because that just defines me. Until very recently, I could not accept an A minus. So, when I came here and started getting A minuses for working my butt off. It broke my heart…In electronics, in my junior year, I got my first B, and I've never worked so hard for a class… Now my perfectionism has faded. But for a very long time I would cry if I got an A minus instead of an A. I got over it though. (White woman, high math persister in bio-sciences)

Seniors also describe how women with high grades may, nevertheless, define themselves as failures and see this as a reason to switch to something else:

> She got one of the highest grades in neuro. It was a B and she was devastated by that. But it was one of the highest grades in the class. It wasn't like it was like a failure. She's still having to wrestle with that, but she's staying with it. I've met women who have switched into econ, history, or other things. (Asian woman, high math persister in bio-medical engineering)

And some women never got beyond their initial shock and self-blame:

> I felt pretty bad about myself. I graduated high school with over a 4.0 and then I got to college and I wasn't doing nearly as well as I thought I would be. So, I felt really bad about myself. I thought I was stupid. I thought that I was going to fail. I thought that I was going to disappoint my family. I thought that I wasn't living up to my full potential and I just didn't really understand why it was happening. (White woman, low math switcher from life sciences to family social science)

A senior explained, from experience, that women often leave to find a major where they can feel good about themselves once more:

> I've met a lot of people who started off doing well and they just, they just lost confidence… They often make the choice to switch to a major that will bring them more confidence--somewhere they will do well consistently. (White female high math persisting senior in Organisms, Ecology and Evolution)

This factor in some switching decisions is discussed in detail in Chap. 9.

How, then, were some women able overcome the feeling they were not good enough, while others were not? This senior struggled with both her own and her internalized parental expectations. In the end, her intrinsic interest in the discipline won out:

> I was definitely the most neurotic person about grades that I've ever met… One of the reasons I gave up my thesis… was I knew it wasn't going to turn out as well as I liked, and… my professor told me I was on track to get like a C. And that freaked me out so much--I'd already given up A's. I could get A minuses, I was okay with that. But a C? No. Not even to save my GPA, just like really for my own pride… Part of it is my mom is very much a perfectionist and has always pushed me to do my best, which means <u>the</u> best. So…I have like a voice inside that's always like, if I didn't get an A it means I didn't work hard enough, I could've done better and I've just been letting myself down. And that's really scary, especially when…I know in some other fields I could have killed it and gotten straight A's and not have worked as hard. It was definitely tempting at sometimes, but science is just way cooler. (White woman, high math persister in life sciences)

Another woman survived, in large part, because a professor took the trouble to put her grades in context and encouraged her to keep going. She was willing to listen because, like the speaker above, she was swayed by her intrinsic interest in the field. However, it was a close call:

> Student: I think it, it definitely took some convincing on my professors' part to really nail it into my head that just because I got a B minus doesn't mean that I shouldn't continue on in the major. Even though that was my first thought. But also, I was so interested in it that I was willing to like latch onto what he was saying and just go with it.

> Interviewer: They could have lost you if he hadn't gotten through to you.
> Student: Yeah, definitely. (Asian woman high math persister in bio-medical engineering)

As in the original study, we heard many stories in which a decision to stay or leave turned (as it did in the above example) on a serendipitous intervention by an instructor or advisor who persuaded a capable student that they should stay. We also note that more experienced, students (e.g., in dorms or sororities) play an important role in explaining how to put low grades in weed-out courses into perspective. This kind of interpretive work is also offered through students' disciplinary and professional organizations such as Women in Science and Engineering (WISE) or the Society of Women Engineers (SWE). However, the chance nature of these encounters means that many women that could have persisted may not have done so. We note how valuable such senior peer "translation work" can be to survival, especially for women with high self-demands whose confidence is undermined by the consequences of curve grading:

> If I hadn't had upper classmen and good study resources telling me that getting a 65 on an exam was a good score for that professor, I would have been like the other people who dropped out of the major because of over-concern about their early grades. They might have stayed if they'd had that help. (African American female high math Persister in Cell Biology and Molecular Genetics)

An entirely different approach to weed-out survival is to weigh the odds. Has so much already been invested that switching involves too great a loss? Or are there advantages to switching after a bruising weed-out course experience? This STEM switcher moved majors in order to ensure graduation:

> My first weed-out class was materials and energy balance which I got through. But it was like, 'Hmm, if I'm struggling this much with it, there's something wrong.' I enjoy a challenge, but if the challenge means utter failure in college, then I need to adjust. (Multi-racial female high math relocator, engineering to food science.)

This persister decides not to take the gamble:

> I remember talking to my parents on the 'phone and thinking,' 'I am so stressed out; is it really worth it? Should I be doing something in the humanities or social sciences instead?' It was just so hard and the professor was so bad. I didn't know if I could handle this. But I also thought, 'I've already taken some prerequisites and it would be a waste to just throw them away.' (White female low math senior in evolutionary biology)

Some women decide that it is "better to bail than to fail," but timing is a critical part of this calculus:

> We lost about half the class at the withdrawal date. I had a C and I thought, 'That's all I need. I just need a C to get through.' And I stuck it out, but I probably should have withdrawn because I ended up failing. It's like a lottery. (Multi-racial female high math switcher, mathematics to hospital management)

Again, we note the theme of "needing to adapt" in order to survive. As we have illustrated, switching may, in itself, be an adaptation—a means to restore damaged confidence and self-esteem. We discuss this theme further in Chap. 9.

### Under-Preparation and Structured Inequality: The Intersection of Race/ Ethnicity and Socio-Economic Status

We cross-referenced the records of the 34 interviewees who reported that they had switched out of STEM majors because of weed-out experiences with those who identified poor high school preparation and problems with transition to college as contributing to their switching decisions. Eighteen of the 34 (53%) reported all three factors in their decision, but only four of these students entered with low math scores. One explanation is that good SAT/ACT math scores may reflect a natural ability to do math but can mislead students into assuming that they are better prepared than they actually are for college–level work:

> I got a 690 for my SAT score and thought I was all set. But my bruising experiences in the Calc 2 and Gen Chem weed-out classes showed me how under-prepared I actually was. I had a heck of a time making up for starting behind and staying in the major. (Hispanic man, high math persister in life science/bio-medicine)

Students who identified themselves as "bright" although critically underprepared include these two women:

> It was freshman year, and they were all hopped up on college with this big, 'I must prove myself' thing. I felt a lot of pressure to be like the smartest one in the room. Like all of us just coming from high school, I usually was the smartest one in the room. So everyone jockeying, but I wasn't in a great position to do that because I hadn't taken the course I needed beforehand. (White woman, high math switcher, engineering to psychology)

> If I had known enough to prepare for the kind of class organic chemistry was, that would have made it easier. But I came into college not knowing about weed-out classes until freshman year and everyone was talking about them.... It helps to know what you are getting into. (African American woman, high math switcher, life sciences to community health)

As the last speaker makes clear, under-preparation includes not being alerted ahead of time to the nature of weed-out courses. And limited prior knowledge of college is exacerbated where advisors fail to steer students away from undertaking several weed-out courses in the same semester:

> I was taking Calculus 2 and Chem 2—which I felt was intended as a weed-out class for chem majors—plus my own biology class. So, I was taking these incredibly challenging courses all at one point in time and it wasn't working. I was failing everything. (White man, high math switcher from life sciences to dance)

The following STEM relocator makes the same point, but also comments (as do we throughout our analyses) on the push-pull nature of switching considerations:

> They're not ready for taking a failure at the same time as success in something else they like. And if they hit another weed-out class at the same time, then they're just like, 'I'm done.' (White man, high math relocator, biochemistry to biology)

As we discuss in more detail in Chap. 12 (on what enables persistence), seeking and finding timely help is a more effective survival strategy than toughing it out alone. Thus, one significant dimension of under-preparation is not knowing that various forms of academic help are available, or that it is important to seek and use them out in a timely way. This is a persistence risk for all students with inadequate college know-how—the kind of social capital that is apt to be missing among working class students of all races and ethnicities:

> I wish I hadn't let one course be the deciding factor of an entire major. I wish I'd known that there's support if you are willing to look, and that it's not a make or break decision in the first course—that if you can't do this, you can't do it at all. And I wish I had known more about tutoring and study groups but I didn't know they existed. (Multi-race woman, high math switcher, mathematics to business)

Again, we also point to the damage that the "constructed hardness" of weed-out teaching methods can do to students who are already struggling to recover from the discovery of under-preparation by their high school:

> Professors do make it harder than it needs to be, either because it's a weed-out class, or because they're not the greatest teachers. Sometimes the professor expects they know things that we should have learned in high school but didn't. So, there's a disconnect right there that makes the class twice as hard as it needs to be. (Asian woman, high math relocator, biology to physiology)

How some under-prepared students survive beyond weed-out courses includes practical help from older students or from advisors who steer these students around such courses until their skills have been built. Some of our under-prepared interviewees had repaired their preparation deficits by taking time out to attend community college or had taken courses required by their major that they could transfer in. A departmental policy of late declaration is also valuable to students who enter under-prepared, giving them time to build missing knowledge and college-navigation skills.

## Are Students of Color at Particular Risk in Weed-Out Courses?

As discussed, the institutional records study identified women of color with below-average ACT/SAT scores as at particular risk of leaving their STEM majors. In the four STEM gateway courses at 36 institutions examined in the Gardner study, Koch and Drake found that African American, Native American and Hispanic/Latino(a) students almost always had DFWI rates that exceeded both the course average and the average for their white peers. In the interview study, of all 85 students who described negative consequences because of weed-out courses, 38% ($N = 32$) were students of color. Among those who moved majors (whether within or outside of STEM) because of weed-out experiences, students of color were 54% ($N = 26$) and white students were 51% ($N = 70$). Of the 18 STEM switchers who cited poor preparation, college transition difficulties and negative weed-out experiences as all having contributed to their move out of STEM majors, nine were students of color—six women and three men. By gender, 14 were women and four were men. Do these data indicate higher risks for students of color in weed-out classes and, if so, why? In the institutional records analysis, students of color do show an observed risk of switching that is 6% greater than white students. However, in the logistic models, this difference disappears because other variables (e.g., GPA and math scores) explain switching rates better than race/ethnicity per se. So, we must ask whether other variables intersect with race/ethnicity to explain the switching risks of students of color in weed-out classes.

In Chap. 5, we outlined the relationship between race/ethnicity, first-generation college status and the enhanced risk of arriving in college under-prepared for introductory STEM courses. Under-prepared students, whether white or of color, commonly attended small or poorly resourced elementary, middle and high schools—many in rural or low-income urban areas—that do not offer advanced math or sciences courses. Students often discover, in retrospect, that the quality of teaching and conceptual challenge in their high schools was lower than that experienced by peers from better schools. Working class parents of all races and ethnicities are also more likely to have limited "social capital" to give to their children. In this context, they have less knowledge about how higher education works, about what career pathways exist, and how to get the most out of college. As we have reported, students of color are over-represented among under-prepared students who switched: 36% are African American, 22% Hispanic, and 16% white. As we discuss in Chap. 11, poorer families are also less likely to be able to provide funds for college, so their children are more likely to have to work to pay for their education and support themselves while in school. Financial constraints and the need to work are, as our iceberg table reveals, a strain that affects 70% of all switchers, 48% of persisters, and are direct contributors to 10% of all switching decisions. Thus, students of color, along with other working class and first- generation students face a set of structured socio-economic and educational disadvantages, including under-preparation for both STEM majors and college, that derive primarily from the limitations of their circumstances. All of these disadvantages are perhaps greatest among immigrant families.

Thus, the playing field is not level for students who enter first year foundational courses, especially courses taught and assessed with weed-out intent. As this Asian STEM senior explains, the weed-out process is blind to variations in socio-economic advantage:

> I've definitely worried that I don't belong here—not in this major; not in this school. I've felt that since freshman year. Just being a minority student, it's harder to connect with people, even professors. But the main thing is that my professors don't understand the circumstances that I and other minority students come from—lack of resources at home. I can't afford a tutor or sometimes even a book. Often, I don't have the book until the third week of class. So, I'm not failing because I want to. (Asian woman, low math persister in life sciences)

As the institutional records analysis reveals, being an under-prepared woman of color adds additional layers of vulnerability. As both the last and next speaker explain, numeric isolation underscores the pervasive weed-out message that, "you don't belong here":

> It's about a sense of belonging. If there's a lot of people like you, it's easier to stay… But that weed-out mentality, if you don't like it and the majority of people are not like you, you are gonna feel it's less attractive. (Hispanic woman, high math switcher, math to business)

This Hispanic woman felt both under-prepared and actively encouraged by instructors to leave:

> The main thing is they just assume we should know some things: 'You should know this already.' And if you don't do well, they suggest you drop the class: 'You should be at this level, and if you're not, don't forget, the last day to drop is Friday.' It's really discouraging. (Hispanic woman, low math switcher, life sciences to psychology)

We received only a few accounts of active discrimination by instructors. This is one of them:

> Weed-out classes definitely affect women and students of color differently from the way they affect white men. One of my women friends and me were the only two black kids in the class. The teacher was always eyeing us like she was making sure we didn't do anything suspicious. She told my friend that she cheated on one of the tests—which she didn't. And there was this huge ordeal—a meeting where she said, 'I am going to keep my eye on you and make sure.' My friend ended up dropping the class because she knew there was no way she could get a passing grade. (African American woman, low math persister in life sciences)

Thus, socio-economic disadvantages, such as the need to work while in schools, that are shared with many white students are often enhanced for students of color by other pressures—social isolation, stereotyping, and discriminatory treatment. Race and ethnicity, per se, may not confer greater risk of switching or relocation from weed-out courses. Nevertheless, problems grounded in inadequate pre-college preparation may be too great to overcome, particularly where weed-out classes are encountered early, and where students find it hard to feel that they belong. There is also some risk of racism in inquiries that define the difficulties of students of color as if they primarily relate to their race or ethnicity. Rather, we argue that the problems experienced by students of color—including problems created by under-preparation—begin with poorly resourced high schools that serve students of all races and ethnicities from working class, financially challenged, families. That said, in the original study, we also found a set of difficulties that become important beyond college entry and that bear exclusively upon students of color. They vary in form by racial/ethnic group, but they all arise from differences in cultural values and socialization, internalization of stereotypes, ethnic isolation and perceptions of racism. All of these problems are exacerbated for women of color by the gendered issues described above—a double burden that has been discussed by other scholars (cf. Ginorio, 1995; Givens, Tassie, (eds), 2014; Manoucheka, Diaz, Ginorio, & Joseph, 2014).

There are two other groups of students that can be discerned in the patterning of losses following weed-out courses. They are students taking these courses as requirements by other STEM majors—sometimes referred to as "serviced majors"; and high-performing and/or multi-talented students.

## Losses Among "Serviced Majors"

As both the interview and institutional records analyses show, weed-out courses enable losses of students in disciplines other than those of particular weed-out courses. In the life sciences, majors are frequently weeded out by courses in the chemistry sequence, and in engineering, majors are often weeded out during the calculus sequence. Students who are weeded out as "a service" that is provided by one discipline to another are highly aware of its apparent illogicalities and are resentful of it:

> It was a chemistry class weeded me out from biology. What's the sense in that? (White female high math switcher, life sciences to psychology)

We also note the problems that servicing creates for well-motivated but under-prepared entrants:

> Failing thousands of students who would otherwise make good contributions to society but who had to show they were ready to be engineers even before they got to college means that you send 70% of them home. (Multi-race female low math relocator, engineering to hydro-geology)

Seniors also struggled with the barriers to progress in their own major that are created by required courses that are geared to weeding students out:

> If you are an engineer or a physicist and you can't do Calc 2 and 3, you are going to struggle and you probably shouldn't be there. But if you are a chemist, biologist, or in lots of other majors who have to take those weed-out courses, it's worthless—even dangerous--to take them that early. And, as a chemist, it's hard to motivate yourself to do something you're not good at, knowing that it's not something you are going to apply, but that you have to get through it or it's going to take you down. (Multi-race female high math chemistry persister)

Thus, students who might otherwise be well-suited to their major by their interest, motivation, and career aspirations are often weeded out in required courses in other STEM disciplines. It is this awareness that has prompted some majors (notably in engineering schools) to provide their own courses in calculus and the physical sciences. This also has the advantage of infusing the content with applications relevant to their major.

## Losing Talented Students Because of Weed-Out Course Experiences

Of the 50 students who switched majors because of weed-out experiences, 35 had high incoming SAT/ACT math scores, 29 had cumulative GPAs in the B range and 14 in the A range. As noted in TAL, high-achieving students may lose their incoming interest through boredom induced by weed-out course approaches to teaching content. As we discuss further in Chap. 10, some students who undertake multiple majors and/or minors, and multi-talented students with viable alternative interests in the arts and humanities as well as the sciences, also switch out of their STEM majors following weed-out courses. Weed-out class experiences also redirect talented students away from STEM disciplines into majors that enable higher GPAs. This increases students' chances of acceptance into competitive graduate professional programs.

In Chap. 2, we reported that 12% of STEM students with GPA's between 3.5 and 4 switched out of STEM, as did 7% with math scores in the highest quartile. While (by definition), only a small group of students with high GPA's receive poor grades in weed-out courses, in the interview study 32 students did so and subsequently left their STEM majors. In addition, receiving an incomplete grade prompted 94 of otherwise high-achieving interviewees to switch out of their STEM major. These findings, while based on small numbers, suggest that high-achieving students who receive (or anticipate receiving) a poor grade or an incomplete in a weed-out courses are another group who are vulnerable to switching. The phenomenon of switching in response to anticipated as well as to actual grades is further discussed in Chaps. 9 and 10.

The clues offered by the interview study are strongly validated by the institutional records analyses. As we have shown, students with the highest levels of incoming SAT and ACT math scores are particularly vulnerable to the effects of DFWIs in first year SF courses. Losses from this highly talented group are greater than those for talented men—a difference that rises to double among women and men with the highest math scores on entry.

## Discussion, Conclusions, and Questions

***What Characterizes Weed-Out Classes?*** This chapter offers a multi-faceted examination of the nature and consequences of what are widely labeled "weed-out" courses or classes in the STEM disciplines. By triangulating findings from concurrent studies—both qualitative and quantitative—we have sought to define and characterize such courses. Drawing on DFWI and course grades data from institutional records, detailed recorded observations of teaching methods in a sample of foundation courses, and student surveys in the same courses, we identify what features distinguish STEM weed-out courses from other STEM foundational courses. These include their grading "severity," and teaching and assessment methods that produce limited student learning gains. In two institutional records studies across 42 colleges and universities of different sizes and types, we investigate risks and consequences for students. Through interviews with students at our six study sites who had left STEM majors following weed-out courses and with STEM seniors who had survived them, we explored how such consequences are created.

Students' problems with aspects of course design, pedagogy, and assessment methods were also reported in Chap. 6 by interviewees in courses across all 4 years of STEM programs. However, it is in those courses that we have identified by multiple measures as "weed-out" in nature and consequences that these problems appear in their most extreme forms. For example, although many courses use a norm referencing (i.e., curved grading) system, in weed-out courses it provides predetermined quotas of students with D and F grades to an extent that depresses STEM department grade averages (Rask, 2010). If the aim is to secure a loss rate of a predictable size, our institutional records analyses clarifies that it does so effectively. As we also observed in the original study, although the intellectual caliber of particular student cohorts taking introductory STEM courses will vary, the loss rates from weed-out courses do not. They are also announced in the traditional warning speech to new students—that at least one-third of them will no longer be in the major by the following year. The same message delivered in the same words was also documented in our original study and its predictive power is clearly longstanding and accurate.

Here (and also in Chap. 9), we discuss how the level of difficulty of any course may be intrinsic in that its hardness derives from the complex and abstract nature of its content. "Hardness" may also be "constructed," that is, made artificially difficult by teaching methods that discourage learning. It may be intended to scare students by inclusion of too much content delivered at too fast a pace for comprehension and assimilation. It may also be caused by poor course design or carelessness—as in test questions that are misaligned with the course material presented. None of these forms of hardness arise from the intrinsic hardness of the material itself. Flaws in course design and delivery also occur in other courses, but constructed forms of hardness are consistent features of weed-out class teaching methods. By confusing, intimidating, and discouraging students, instructors effectively convey the message that the major is "too hard for them." Even when interviewees are aware of this manipulation, they may leave because they have become disillusioned or have other options. They have negative consequences also for students who stay: seniors described how the combination of poor teaching and assessment methods experienced in weed-out courses had failed to equip them with the knowledge and skills that they needed in order to tackle genuinely difficult content in later STEM courses. This is one of the many ironies that we encountered. Another is that, dullness of presentation perhaps intended to discourage those thought less fit to continue, also undermines the interest of some high-performing and multitalented students who leave for majors that offer them creativity, engagement and intellectual challenge. Dissipation of incoming interest in the sciences among students of this caliber, and their reasons for moving to other disciplinary fields, are discussed in detail in Chap. 10.

***How May the Persistence of Weed-Out Practices Be Explained?*** The evidence that we have presented from several sources in this chapter begs the question "Are the instructor practices and attitudes, and the risks to persistence that they create, intentional?" As we have illustrated, many students—both switchers and persisters—believe that weed-out courses are indeed intentional because of the consistency in weed-out instructors' classroom behaviour. Seniors also recount a startling improvement in pedagogy and attitudes towards students by instructors whom they encountered in weed-out courses, then later in their majors. They also appear intentional because of the size and nature of student losses they consistently produce.

The course design, teaching methods, and assessment practices that we have documented do not reflect what is now widely known about how to teach such that students learn well. Such disattention might reflect individual choices by instructors. However, in that case, the incidence would be random and variable, whereas we have documented distinct pedagogical patterns that are consistent over time.

The persistence of weed-out courses might still not arise as the result of deliberate departmental decisions. Rather, it may reflect deeply socialized beliefs and attitudes about approved professional practice and what enables persistence that have been recorded among STEM seniors and graduate students (Seymour et al., 2005) and, in this study, among STEM faculty in foundation courses by our collaborators, Ferrare and Miller (2019). Gradually assuming the beliefs, and replicating the practices, of successful members of the disciplinary community is a normal process of apprenticeship (Laursen et al., 2010). It is also a powerful way by which traditional practices are rewarded and perpetuated and attempts to introduce improvements in teaching practices (however research-grounded) are discouraged and discounted (Seymour & DeWelde, 2016; Foertsch et al., 1997). Thinking about the perpetuation of weed-out course practices as products of professional socialization into the STEM academy may explain why undergraduates were puzzled that graduate TAs seemed not to understand their complaints about weed-out course teaching and testing and defined them as "just normal." It would also explain why graduate student teachers used the same ineffective teaching methods as their departmental instructors. Thus, professional socialization could explain why some instructors go on teaching in particular modes. However, it is not enough to explain the consistency, dysfunctional character and patterned outcomes that we have documented.

Another way to make sense of the system-like patterning of weed-out courses that is offered by classic structural functionalist theory (cf.Merton, 1957 ; Parson, 1968). These theorists proposed that any set of practices, beliefs, and attitudes that consistently maintains its form over time does so because it is providing a service that is valued and supported by the system of which it is a part. In this case, STEM departments, each nested within their own disciplines, may be considered as "systems." Some departmental practices have an official recognized status—formal positions, committees, tenure and promotion rules, etc. However, some approved practices are "latent" in that they are not officially recognized. Indeed, if questioned, group members may deny that they exist to ward off the scrutiny of non-members. Nevertheless, they exist and persist precisely because they serve purposes valued by the group. Thus, weed-out courses use methods that system members see as appropriate in securing valued outcomes. They may also continue, somewhat under the radar of host institutions, because they develop a taken-for-granted status that is tacitly recognized by wider members of the institution such as deans, advisors, and faculty in other departments. They are, in effect, hidden in plain sight. Their tacit, taken-for-grated status may explain why there is so little published research on weed-out practices, notwithstanding the widespread nature of the phenomenon. We argue, therefore, that weed-out courses do not have to be formally "designed" in order to be intended. They can simply have evolved to fit their purpose and, so long as departmental members see them as performing a necessary function in a manner that produce required results, they are likely to continue. As phenomenologist, Alfred Schutz explains, longevity per se ensures for taken-for-granted place practices that have stood the test of time and are accepted as the normal way to accomplish something of value (cf. Schutz, 1971). As Schutz further explains, such practices are unlikely to be questioned unless a circumstance arises that their hitherto unquestioned rationale cannot explain.

***What Purposes Are Weed-Out Courses Intended to Serve?*** So, what are the purposes of weed-out courses that can explain their latent functionality, intentionality, and taken-for-granted status by departments and in the wider institution? The evidence available to us comes, inferentially, from our qualitative data sources and from our analyses of weed-out course consequences derived from institutional records and student transcript data. What we and other researchers lack are direct statements from members of departments about what they perceive as the purposes of weed-out courses. Like all inquiry into practices that are not formally recognized, this information is difficult to obtain. However, it would be a highly desirable to pursue such an inquiry. One of our most informative sources are the focus groups of seniors from multiple STEM disciplines. Seniors, who often worked in proximity with faculty and graduate students, were well-informed as to their attitudes and expectations. Some, who aspired to a career in academe, expressed sympathy with the needs of the profession and the discipline that the weeding out process was presumed to serve. From these discussions, and from other student observations, we infer that what STEM faculty are seeking to ensure is twofold. First, they want only the most intellectually able, committed, hard-working, and interested students to stay in their major. Second, they have limited capacity to teach their majors and must reduce their entering numbers in order to accommodate them. They view the challenges of weed-out classes as tests of intellect, determination, and interest, and seek to identify students with sufficient stamina to handle the challenges that lie ahead. Indeed, they see it as no kindness to students to let them continue if they cannot surmount that challenges of what they often refer to as disciplinary "rigor." Seniors tended to applaud these intentions but also questioned whether faculty have merged the distinction between intrinsic and constructed forms of hardness in defining "rigor." Although, as survivors of the weed-out system they could claim some of the attributes for which they have apparently been selected, they also offered accounts of friends and former peers who, notwithstanding their talent, interest, and capacity for hard work, were inappropriately diverted by weed-out experiences into other paths.

Clearly, departments secure enough majors who possess desired attributes to be continually reassured that the system works. However, whom may have lost by this process that they might have preferred to retain? The reason that departments may not be able to answer this question is because they do not collect and analyze data that could address it. They could (as we have done) draw data from the institutional records and student transcripts. However, to the degree that the system appears to be working, they would have no reason to do so.

***Which Groups of Students Are Lost Through Weed-Out Practices?*** With this in mind, we draw attention to our summary of findings that establish which groups of students are at most risk of leaving both STEM and STEM-serviced majors following weed-out course experiences. Students that are lost because of weed-out course experiences include high-performing students—notably women, especially those accustomed to getting good grades. As described in Chap. 2, women with the same SAT math scores as men get higher grades—suggesting perhaps that they work harder. Women also enter and leave STEM majors with higher average performance scores than both male switchers and persisters—a phenomenon that we first reported in the 1997 study. Also lost are multi-talented students of both sexes whose interest was dissipated by lack of intellectual challenge, engagement with authentic science, and exploration of theory in the limited "school science" presented. Some students who undertake multiple majors and/or minors, and multi-talented students with viable alternative interests in the arts and humanities as well as the sciences, also switch into them. (These groups of high-performing switchers are further discussed in Chap. 10.) Even though students with high ACT/SAT scores are less likely to switch, they are still vulnerable to the effects of DFWI. Losses from this talented group are high among both women and men with the highest math scores on entry. A subset of these students make pragmatic moves from STEM disciplines into majors that enable them to improve their GPAs and, thus, increase their chances of acceptance into competitive graduate professional programs.

Also weeded out are high proportions of students from families with low socio-economic status (indicated in the institutional record by PELL eligibility and awards) and/or who are first-generation college entrants. At particular high risk are all students who enter under-prepared by under-resourced high schools—a risk that is enhanced for women of color. Even good SAT/ACT math scores reflecting natural math ability can mislead students into assuming that they are better prepared than they actually are for college–level work. Among these bright, but critically under-prepared students, are many students of color. Under-preparation also includes college entry with insufficient "social capital" to know how to navigate undergraduate education. Critically, this includes what sources of academic help are available and that it is important to seek and use them out in a timely way. We point to the damage that the "constructed hardness" of weed-out teaching methods can do to students who do not realize they are under-prepared until entering their first foundation courses. These students are unable to make up the deficiencies in their high school preparation before they are hit with weed-out courses—often (if poorly advised) two or more at once. Under-prepared students are very quickly overwhelmed by the difficulty posed by working hard to catch up while also tackling new concepts and assignments. We have described the high losses of working class youngsters of all races and ethnicities from under-resourced high schools as a form of structured inequality. It is another irony of the unexamined weed-out tradition that, although STEM instructors and departments do not knowingly target this student demographic, weed-out class methods consistently ensure high (and early) loss rates among bright, but socio-economically and educationally disadvantaged, students.

As Koch and Drake demonstrate, these losses are not only of switchers to non-STEM majors; they are also of students in otherwise good academic standing who contribute to the 20% of STEM majors who leave college altogether.

To repeat Schutz's proposition, taken-for-granted practices and the beliefs that support them are unlikely to be questioned unless a novel experience turns up that their unquestioned frame of reference cannot explain. If the tacitly approved purpose of weed-out courses is to secure enough—but not too many—students with the "right" attributes to continue in STEM majors, but STEM department do not know from data how many students who possess these attributes they are actually losing, or why, what difference should such knowledge make? When people of good intention support a set of practices that turn out to be highly disadvantageous to women—especially high-performing women and women of color—and to students of all races and ethnicities from poorer families who are under-prepared by under-resourced high schools, how should they respond? Furthermore, when practices thought to dismiss only those students with less ability, interest, motivation, and willingness to work hard, are called into question because patterned losses consistently include hard-working students of high ability, interest, and motivation, what adjustments of attitudes and practices might STEM departments, Colleges of Engineering and their faculty make? Is the continuance of weed-out courses in their present form likely once their dysfunctional outcomes are known? We await responses to these questions with interest.

## References


Breussoud, D. (2015). Insights from the MAA national study of college calculus. *The Mathematics Teacher, 109*, 178–185.261.

Eagan, K., Hurtado, H., Figueroa, T., & Hughes, B. (2014). Examining STEM pathways among students who begin college at four-year institutions. *Commissioned paper prepared for the committee on barriers and opportunities in completing 2- and 4-year STEM degrees*. Washington, DC: National Academy of Sciences. Retrieved from: http://sites.nationalacademies.org/cs/groups/dbassesite/documents/webpage/dbasse_088834.pdf

Ferrare, J. J. (2019). A multi-institutional analysis of instructional beliefs and practices in gateway courses to the sciences. *CBE Life Sciences Education, 18*(2), ar26.

Ferrare, J. J., & Miller, J. M. (2019). Making sense of persistence in scientific purgatory: A multi-institutional analysis of instructors in introductory science, technology, engineering, and mathematics (STEM) courses. *The Journal of Higher Education*, 1–26.

Flanders, G. R. (2017). The effect of gateway course completion on freshman college student retention. *Journal of College Student Retention: Research, Theory & Practice, 19*(1), 2–24.

Foertsch, J. M., Millar, S. B., Squire, L., & Gunter, R. (1997). *Persuading professors: A study in the dissemination of educational reform in research institutions*. Report to the NSF education and human resources directorate, Division of Research, Evaluation, and Communication. Washington DC: University of Wisconsin-Madison, LEAD Center.

Freeman, S., Eddy, S. L., McDonough, M., Smith, M. K., Okoroafor, N., Jordt, H., & Wenderoth, M. P. (2014). Active learning increases student performance in science, engineering, and mathematics. *Proceedings of the National Academy of Sciences, 111*(23), 8410–8415.

Ginorio, A. (1995). *Warming the climate for women in academic science*. Washington, DC: Association of American Colleges and Universities, Program on the Status and Education of Women.

Givens, S. M. B., & Tassie, K. E. (Eds.) (2014). *Underserved women of color, voice, and resistance: Claiming a seat at the table*. New York, NY: Lexington Books.

Grave, N. D. (2018). *Demographics and the demand for higher education*. Baltimore, MD: Johns Hopkins University Press.

Greenwood, M. R. C., & North, K. K. (1999). Science through the looking glass: Winning the battles but losing the war? *Science, 286*(5447), 2072–2078.

Hrabowski, F. A. (2003). Raising minority achievement in science and math. *Educational Leadership, 60*(4), 44–49.

Kardash, C. A. M., & Wallace, M. L. (2001). The perceptions of science classes survey: What undergraduate science reform efforts really need to address. *Journal of Educational Psychology, 93*, 199–210.

Koch, A. K. (2017). It's about the gateway courses: Defining and contextualizing the issue. In A. K. Koch (Ed.), *Improving teaching, learning, equity, and success in gateway courses* (pp. 11–17). *New Directions for Higher Education* (180). San Francisco, CA: Jossey-Bass.

Koch, A. K. (2018). Big inequity in small things: Toward an end to a tyranny of practice. *National Teaching and Learning Forum, 27*(6), 1–5.

Koch, A. K., & Drake, B. M. (2018). *Digging into the disciplines: The impact of gateway courses in accounting, calculus, and chemistry on student success*. Brevard, NC: John N. Gardner Institute for Excellence in Undergraduate Education.

Koch, A. K., & Rodier, R. (2014). *Gateways to completion guidebook*. Brevard, NC: John N. Gardner Institute for Excellence in Undergraduate Education.

Laursen, S. L., Hunter, A.-B., Seymour, E., Thiry, H., & Melton, G. (2010). *Undergraduate research in the sciences: Engaging students in real science*. San Francisco, CA: Jossey Bass.

Manoucheka, C., Diaz, S., Ginorio, A., & Joseph, R. (2014). A resistance story: Negotiating the institutional and material through collectivity. In S. B. Givens & K. E. Tassie (Eds.), *Claiming a seat at the table: Feminism, underserved women of color, voice and resistance* (pp. 91–106). Blue Ridge Summit, PA: Lexington Books.

Merton, R. K. (1957). *Social theory and social structure* (revised and enlarged edition). New York, NY: The Free Press

National Action Council for Minorities in Engineering, Inc. (2013). *2013 NACME data book: A comprehensive analysis of the "new" American dilemma*. White Plains, NY: Author.

Parson, T. (1968). *The structure of social action*. New York, NY: The Free Press.

President's Council of Advisors on Science and Technology (PCAST). (2012). *Engage to excel: Producing one million additional college graduates with degrees in science, technology, engineering, and mathematics*. Retrieved from www.whitehouse.gov/sites/default/files/microsites/ostp/pcast-engage-to-excel-final_2-25-12.pdf

Prosser, M., & Trigwell, K. (2013). Qualitative variation in approaches to university teaching and learning in large first-year classes. *Higher Education, 67*(6), 783–795.

Rask, K. (2010). Attrition in STEM fields at a liberal arts college: The importance of grades and pre-collegiate preferences. *Economics of Education Review, 29*(6), 892–900. https://doi.org/10.1016/j.econedurev.2010.06.013

Schutz, A. (1971). *Collected papers: The problems of social reality*. The Hague, The Netherlands: Martinus Nijhoff.

Seymour, E., & DeWelde, C. (2016). Why doesn't knowing change anything? Constraints and resistance, leverage and sustainability (with K. De Welde). In Weaver, G. C., Burgess, W. D., Childress, A. L., & Slakey, L. (Eds.), *Transforming institutions: Undergraduate STEM education for the 21st century*. Purdue, IN: Purdue University Press.

Seymour, E., Melton, G., Pedersen-Gallegos, L., & Wiese, D. J. (2005). *Partners in innovation: Teaching assistants in college science courses*. Boulder, CO: Rowman and Littlefield.

Suresh, R. (2006). The relationship between barrier courses and persistence in engineering. *Journal of College Student Retention: Research, Theory, and Practice, 8*(2), 215–239.

Tinto, V. (2000). Linking learning and leaving. Exploring the role of the college classroom in student departure. In J. M. Braxton (Ed.), *Reworking the student departure puzzle* (pp. 81–94). Nashville, TN: Vanderbilt University Press.

# Chapter 8
# Students' Perceptions of Good STEM Teaching

**Raquel P. Harper, Timothy J. Weston, and Elaine Seymour**

## The Experience of "Good Teaching" in STEM Foundation Courses

We begin this chapter with an account of findings from the Student Assessment of Their Learning Gains (SALG) survey. These respondents were at an earlier stage in their education than the STEM switchers and persisters in the interview study.

R. P. Harper (✉)
Ethnography and Evaluation Research (E&ER), University of Colorado Boulder, Boulder, CO, USA

Center for STEM Learning, University of Colorado Boulder, Boulder, CO, USA
e-mail: Raquel.Harper@Colorado.EDU

T. J. Weston
Ethnography and Evaluation Research (E&ER), University of Colorado Boulder, Boulder, CO, USA

National Center for Women and Information Technology (NCWIT), University of Colorado Boulder, Boulder, CO, USA
e-mail: westont@colorado.edu

E. Seymour
Ethnography and Evaluation Research (E&ER), University of Colorado Boulder, Boulder, CO, USA
e-mail: Elaine.Seymour@colorado.edu



E. Seymour, A.-B. Hunter (eds.), *Talking about Leaving Revisited*,
https://doi.org/10.1007/978-3-030-25304-2_8

## *What Learning Gains Did Students Make in Introductory STEM Courses: Findings from the Student Assessment of Their Learning Gains (SALG) Survey*

In contrast to negative accounts from students in weed out courses, most students in STEM introductory courses in the survey sample assessed their teachers and courses positively. This can be seen in average student ratings for the "core" section of the SALG survey. The first core question asks students to rate the extent of the learning gains they had made in five key areas of their course. The results are presented in Table 8.1.

For most items (scaled 1–4), more than half of the students responded that they had made learning gains in the two higher categories (moderate gain and great gain). However, some lower ratings were given to "applying what I learned in this class in other situations" (average = 2.40) and "*using concepts from this class to address real world issues*" (average = 2.40).

A set of core questions asked students to identify, in some detail, the degree to which particular aspects of their courses had helped them to learn. Their answers, which are grouped by class characteristics and activities, are presented in Tables 8.2, 8.3, and 8.4 and in also Appendix Y. (These additional tables may also be of interest to readers who wish to learn about the kinds of student feedback that can be gained by the use of the online SALG survey.)

It is notable that all of the first group of pre-coded class aspects (shown in Tables 8.1 and 8.2) proved to be those spontaneously raised by students in the interviews and focus groups as matters of primary concern (as discussed in Chaps. 6 and 7). The items in Table 8.2 focus on their instructor's class design (in terms of content, pace, and alignment of class elements), the teaching methods used, and the class atmosphere and level of peer civility established by the instructor.

The set of items shown in Table 8.3 reference the degree to which students made gains in their interest and enthusiasm for the course subject, sense of mastery of

**Table 8.1** STEM foundation course students' responses to the question "As a result of your work in this class, what gains did you make in your understanding in each of these areas?"

| Type of response | No gain (%) | Little gain (%) | Moderate gain (%) | Great gain (%) | Average | *N* |
|---|---|---|---|---|---|---|
| The main concepts explored in this class | 9 | 23 | 40 | 28 | 2.88 | 1431 |
| The relationships between the main concepts | 11 | 25 | 41 | 23 | 2.76 | 1431 |
| Connecting key ideas from this class with other knowledge | 14 | 29 | 37 | 20 | 2.62 | 1431 |
| Applying what I learned in this class in other situations | 21 | 32 | 31 | 16 | 2.42 | 1431 |
| Using concepts from this class to address real world issues | 22 | 22 | 22 | 22 | 2.40 | 1431 |

**Table 8.2** STEM foundation course students' responses to the question "How much did the following parts of this class help your learning?"

| Type of response | No gain (%) | Little gain (%) | Moderate gain (%) | Great gain (%) | Average | *N* |
|---|---|---|---|---|---|---|
| The teaching methods used in this class | 12 | 24 | 33 | 31 | 2.83 | 1349 |
| How class activities, readings, assignments (and labs, if relevant) fit together | 7 | 24 | 44 | 25 | 2.86 | 1347 |
| The pace of this class | 11 | 26 | 43 | 20 | 2.73 | 1343 |
| The amount of content presented in this class | 8 | 23 | 42 | 27 | 2.89 | 1341 |
| The instructor's explanation of how the class activities, reading, labs (if relevant) and assignments related to each other | 14 | 27 | 30 | 28 | 2.72 | 1341 |
| The expectation of courteous student-student behavior established by the instructor | 12 | 22 | 35 | 31 | 2.84 | 1346 |
| The atmosphere of this class | 12 | 25 | 33 | 30 | 2.81 | 1332 |

**Table 8.3** STEM foundation course students' responses to the question "From your experience in this class, rate the following"

| Type of response | None (%) | A little (%) | Moderate amount (%) | A Great amount (%) | Average | *N* |
|---|---|---|---|---|---|---|
| Your enthusiasm for the subject | 14 | 27 | 38 | 22 | 2.68 | 1293 |
| Your interest in taking, or planning to take additional classes in this area | 25 | 23 | 27 | 26 | 2.53 | 1295 |
| Your interest in discussing the subject area with friends or family | 23 | 26 | 29 | 22 | 2.49 | 1294 |
| Your confidence that you understand the materials | 8 | 24 | 44 | 24 | 2.84 | 1293 |
| Your willingness to seek help from others (TA, teacher peers) when working on academic problems | 8 | 21 | 40 | 31 | 2.95 | 1291 |
| Seeing this as a welcoming, inclusive field in which to start a career | 29 | 27 | 26 | 18 | 2.34 | 1292 |
| Confidence that you could succeed in a career in this field | 26 | 27 | 27 | 20 | 2.42 | 1294 |

course content, willingness to seek appropriate help with academic problems, and appraisal of the field as a career prospect. Taken together, they explore the extent to which particular student attitudes had moved in a positive direction by the end of the course.

**Table 8.4** Student responses to SALG questions about class climate

| Type of response | None (%) | A little (%) | Moderate amount (%) | A great amount (%) | Average | *N* |
|---|---|---|---|---|---|---|
| I was comfortable asking questions in this course | 7 | 20 | 50 | 23 | 2.88 | 1154 |
| I was comfortable approaching the teacher out of class | 7 | 17 | 48 | 29 | 2.99 | 1153 |
| The content in this course was difficult | 2 | 21 | 52 | 25 | 3.00 | 1149 |
| The content of this course was not that difficult, but the way the course was taught made it difficult to learn | 22 | 35 | 22 | 21 | 2.42 | 1151 |
| This course was very competitive | 9 | 44 | 35 | 13 | 2.51 | 1154 |
| This course was a "weed out" course for the major | 16 | 39 | 26 | 19 | 2.49 | 1153 |
| Students worked together in this course | 6 | 20 | 54 | 21 | 2.92 | 1150 |

*Note:* Students can choose "non-applicable" for items with lower counts

Two items in this group of responses were rated more negatively than positively. They are seeing this as a welcoming, inclusive field in which to start a career (2.34) and confidence that you could succeed in a career in this field (2.42).

Using a four-point agree/disagree scale, a final set of core questions probed the extent of negative and positive student appraisal of four course characteristics that were also found to be a source of concern in student interviews. They are shown in Table 8.4 as the students' level of comfort in asking questions in class and approaching the instructor out of class; the intrinsic or constructed nature of course "hardness"; the degree of collaboration and competition among students, and whether they assessed the course to be "weed out" in intent.

Overall, the aggregated results of the survey show that, across courses, students rate their learning gains as moderately high. For many activities and teaching methods, 60% or across courses more of the respondents gave ratings of "moderate gain" or "great gain" to basic teaching activities such as lecture, homework, and the overall teaching methods used in the class. Higher ratings (sometimes over 70%) of "agree" or "strongly agree" were also given to the overall course climate in areas of comfort in approaching the teacher and working together with other students. Slightly lower ratings occurred for assessment activities (i.e., *the number and spacing of tests*) and seeing the course as a gateway to a career. The latter ratings are most likely a function of students taking courses outside of their own major.

The other important finding from the survey is found in items with missing or "non-applicable" responses. Questions where few or very few students gave ratings indicate that these students' instructors do not include activities such as projects, presentations, or group work in their courses. Students may also be inadvertently making a comment about the quality of instruction in the relatively low numbers of non-missing responses to questions: *Interacting with the instructor during class* and

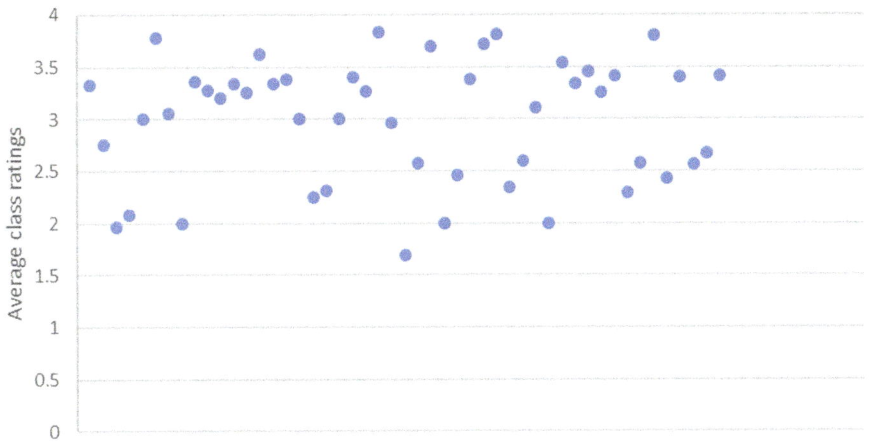


**Fig. 8.1** Course averages of student responses for the question "What gains did you make in your understanding in each of these areas? The main concepts covered in this class"

*Interacting with the instructor during office hours*. For these two questions, fewer than three-quarters of respondents reported interacting with the teacher in class and less than half indicated they met with teachers during office hours.

While aggregated results can show important patterns in how students view their teachers and their own learning, they inevitably mask dramatic differences between teachers and courses. Thirty percent of variation in SALG ratings of the course is due to students' reactions to particular teachers. For example, when asked about gains in *the main concepts explored in this class*, the average student rating was 2.88. However, when looking at the distribution of averages by teacher (shown in Fig. 8.1) averages ranges from very high (3.86) to very low (1.7). This means that a substantial number of students are very dissatisfied with their learning in specific courses.

## How Students Characterize Good Teaching

In interviews, students tended not to make a sharp distinction between the qualities of attitude and behavior that, for them, defined "the good teacher," and the teaching strategies and aspects of course design that best enabled their learning and engagement with the course material. They were, however, very clear about what constitutes good teachers and good teaching and how each of these benefited them. There was also a high degree of agreement about these characteristics: Switchers and persisters offered similar descriptions, and, as we discuss later in this chapter, there were no strong differences between men and women or between students by race and ethnicity. Table 8.5 summarizes what both switchers and persisters considered

**Table 8.5** Characteristics of good teachers and teaching by switchers' and persisters' frequency of descriptions

| Characteristic | Switchers (%) | Persisters |
|---|---|---|
| Teacher is open, approachable, and shows concern about student learning | 42 | 43 |
| Delivery is engaging | 40 | 44 |
| Organized, coherent course structure | 31 | 31 |
| Interactive and inquiry-based teaching methods | 25 | 15 |
| Provides examples and applications; makes connections | 24 | 36 |

to be the most important aspects of effective teachers and teaching. In many ways, these are the obverse of what students reported (cf., Chap. 6) as types of instructors and instructional methods that caused them problems.

## *Most Valued Qualities: Teachers Who Are Open, Approachable, and Show Concern for Students' Learning*

As is evident in Table 8.5, the qualities most frequently described and valued (by 42% of switchers and 43% of persisters) were teachers who showed in their teaching methods and interactions with students that they wanted students to succeed and treated them with respect and encouragement when approached with questions or requests for individual help. Students most commonly encapsulated these desired attributes as teachers showing that they cared about students' learning in their courses. SALG respondents also wrote that they liked teachers who were approachable and seemed to care about their students. Caring was manifested in encouraging questions and interaction in their classes, and being open, approachable, available, and helpful to individual students. For persisters, these attitudes and behaviors were as highly prized as course content delivered in engaging ways. Indeed, students seemed to yearn for their teachers to care about them, both as learners and as people, and for them to make it evident that they wanted their students to succeed. As illustrated in the following selection of quotations from our text data, students commonly used both "caring" and "friendly" as glosses for these valued teacher attributes:

> I feel that teaching well is when you actually care about the students. You want them to learn the material ... When they teach well, they're like 'I want you to remember this and carry this on for years to come.' They actually care about your process, and whether you're doing well or not doing well. (Pacific Islander woman, biological sciences persister)

> The professor is phenomenal ... She understands that I'm somewhat intelligent. She wants me to pass her class and she wants my help figuring out what would make me more engaged in her class...I know that she wants us to do well. (White man, switcher, biological sciences to film studies and production).

How instructors approached answering students' questions—whether inside or outside the classroom—was key in establishing whether they cared that their students were understanding the material and making progress. To ensure this, the "good teacher" encourages students to ask questions when confused or curious and sets aside class time to address them:

> He would always stop the lecture and ask if you had any questions. ... After the lecture, he stayed for a while so if you had questions, but you didn't want to raise your hand, you could go and ask him ... He always opened up his time for questions. (African-American woman, switcher, biological sciences to community health)

> One of the reasons why he is my favorite professor is because he humors hypothetical questions about research. I think that's really good for young scientists--learning to ask the right questions. (Multi-racial woman, biological sciences persister)

> The professor was very helpful, understanding and outgoing. The way in which he approached teaching was helpful for myself and I believe the majority of the class as it allowed us to openly ask questions and explore physics (SALG comment from a white, female nursing major)

Encouraging questions also required creating a safe, comfortable atmosphere where students were not afraid to speak up or worry that their question was "stupid":

> Whenever people ask stupid questions, which happens often--I did it too--he would always pause for a moment and say, 'That's a very good question,' and then explain it. (White man, switcher, physics to journalism)

Teachers who showed concern that their students were coping with difficult material were those who acknowledged that it was hard to grasp and who used a variety of ways to help students gain an understanding:

> Good professors acknowledge that some things are hard ... They look you in the eye and they're like, 'This is hard stuff. This is hard to understand...I'm open to talk about this after hours.' They have open office hours ... and they make sure everybody is on the same page... They're invested in your success. (White woman, geosciences persister)

Thirty percent of SALG commentators also praised instructors who explained concepts clearly and simply:

> My professor and TA were both really approachable and helpful. They presented subject matter clearly and were willing to slow down and help students if need be. (White, male engineering major)

> Topics were discussed in a manner that supported learning and they were simplified enough to be comprehensible. (White man, horticulture major)

Checking whether students were understanding the material as they went along, and being approachable and available outside of class time when they did not, were both seen as critical:

> He really took teaching to heart in the sense that he never shied away from questions after class or questions in office hours. And he set up review sessions outside of class. He was available more than anyone else. (Asian man, biochemistry persister)

> He had a very open-door policy ... 'Come in and talk to me if you need to. Send me an email. Schedule a meeting.' He has a weekly meeting with every single group of students in

> his class, outside of class time. His policy was, 'If you need to talk to me, send me an email and I'll meet with you.' (White man, chemical engineering persister).

Students also described "caring" in terms of having compassion for their difficulties and outside pressures that were unrelated to the course. They appreciated teachers who showed concern for their well-being. For example, one young man who appreciated his professor's understanding of his difficult family situation and her willingness to work around his challenging personal schedule, offered this testimony:

> I took a biology class my freshman year in the chemical engineering department … and that professor really cared. … My dad was going through chemo and I would go down and see him, probably once or twice a month, and sometimes I'd have to miss class or a test or something. And she was like, 'Y'know, go ahead, we'll figure it out later.' … I didn't really care about the grade I got in that class. And I think she knew that there were more important things, and it was cool to see that in a teacher. (White man, switcher, engineering to economics)

A teacher's willingness to take account of students in personal terms also brought benefits. The more students felt their teachers gave them personal attention, the more they were motivated to work hard and show that they were committed to their studies. It was also an important factor in students' determination to persist:

> The instructors that I've had a good personal relationship with …I guess it can be pretty motivating … like you want to show them that you're dedicated to learning everything you can from their course. (White woman horticulture persister)

> He consistently stayed after class with me for… 10 or 15 minutes … to talk about everything from his research to my life … So, he's been a really strong support. (White man, chemical engineering persister)

> When I feel the professors are really invested in our success--like they really care about how we do in their class, how we do in life, like, if we're having trouble finding a job and stuff, the more committed I feel to do my best. (White woman, relocator, chemistry to physical sciences)

Instructors who learned students' names also helped students feel connected, not only to that teacher, but also to their major and the department. It was a courtesy that made them feel respected and part of a community:

> I hate it when my teacher doesn't know my name. I think that it's a little thing, but the electronics class professor made an effort to get to know the 80 people in his lecture. He made the effort. If you're there every day, he'll know your name. And that makes me respect him more, because he respects me. (White man, switcher, chemistry to theater)

> I switched to material science. I think we have 40 people in our major, and all of the professors know all of our names…We've formed really tight-knit communities. (White woman, relocator, chemistry to physical sciences)

Being open to students and showing interest in their learning and aspects of their lives is clearly important to students, and (as we described in Chap. 6) failure to connect with them has consequences for their persistence. Some students switched out of the sciences, and some relocated to different STEM majors, in part because they had found among teachers in other departments greater openness, approachability, and concern for student learning and welfare:

> In my communication classes… I had professors that got to know me better….They actually took some interest in my career. They said, 'Why are you majoring in the sciences, when you have such a great ability in writing….' And I was thinking, 'Nobody's ever said I'm good at science, so why am I doing it?' (Asian woman, switcher, biological sciences to communications)

> This is definitely one of the major reasons why I switched out of chemical engineering … The professor would constantly remind us how half of us weren't going to make it …that we should just give up … I switched after that semester (to the physical sciences). The physical thermodynamics professor was very outgoing, and very upbeat. He encouraged us to work together and not to try and compete against each other. So that was definitely a breath of fresh air. And I was very glad that I did switch from chemistry after that. (White woman, relocator, chemistry to physical sciences)

## *Engaging Delivery*

Both switchers and persisters want their teachers to be passionate about the material that they teach, project their enjoyment in teaching it, and make an effort to engage, motivate, and inspire their students. This was the most-often described characteristic of high-quality teaching offered by persisters (44%) and the second-most described by switchers (40%). These qualities were also highly valued by the SALG respondents (30%):

> It helps that my professor is very engaging and passionate about what she does. It really helps facilitate my learning. (White woman, psychology major)

We also learned how important this is to students from interviewees who looked for and did not find these qualities in their STEM teachers:

> I need the teacher to be very passionate, which seems like a given, but I've had so many teachers where … words just kind of flow out, and they don't care what they're teaching about, so I don't care what they're teaching about. (White man, switcher, physics to theater)

Especially when the material is complex and difficult to understand, both switchers and persisters wished their instructors would present it with more enthusiasm. Students explained that this helped them understand the material and motivated them to work hard to understand it:

> If you have a good teacher who is really passionate about the subject, that can make the class, even though it's a very hard subject. It can be so much fun because you enjoy the teaching and you enjoy the material. (Asian woman, biological sciences persister)

> If they're enthusiastic about it, and like making jokes or being entertaining, keeping it upbeat, as opposed to just standing and lecturing, that helps a whole lot. Classes I've had where the teacher just talks at you and then leaves, those are no fun at all. (White male, switcher, engineering to psychology)

When teachers are excited about what they are teaching, students seem to absorb that excitement. They describe feeling more attentive, encouraged, inspired, and interested in the material. Although all of the three students quoted below switched or relocated, they had carried with them memorable examples of good science and math teaching:

> His pure raw passion for math … He could tell you all these cool things and you felt like, 'Oh! It's so awesome!' (White woman, switcher, mathematics to economics)

> He was very animated … He just made things exciting and was very good at involving the class and keeping us on our feet, doing problems all around the room, and like getting everyone excited about getting things right. (White man, switcher, environmental science to economics)

> He was always super excited. In the chemistry halls, we have this huge chalkboard. There's like three panels so you can move it up and down. And he would always get his chalk and throw it at the board … It was so great because he would always hit the word he was talking about like he had been doing this for years … It was entertaining and it kept us awake. And it made you more interested and more excited to be there. (Asian woman, relocator, mathematics to life sciences)

The following descriptions by persisters also convey that engagement, enthusiasm, and sense of fun deployed by "good" instructors have the potential to bond students both to the teacher and to the discipline:

> My CALCULUS 2 class … that professor was really good. … He was always happy, and he had a really charismatic way of conveying his interest in the topic. I felt that students felt more connected to him. And because of that, we were able to pay better attention. (Multiracial woman, computer science persister)

> I just took O-chem, and I loved it. My professor was – you just knew she was pumped about it. Every day we'd have competitions of pulling pipettes or whatever with fire … like 200 people in a lecture and everyone paid attention because she was so excited and loved to blow things up and stuff. (White woman, chemical engineering persister)

> He is a professor of quantum mechanics … He was amazing, he truly inspired me. … He made physics concepts accessible to everyone … I mean this is just beautiful, and I, there was a point where I never slept because I watched U-tube videos on physics and actually enjoyed it … He was a goofy guy, and just made silly jokes, that helped. (Asian man, mechanical engineering persister)

As some SALG respondents also explained, the use of humor also has learning benefits. Three students in the interview study explained how humor can hold attention, ease students through hard material, and clarify or illustrate concepts to help ensure that students remember them:

> Some of my professors stick out. … My chemistry professor would make jokes about the molecules. And even though they were always cheesy, I would always listen in lecture because I was waiting for them to happen. (White woman, switcher, mathematics to hearing and speech science)

> His sense of humor was really funny… He would always make these awkward jokes to kinda lighten the mood. He would tell math jokes sometimes. After we would do a really hard problem, he would make a joke … the personality that he put into the teaching really made that class. (White man, computer science persister)

> In Calculus 2 … He uses a lot of visual tools … There is this one topic … the bi-rule factor … it's the factor that sticks out completely orthogonal … it's always sticking up … One day he brought in a Mickey Mouse hat and had two chopsticks sticking out of his head…He was

> trying to show us, and I thought that was a really good representation. He did it in a humorous way so that it wasn't boring. (Asian man, aerospace engineering persister)

Other scholars have also found that an instructor's enthusiasm for the material being taught is one of the most important characteristics of good teaching identified by students (e.g., Lee, Kim, & Chan, 2015). Excitement and passion are contagious, and students wish that more instructors would show them how amazing, interesting, connected to other concepts, and applicable to the real world the subject they are teaching can be. As we have illustrated, this also has benefits in promoting student engagement, motivation, learning retention, and bonding to the discipline. Switchers also can take away memorable aspects of good STEM learning experiences.

## *Organized, Coherent Course Structure*

A third of both switchers and persisters included good organization and structural coherence in their descriptions of effective teaching. For students, these course attributes include: Good alignment between course elements; clear objectives, expectations, syllabi, and lesson plans; logical sequencing in presentation of concepts; and a consistent teaching style throughout the course.

Students identified the best teachers as those who made sure that all aspects of the course aligned well. They identified as important in enabling their understanding and progress alignment between course and lab content, textbooks, and tests:

> My genetics class ... it was very organized, there was a reading before each lecture ... and after the lecture there would be assigned problems out of the book that were very relevant ... I feel like I learned everything so well, it was everything that I needed as a learner to be successful. (White woman, applied plant science persister)

SALG commentators also praised teachers organized their courses well (13% of comments). They appreciated courses where different course elements supported each other and where assessment was linked to content:

> This class content was organized very well by the teacher. It followed the readings very well, and the assignments connected with both very well. Overall the course made sense and was connected in every aspect. The environment of the students was intense, which made for a productive, if stressful, environment. Our discussion section professor was immensely helpful in breaking down concepts and making sure we understood how to go about the homework and exam problems. (White woman, speech and language major)

Students appreciated a thoughtful choice of texts to support course content so that each supplemented and reinforced the other:

> It helps a lot when professors take time to find a good text book that aligns. I think a lot of professors just throw whatever they think fits. In our reactor engineering class, our professor actually took the time to find a book that was good, enjoyable to read, and clear in its examples. (White man, chemical engineering persister)

Good alignment also includes well-thought-out intellectual links and sequential organization of lectures and labs, such that what they learn in one context makes intellectual connections with, and reinforces, what they learn in the other.

> He was a really good teacher … really thorough in presentations and good about answering questions … then we had a lab hour to build these models and practice … we had TAs teaching us how to do them … And the professor would stop by the labs, which I don't think you see very often … and he had a really detailed handout and trouble-shooting guide … That was probably the best engineering class. (White woman, switcher, civil engineering to finance)

> I always respond better to teachers who obviously have an interest in the material and can also relate it to something interesting. (White woman, biological sciences persister)

Students also viewed as important to their progressive understanding that course material and activities be logically-sequenced, well organized, and clearly outlined in the syllabus. Talking through course objectives with the class allowed them to understand what they were going to learn, the significance of particular course topics and relationships between them. It provided a road map that students could refer to as they progressed along it. Good teachers also referred to the map as they went along and sign-posted where they had been, where they were now, and where they were going next. Making the structure of lessons patent and consistent allowed students to understand their own intellectual journey and feel clear and confident about their progress:

> She went full-on through all the objectives on what you're going to learn … And even a year or two down the road, if you still have the syllabus, you could look back and be like, 'Oh yeah.' (White woman, switcher, biomedical sciences to sports medicine)

> My linear algebra teacher, I really admired. … He had a lesson plan that was beautifully bullet-pointed. Here's an example, here's a theorem, here's a definition. …Just really straight forward. (White woman, switcher, engineering to psychology)

> In my O-chem class, my professor was very consistent about what he taught. Even if you missed something, you could go back from the examples he gave before and you could figure out where he was going. (White man, biochemistry persister)

> This class was very well organized and taught and increased my knowledge of the subject exponentially. I wouldn't call me an expert on the subject, but I would feel comfortable answering a question on certain topics, if asked. (SALG write-in comment, white woman, biology major)

Again, we underline that many switchers gave full credit to particular instructors in their original majors whose courses they recalled with pleasure. A sense of continuity in the overall structure of the course was enhanced where instructors reviewed prior class work before launching into new material that built upon it:

> One of the most helpful things for me was his taking ten minutes at the beginning of each lecture every class to go over what we went over in the last lecture, so it was more of a continuation. (Asian man, biochemistry persister)

The value of well-organized and aligned course content based on clear learning objectives that are made patent to the students is also attested in research findings (e.g., Akiki, 2014; Lee et al., 2015). Students' positive appraisal of coherent course structure was also affirmed (and, where missing, lamented) in the original study.

## *Active, Interactive and Inquiry-Based Teaching Methods*

As described, most switchers (99%) and over half of persisters mainly experienced straight lectures in lower-division courses. Most switchers had left before experiencing the more interactive teaching strategies (and smaller classes that enabled them) that were experienced by persisters in later courses. However, 33% of persisters and 26% of switchers described, with approval, their experiences with various forms of active, interactive, and inquiry-based teaching across STEM courses of varying levels. They described several benefits: enabling better understanding of concepts, making the material more interesting, holding their attention, and allowing them to make solid connections between ideas. These were the fifth most-mentioned types of effective teaching.

Students answering the SALG praised many of the same characteristics of good teaching described in interviews: 29% of respondents praised active and interactive teaching methods and the use of non-lecture-based teaching methods. We conducted a simple cluster analysis using the types of teaching methods reported by respondents and found two groups of courses (as represented in Fig. 8.2). In the first group of courses ($n = 15$), instructors used clickers (audience response systems)—a practice associated with group discussion, short in-class reviews, and the integration of simulations and demonstrations into lecture. The other group ($n = 13$) used mainly lecture with fewer associated methods and slightly more group and individual projects.

While many written comments by SALG respondents were positive about courses incorporating clickers, overall the ratings for students' learning gains were lower in courses with clickers. Part of the difference may be that many courses incorporating clickers are larger, lower-division courses whose high rates of D or F grades,

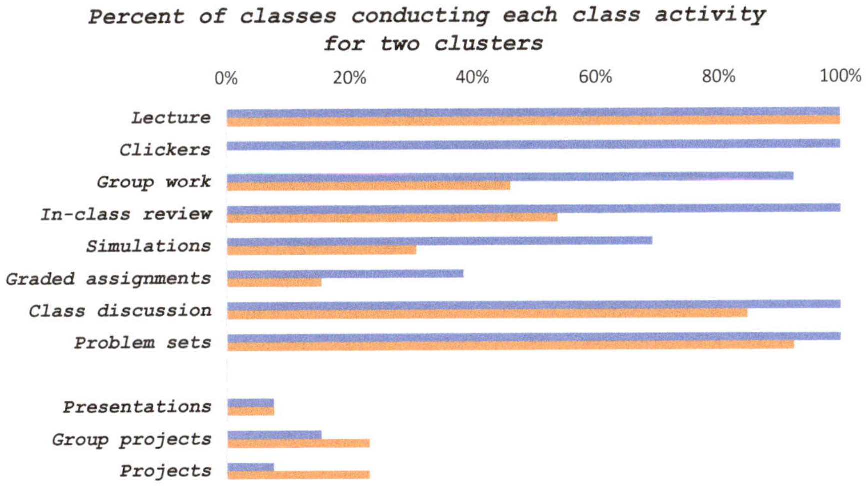


**Fig. 8.2** Percent of classes conducting each class activity for two clusters

withdrawals, and "incompletes" (DFWI) suggest that other aspects of these courses undermine whatever benefits clicker questions might offer. By contrast, both switchers and persisters found the use of clickers especially beneficial in large classes where other forms of interaction were limited. The apparent discrepancy is explained by commentary from both studies in which students clarify that clicker use is beneficial only when it is appropriately done and as a complement to good quality teaching not at its expense.

The following interview extracts illustrate the benefits to students of successful deployments of the technology:

> This teacher asks questions in the middle of class and you answer with clickers. The questions are really helpful…You can quiz your learning in the middle of lecture and make sure you understand the concepts. And I feel really confident after I get them right and I understand what I need to learn if I don't get them right. (Asian woman, relocator, chemical engineering to chemistry)

> He's just so interesting. That class is interactive all the time. So, he'll teach a little bit, go over some notes, and then you'll have 50 clicker questions through the whole lecture. (African-American man, aerospace engineering persister)

> I tend to understand a lot better in classes where they use clickers. I remember stuff better. (Native American woman, astronomy persister)

> In Organic 2 I thought the teacher was really effective… He had clicker questions – it made it helpful to have a benchmark of where I should be so I know I won't fall back.… The teachers that had checkpoints with clicker questions could evaluate where the class was – and when they realized they might be leaving a lot of people behind they would review the material again. Those teachers were helpful. (African-American woman, switcher, food science to theater)

Some SALG respondents who cited the benefits of the value of instructors' use of clickers also explained how effective use of this technique was helpful when used to check for understanding, correct misconceptions, facilitate small group student discussions, and build students' confidence in their progressive mastery of concepts:

> Discussing the clicker questions and recitation were especially helpful for my learning. It forced me to actually sit down and think through the problems and how to solve them. It made understanding the concepts easier. (White man, biology major)

> The use of clicker questions really helped me keep up with the concepts we learned that day. (White man, mechanical engineering major)

> I was able to understand why our answers were wrong right there with the teacher from the clicker questions (African-American woman, biology major)

> The teaching style for this class was enthusiastic and full of motivation. Clicker questions were provided every lecture, and this technique allowed me to develop a sense of confidence in my answers. I really enjoyed this course because it was very challenging—much of the material challenged what I had previously believed to be true/false. This was helpful for my expanding my knowledge and furthering my learning. (Multiracial man, biology major)

Other SALG respondents described their frustration with the use of clickers. As this student explained, the method only enables learning when instructors have adequately taught the material and what clicker questions explore is consistent with the instructors focus in tests. This student also cites the same conditions for effective group work:

> Clicker questions were extremely unhelpful, as was attending lecture. The material was not taught in lecture so often I felt more confused. Working with others only helped if they had taken a physics class before this one and knew the material. Otherwise, no one around me knew the correct answer, which was frustrating. Discussing clicker questions often felt like a waste of time because I would have *much* rather have spent that time listening to a lecture where the material was taught. The key to success was memorizing information for the tests. The tests asked extremely specific problems, so even though we were not taught equations in class, memorizing the equations was vital. (White woman, biology major)

Clicker questions also frustrated students when they came at the expense of direct explanations of material, particularly in fast-paced classes with a high content level:

> The pace of the class was very fast and yet we still managed to waste time answering clicker questions. [When we answered questions] we just talked about other things, we should have had more time to go over why an answer was right or wrong, how to do the problem, why we set it up the way we did, etc. (White woman, environmental studies major)

Both interviewees and SALG respondents explained that clickers are beneficial where instructors used them to check whether students are understanding well-taught concepts by posing carefully crafted questions. They cited examples of trivial usage and particularly disliked their use to check attendance.

Students described other interactive techniques used by some STEM instructors. One simple strategy that students found effective was teachers' pausing in their presentation to ask them questions to make sure that they were understanding the material:

> My thermo professor, he really enjoyed teaching us because he was so into what he was doing, and he would always ask us questions to see if we understood. He would take the time. And I really liked that, because he made the class lively all the time. (Asian woman, biological engineering persister)

One of the benefits of this method was that students came to expect they would be asked questions. This motivated them to read the assigned material and review the content of the previous lecture so that they felt ready to offer answers. Because they felt rewarded when answering correctly, this further reinforced pre-class preparation:

> He would ask us questions … I think it was very helpful because towards the end, some people felt very confident and rewarded when they would be able to answer his questions. And that format -- being forced to really up your game so you were ready to answer his questions -- I felt that was very valuable. (Asian man, biological sciences persister)

A persister described how his mechanical engineering instructor ensured that students stayed alert and active and were absorbing the material by getting them to relay back to him in their own words what he had told them. This also had the benefit of reinforcing what they had learned. Having students both ask and answer questions in a discursive format was also effective for keeping students alert and active. One engineering persister described teachers who encouraged back-and-forth student interaction by awarding participation points. Every student was expected to participate by asking or answering at least one question.

Persisters who had experienced structured peer-to-peer interaction in small upper-division classes reported on its value for their learning. It took several forms; for example, instructors asking students to explain concepts to those sitting next to them or having students work in small groups on practice problems. In the following extracts, students described these strategies and their benefits:

> He'd have us turn and talk to someone next to us and discuss things ... it forces you to teach someone else the material during class and it was a really great way to learn. He did that on purpose to help us process the material and not just memorize it. (White woman, relocator, biological engineering to microbiology)

> He would do all the theory behind a certain concept, and then he would get us to come up with a proof. We would work in groups of three or four and it was a creative way to help that proof make more sense rather than giving it straight to us. (White woman, relocator, mathematics to computer science)

> So the best teachers I've had, as they're lecturing to the students, they have practice problems and then they say, 'Hey, now that I've taught you this, why don't you practice this and then discuss it with your neighbor.' (Asian woman, biological sciences persister)

Many SALG respondents also reported that they gained understanding and confidence in their mastery of the material from working in groups, both inside and outside of class. Many students participated in study groups or asked other students questions via email. Most group work in class was structured around working on problem sets or finding answers to clicker questions:

> I really liked the peer learning, and how we learned through doing and problem solving as a group. Although I would be very challenged in a conventional exam-type situation with this information I have learned, I feel that I have learned to apply and do biology through this class effectively. (White woman, biochemistry major)

> With this being an interactive learning class, working with my group members and teachers on understanding the knowledge was very beneficial. It was easy to approach them, ask questions, and receive helpful explanations. (White woman, genetics major)

Guided discussions in groups were sometimes structured by discussion questions that required out-of-class meetings:

> The class had many discussion questions that allowed students to talk to each other. Also, the homework also allowed students to meet up and discuss the questions. Overall, it was a good learning atmosphere and it allowed me to build on my social skills with my classmates. (Asian-American woman, genetics major)

Again, we note the confidence-building potential of focused collaborative work.

Persisters described the use of other active learning techniques in upper-division courses. Although the forms of these strategies varied, students described their benefits as making it easier to learn, enabling deeper understanding, and making concepts more accessible to students with different learning styles:

> My physiology teacher was amazing...Every single class he was at the light board drawing it from the inside out, making you understand the full story. (White woman, biomedical engineering persister)

> One of the things that made it easier to learn from the more intensive classes here was this active learning style in my foundations biology class. It was a lot easier for me to pick up

> on things where there are multiple learning styles being integrated into a course.... For me, active learning is more in-depth. (White man, microbiology persister)

> My invertebrates' professor, he did a lot of demonstrations and had a lot of stuff to look at... Instead of just a picture on a slide, he actually made it so that we could interact with something. I'm a kinetic learner; I need more than to just like sit there and listen...so if I can like apply it to something it works—and he did that really well. (White woman, zoology persister)

Some instructors found other creative ways to engage their students. For example, a white female engineering persister described how a technique used by one of her engineering instructors engaged students: "She would play music at the beginning of class and that brought life into the classroom." As noted by several teaching scholars, regardless of the format, interaction, and activity-based teaching helps students to stay engaged and involved throughout class and increases student retention rates (Ambrose et al., 2010; Ejiwale, 2012; Gao & Schwartz, 2015; Svinicki & McKeachie, 2010).

### *Providing Examples, Showing Applications, and Making Connections*

Some of the most effective instructors described were those who helped students to connect theoretical material with real-world phenomena. As reported in Chap. 6, many STEM instructors failed to do this. However, nearly one-quarter of switchers (24%) and over a third of persisters (36%) described how instructors who provided examples and applications had helped them to apply what they were learning and had increased their depth of understanding and retention of new knowledge:

> Every abstract or new concept that he taught, he would have a real-world example for every single thing. That was really, really helpful. (Multi-racial man, mechanical engineering persister)

> My favorite professor always incorporates the current events of science into his lectures. He puts in even more effort in updating them. Instead of following a slide about a certain technique, he'll follow it with a news article about how they're using that technique in Sweden to something with genetics. (Multi-racial woman, biological sciences persister)

> He would bring in outside things rather than just lecture. He would start his lecture with a chemical that you know came from organic chemistry, and he would talk about it and show the application to the real world, and then he would start his lecture. So, it wasn't just a lecture ... when you see something that relates to the outside world, you're like, "OK." (African-American woman, switcher, food science to theater)

Some instructors, as above, used the strategy of beginning with something familiar to students, then progressively working backwards into concepts that students needed in order to explain it. This method, pioneered in the 1990s in ChemConnections chemistry teaching modules (Anthony, Braun, & Mernitz, 2012) proposed that, "students learn best when they can build on past experience, relate what they are learning to things that are relevant to them, have direct 'hands-on' experience, construct their own knowledge in collaboration with other students and faculty, and communicate their results effectively. The modules, therefore are based on a question from the student's surroundings such as: *What should we do about global warming?* and *How can we purify our water?* Each module centers on an interesting question that provides a context for understanding and applying specific chemistry concepts. The module question, and its accompanying story line, provide a contextual framework and springboard for guided inquiry and exploration." (Anthony et al., 1998, p. 322). Interviewees in the study sample described the effectiveness of teachers who deployed this method:

> I have a really good teacher … He will start with a real-world example of a math theory and then move on to showing the mathematics actually behind it and then do a visual exercise where you could see graphs or run through a simulation and then go back to the real-world example. (White woman, computer science persister)

> My organic chemistry teacher would bring in outside things….He would talk about an idea, show its application to the real world, and then start his lecture…. It's very effective. (African-American woman, switcher, life sciences to theater)

Students reported that illustrations, real-world applications, and making connections to other areas of knowledge all helped them to understand and remember the material:

> In Organic Chemistry 2, the style was 'OK, here's a mechanism, but here's how it's used in the system.' So, we learned about isomerization when light hits a certain double bond, and then my professor showed us what actually happens in your eye cells, and that's how you're able to see. And I mean, I still remember that, even though it was a year ago. (Asian woman, biological sciences persister)

> Our professor started off the lecture by introducing how the magnetic waves aligned with the orbitals and pulled electrons, and then you got signals from those electrons based on polarity. That helped us really define what it meant to do spectroscopy, so we had a much easier time understanding. (White man, physics persister)

> He had baked potatoes, and he would stick little thermometers inside to measure the temperature at different depths into the potato, and then talked about 'What's happening when you bake a potato?' That was very memorable. (White man, chemical engineering persister)

SALG respondents also wrote that lectures were more compelling when teachers used examples and applications from real life:

> The concepts learned in this class have many real-world applications. I have learned not only how to perform the calculations and solve problems on tests, but also how to apply the concepts to situations outside of the classroom. (Asian-American woman, biology major)

> I never studied physics before, so I was able to connect many classroom concepts to my day-to-day life. (White woman, animal science major)

Again, using techniques developed for teaching introductory chemistry (c.f. Hill & Holman, 2011, sixth Edition) students across STEM disciplines described how the use of aides-memoires, anecdotes, and explanation of how concepts can be used to explain things that are familiar in the student's world, provided them with context, enabled their understanding of complex topics, and helped with recall:

> I still remember all the names of the bones. My biology teacher, she would tell a detailed story about each bone that would help you remember. So, on the test when you have to remember 206 bones, you're able to name all of them pretty easily. (White woman, switcher, biology to psychology)

> The side stories really help me … We learned about this molecule and this is how you can eat this molecule in food and how it affects you. So now I remember that an apple has X, Y, and Z chemicals and that's what processes them in my body. (White woman, switcher, biology to economics)

*How* teachers give examples also makes a difference. Students especially appreciated active examples provided in real time. For instance, some teachers would write out or draw the conceptual material while talking the students through it. Helping students follow the teacher's thought processes or methods step-by-step was considered far more effective than reading the same steps from PowerPoint slides. Some of these interactive forms of teaching were done using sophisticated equipment; others worked well using simple, low-tech means:

> My reproductive teacher had Smartboards, where you write on the screen and then it shows up on the slide. He does it solely from memory of what he knows and he writes it down every single day. Every slide is different, he's writing it as he goes, and that was really awesome because he's explaining it as he's writing it down. He's drawing diagrams. He's explaining the diagram. It really helped me to see the process of how he's thinking and how he's going through it. (Hispanic woman, equine science persister)

> He had an old-school projector--the kind that you have to spray and erase with a paper towel and all these colored markers. He projected it into the lecture hall and he would write out the mechanisms and circle and draw arrows and stuff rather than just talking about it. This really helped for orgo, for knowing the mechanisms. (African-American woman, persister, biochemistry to biology)

> He'd have a projection of watching him type. So, you'd have the projector watching him develop the code so you could see it on your own screen and how you were doing it. (White man, computer science persister)

> Watching her do things step-by-step. The teacher was awesome and she worked out every single mechanism and did it herself. And then it just seemed like a puzzle that you were solving--trying to figure out where the electrons were moving, and that helped a lot. (White woman, relocator, environmental engineering to chemical engineering)

Even in more traditionally taught lecture classes (notably chemistry and physics), SALG respondents reported that they learned more when demonstrations or large-screen computer simulations were offered:

> The demos used in class as well as the molecule models particularly helped me to visualize what we were currently learning and focusing on. (White woman, chemistry major)

> I liked whenever professor [name] brought her molecular structures to class to show us examples. (White woman, human development major)

Drawing on a variety of these methods that placed concepts in an accessible context these instructors were able to carry their students along with them in thinking through the material. Their students' testimony also makes it clear that the extra effort that these instructors made to ensure that their students got the most out of their classes was also part of what motivated and encouraged them.

Over the last two decades, a significant and growing body of empirical research underscores what students in this study reported, namely, that student learning is enhanced when teachers employ active, interactive, and inquiry-based methods of instruction (see, for example, Akiki, 2014; Armstrong, Chang, and Brickman, Henderson & Dancy, 2007; Ejiwale, 2012; Freeman et al., 2014; Gao & Schwartz, 2015; Hanson, Paulsen, & Pascarella, 2015; Holdren, Lander, and the President's Council of Advisors on Science and Technology, 2012; National Research Council (2012); Mulnix & Vandergrift, 2014; Sawyer & Alper, 2014; Watkins & Mazur, 2013). These scholars emphasize the importance of engaging students as active partners in their own learning process in order to improve their depth of understanding and ability to apply, extend, and transfer their knowledge (Seymour, 2007). Enabling students to engage with the development of ideas through active forms of learning (such as discussion, small group work, peer learning, and hands-on inquiry) is found to deepen their understanding of concepts. Activity-based teaching also keeps students engaged and motivated. Ejiwale (2012) argues from data that using practical illustrations and real-world applications is highly effective in increasing students' depth of understanding. Students also remember what they learn when they are enabled to interconnect concepts and use them to understand real world phenomena and processes. The more students can identify with the material, the more willing they are to invest time in understanding course content and the more likely they will stay engaged and motivated (Gao & Schwartz, 2015). Gao also reports that active forms of learning can also increase student retention from 10–30% to 80–90% levels.

The National Science Foundation, the National Academies of Science, many private foundations, and STEM disciplinary and professional societies promote the use of research-based instruction strategies (RBIS) that incorporate interactive, active, and inquiry-based pedagogies and are important repositories of resources for teachers. Disciplinary-based and also cross-disciplinary networks of STEM instructors share, spread, and support each other in learning these methods and many instructors publish the results of their work in a growing scholarship of teaching and learning. Some of the many active research-grounded teaching strategies promoted include acting as classroom facilitator, explaining concepts and procedures from multiple perspectives, organizing lessons around outcomes-based planning, teaching collaborations with colleagues both inside and outside of the department, frequent interactions with students, and providing scaffolded challenges (e.g., Akiki, 2014; Ejiwale, 2012; Mulnix & Vandergrift, 2014; Gao & Schwartz, 2015; Hanson et al., 2015; Lee et al., 2015).

None of these teaching methods were evident in STEM classrooms in the original study. However, various active and interactive pedagogies were reported in the current study by 26% of switchers and 33% of persisters. They were also observed in 16% of introductory STEM classes in Ferrare and Miller's (2017) class observation study that was conducted in the same 80 foundation courses that were surveyed by the SALG instrument. Small group work was reported as the most common form of innovation in all three studies. In 9% of all courses (in physics, chemistry, biology and engineering classes of varying sizes) Ferrare observed frequent facilitation of peer interactions. Students were engaged in hands-on activities that required creation and problem-solving, and instructors spent half of their class time moving around the classroom answering questions and talking with groups of students. Ferrare also found that 8% of instructors (mostly in large classes) combined a variety of lecture methods with small group work. As we observed in Chap. 6, these findings indicate a somewhat modest uptake of the array of RBIS now available to instructors. However, these findings will be encouraging to instructors and departments that are actively seeking to implement teaching methods that meet students' expressed preference for greater engagement with their instructors in the learning process.

## Students' Positive Accounts of Their Experiences with Graduate Students as Teacher and in Learning Support Roles

As described in Chap. 6, persisters reported more positive than negative experiences with their teaching assistants (viz., 33% positive, 10% negative). Switchers, however, were more likely to comment on their experiences—good or bad—with teaching assistants when asked about the kind of instruction they received. This is not surprising as most switching occurs following introductory STEM courses that are heavily staffed by teaching assistants in accompanying laboratories and recitations, and in some instances, with graduate TAs as class teachers. Switchers' reports of their learning experiences with teaching assistants were 47% negative (as discussed in Chap. 6) and 53% positive. Taken overall, our sample of STEM undergraduates reported that their TAs—whether in teaching or learning support roles—were more accessible, more supportive, and more enthusiastic about the subject and also about teaching, than were the instructors who lead their STEM courses. This finding also mirrors that reported in the original study.

### *TAs as Approachable and Available Sources of Help*

Both interviewees and SALG respondents cited their TAs as easier to approach, talk to, and request help from than their instructors—an experience that partly derives from the contexts in which students encountered them. TAs often worked with students in labs and recitation sections where the numbers were small enough for natural interaction and ease in getting to know one another, or in learning centers where they worked one-on-one with students:

> The TAs definitely you talk to a little bit more just 'cause you see them in a smaller class--like thirty people. So, you would get more one-on-one time with the TAs. …I didn't really talk to my professor at all. (White woman, switcher, chemistry to French)

Persisters, especially, reported that their TAs were more approachable and supportive than their primary instructors, took regular advantage of this opportunity, and credited their part in enabling them to survive difficult courses:

> I'd say TAs are more helpful at office hours than professors are. Like, if I have a TA that has office hours on Wednesday, and a professor that has office hours on Thursday, I'll go out of my way to make sure I start my homework before Wednesday so that I can ask my TA questions rather than my professor. (White woman, civil engineering persister)

> I never really interacted much with professors …I interacted more with the TA, and that was kind of what got me through some of the more complicated subjects. (Asian man, aerospace engineering persister)

Teaching assistants were also seen as far less intimidating and easier to talk to than instructors, especially for first-year students. Being able to ask the TAs questions could also help freshmen eventually feel more comfortable approaching instructors with their questions:

> [Faculty] are really intimidating and so the only office hours I went to as a freshman were TA office hours. 'Cause they're younger, and they know you from a smaller discussion or lab section. I learn best by talking out things with people, like I ask you a question about something I'm confused about and then you ask me a question back. …But since then I haven't hesitated to go to any professor's office hours … now I go all the time. (Asian woman, biochemistry persister)

Establishing working relationships was enabled by contact through recitation and lab meetings, by being close in age, and by TAs' more recent experience with learning the same bodies of knowledge and, thus, their appreciation of the difficulties entailed and how best to approach them. Switchers who were struggling also appreciated how responsive and available their TAs were compared with their primary instructors:

> My chemistry TA was super cool. If I had any questions I could go talk to him, or I would send him an email, and he would respond to it within like twenty minutes. (White woman, switcher, chemical engineering to accounting)

> You can always take your homework or tests back to the TA, the teacher is not always available. But the TAs are available most of the time, so you can take that work back and get a better understanding of what you did wrong. (White man, switcher, engineering to psychology)

However, TA accessibility and responsiveness reflected willingness to help first-year undergraduates particularly that went beyond the benefits conferred by simple proximity. It was their attitudes and behaviors towards both students and their teaching role that made TAs a more available, and thus more helpful, learning resource to students than many STEM instructors.

## *Establishing Teacher-Learner Relationships*

Students described how much easier it was to develop a working relationship with their TAs than with their instructors. They felt more comfortable asking TAs questions and experienced them as more understanding of their learning needs than were instructors. Students explained this difference in terms of their shared experiences as students and, thus, an understanding of the sources of particular difficulties arising from their own experiences in the same or similar courses:

> They understand what it's like to be a student, 'cause most of them are grad or doctoral students, so they understand where the problems are, and I feel like some of the professors just don't remember what it's like to be a student. (White woman, switcher, chemistry to French)

> I felt like the TAs understood us more [than the professors] and were kind of in the whole struggle with us. (Asian woman, switcher, biological sciences to strategic communication)

> I like having TAs, 'cause they've just recently taken the course, so they understand what you're talking about … I think it's easier for someone to explain stuff when they're more on your level. Whereas the teachers are a lot smarter than we are, so it's hard for them to explain things at a simple level. (White woman, switcher, computer engineering to health and wellness)

They also found that TAs were more apt than instructors to see it as part of their job to address students' learning difficulties and find ways to surmount them:

> The TAs are usually graduate students or older students and I always find it helpful to talk to them because they have taken the class before and they're not talking from the professor's perspective – where the professor always thinks, 'Why don't you get this?' The TA will say 'OK, let's go through this step by step.' (African-American woman, biological engineering persister)

We also note (as in the original study) the importance to some students—often women—of learning in the context of a personal dialogue with their teachers. The possibility of a more personal teacher-learner relationship was seen as much greater with TAs than with instructors, especially in large lecture classes:

> The TAs are good. That's how I passed some of my classes -- going to the recitations and the TA's office hours. They were all helpful in having a more personal relationship with you, since the class size was so big in chemical and biology engineering … there could be up to a hundred students. There wasn't as much interaction between you and the instructor. But the recitations were smaller, so you could still get the help you need from the TA. (White woman, chemical engineering persister)

## *Teaching Assistants as Effective Teachers*

Notwithstanding some complaints about the poor quality of teaching by particular graduate assistants (discussed in Chap. 6) there were also many appreciative appraisals of TA teaching, especially from persisters. SALG commentators also wrote that TAs in labs and recitations had enabled their learning. TAs were cited as instrumental in clarifying complex lecture material, and some TAs were described as more interested in teaching, in active pedagogical methods, and as more invested in the students' learning needs than were many instructors:

> I feel like a lot of the teachers that are here don't really like teaching. And a lot of the TAs are the best teachers I've had. (White man, mathematics persister)

> Professors are annoyed when you come to their office hours 'cause I feel like teaching isn't their priority. It's more like it's their research. … The TAs were more willing to help. I think they were better at explaining the material. (White woman, life sciences persister)

> I love TAs, I think they're super great. Especially lab TAs, because a lot of them have fun with it because they love the subject. And they love teaching, so it's much more personal. They actually learn your names and know who you are, so I like them. (Asian Woman, biological sciences persister)

> The TAs are the best part of my education.… They're there because they want to help or because they want to give back. TAs – even more than professors – if you show up to their office hours they get excited, and that passion is what really makes the difference. They're the ones I always go to. They have more time for you than the professors do. (White woman, biological sciences persister)

Persisters in particular expressed appreciation for the quality of teaching provided by graduate TAs, and some appraised it as superior to the instruction they received from instructors:

> The TA has a lot more responsibility than the professor really, especially in my lower biology classes. There was a huge disconnect between the professor and the student. The TA was the one who taught you, who graded all of your tests. And if you had a question, you went to the TA office hours. The TAs were definitely your teachers. (African-American woman, biological sciences persister)

Students were thankful for TAs who were patient and willing to spend time to explain and clarify difficult concepts and other complex material presented in lectures:

> The TAs didn't mind taking the time to explain the problem and would even go over it multiple times if we weren't getting it. (White woman, switcher, mathematics to landscape architecture)

> I like also when the TA can simplify what the professor has been talking about in class. My O-Chem professor would go off on these tangents, and it would take him an entire class period to say something that the TA could say in twenty minutes. So, when I went to the recitation, the TA would re-explain it and that was very helpful in clarifying the material. (Asian woman, biochemistry persister)

> I had a really good TA who was very helpful and explained things pretty well. It really makes a difference when you have a helpful TA. I operate best when I have multiple sources of help. So, I could go to him and ask, 'Hey Jasper, can you clarify something that I learned in lecture?' And he would. (White woman, switcher, civil engineering to geography)

## *Seeking Help from Teaching Assistants*

The last speaker got the help that she needed because she took the initiative to seek it out. As we reported in Chap. 5, significant factors in students' transition into both college and STEM majors were awareness that help was available and an understanding that it was important (and legitimate) to use it. As we also reported in the original study, failure on both counts left students open to switching risk. We found that the persisters more frequently than switchers took advantage of the help offered by teaching assistants—typically during recitations and laboratories, office hours, and help rooms. Indeed, they often saw it as a critical factor in their survival in the major. As one electrical engineering persister attested, "There was no way I would've gotten through the course without my TA." Like many other persisters, he regularly went to her office hours, sought her out when he needed help to understand something particularly challenging, and to make sure he was on the right track with assignments:

> I've had really great TAs, and I would meet with them before I would do papers or lab reports and they helped me so much. (White woman, physiology persister)

TAs also acted as an important hidden force in enabling students who entered college under-prepared by their high schools to bring themselves up to speed with their peers:

> I took as much advantage of TA office hours and tutoring resources that I could in those first couple of years. I realized that I had gotten into something that I wasn't necessarily prepared for, especially with organic chemistry and advanced math classes. (White man, materials science persister)

Taking advantage of available help was a key difference between persisters and switchers who were more likely to express regret that they had not sought the help of teaching assistants. In the original study, we reported that failure to seek help—whether from a TA or from any other available source—was particularly marked among young male switchers. (Women were much more comfortable in asking for help but were more devastated when it was not forthcoming.) When asked why they did not seek help, the reasons that these students offered echoed those of the earlier study: They had thought of going to TA office hours but didn't because they assumed it would be "another waste of time," or that they would have to wait a long time because the help rooms "were always super packed." What these insubstantial rationales share is a sense that it is difficult for some students—especially young men—to accept that they may not be able to do everything on their own, and to give themselves permission to accept help without which they may not survive in their chosen major.

## *Graduate Teaching Assistants in Context*

Graduate teaching assistants are known to play a significant role in undergraduate students' learning experiences, especially in STEM courses. At research universities, especially, most undergraduate STEM students will receive instruction from teaching assistants in their laboratory courses and recitations (Dotger, 2010; Gardner & Jones, 2011). Undergraduates commonly have more contact with teaching assistants than with faculty (Gardner & Jones, 2011). This was especially evident in our switcher sample, who commonly associated their learning experiences in STEM courses with the teaching assistants they encountered in recitations, laboratories, and sometimes lectures.

Although graduate students play such a key role in undergraduates' learning experiences during their first two years in a STEM program, most TAs receive very limited preparation for their teaching and learning support work (Gardner & Jones, 2011; Dotger, 2010; Dudley, 2009; Seymour, Melton, Pedersen-Gallegos, & Wiese, 2005). Lacking adequate tuition in learning theory, pedagogy, assessment, or classroom management, teaching assistants (like young faculty) draw upon their own experiences as students or model their practice on that of departmental faculty—thus perpetuating the cycle of less effective teaching methods into the next generation (Dudley, 2009; Seymour, Melton, Pedersen-Gallegos, & Wiese, 2005).

Other studies report that graduate teaching assistants frequently feel overwhelmed and under-supported by their departments in their teaching assignments (Bomotti, 1994; Dudley, 2009). Dudley (2009) found that graduate students wanted the following forms of preparation in order to do their work effectively: More communication with, and instruction from, the instructors with whom they were working to help them understand how the laboratory or recitations that they taught aligned with the lectures, what were the instructors' learning objectives for the students; what subject knowledge they needed to know well in order to help students prepare for assignments and keep lab activities up to date and enliven them with real-life examples; and an agreed set of policies for both grading and classroom management. Not only do graduate teaching assistants wish for more training by departments and guidance as part of an ongoing working relationship with their instructors, but, as is evident from our interview study results, undergraduate students are often critically aware when graduate students lack sufficient knowledge of the discipline or training as teachers, and of the effects of poor communication and alignment between TAs and their instructors.

## What Has Changed Since the Original Study in How Students Characterize 'Good' and 'Bad' Teaching and How These Vary by Gender and Race/Ethnicity?

In *Talking about Leaving: Why Undergraduates Leave the Sciences* (Seymour & Hewitt, 1997), we contrasted young men's predominantly instrumental expectations of their STEM instructors with the more affective expectations of young women. These were evident, for example, in their reasons for disliking large classes. Men emphasized the assignment of less experienced instructors to these classes, and the depressing effect on grades of the high level of competitiveness that large classes encouraged. Women complained that the large classes made learning more impersonal: It was harder to establish a relationship with their professors and easier for them to ignore you. In their definitions of "good" and "bad" teachers, women more often than men stressed the importance of their teacher's personal attention to them.

For women, "good" teachers were approachable and readily available for consultation, and friendly in their manner of teaching and towards students as individuals. They were patient and willing to explain difficult material, courteous and supportive, and showed that they cared about students' learning. Men were less concerned with teachers' openness to personal contact with them than with their effectiveness in presenting the course material. Good instructors were enthusiastic, engaging, and explained concepts well. They gave good examples, demonstrations, and real-world applications, and encouraged questions; and they checked to see that student were understanding. Men also appreciated teachers who challenged and motivated them and pushed students to work hard.

We offered strong textual evidence that "failure to establish a personal relationship with faculty represents a major loss to women, and, indeed, to all students (including men and women of color) whose high school teachers had given them considerable support and fostered their potential." For these students, "engaging the teacher in a personal dialogue appears to be critical to the ease with which they can learn and to their level of confidence in the adequacy of their performance. The abrupt withdrawal of a special teacher-learner relationship and its replacement with the 'impersonality' of college classes was reported to be extremely disorienting." (p. 267) We offered this finding as a partial explanation for the loss of incoming confidence among white women, and students of color of both sexes, that were the focus of a large number of research studies at that time (e.g., Arnold, 1987; Astin, 1993; Ginorio, Brown, Henderson, & Cook, 1994; Manis, Sloat, Thomas, & Davies, 1989; Rosser, 1990; Ware & Dill, 1968). For a review of this literature, see Kimball (1989) and Oakes (1990). Our findings offered an explanatory link between the then high loss rates among high-ability women entering STEM majors and their unmet need for personal connection with their teachers as an important contributor to their loss of incoming confidence (see also, Seymour, 1995).

The gendered dichotomy in definitions of "good" and "bad" teaching was much less evident among men and women in the four main groupings of students of color in our TAL sample, namely, African-Americans, Hispanics, Asian-Americans and Native Americans. Indeed, across both sexes in all four groups, half as many students of color as white students (i.e., 21% vs. 42%) complained about "poor teaching". Rather, we found a distinctive tendency among all students of color to blame themselves, rather than their instructors, for their learning difficulties. However, as with white women, the decision to leave a STEM major was often precipitated by loss of confidence in their ability to do science. Also, like white women, many students of color were accustomed both to a high degree of individual support from high school teachers and to rewards for effort as well as performance—expectations that they carried into college. They were, thus, ill-equipped to deal with the impersonality of traditional STEM teaching, and their first experiences of "objective" grading could seem unfeeling or discriminatory. As they were often the best students in under-resourced high schools, many students of color also experienced the devastating reality that they had entered college under-prepared but over-confident. In addition, unlike white women who complained that instructors did not make themselves available in office hours, students of color of both sexes often did not ask questions or seek help. Having worked intensively with their high school teachers, they also had less experience of working with peer study groups. Thus, for all these intersecting reasons, a history of high reliance on learning relationships with teachers, low assertiveness, and a tendency to self-blame left many students of color at high risk of switching. African-American women, however, stood out as exceptions to the pattern of self-blame and lowered confidence and were among the most independent and self-assertive students in our samples (c.f. Chap. 6, *Talking about Leaving*).

In the current study, what students defined as "good" and "bad" in the quality of the teaching they experienced was more homogenous by gender, race/ethnicity, and switcher status than in the original study. We found that all student groups were now concerned about the quality of their personal relations with teachers, such that "friendliness" and caring about their students were the commonest defining features of what students described as 'good' teachers. As explained earlier in this chapter, these were shorthand terms for teachers who were approachable and available, invited questions inside and outside of class, explained that the material, although hard, was within their grasp, encouraged them to come to office hours, showed compassion for their difficulties, knew their names, and took an interest in their learning and their lives. Other characteristics of 'good' teaching focused on teaching competence and effectiveness, but they also included attitudes towards students and their teaching role. In frequency order, these were: Delivering course material in an engaging way; being organized and coherent; using interactive instruction methods; and providing helpful real-world examples.

Students overall defined 'bad' teaching (also in frequency order) as: Presenting material in a disorganized manner, pitching material at an unsuitable level for the course; behaving in a distancing or intimidating manner; presenting material poorly; showing little engagement with either the material or the students; and conveying research as more important than teaching. Again, as described earlier, worst-case descriptions glossed as "unfriendliness" included: Avoiding questions or answering them begrudgingly; mocking and belittling questioners; treating difficulty in understanding with incredulity or annoyance; discouraging office visits by rude or aggressive behavior or referring students to the TA, and not learning their students' names.

The convergence, over time, of all students on what they wanted and did not want from STEM teachers is marked. In contrast to the findings of the original study in which men were described as more "instrumental" in their concern for effective presentation of materials and women more "affective" in their focus on the student-teacher relationship, in this study, similar proportions of men and women (across all races and ethnicities) expressed concerns about how material was presented (viz., 29% of women and 31% of men). "Engaging delivery" was the second most-commonly cited characteristic of 'good' teaching. About one-third of all men and women (37% of women and 31% of men), regardless of race/ethnicity and switcher status, defined 'good' teaching as engaging and enthusiastic.

That said, there were some variations, but with far less distinct patterns, whether by gender or race/ethnicity, than were found in the original study. For example, male switchers of color were particularly likely to describe teachers' attitudes and behavior toward them when discussing teaching quality. Sixty percent of these men (in contrast with 30–40% of all other students) defined 'good teaching' as being "approachable," "treating students fairly," and "caring about students." They were similarly likely to define "bad" teaching as being "unfriendly," "unapproachable," "impersonal," and "rude."

An entirely new phenomenon was student awareness of alternative forms of teaching to 'straight lecturing' and it was more evident among switchers than persisters and among white switchers than switchers of color. Among white switchers, 23% of women and 20% of men described 'good' teaching as including active, interactive and inquiry-based instruction methods, whereas, only 14% of female switchers of color and no male switchers of color did so. White switchers of both sexes included reference to these techniques in defining 'good' teaching and wished that more STEM faculty would use them. Whether this difference reflects the wider experience of white students as learners in high school and non-STEM disciplines, or respect for traditional teaching by students of color is a matter for speculation. However, as in the original study, students of color tended to be less judgmental about their teachers than white students. Although research-grounded teaching methods were valued by some persisters, fewer persisters (15%) than switchers (25%) included them in their characterizations of 'good' teaching.

Also among switchers, more white women (38%) discussed their frustrations with poor delivery of lecture materials than did either women of color (10%) or men overall (20% of white men, and zero men of color). Although, as above, persisters judged their teachers less than did switchers about their failure to use active and interactive methods, they complained more than did switchers about STEM instructors' poor presentation skills (38% of persisters compared with 30% of switchers). This was the persisters' second-most common definition of poor quality teaching and there were no differences by race/ethnicity or gender on this issue. Thus, persisters were, as a whole, more conventional in their evaluation of teaching methods than were switchers. They accepted lecturing as the norm, but expected lecturers to be effective and engaging in promoting understanding.

We noted several other distinctions with respect to white men. White male switchers were more likely than any other group of switchers to complain about STEM teachers who delivered course material at an unsuitable level. (Among switchers, 47% of white men, zero men of color, 28% of white women, and 19% of women of color complained about course level). These men were frustrated by instructors who failed to review essential material before launching into new, complex material. They were also more likely than other switchers to sense that STEM faculty valued research more than teaching (viz., 27% of white men, zero men of color, 13% of white women, and 10% of women of color). Thus, as in the original study, among switchers, white men, more than white women or students of color of either sex framed their complaints in terms of flaws in instructor competence, efficiency, and commitment to teaching. Twenty-two percent of persisters also found these issues problematic, but there were no differences by gender or race/ethnicity in the level of complaints.

We, thus, observe some significant changes in what students want from their STEM teachers. First among these, is a convergence across all student groups in their characterization of "good" teaching. The expectation that teachers should be engaged in their students' learning and treat students with respect and encouragement is now widely endorsed by white men as well as by women and students of color. Women, and to a lesser degree, students of color, now balance this preference with more (instrumental) critical appraisals of classroom teaching techniques while retaining their (more affective) concern for the quality of individual relationships with teachers. However, this falls short of the level of teacher-dependency that prompted able women to switch following a loss of confidence that was so marked twenty years before. Switchers also emerge as somewhat more aware of new teaching trends, and more consumerist about the quality and value for money of their university education than persisters in their expectations of research-grounded teaching methods. Although they also register similar values to those of white peers, students of color remain somewhat more reluctant to criticize teachers and are more conservative in their appraisals of teaching quality. White men are still more instrumental in their teaching preferences than other students and male students of color are more critical of teachers' personal attitudes and behavior towards them. Notwithstanding these variations, we record a convergence towards gender and race–ethnicity neutrality in student assessments of teaching quality in which more is expected of teachers, both as enablers of learning in the classroom and as supporters of individual student progress.

## References

Akiki, T. K. (2014). A review on effective teaching and learning in higher education. *European Scientific Journal, 1*, 159–164.

Ambrose, S. A., Bridges, M. W., DiPietro, M., Lovett, M. C., Norman, M. K., & Mayer, R. E. (2010). *How lerning works: Seven research-based principles for smart teaching*. San Francisco, CA: Jossey-Bass.

Anthony, S., Braun, K. L., & Mernitz, H. (2012). *ChemConnections activity workbook*. New York, NY: W. W. Norton & Company, Inc.

Anthony, S., Metnitz, H., Spencer, B., Gutwill, J., Kegley, S. E., & Molinaro, M. (1998). The ChemLinks and ModularCHEM consortia: Using active and context-based learning to teach students how chemistry is actually done. *Journal of Chemical Education, 75*(3), 322. https://doi.org/10.1021/ed075p322

Armstrong, N., Chang, S.-M., & Brickman, M. (2007). Cooperative learning in industrial-sized biology classes. *CBE-Life Sciences Education, 6*, 163–171. https://doi.org/10.1187/cbe.06-11-0200

Arnold, K. D. (1987). *Values and vocations: The career aspirations of academically-gifted females in the first five years after high school*. Paper to the Annual Meeting of the American Education Research Association, Washington, DC.

Astin, A. W. (1993). *What matters in college? Four critical years revisited*. San Francisco, CA: Jossey-Bass.

Bomotti, S. M. (1994). Teaching assistant attitudes toward college teaching. *The Review of Higher Education, 17*(4), 371–393.

Dotger, S. (2010). Offering more than "Here is the textbook": Teaching assistants' perspectives on introductory science courses. *Journal of College Science Teaching, 39*(3), 71–76.

Dudley, M. (2009). Jumping out of an airplane: A TA's perspective on teaching effectiveness. *Eastern Education Journal, 38*, 1–10.

Ejiwale, J. A. (2012). Facilitating teaching and learning across STEM fields. *Journal of STEM Education, 13*(3), 87–94.

Ferrare, J. J., & Miller, J. (2017, February 25). *Cultural models of persistence and inequality in introductory STEM courses*. Annual Meeting of the Sociology of Education Association, Monterey, CA.

Freeman, S., Eddy, S. L., McDonough, M., Smith, M. K., Okoroafor, N., Jordt, H., & Wenderoth, M. P. (2014). Active learning increases student performance in science, engineering, and mathematics. *Proceedings of the National Academy of Sciences, 111*(23), 8410–8415.

Gao, X., & Schwartz, B. (2015). Classroom implementation of active instructional strategies for undergraduate STEM education. *INternational Journal of Information and Education Technology, 5*(9), 688–692.

Gardner, G. E., & Jones, M. G. (2011). Pedagogical preparation of the science graduate teaching assistant: Challenges and implications. *Science Educator, 20*(2), 31–41.

Ginorio, A. B., Brown, M. D., Henderson, R. S., & Cook, N. (1994). *Patterns of persistence and attrition among science and engineering majors at the University of Washington, 1985–1991*. Sloan Foundation: Report to the Alfred P.

Hanson, J. M., Paulsen, M. B., & Pascarella, E. T. (2015). Understanding graduate school aspirations: The effect of good teaching practices. *Higher Education, 71*, 735–752.

Hill, G. C., & Holman, J. (2011). *Chemistry in context* (6th ed.). Oxford, England: OUP.

Holdren, J. P., Lander, E., & The President's Council of Advisors on Science and Technology. (2012). *Report to the President: Engage to Excel: Producing one million additional college graduates with degrees in science, technology, engineering, and mathematics*. Washington, DC: Executive Office of the President.

Kimball, M. M. (1989). A new perspective of women's math achievement. *Psychological Bulletin, 105*(2), 198–214.

Lee, H. H., Kim, G. M. L., & Chan, L. L. (2015). Good teaching: What matters to university students. *Asia Pacific Journal of Education, 35*(1), 98–110.

Manis, J. M., Sloat, B. F., Thomas, N. G., & Davies, C.-S. (1989). *An analysis of factors affecting choices of majors in science, mathematics and engineering at the University of Michigan*. Ann Arbor, MI: Center for Continuing Education of Women.

Mulnix, A. B., & Vandergrift, E. V. H. (2014). A tipping point in STEM education reform. *Journal of College Science Teaching, 43*(3), 14–16.

National Research Council. (2012). *Discipline-based education research: Understanding and improving learning in undergraduate science and engineering*. Committee on the Status, Contributions, and Future Directions of Discipline-Based Education Research. S. Singer, N. R. Nielsen, & H. A. Schweingruber (Eds.). Board of Education, Division of Behavioral and Social Sciences and Education. Washington, DC: National Academies Press.

Oakes, J. (1990). Opportunities, achievement and choice: Women and minority students in science and mathematics. In C. B. Casden (Ed.), *Review of educational research* (Vol. 16, pp. 15–222).

Rosser, S. V. (1990). *Female-friendly science: Applying women' study methods and theories to attract students*. New York, NY: Pergamon.

Sawyer, K., Alper, J., National Research Council, Chemical Sciences Roundtable, Board on Chemical Sciences and Technology, & Division on Earth and Life Studies. (2014). *Undergraduate chemistry education: A workshop summary*. Washington DC.

Seymour, E. (1995). The loss of women from science, mathematics, and engineering under graduate majors: An explanatory account. *Science Education, 79*(4), 437–473.

Seymour, E. (2007). *The US experience of reform in science, technology, engineering, and mathematics (STEM) undergraduate education*. Paper presented at the Policies and Practices for Academic Enquiry: An International Colloquium, Marwell Conference Centre, Winchester, UK.

Seymour, E., & Hewitt, N. M. (1997). *Talking about leaving: Why undergraduates leave the sciences*. Boulder CO: Westview Press.
Seymour, E., Melton, G., Pedersen-Gallegos, L., & Wiese, D. J. (2005). *Partners in Innovation: Teaching assistants in college science courses*. Boulder, CO: Rowman and Littlefield.
Svinicki, M., & McKeachie, W. J. (2010). *McKeachie's teaching tips: Strategies, research, and theory for college and university teachers*. Belmont, CA: Wadsworth Publishing.
Ware, N. C., & Dill, D (1968). *Persistence in science among mathematically-able male and female college students with pre-college plans for a scientific career*. Paper presented at the Annual Meeting of the American Educational Research Association, San Francisco, CA
Watkins, J., & Mazur, E. (2013). Retaining students in science, technology, engineering and mathematics (STEM) majors. *Journal of College Science Teaching, 42*(5), 36–41.

# Chapter 9
# The Struggle to Belong and Thrive

**Dana G. Holland**

## Struggles with STEM Curricula and Conceptual Understanding

A substantial, and growing, body of literature addresses students' conceptual problems in understanding STEM content and the process of developing expert competence in a STEM field. The focus of much discipline-based research is how to increase students' conceptual understanding, address common misconceptions, and build curricula and assessments based on learning objectives (Talanquer, 2014). Though often tied to specific STEM knowledge (e.g., Montfort, Brown, & Pollock, 2009; Stetzer, van Kampen, Shaffer, & McDermott, 2013) some studies examine the processes by which scientific reasoning and meta-cognitive practices are built (Koenig, Schen, Edwards, & Bao, 2012). This line of research generally follows a cognitive theoretical model by which problems with conceptual understanding are seen to reside primarily in students' mental processes. More rarely addressed are questions about how the pedagogies, assessments, and learning spaces used by instructors influence the development of conceptual understanding (Smith & Wood, 2016; Streveler, Litzinger, Miller, & Steif, 2008; Talanquer, 2014). The findings that we present here make a strong case for consideration of both instructional context and interpersonal dynamics in improving students' content understanding and thereby reducing the contribution of conceptual difficulties to STEM attrition.

D. G. Holland (✉)
Ethnography and Evaluation Research (E&ER), University of Colorado Boulder, Boulder, CO, USA

Center for STEM Learning, University of Colorado Boulder, Boulder, CO, USA
e-mail: Dana.Holland@Colorado.EDU



E. Seymour, A.-B. Hunter (eds.), *Talking about Leaving Revisited*,
https://doi.org/10.1007/978-3-030-25304-2_9

From the "iceberg" table presented in Chap. 3, we note that a number of students' academic struggles were rooted in curricular design and content. Table 9.1 shows that curricular issues were cited as contributing to decisions to leave their STEM majors by 31% of switchers. They were also of concern to over half of the persisters.

Students' conceptual difficulties with STEM related both to aspects of the curriculum and to instructional practice (which is the focus of Chap. 6). Table 9.2 shows that conceptual difficulties contributed to decisions to leave STEM programs for 16% of switchers overall. Although persisters seldom cited grave conceptual difficulties with STEM content, nearly half described some conceptual challenges.

These findings are very similar to those reported in the original TAL study, in which curriculum issues factored into decisions to switch out of STEM for 35% of students and conceptual difficulties with STEM content were significant for 13% of switchers.

Of foremost concern to students in the current study was the pace of curricular coverage and content load of courses, and the resulting speed at which they were required to comprehend material. Students who experienced conceptual difficulties often said that they had developed only *partial* understanding or had thought that they understood course material until test results indicated otherwise. Compounding these problems, taking multiple STEM courses simultaneously over a relatively short duration of time, as is typical early in STEM majors, required students to juggle both their time and focus in order to gain understanding across multiple

**Table 9.1** Proportion of students negatively affected by aspects of their STEM curriculum

| Switchers | Contributed to switching (%) | Negatively affected (%) | Persisters | Negatively affected (%) |
|---|---|---|---|---|
| Overall | 31 | 86 | Overall | 56 |
| Students of color | 19 | 92 | Students of color | 74 |
| White students | 36 | 84 | White students | 46 |
| Men | 34 | 94 | Men | 55 |
| Women | 30 | 82 | Women | 59 |

**Table 9.2** Proportion of students negatively affected by conceptual difficulties with STEM

| Switchers | Contributed to switching (%) | Negatively affected (%) | Persisters | Negatively affected (%) |
|---|---|---|---|---|
| Overall | 16 | 80 | Overall | 42 |
| Students of color | 12 | 92 | Students of color | 54 |
| White students | 17 | 76 | White students | 35 |
| Men | 20 | 83 | Men | 50 |
| Women | 13 | 79 | Women | 38 |

curricular areas. In addition, many STEM courses involve not only lecture classes, but also laboratory sections. Lab sections often felt like a separate class rather than an extension of an associated course. This issue is an example of a second contributor to conceptual problems, namely misalignment of course components. Each of these issues is discussed in detail below.

## *Problems with the Pace and Content Load of STEM Courses*

Generally, switchers and persisters experienced the same problems with course load and pace. However, one notable distinction between them arose from course pacing. Where courses were paced too slowly for some persisters, they became bored and disinclined to attend class. More commonly, students struggled with the rapid pace of courses, which they characterized as content being "thrown at them." Tough course material, especially in physics and calculus, was, in effect, made tougher when presented at a rapid pace. The reported result was *incomplete understanding* that undermined students' ability to properly comprehend ensuing course content and caused ongoing academic struggle:

> If you didn't completely understand one thing, if [the professor] sort of blazed through it and just kept going, and then [the concept] would keep coming back. That was the main problem. That happened with the calculus course especially. [For example], there was one thing early on that I didn't completely understand that just kept reappearing. And because I never got a full understanding of it, I was suffering trying to do it. (White man, switcher, aerospace engineering to studio art)

As noted in the original TAL study, even STEM seniors nearing graduation could feel insecure due to partial or incomplete understanding of important content.

The fast pace of STEM courses was challenging for some students because it contrasted with a slower pace of curricular coverage in high school, where (as discussed in Chap. 5) there was often less emphasis on conceptual understanding or demonstration of conceptual knowledge through application. Challenges associated with the fast pace of courses were compounded where professors spoke with accented English or wrote illegibly. As one student noted, veteran professors seem particularly adept at "writing and talking faster than anyone can ever listen to them, let alone write down what they are writing." Students often felt that they had no time to process information during class and that "nothing was sticking." As indicated in the discussion of weed-out classes in Chap. 7, this was more problematic in introductory courses that surveyed a wide breadth of information compared with upper-division courses that emphasized in-depth mastery over a more distinct and delimited area of content:

> They try to teach an entire textbook too quickly, and you don't have a chance to grasp the things that they test you on. At the point that they're teaching, you just grasp the general concept; [however], when they're testing you [it is more] specific, things that you never got to really figure out—things that you never had the time to do. I think sometimes it's almost impossible to figure out everything that they're trying to tell you. It's just way too much. If

> you want to master something, you need to have the time to master it, which is what I like about my upper-division science classes. Since you're doing fewer chapters, it is harder because you're going so far into [the material]. But you do have the chance to really master it instead of [trying to] master so many things at one time that it's overwhelming and you don't know what to do or what to study and you just try to do your best. (White woman, switcher, biology to human development)

Some conceptualization problems were grounded in insufficiencies in high school preparation on the one hand (as described in Chap. 5) and, on the other, material that was pitched too high for a foundational course. Other problems were rooted in instructors' unwillingness or inability to explain concepts in alternative ways or explain how concepts were related or relevant to subsequent courses. Some students described feeling "lost" in the curriculum, unable to bridge what they saw as illogical gaps in instructors' presentation, and becoming so confused that they did not even know what questions to ask to improve their understanding. Students could, through trial and error, often do a prescribed task, such as a lab that was associated with a concept, but be unable to explain its underlying dynamics and scope in an assessment. Conceptualization problems were compounded by the apparent divergence between how concepts were presented in lectures and how they were used in applications or appeared in assessments. A concept could be perceived as making sense in lecture, but be lost to students when attempting to demonstrate their understanding:

> One thing we kind of joked about when I was a STEM major is that the lecture makes sense, the homework is hard, and the exam is as though you've never seen the subject—essentially an escalating difficulty. So, even if I got everything in the lecture, which I usually did because the professors I had were very good, it just got worse and worse as soon as I left the classroom. (White man, switcher, chemical engineering to psychology)

As indicated by this speaker, (and elaborated in Chap. 6), when courses were designed so that lecture-style classes alone supported student learning and lacked opportunities to build and reinforce understanding through discussion, application, and experiential consolidation, students could struggle to fully understand and apply the concepts presented in the course. Waning interest could compound comprehension difficulties, especially where students felt that they were exerting considerable effort to no avail.

## *Problems Due to Misalignment of Course Components*

How well course components were designed to relate to each other had consequences for students' conceptual understanding of course content. Misalignments created extra challenges. They included misalignments between classes (typically lectures) and homework assignments, homework and exams, and between class and laboratory sessions. Students expected a lab session to complement the class that it was associated with. Problems ensued when labs seemed unrelated to a class or when students were expected to know the content that was dealt with in lab prior to its being covered in a class session. Labs were not uniformly experienced as conceptual and were often viewed as "cookbook-like" sessions. As one student described them:

> The concepts were there, but they weren't explained as well in labs. Instead, they were just kind of, 'mix this with this and this and hopefully you get this, and I'm not going to tell you how that applies to the concept that you were discussing in lecture'... So [labs] were related, but very loosely. (White woman, switcher, chemistry to French)

As this student points out, the mere inclusion of a hands-on application in a course (such as a lab) did not mean that an underlying concept was further consolidated for students.

## *Problems Negotiating Multiple Courses Simultaneously*

A final source of conceptual problems that also contributed to switching was trying to juggle multiple courses simultaneously. An uptick in the pace of one course, especially when the content was novel, could lead students to focus on that course more so than others. This precipitated problems of keeping up in the neglected courses. This was most problematic for students during their first year when STEM courses may not align conceptually across a program as well as they do in later years:

> As the semester progressed, things were getting really difficult really fast. It kind of threw me into a panic. I would end up spending all of my time studying for one class and then the other two would start to slowly drift into the background. And then I'd realize, 'Oh, these two need some focus.' So, I'd go over and rescue one of them. And then the other one that I had been working on before would start [becoming a problem again], and it was just a slippery slope. A slippery, slippery slope. (White man, switcher, biology to dance)

> Well first semester freshmen year I took Introduction to Chemistry and Biology and those two together weren't too like terribly hard and they seem to go into each other—to coincide just a tad here and there. So, they weren't too hard to manage. But the next semester I took calculus, Chemistry 1, and genetics and that was harder to keep track of. The genetics class wasn't too bad until the last unit when things kind of went out of whack and a lot of students felt the material at the end of the class wasn't taught as well as it was at the beginning of the class. And then calculus was just a disaster the entire semester, so it was hard to keep up with. Chemistry wasn't bad. But with calculus being so crazy and hectic, and genetics being alright until the end of the semester. … Then, at the end of the semester, chemistry also was a little harder for me to handle because everything sort of hit [at once] and I just lost control of everything. (African-American man, switcher, physiology to psychology)

Students' progression through STEM degree programs essentially requires that they take multiple STEM courses at once. However, there is a high risk that juggling particular combinations of courses in certain time frames is a bar, both to solid conceptual understanding and to survival in the major. Whether and how persisters overcame challenges associated with this *juggling process*—including sheer luck—are described in Chap. 12.

## *Conclusion: Conceptual Problems and STEM Switching*

While some degree of academic struggle with the curriculum and with STEM content is likely evitable, and arguably, desirable, these results suggest that attention should be paid not only to *what* is included in the curriculum, but to *how* it is paced and aligned, both internally within a single course (in class sessions, homework, labs, quizzes, etc.), and across courses in a disciplinary program. Content challenges can be good and motivating for students when they are thoughtfully devised *as part of* the curriculum. However, as these findings show, they can also create unnecessary struggle, confusion, and low levels of comprehension and may confound students' commitment to persistence when they occur *because of flaws in* the curriculum structure.

## Problems Related to "Hardness" in STEM

One way to think about the conceptual difficulties that STEM students' experience is informed by the colloquial notion of a subject or task "being hard." In the original study, when asked what makes STEM "hard," students indicated that conceptual difficulties were sometimes inevitable and intrinsic to STEM subject matter. At other times, however, students saw their difficulties as contingent on curriculum and instruction and thereby, in effect, *situationally constructed* by the instructional context.

Our findings echo these earlier results and similarly reveal multiple dimensions of STEM students' experiences with hardness. Both switchers and persisters attributed the hardness of STEM majors to malleable aspects of its instructional culture that made STEM majors harder than they need to be. These include weed-out practices, curve grading, competitive classrooms, and incoherent curricular design among other issues. As described in Chap. 6, a substantial number of both switchers (26%) and persisters (25%) described from experience how poor quality teaching makes STEM disciplines more difficult than they need to be. Students were troubled by inconsistencies among instructors teaching the same course, with some making the same course content more accessible than others. Instructors' attitudes and dispositions also mattered, and it was clear to students when faculty did not want to be teaching them. One notable contrast between persisters and switchers in how they experienced hardness as an element of STEM instructional culture was that persisters tended to describe hardness challenges as a characteristic of particular instructors or courses but not of STEM majors per se. Switchers, in contrast, tended to describe these challenges in more general terms as typifying their STEM higher education experience.

Consistent with the discussion earlier in this chapter, both switchers (12%) and persisters (8%) associated hardness with the pace and curricular coverage of STEM courses. Students complained about the large amount of information that they were expected to memorize in a short amount of time—especially in biology and chemistry—which were said to require the most memorization. Students also contrasted memorization of STEM knowledge ("learning the facts") with the development of conceptual understanding that required understanding the underlying logic of the system or phenomena under study rather than just absorbing content. Time management skills were considered necessary to cope successfully with the fast pace of courses. Pace and overload issues were especially acute in institutional settings that follow a quarter system or block plan, since in comparison to semester systems, more course content must be covered in a shorter time period. Students also faced particular difficulties keeping up with the pace of curricular coverage where they were simultaneously pursuing non-STEM interests and obligations, such as sports, church, family, work, and hobbies.

Switchers and persisters also varied in their views about what makes STEM hard. Although considerably more switchers (24%) than persisters (12%) associated STEM hardness with the amount of time and effort required in STEM course, students uniformly agreed that success in STEM requires time and effort. There were, however, substantive differences in how each group associated expenditure of time and effort with hardness. Persisters described hardness as requiring effort and time to understand concepts, cope with heavy workloads, and apply oneself, that is, by "putting effort into it." For persisters, this effort was internally driven and not associated with time spent in working with external sources of help, such as tutoring or instructors' office hours. Persisters also regarded all STEM content as comprehensible so long as they put in sufficient effort. They saw nothing that they "couldn't wrap [their] minds around" or that was "beyond the scope of understanding," given sufficient time and effort to grasp the material. This could include substantial time doing homework problems, particularly in math and physics, or time memorizing information, particularly in chemistry and anatomy.

In contrast, the effort that switchers put into overcoming hardness tended to be externally supported by supplemental help rather than internally sourced by applying oneself to understanding the material. Some switchers mentioned deficits in their incoming knowledge as a result of poor high school preparation, or as challenges arising from their inadequate content knowledge in contrast to professors' assumptions about what they should already know. Also, in contrast to persisters, switchers had conceptual difficulties with STEM content that they were unable to overcome, including difficulties in demonstrating their understanding in assessments. STEM grades were, therefore, an issue for these switchers and they were often dismayed that the amount of time and effort that they put into a course did not result in a desirable grade—an issue taken up more fully in the following section.

A small number of students overall (6%) commented on how the apparent hardness of a STEM course could arise through hearsay from other students. There were important differences, however, in how switchers and persisters reacted to hearsay. Persisters heard about the reputation of a specific course as being hard and interpreted it as intended to "scare" or "intimidate" them. They tended to experience the course as less hard than second-hand assessments had led them to expect. Persisters speculated that students who did not take courses seriously perhaps generated false stereotypes about them, or succumbed to such tales by becoming unnecessarily anxious and intimidated. In contrast, switchers tended to take what they had heard about specific courses as legitimate information and reacted with apprehension. Student hearsay about the hardness of courses could, thus, create anticipatory *stressful emotions*, such as anxiety, apprehension, and fear, especially for those switchers who knew other students who had failed these courses. Some switchers had reacted to hearsay about the hardness of a particular course by postponing taking it, or withdrawing from programs that required it.

A modest number of switchers (9%) and a substantial number of persisters (23%) described hardness as an intrinsic part of STEM education. Related to this, the most notable distinction between switchers and persisters was the negativity implicit in switchers' characterizations of STEM majors. Switchers described STEM education in terms of coldness, abstraction, exactness, and leaving "little room for humanity." They sometimes rejected having to explain their thought processes, preferring a more technical, clear-cut approach to STEM instruction. Not surprisingly, persisters were much more positive about the nature of STEM education. Many contended that STEM requires abstraction and visualization, and characterized STEM as being both abstract and concrete depending on the field of study. Physics, more so than other STEM fields, was viewed as requiring creativity, critical thinking, and the ability to think abstractly. Engineering and computer science were associated foremost with using logic, problem solving, and the ability to apply facts. Chemistry and biology diverged from these characterizations in students' emphasis on their inclusion of a lot of information. Biology, in particular, was said to require understanding of "the facts," and to necessitate a great deal of memorization and identification. These disciplinary differences were associated with distinct types of hardness, as depicted by a neuroscience student:

> I think that different types of sciences are difficult in different ways. For neuroscience and for biology, and kind of for psychology, those are much more memorization-based classes, at least in my experience. Especially like molecular biology. So students who don't like to spend a lot of time memorizing terms and processes find it more difficult. But then for chemistry, physics, and math, it's very much a problem-solving based rhetoric. I have friends who are neuroscience majors, and they love neuroscience and didn't find it very difficult because they can memorize. But organic chemistry was really a big struggle for them because it was a problem-solving class and they weren't used to thinking and manipulating things in their mind. (White woman, neuroscience persister)

Drawing on these ideas about the nature of STEM education, a small number of both switchers (9%) and persisters (7%) considered STEM majors to be hard because they require innate abilities that some people have more than others. Innate ability was generally defined as the way someone reasons—a "thinking style," how the "brain is wired," and other inherent mental capacities, such as being a "math person" or a "biology person." By these theories, if a person has particular innate capacities, then STEM majors should be less hard for them. Among those switchers who associated hardness with innate ability, good teaching was seen as critical in making STEM comprehensible because it could compensate for lack of particular innate abilities.

As with ability, interest in STEM disciplines was also considered to be largely innate (although amenable to change) and to influence how hard STEM content was judged to be. While a very small number of switchers (3%) associated hardness with their lack of interest in STEM, persisters (8%) were more inclined to believe that their interest helped them to overcome their academic struggles. For persisters, "passion" for a subject and "the desire to understand" shaped what they got out of a challenging course. It was interest that largely propelled persisters' commitment to STEM courses and majors, and spurred their "personal investment" and motivation to "try harder." When persisters encountered a course or substantive area of their discipline in which they were less interested, they nevertheless persevered in order to "at least pass," or treated the course "as a barrier" while focusing on "the end goal." They reasoned that a single challenging course was a small part of a large major and assumed that it would prove to be useful. Indeed, some persisters contended that when students say that a math or science course is hard, what they are actually saying is, "They don't like it and don't find it interesting."

## *Conclusion: What Makes STEM Hard?*

As in the original TAL study, our current findings show considerable variation in how students understand what makes STEM majors challenging. Students viewed the STEM disciplines both as intrinsically hard because of the nature of the subject matter and for extrinsic reasons that are amenable to change. The latter included aspects of STEM instructional culture that perpetuate poor teaching practices. Some students grounded their views about STEM content hardness in assumptions about ability as being natural and fixed rather than as something that grows with interest and effort (Aronson et al., 2002; Blackwell et al., 2007). Some clarified that particular STEM disciplines are hard for different reasons. The hardness of STEM courses was also argued to be socially constructed as student hearsay that persisters were more likely to ignore than switchers. The considerable amount of time and effort required by STEM courses (which was prominent in many students' views about what makes STEM hard) often did not translate into success for those students who switched out of STEM due, at least in part, to curricular or conceptual difficulties. At the same time, students' appraisals of their success in terms of the grades they received in STEM courses were quite variable. As the next section explains, there was surprising diversity among students in what constituted good and poor STEM grades.

## Struggles with Poor STEM Grades and Grading Practices

STEM persistence research generally uses student grades and GPAs as predictors of STEM retention or attrition (e.g., King, 2015; Ost, 2010; Rask, 2010), or as ways to operationalize student performance (e.g., Chen, 2015). In these studies, grades are assumed to be objective, standardized, and reliably reflective as measures of students' conceptual understanding. They are also assumed to be comparable across classes, courses, and institutions. However, results about the relationship of GPAs and conceptual understanding to students' STEM persistence are contradictory: some studies show that GPA can predict STEM attrition and others show no meaningful differences between the GPAs of those who persist and those who opt out (Geisinger & Raman, 2013). A partial explanation for these discrepancies is that, while grades may function as objective standards of performance, they are also *subjectively experienced.* As noted in the original TAL findings, grades are labels to which people react emotionally and behaviorally. These subjective experiences are then *responded to* as real and significant by students. Moreover, as grades (and by extension GPAs) may not be as standard and comparable as assumed, efforts to understand their impact and associations with STEM persistence are necessarily complicated. Our results provide additional insight into the significance and effects of STEM grades on persistence in STEM programs.

The academic struggles reported by students in our interview sample were often extended to include struggles with STEM grades. Low grades put students at particularly high risk of switching out of STEM, *even when the grades in question were sufficient for them to continue in a program.* The original TAL study found that almost one-quarter (23%) of switchers reported that discouragement and loss of confidence created by low grades, especially in early classes, was a factor in their decisions to leave a STEM major. Similar low grade-related discouragement was also reported by 12% of persisters. As described in Chap. 3, in our current study switchers' reports of difficulties with STEM grading have increased over time, and persisters continue to report grade problems with the same frequency as two decades ago. In Table 9.3, 61% of switchers cite grades as a factor in their decisions to leave, while 44% of persisters also cite grades issues. Grade-related problems were found to be especially challenging for students of color, affecting 92% of switchers and 59% of persisters. Men of color and women of all races and ethnicities were both more likely to report that grade-related problems had contributed to their switching decisions (69% and 67%, respectively) than were men overall (51%).

Chapter 7 describes how learning problems and poor grades in early STEM courses influenced switching decisions. Here, we focus on how STEM grading practices influenced students learning and decisions to persist over the entire course

**Table 9.3** Proportion of students negatively affected by STEM grades

| Switchers | Contributed to switching (%) | Negatively affected (%) | Persisters | Negatively affected (%) |
|---|---|---|---|---|
| Overall | 61 | 79 | Overall | 44 |
| Students of color | 69 | 92 | Students of color | 59 |
| White students | 59 | 74 | White students | 35 |
| Men | 51 | 71 | Men | 37 |
| Women | 67 | 84 | Women | 49 |

of their degree program. At the extreme, we found that 12% of students in the interview sample, including comparable numbers of persisters and switchers, failed or had to repeat classes, sometimes multiple times. Perhaps counter-intuitively, students with high incoming math backgrounds were not immune to failing and repeating STEM classes. While students with low-math background were more likely to fail and repeat classes (34% of the low math sample), 11% of students with high-math backgrounds also failed or repeated STEM classes. Only nine of the 189 persisters in the sample considered failing a course to be a psychological crisis, and only six saw it as threatening their continuance in the major. However, women had less tolerance for receiving low grades and failing classes than did men and were over twice as likely as men to report demoralizing and psychologically traumatic effects of low grades (viz., 24% of all women, and 10% of all men). Among students, who switched out of STEM, half of the women said they were demoralized by receiving low grades, while less than a third of the men expressed this concern. Among students who stayed in STEM programs, men were over twice as likely as their female counterparts to fail and/or retake courses. Also of note, although only nine persisters expressed a critical level of anxiety about low grades, these were nearly all women ($n$ = 8 of 9). In sum, reactions to poor STEM grades were gendered and women were considerably more subjectively affected by grades than men. These findings are further elaborated below.

## *Transition from High School to University STEM Grading*

It was not uncommon for switchers to have experienced STEM "grade shock" as part of their transition from high school to higher education, especially during the first few semesters. Typically, grade shock affected those who had been "top students" in high school and had seldom, if ever, received grades below a B. Grade shock was equally likely to be experienced by students with high incoming math scores:

> I never got a grade lower than a B+ in a high school class. The lowest grade on a test I ever got might have been the very rare 79%. So, it was just kind of shocking [when I started college]. Like I remember my first honors math class, I got less than 50% ...and then my first physics test I think I got a 70%. And the next one I got less than 50%. (White woman, high math, switcher, general engineering to psychology)

Several consequential tendencies were associated with grade shock. First, as described in Chap. 7, the shock of receiving low STEM grades often occurred relative to past experiences with high school grades rather than as prompted by the absolute value of a grade. Receiving a C or even a B could be experienced as personal—if not objective—failure because of its disconnection with past experience. Secondly, low STEM grades were provocative, prompting introspection and (as elaborated in Chaps. 10 and 12) sometimes prompting thoughts of switching among those who did eventually switch, as well as among those who did not. Low grades could also provoke emotional stress, both in real time, and in disquieting projections of an imagined future in a STEM program:

> Even now I'm still kind of struggling with that. In high school I was used to getting straight As. I was always at the top of my class, one of the smartest kids in the school. Then coming here it was a complete shock to actually get a C, and to have to work more for that A. But even when I did work hard, I would still be getting Cs...And honestly, I think that's a big part of why I switched out. It was after I failed a test in physics, and I was really worried about passing the class. I called my mom and said, 'I can't keep doing this. I can't do this every semester, where I'm worried about failing a test, failing a class, and then come so close to failing it.' I just couldn't psychologically handle that. (White woman, high-math switcher, civil engineering to psychology)

Third, grade shock affected some students particularly acutely because low grades undermined their identity as good and competent individuals, thereby putting them at particular risk of switching into a field where they could regain a sense of successful selfhood:

> That experience in math was one of the many low points I had. It was second semester last year and it came at a time when I was really questioning a lot about me as a person and finding out things about me. So, it left me feeling like I couldn't offer much really. Math had always been one of my strong subjects so when I ended up getting a D in it, it was just heartbreaking for me. So, honestly, to this day I'm still trying to make things better and build myself up. But that was a huge knock for me. (African-American man, low-math switcher, physiology to psychology)

> I was so motivated during high school. ... I was such a type-A, really go hard, you know, 'I want to be the best: the best or nothing at all,' [kind of person]. And then to just kind of waddle my way through this major that I [was trying] my hardest yet I have no chance at being the best. [It just] sucked...That's critical for me because I never want to go through life being mediocre. I would rather do something that is less prestigious...I don't want to sound that pretentious, but you know what I mean? [I want to do] something that I can be competitive, that I can hold my own. Whereas, with engineering, I try so hard and yet I'm completely mediocre. (White man, high-math switcher, chemical engineering to psychology)

For both of these students, poor STEM grades were stigmatizing, and represented a reversal both of past recognition symbolized by good grades and of prior identities as good STEM students. For such students, switching was a means of recovering emotionally and psychologically from grades that were not in line with past experience.

## *Pre-emptive Switching Due to Low Grades*

As subjectively experienced, the power of grades transcended their objective value for many students, blurring and confounding the distinction between good and poor grades as well as between real and imagined future grades. For a number of switchers (16%) doing poorly in a class, or sometimes on a single exam, led them to fear that they would do equally or more poorly in later STEM classes or compromise their overall GPA. Importantly, these students tended to predict from poor grades in one class or on one exam—typically very early in their STEM coursework—to a likely future poor overall performance. Perceived poor performance in one class led the following student to presume that she could not be successful in a STEM program overall:

> If I had to say one reason why I shifted [out of STEM] it was my feelings from my performance in that first class that I was not capable of being successful in that major. And that's what it comes down to for me. That I couldn't be successful in that major. (Black woman, switcher, math to management)

These students in effect *pre-emptively* switched out of STEM in part due to grades. That is, they switched less because of actually receiving poor grades that compromised their continuing in a STEM program, but because of the *perceived possibility* of failure and a forced exit happening in the future. In other words, pre-emptive grades-related switching was motivated less by failure itself in terms of receiving low or failing grades, than by *fear of future failure*. This was particularly worrisome for students who had aspirations of professional school entry, especially medical and veterinary schools:

> To get into veterinary school I feel like I probably won't be as competitive as others so my focus is to get [my GPA up] by the time I graduate. … The fact that I have to be the top student to get into [veterinary school] somewhere is kind of a big deal, [it's] a problem. (African-American woman, animal science persister)

Perhaps not surprisingly, life science persisters were more likely than persisters in other STEM disciplines to spontaneously describe grades as important for their graduate school and career aspirations.

## *Adjusting for Relative Success*

One difference between switchers and persisters who experienced low grades was an adjustment process in which, over time, some students became more comfortable with receiving low grades, as well as more familiar with common STEM grading practices. While a number of switchers also made these adjustments, a subset did not, and the *absence of adjustment* to STEM grading directly contributed to their leaving STEM programs. Ironically, some switchers appeared to care too much about keeping their grades high to continue in STEM programs, while some of those who persisted despite poor STEM grades did so by cultivating an attitude of not caring about their grades. However, as mentioned, those students whose ultimate educational goal was professional school had, arguably, good reasons to care about their grades and their GPA and to resist adjusting their expectations downward.

The process of adjusting to poor grades could take different forms, including changing attitudes and study habits, becoming more knowledgeable about STEM grading practices, and taking particular actions. One attitudinal aspect of adjusting to STEM grades was normalizing failure, that is, learning to take a single poor grade as a setback, not a deal-breaker, and thus limiting its psychological ramifications. In the private university in our sample, a woman persister described faculty as having been instrumental in teaching her to "calm down more about grades and take care of myself." For her, earning a B seemed "poor" because she had been accustomed to earning As in high school. The experience and advice of others did help many STEM students to adjust their attitudes such that their academic struggles, including poor grades, did not come to signify a lack of talent or ability. This computer science student, for example, recounted how talking to older peers and her parents helped her interpret her academic struggles in a way that offset her inclination to fear future failure and to pre-emptively switch out of STEM:

> I decided to stay [in STEM] mostly because I talked to a few people older than me: my parents, a girl who was a Junior …and a guy that I met who was in my major. … They helped me see things more in perspective, in terms of, 'This is one class in a big major.' And they helped me understand my strengths, like, 'You are gifted in this and just because you're struggling doesn't mean it's necessarily gonna be the end.' And I think that helped me put it into perspective [that] the world's not ending, 'You're gonna be okay.' I tend to get really wrapped up in the moment a lot of times. (Asian woman, computer science persister)

The first semester of freshman year was considered to be an especially crucial time in adjusting attitudes after post-high school grade shock, when the experience of low grades prompts self-reflection about how much and what kind of effort is necessary to succeed in a STEM major:

> I think it really takes that first semester [to adjust]. It's funny 'cause I mentor kids now [who are] in their first semester of college…Until you get your grades back you don't really realize like, 'Wow, I should have [studied more or differently]. This wasn't as easy as I thought.' So really, I think after you get your first semester grades back and you [tell yourself], 'Okay I need to buckle down and try harder,' [then you] realize this is different from high school. (African-American woman, civil engineering persister)

Even when students were warned by others that STEM grading in higher education is more stringent than in high school, this could be hard to take in until it was personally experienced and dealt with:

> I was really disappointed in myself [because of poor grades]. I thought about whether this is what I should be doing... After the semester was over, I reviewed what I did and realized that I just didn't have the right work ethic. I wasn't prepared. My brothers had told me what I should expect, but it's different when you're actually the one going through it. They told me that you'll need to study more. I thought studying was just reading my book, not actually going back through my work, through my notes. I realized that I definitely needed to change the way I had been studying. So, I changed my study habits. (African-American man, biology persister)

Gaining knowledge and understanding of STEM grading practices facilitated the adjustment process for some students. Disparities in grading practices between STEM and non-STEM courses caused some concerns for students. However, knowing the overall GPA or typical level of performance in a STEM program helped them adjust expectations of their own performance:

> In my liberal arts classes like French or philosophy I would be getting easy As and then in my astronomy class I would be getting maybe a B or B-minus. That was very, very difficult. Because it [made me] feel like I wasn't cut out for the work that I was trying to do. But I realized that there are lots of different parts of the field and there's more to it than just doing an exam, and that a lot of other people are doing poorly as well. And that what felt like doing poorly to me really wasn't all that bad. (White man, astrophysics persister)

Coming to understand the practice and consequence of curve grading was particularly important. For most students, not being assessed in absolute terms by demonstrating conceptual understanding but relative to your classmates was novel and alarming:

> I was used to getting As. So [when] I'd get a C when I first started [college] it was, 'Oh no!?!' But now, I've gotten more, not used to getting lower grades, but I have been okay with not getting As on everything. So that's been less hard on me, getting a bad test grade. And then also, if I get a lower test grade, it could be curved and I know that now as opposed to when I started. And [I know that] there are a lot of other grades in the class that also get curved or points readjusted. And so, after a bad test I'm not as worried anymore. (Multiracial man, engineering persister)

> My first semester was actually kind of tough. I was about to change, to switch out of biology, because I wasn't doing as well as I wanted to. But that was because I realized that in high school everything is, '[You get] 90%, you're an A; 80%, you're a B.' But in college everything is based on a curve. I didn't realize that I was doing above the curve. It turns out that I was actually doing really well, but I didn't realize that…I'll give you an example. So my first freshman biology course, I got a 66% on my first exam, and that was really frustrating for me because I studied very hard and prepared for it for several days. And so I didn't understand why that was a 66%. But then when I went to the next lecture, and it turns out that the average was a 55%. So, in terms of the grade distribution all that really mattered was doing above the average. And I didn't understand that concept [initially]. I just thought, 'Okay your grade was the number that you got and that was it,' because that's how it was in high school. (Asian woman, biology persister)

Assessment relative to peers in some classes could occur in a literal way in the practice of ranking students according to performance on an assessment, then making these rankings public:

> Like my Introduction to Evolution class, it was a ranked class. So what they did is they ordered you in terms of your rank in that class, from 1 to 250 or something [depending on how well you do on tests]. So that was kind of a shock to me, because I wasn't used to being ranked against my peers. (Asian woman, biology persister)

However, students' understanding of curve grading was not always complete and some assumed, in effect, that a criterion-referenced testing logic was being used rather than a norm-referenced one. In other words, many students thought that testing and grading results on a curve indicated both how much they had or had not learned of the curriculum as well as how well they had performed relative to their classmates. They may well have thought this because most of their non-STEM classes relied on criterion-referenced testing that uses assessment to show the extent to which students comprehend the curriculum—regardless of how many do so (e.g., King, 2015). Consistent with TAL findings, in the current study curve grading methods had negative consequences for some students beyond their grades per se because it disconnected students' appraisals of their effort from any corresponding reward:

> Going into college with the curve and getting less than 50% on exams, but actually it could be translated into a C or better, the concept was mind blowing. I never knew where I should shoot for. I knew I wasn't going to get higher than a 90, so I didn't even think about that. My goal was to get Bs. It was also stressful seeing bad scores even though they didn't mean that you were failing because there was such a big curve. And then the fact that it differed in different classes so there was no consistency, it really just depended on the class as a whole. (African-American woman, switcher, food science to theater)

As this same student further explained, curve grading could undermine incentives to work hard and do well because of limits on the proportion of a class that could earn high grades:

> If everyone was doing well, if everyone was getting 90s and the class was based on only letting [a certain] percentage of people pass, that still means that there was going be a percentile of people who compared to everyone else were the worst. And so grading was [the result of] your work based on everyone else and not based on your work. (African-American woman, switcher, food science to theater)

Curve grading also created disconnects between students' appraisals of their conceptual understanding and corresponding reward. Perhaps, most perversely, curve grading could cause students to question their competence in understanding course content even when they had performed well relative to the curve:

> You hear a lot about science classes, people might say, 'Oh I got a 52 on this exam.' And you say, 'Okay, well, what was the average?' And they'll be like, 'Oh the average was a 48, so it wasn't that bad.' I think that's really common in a lot of the STEM courses… They kind of set it up so everyone is going to fail and then eventually there is a really big curve at the end, which I think is discouraging. I would hate it if I would get back an exam that says I got a 52 on it, and technically, yes, I did still get above the average if the average is a 48, but still, a 52 is a 52. And that means you missed almost half the material, which means you don't know it well, and you don't understand it, and you can't perform well. (Hispanic woman, switcher, math to management)

These problematic aspects of curve grading appeared to affect those who ultimately switched out of STEM more so than those who persisted in STEM, as Chap. 12 will further explain.

Also implicit in students' interpretations of curve grading was buying into the standard view of grades as objectively representing conceptual competence. As a result, a number of students who did not fully adjust to STEM grading practices had trouble accepting what they assumed to be sub-par performance and corresponding lack of competency. This particularly affected some students who earned Cs in STEM classes, despite the objective adequacy of this grade:

> Going into the final exam and feeling like I did not do well and just getting a C—I felt like that was a clear indicator that I should try something else. (Asian woman, switcher, computer science to accounting)

As the findings discussed in the next section will show, female students had a particularly difficult time accepting earning grades of C in STEM courses.

## *Women's Differential Sensitivity to Earning Grades of C*

As with the student quoted above, women were especially prone to switching as a result of earning Cs because this fed into concerns that they were not as competitive in a STEM major as they wanted to be and that they were unlikely to be successful in the future. Women's disappointment with earning Cs was expressed as *fear of future failure:*

> I was getting Cs and they always had curves. They always had curves because everybody doesn't do that well. And so that was kind of what saved me, are the curves. But I would still get Cs on the test, especially the last quarter... That kind of killed me and disappointed me because I knew I wasn't going to do well. And I knew I had to take organic chemistry and inorganic chemistry, maybe two years of chemistry. And that scared me half to death, because I didn't want to fail. (Asian woman, switcher, biology to strategic communication)

Women also characterized this as fear of lacking competence, which, as described above, was especially hard to judge with curve grading:

> It wasn't for the grades that I switched, but I can't say the grades didn't make me reconsider the major. And it was not even the grades that I got at the end [of a course], but the grades I got throughout the course. For me when I fail an exam that tells me that I didn't understand the material. So then to get a B at the end of the semester, which says I understand it pretty well, that would seem like a real disconnect to me. [I thought] 'How did I end up with a B when I completely failed?' Or maybe not completely failed, maybe, Ds. I guess that's not completely failing. But when I didn't even have average understanding?! So I really didn't like that because I do know chemistry. I know organic chemistry, but I never really felt like I understood it, honestly, because I never had a grade that reflected, 'Yes, you do [understand] this,' until the final grade. And then I [wondered], 'Do I know? Why did I get a B? Because I don't feel like I understood this.' ...I put myself through all this stress, all this hard work. I did end up with a good grade, but I didn't like it at all. (White woman, switcher, microbiology to dietetics)

Women took grades as serious indicators of their competence in a subject, both in their present coursework and in future work contexts. For example, this woman who switched out of engineering left little space in her self-assessment for what she might bring to an engineering career other than what had been measured through course examinations:

> The grades that I got were lower than I was used to. So, for me, I had to say, 'According to my standards that is failure.' So, you know, getting a C on my first exam in college was like, a huge shock. And while I was determined to do better on my next exam, it was still a rude awakening. But I thought to myself, 'You know, throughout the semester, maybe this isn't for me. Maybe I'm not as good as I thought.' And, again, the amount of work that a lot of engineering students put in is extensive...then you take the exam, you get a bad grade and you say, 'How can I put in so much time and so much effort and feel pretty good about my knowledge, and then perform so poorly?'...I place a very high value on being very competent in what I'm doing. And I felt zero competence [in engineering]. So, I said [to myself] I wanted to do something in my life that I feel good about my work and find personally satisfying. And I said [to myself], 'Even though I like these subjects, I don't want to get to my job and not know how to do it'...You hear people saying, 'I got a 2.8 GPA, but I got hired right out of college as an engineer.' And I always thought to myself, 'How much are you

> learning, necessarily?' A lot of our exams were based on application of concepts and, essentially, that's what you're doing in the engineering field. You're applying the concepts, and I thought to myself, 'If I can't do that, then will I be competent in what I'm doing?' Which, if I'm not, I will graduate with a degree that I physically can't take to a job market because I still don't know what I'm doing. (White woman, switcher, engineering to management)

Again, poor, as well as objectively moderate, grades in STEM that were earned by women—that is, C-grades—could provoke fear of failure and corresponding concern about competence. This pattern was especially evident among women in life science programs. In our interview sample, this included: a white woman switcher who "legitimately thought [she] would fail neuroscience," provoking so much anxiety and stress that she visited a medical counselor and dropped the major mid-way through a semester; another white woman who switched out of biology due to "the chance of failing or not getting the A or B"; and a multi-ethnic woman who switched from biology to psychology because she felt like she was "not succeeding in the sciences" and was not sure she could succeed after having failed one organic chemistry test.

Female students' sensitivity to middling grades was also particularly acute among high-achieving women. As detailed above, in interviews they described how receiving just one C-grade in a STEM course contributed to their decisions to leave a STEM program. Similarly, in his transcript analysis of the larger student population at our study sites, Tim Weston provides findings that support and expand upon those from the interview study. Although Weston found that that students with higher GPAs switched less, higher achieving students who received just one C-grade switched more than students with slightly lower GPAs (3.0–3.5 GPA). Figure 9.1 shows that this trend affected both men and women in the highest (of four) GPA blocks (i.e., 3.5–4.0 GPA), but was most dramatic among women. High-achieving women who received just one C-grade were almost three times more likely to switch out of STEM than women with comparable overall GPAs who received no Cs (41% versus 15%). This disparity in sensitivity toward receiving Cs was evident, but less marked, among students in the three lower GPA blocks.

Female students who stuck with their STEM major despite anxiety over their grades sometimes recognized the liability of their own high expectations and the need to moderate them as a way to adjust and succeed. It was also evident that female students were aware of this gender disparity, with women holding themselves to higher expectations for grades than their male peers:

> I think the fact that we as females feel the need to prove ourselves in these fields shows the inequality that exists. Like the fact that I'm, you know, mad at myself for getting less than a grade in the A-range, where the guys do not give a crap [about their grades]. (White woman, environmental science persister)

Consistent with transcript analysis findings, interview data showed that male students generally had an easier time both adjusting their performance expectations, in light of the norms of STEM grading practices, and becoming comfortable with receiving Cs. Men were less likely to question their competence in a STEM field as a result of poor or moderate grades. Male students also expressed less concern than

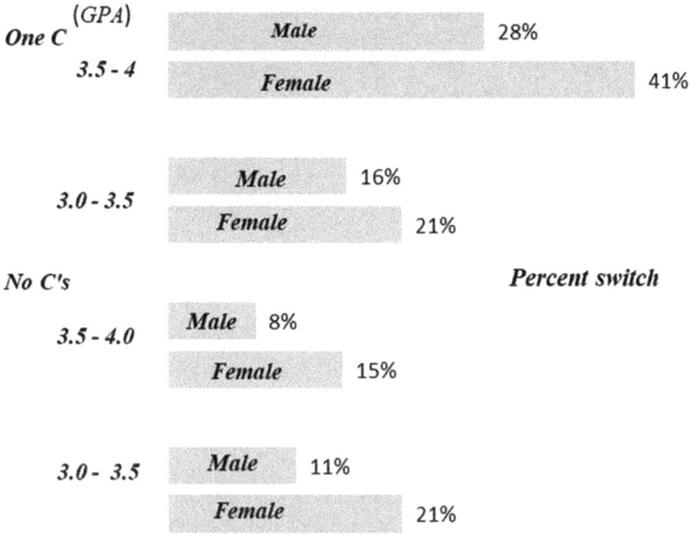


**Fig. 9.1** Percentages of male and female students in four GPA blocks in the institutional student sample who switched out of STEM majors following receipt of a C grade in one course

female students that low grades or a modest GPA would impede their future success in a STEM career:

> I guess biology was tough...and I struggled, but the curve came along and I got an A. I thought I was going to get a C. I've never gotten lower than a C in any course. So, no horror stories for me. I didn't even withdraw from anything…I wasn't stoked about getting a C. But I also wasn't like, 'My world is over,' you know? People focus too much on the grade. (White man, relocator, computer science to natural resources)

> [My grades] started out better. I was getting As and Bs and occasional Cs, like my math classes in the beginning were all Cs. And I'm not especially sad about that. I mean, the [average] among engineers is Cs…[Getting] an A in a difficult class is just unrealistic. In different majors that might be possible, but not in engineering necessarily. So at the beginning, in the first two years, I got As and Bs and some Cs. Now it's mostly Cs and Bs. I hope it doesn't slide down too much further. At the moment [my GPA is] 2.78. I'd like to get back to 3.00 by the time I graduate. We'll see if that happens. (White man, mechanical engineering persister)

Male engineering students, in particular, were comfortable with getting grades that though they were not stellar, were *good enough* to formally continue in a program. Moreover, male engineering students anticipated further learning on the job that would eventually obviate their middling STEM program performance:

> In chemical engineering, there will [for example] be an average of the 30 out of 100. And so the curve just gets bumped up a bit, 80 from 30...I may have felt horrible, may have felt like a failure, but I came out with a B or even coming out with a C in engineering [is good]… it is relative between the people who got 20, and the people who got 35…That's what I've told younger chemical engineers. I just say just keep doing it because if you like the subject and you like its application that's what matters. You'll be able to do it. The career

> world will get you even if you have a 2.2, even though you're not going to be able to put your GPA on your resume. You know, 'Get out there and work in the field you want to'...If you get a 2.1, it will be hard for them to say, 'Okay, we're going to hire you over any competition,' unless you have the exact work experience probably...[they may have to] do something lower level that doesn't necessarily involve engineering, [but just] have your degree and they'll move you [eventually]...Anyhow, it's not about transferring to business or transferring to something that you can do and get As in easily. It's about the major, and the degree itself does mean something. And I hope to be an employer one day who's able to say that. (White man, chemical engineering persister)

The findings presented here and in earlier chapters about gender differences in students' subjective views of grades show how assessments of course and program performance are often situated in students' projections about being sufficiently prepared to succeed in the workforce. While there were disciplinary distinctions in these results—with life science majors intent on professional school showing heightened sensitivity toward grades—women generally appear more inclined to equate their competence both in the present and in future work settings with their grades than do men. Our related findings align well with the theory of *professional role confidence*, in that relatively poor grades led female students to lose confidence in their future ability to successfully fulfill professional roles, competencies, and identities, more so than many male students (Cech, Rubineau, Silbey, & Seron, 2011).

## Actions Taken to Adjust to Poor Grades

In addition to changes in attitude, knowledge, and expectations about STEM grading, receiving low STEM grades could provoke reactions of different kinds. In keeping with the standard view of grades as objective reflections of comprehension, some students responded to poor grades as a signal that they should adjust their study habits. (This was sometimes part of their general transition to higher education.) Low grades could also prompt students to work harder, bolster their determination to succeed, and become more effective and efficient in their study practices:

> I definitely treat every class as if there was no curve. When I get low grades I go through [the test] and try to find each question in the book or in my notes, to make sure the answer is there somewhere. And so that tells me they're not just making it up. So I know it's there and that either I didn't look over it or I forgot [so I got it wrong]. And then I highlight those sections because usually my exams will be cumulative, so I know it's going to show up again somewhere. (Hispanic women, relocator, biomedical science to zoology)

> I don't think I learned how to really hard-core study until I got to my spring semester Junior year. That might have been the worst semester [in terms of how] I performed, but I learned how to study. I say that because I was taking four upper-division material science courses and they were really hard. Unfortunately, I only got average grades in all those classes because I was trying to balance them all out, but I really learned how to study a lot. It took me that [length of] time and it [took] me taking those hard classes to get to that level. But now, once I got to that level, this year has been so easy, just like a breeze with studying. And

> I've been able to perform really well from shorter amounts of study time but efficient amounts, efficient study time. Before I would study, study, study, but I wouldn't be efficient, so I finally learned that. And now as a senior, I finally have it down after three years. (Hispanic woman, relocator, chemical engineering to materials science)

Out-of-class STEM experiences that were positive for students, such as undergraduate research, could also compensate for the negative psychological effects of poor grades. As part of assessing their effort in light of low grades, students could also determine that they needed additional help.

Students whose grades were so low in STEM courses that they had to retake them—especially chemistry, calculus, or physics—sometimes did this at local community colleges and then transferred the credit back to their 4-year program. Alternatively, either failing a course or earning a grade that was too low for program credit (typically a D) sometimes precipitated relocation to another STEM program, rather than a switch out of STEM altogether. For example, one student moved out of a STEM B.S. program in geology and into a B.A. program in earth science because it required less mathematics, which was a challenging discipline for her. Students' overall GPA could also be too low to enter some programs, which also precipitated relocations within STEM. The courses most often involved in adjusting to grades by switching within STEM were chemistry and calculus. The following student, for example, shifted out of a double major in order to avoid chemistry:

> I ended up choosing biomedical and mechanical engineering and then I failed Chemistry 2. I decided I did not want to take chemistry for another three years so I switched to just mechanical engineering. (White woman, mechanical engineering persister)

## *Adjusting to Extreme Problems with STEM Grades*

As detailed in Chap. 7, one of the biggest grade-related traumas that students dealt with was failing or being required to retake a course. In terms of STEM persistence, grades objectively matter most when students' continuance in a program is formally contingent on passing particular courses or maintaining a particular GPA. Related to this, STEM programs often have minimum grade requirements. Most commonly, programs require a grade of C or better in particular courses or for their program GPA overall. The consequence is that receiving a C-minus or a D grade is, in effect, failing—for program, if not university, credit purposes. Not passing a class and having to repeat it could itself provoke switching, as happened for one Hispanic woman who switched from physiology to psychology rather than take physics a third time. Other program policies that affected students' STEM persistence were restrictions on the number of times that students could retake courses and deletion of failing grades once a class was passed to prevent being placed on academic probation. Because some scholarships also have minimum grade and GPA requirements, some students switched to ensure that they did not lose their scholarships. Grade-related policies could cause both switchers and persisters to struggle over what they viewed as the arbitrary nature of the rules. One student, for example, persisted despite often receiving grades that were a few percentage points too low for program credit:

> I think there should be an exception. For a lot of our classes it's the C minimum requirement and I always get the D. But I'm 2% away from that C and I'm just, 'Can you make an exception for those people that work their asses off and really tried but couldn't quite cut it because they don't test well or something?' (African-American man, physics persister)

Another switched because he could not break through the D threshold in one required course and his appeal to repeat the course a third time was denied:

> I didn't make it through one of the programming courses. I got a D one semester, and I got a D the next semester. And you're capped at taking a class twice. I really wanted to take one more shot at it, but they didn't allow my appeal. But that is the class where the grade denied me from continuing in the major. (White man, switcher, computer science to economics)

Persisters' adjustment to STEM grades included normalizing receipt of relatively low grades. This logic extended to more extreme problems with grades, such as having to retake classes, which was sometimes not only accepted, but seen as fortuitous. Retaking a course could give students more exposure to important content matter and to more inspiring professors:

> I failed Physics 3, which is the electricity and magnetism…I thought I would scrape by with a C-minus, but I failed. I'm so lucky that I failed that course…because I took it again, I took Physics 3 the second time, and I had this one professor that I just loved in that class. I just fell in love with physics, at that point I really wanted to switch into physics as my major, that's how much I love physics…He was amazing, he truly inspired me right from the first lecture of the course. (Asian man, mechanical engineering persister)

Retaking a course could also motivate students to master course content through various means as well as to see the larger *life lessons* and connotations of their personal struggle:

> Organic Chemistry, for example, was a very difficult class for me. To the point that I've taken the two-semester organic chemistry sequence, but the second semester I took twice, and I passed the second time, though [I did not do] great. So now this semester, I'm supplementing it with a different organic chemistry class and doing a lot better. It's one of those things—you have to try different things. And I think I learned that from doing [undergraduate] research. I went into research with the idea of success because that's all that you see. But you talk to the people doing the real research and they are like, 'I failed this experiment for three years before I got a steady experiment to pull out data.'…So it's things like that that also influenced how I see it too. It's a setback, but there are so many things to push yourself forward. (Asian man, zoology persister)

These lessons could include becoming more aware and comfortable with one's own learning style, as was the case for an African-American persister in a math program:

> Because I didn't do as quite as well the university wanted me to I had to retake that. So, sometimes I didn't make it. But I learned from that...it is all learning experiences. University is a great playground for learning …Even the work that we do, the jobs that we have, everything that we do, all of those things are learning experiences ..And so, if I don't make [a particular grade] then that doesn't mean that I failed. I just need a little bit more time, which is usually the case for me. (African-American man, math persister)

## *Student Evidence that STEM Grades Are Not Standardized*

By the standard view of grades, a student's grade is contingent on his or her own performance and not that of an instructor. Students in our interview sample who did not receive passing grades the first time that they took a course, and then had to retake it, were able to compare class experiences across instructors, and sometimes between departments. Not infrequently, their experience was contrary to the standard view of grades. They found that their own performance and the grade that they received were contingent upon the quality of the instructor, rather than the sole outcome of their own effort and understanding. This caused students to question the link between effort and reward as well as the objective fairness of grades:

> Grades don't actually reflect how much effort goes in. For example, I took a fluids class through the chemical engineering department, and I failed it. I've never failed a class quite as hard as I failed that class because I just had no idea of what was happening. Then I took the same class again through the mechanical engineering department, just trying to pass the class, and I got an A in it. And it was shocking [because] I didn't have to put any more effort in the second time, but I got an A. So, you know, the grades don't actually show everything that goes on behind the scenes. There's a lot more to it. (White man, mechanical engineering persister)

Contrasting experiences with different professors for the same course was in effect a natural experiment showing that instructor performance played a significant role in student outcomes:

> For calculus, I ended up failing my first quarter. I feel like a lot of it has to do with the professor because he's one of these professors that I guess he's a kid genius or something…He would just stare at the board and go through things really fast. And it wasn't enough time for me to pick up the things that he was going over. Then I retook it and I got a B. That professor was a lot more helpful. He went through things and I understood all the material better. That's when I was still thinking of doing engineering because I was like, 'Okay, I can understand this. I just need the right professor.' (Hispanic woman, switcher, general engineering to finance)

> [After failing] I repeated the same [physics] class the next year with a different professor and the way he taught it was completely different. He had different homework and different assignments for class and I got a B in that one—-it was easier to understand…I don't understand the inconsistency. (White man, switcher, mechanical engineering to sports management)

Some students, however, were hesitant to attribute differences in their performance to different instructors, despite evidence that this was the case:

> Well calculus was the class where I really had the most difficulty understanding what was going on. When I took it the second time…I realized that it was exactly the stuff I learned in high school. So that one semester that I took it I don't know what happened. It was, I guess mostly the teacher and the way he was trying to teach it that was most of the problem. Because I went to all of the classes and I took notes and I studied…It's just weird that I had been only one semester without math and then I came into it and I didn't understand what was going on [and failed]. Then after a three-month summer break, [I took it again] and I remembered everything from my senior year and everything just made sense. (African-American man, switcher, physiology to psychology)

In addition to failing and retaking courses, students questioned the fairness of grading created by curve grading practices. As noted in Chap. 7, and found in the original study, curve grading could detach effort from reward in students' assessment of the adequacy of their performance. Confounding the standard view of grades, curve grading undermines the role of grades as formative feedback during a course since the outcome of student efforts is sometimes not revealed until the end of a course. There were also indications that instructors were not always adept at gauging the difficulty or alignment of their exams with course content, leaving students confused and frustrated when they expected a curve that was not forthcoming:

> There were times that they wouldn't curve properly. I mean there were classes where they didn't curve at all and we were all sitting there going, 'Are you freaking crazy? We all got Cs and Ds on your exam. There's a problem with your exam not a problem with us'…Some of my freshmen teachers [seemed to have the attitude of], 'You need to fail.'...But there's something wrong with the teaching method not the students when half the class has Cs and Ds in your class. That should be an indication that you need to re-write your exams or re-write your syllabus or find a different way to teach it. (White woman, switcher, biology to sociology)

### *Conclusion: Grades and STEM Switching*

As this section has detailed, STEM grades are consequential to students in many ways that have direct bearing on decisions to switch. Some of the ways that grades affect students—such as receiving grades too low to meet program standards—are in line with the standard view of grades as objective appraisals of performance. However, other experiences with STEM grades are more subjective, such as interpreting a C- grade as "failure" and as indicating a lack of capacity and competence. The next section picks up the themes of competence and subjectivity to address how institutional climate and belonging shape students' persistence in STEM programs.

## Belonging, Institutional Climate, and STEM Persistence

*Belonging* in higher education research is generally defined as seeing oneself as socially connected and part of a particular social group (Hausmann, Ye, Schofield, & Woods, 2009). In practice, it translates into students feeling that they have made the right choice of major and continuing to pursue their choice to the point of graduation. As outlined in Chap. 1, much of the research on belonging has evolved from Tinto's influential model of undergraduate student retention which theorizes that, students who socially integrate into campus communities increase their commitment to the institution and are more likely to graduate (Berger & Lyon, 2005; Demetriou & Schmitz-Sciborski, 2011; Tinto, 1975). Related models emphasizing "institutional fit" (Bean, 1980) and "student involvement" (Astin, 1993) further elaborate the idea and importance of a sense of belonging to students' persistence to graduation.

A *sense of belonging* connotes student perceptions of acceptance, fit, and inclusion, while *lack of belonging* is associated with isolation, marginalization, differential treatment, and stereotypes (Museus, Yi, & Saelua, 2017; Wilson et al., 2015). Belonging has both cognitive and affective dimensions and is influenced by institutional *climate*, which is typically characterized as hostile and chilly or welcoming (Hurtado & Carter, 1997). Early research on institutional climate focused on campus-level cultural dynamics, while more recent research has focused equally on classroom-level environments. With regard to STEM disciplines, Wilson et al. (2015) found that belonging at the class level, more so than the university level, was associated more strongly with engagement and commitment to persist.

The effect of a sense of belonging on students' persistence in degree programs can be complicated to discern. A sense of belonging exerts independent effects on student persistence (e.g., Wilson et al., 2015), yet it is also an outcome of other factors that can both confound and bolster belonging (Seymour & Hewitt, 1997). Indeed, a large body of research has shown that prejudice, feelings of alienation, and negative racial experiences all compromise development of a sense of belonging and commitment to degree completion for students who are underrepresented in STEM (Chang, Eagan, Lin, & Hurtado, 2011; Hurtado et al., 2007; Locks et al., 2008; Smedley, Myers, & Harrell, 1993). Factors that can mitigate negative experiences include good academic performance, parental and peer support, and programs that promote integration (Chang et al., 2011; Hausmann et al., 2009; Nora & Cabrera, 1996). Interactions with faculty, both in and outside of the classroom (Christe, 2013), have also been found to play a role in influencing students' sense of belonging, with the nature and extent of these interactions themselves influenced by institutional selectivity and other structural features of higher education institutions (Hurtado et al., 2011). The campus climate in highly selective institutions with a predominately white student body, for example, has been found to undermine the sense of belonging for students of color (Chang, Cerna, Han, & Saenz, 2008; Espinosa, 2011). Hurtado and Carter (1997) found that peer interactions influence students' sense of belonging more profoundly in the early years of an undergraduate program, and that interactions with faculty become more influential in later years. Taken together, these findings indicate that development of a sense of belonging depends upon multiple influences and has different consequences at various points in students' academic pathways.

In STEM specifically, Museus et al. (2017), among others, have found that sense of belonging predicts intent to persist. Co-curricular opportunities that cultivate a sense of belonging, such as academic peer relationships, participation in undergraduate research, and involvement in STEM-related clubs and organizations, can off-set problems with poor academic preparation and contribute to STEM persistence for underrepresented minority students (Chang, Sharkness, Hurtado, & Newman, 2014; Espinosa, 2011). As noted above, the effects of climate on belonging, and thus, persistence, have been found to be most salient at the classroom rather than campus level (Wilson et al., 2015). Along similar lines, positive everyday interactions in STEM programs and departments, both between students and faculty and among students, have been shown to create an encouraging climate for instruction and commitment to persist among student groups that are underrepresented in STEM (Thiry, 2017).

In addition to the interactional aspects of belonging, the *culture of STEM*—its ethos, implicit rules, and frames of meaning and priorities—has been found to influence students' sense of belonging. Research has consistently shown that women and other students who are underrepresented in STEM are disadvantaged by STEM culture. For example, Smith, Cech, Metz, Huntoon, and Moyer (2014) found that the prevailing view of STEM education as supporting individualistic goals contributes to *belonging uncertainty* among Native American students who value communal goals and opportunities to give back to their home communities. In computer science, which generally has a high level of gender disparity, aspects of disciplinary culture have been found to bring about disaffection and disinterest among girls and women. This includes reactions to "geek" stereotypes about the field, poor accommodation of non-technical interests, and failure to incorporate social implications into computer science curricula (Margolis & Fisher, 2003). Female computer science students frequently question their likely future success in the field after their sense of belonging in the discipline has been undermined (Benbow & Vivyan, 2016). Similarly, numerous studies have shown that the individualistic culture in engineering programs compromises sense of belonging for students who value working with others (Bonous-Hammarth, 2000; Geisinger & Raman, 2013). Women, in particular, appear to be negatively affected by representations of STEM fields as incompatible with communal goals and altruistic motivations (Clark, Fuesting, & Diekman, 2016; Brown, Thoman, Smith, & Diekman, 2015; Diekman et al., 2010, 2011; Espinosa, 2011). The view of STEM education as contrary to communal values also appears to have some grounding in STEM program curricula. Specifically, Cech (2014) found that engineering programs promote a *culture of disengagement* from societal concerns in showing that students' interest in how engineering relates to public welfare declines over the course of progression through degree programs.

## *STEM Students' Sense of Belonging and Experience of STEM Program Climates*

Students described a number of problems that arose from the climate of their STEM programs and their consequent sense of belonging that, for some, contributed to their decisions to switch or relocate. These included issues with competitive culture, which in the original TAL study played into switching decisions for 15% of students. In the current study, as indicated in Table 9.4, over half (52%) of switchers were negatively affected by problems related to the climates they encountered and

**Table 9.4** Proportion of students affected negatively by STEM climate and belonging

| Switchers | Contributed to switching (%) | Negatively affected (%) | Persisters | Negatively affected (%) |
|---|---|---|---|---|
| Overall | 52 | 81 | Overall | 42 |
| Students of color | 62 | 88 | Students of color | 60 |
| White students | 49 | 79 | White students | 32 |
| Men | 46 | 74 | Men | 30 |
| Women | 56 | 85 | Women | 52 |

with their difficulties in developing a sense that they belonged in a STEM program. These issues were slightly more likely to affect women rather than men who switched, and proved even more damaging to students of color—whether or not they switched—than to white students. Indeed, the difference on these measures between both all women and students of color who persisted and their white and male peers was even greater than that for switchers. This suggests that these problems do not necessarily diminish as students move towards graduation.

The commonest problems with belonging were experienced by students of color (62%), especially those who were also first-generation college students and/or had low-income family backgrounds, as well as women (56%) of all races and ethnicities.

As the following sections describe, our findings show that students' sense of belonging is an outcome of five phenomena. These are:

1. Students' assessments of their own and their peers' *competence in STEM*, including the adequacy of their preparation in high school to succeed in a STEM program;
2. Students' encounters with, and reactions to, *competitive individualism* among their STEM peers;
3. The *formation of cliques* among STEM students and related interactions with peers that affect access to, and dynamics within, *study groups*;
4. The nature and extent of *connections and relationships* built into STEM curriculum and instruction;
5. *Interactions* with instructors.

These findings, taken together, show that belonging problems are not just a "sense" or feeling that resides as a psychological phenomenon within students. Instead, problems with climate and belonging often manifest from *status competitions* that are instigated by some students to gain advantage, and assert their own superior belonging, by stigmatizing and excluding others. Despite the well-documented importance of peer-to-peer relationships for STEM persistence that is elaborated in the literature (e.g., Chang et al., 2014; Espinosa, 2011; Ong, Wright, Espinosa, & Orfield, 2011) as well as shown in Chap. 12, our findings also show that peer interactions can create stress and isolation, leading some students to question their commitment to persist.

## Assessments of Competence in STEM and Belonging

Perceived competence in discipline specific content was an important determinant of whether students developed a sense of belonging in their STEM major. Grades, as well as formative feedback on performance, served as signals both of self-competence and *recognition of competence by others*. For many persisters, it was recognition of competence that confirmed their sense of belonging in STEM:

> Writing long lab reports for classes that you put tens and possibly hundreds of hours into, and then getting quality feedback on those reports, has been extremely helpful and validating. It really shows me that I'm in the right place. (White man, high-math astrophysics persister)

Not surprisingly, given the importance of pre-college preparation in STEM persistence (described in Chap. 5), competency-based appraisals of students' belonging in a STEM program could be contingent on their assessment of whether they had been adequately prepared in high school. Those who were well-prepared and experienced a smooth transition to college-level STEM courses (such as the biology student quoted below) had a correspondingly easy time validating their sense of belonging within a STEM program:

> I felt [a sense of belonging] pretty early on probably because I had that preparation...My first science class, freshman year was Biology of Vertebrates, which is traditionally pretty hard…I did well and I really enjoyed it. If I had done poorly or had not been interested, that would have been a clear sign that something was amiss, but right off the bat, I felt good about it. (White woman, high-math biology persister)

In contrast, students who were less well prepared in high school, and who assumed, therefore, that they struggled more than other students and were less competent, were more likely to experience *belonging uncertainty* and eventually to switch out of STEM. One indicator that students used to appraise their preparedness was whether they knew particular curricular content that STEM faculty assumed that incoming students should know. Not knowing this content could result in being stigmatized:

> In the biology department, they expected a certain [level of] knowledge from you. They expect you to know all these things, and then if you didn't know them, you're looked down upon. (White woman, high math, switcher, biology to psychology)

This same student's experience after switching into psychology contrasted with biology in that faculty had less singular expectations about students' incoming knowledge base and were, as she characterized it, "kind of excited that they get a chance to explain it," rather than tacitly making pejorative assumptions.

*Belonging uncertainty* was often a product of interactions within programs and classrooms, especially among peers. These included students judging their peers according to assumptions about their backgrounds. Students could deny being biased or "discriminatory," yet question the legitimacy of other students' belonging based on appearance and behavior. Students could judge peers as *not belonging* based on, for example, their age (as a non-traditional student) or apparent lack of competence (evidenced by questions asked in class):

> I definitely feel there are some people who don't belong. Not in any discriminatory way, but if I hear someone in class ask a really simple question, it does kind of make me think, 'What are you doing here?'… There's someone in my Advanced Organic Chemistry lab. She's a non-traditional student, and she's got grandchildren probably. I've never asked her, but I'm [thinking], 'What are you taking this course for? Are you really trying to go back to school and get a degree in chemistry right now?' It's just kind of confusing. If it's really interest, then more power to her. But I'd say personally I'm a chemistry geek. I love chemistry. I love talking about chemistry. I read about chemistry in my free time sometimes. So, I feel like I definitely fit within the department. (Hispanic man, high-math chemistry persister)

As this extract indicates, competence and belonging were rooted in students' perceptions of their performance and reasons for choosing a major in comparison to their peers. Whether students saw themselves as competent in STEM and therefore "rightfully belonging" in a STEM program was seldom assessed solely in intrinsic, personal terms. Instead, as the next section explains, it was more commonly assessed through comparisons with peers.

## Competitive Individualism

Comparing oneself to peers—by various metrics of competency, such as degree of struggle, grades, and pre-college preparation—often formed the basis of student appraisals of their own competence as STEM majors. While this was sometimes a neutral self-assessment based on other students appearing to struggle less or understand more, it was also the outcome of students who asserted higher status than their peers. Women had more complaints about this form of competitive jostling than did men, and women's sense of belonging in STEM was more often damaged by it. Ten percent of switchers, all women, reported that competitive behavior by their peers had contributed to their decisions to leave STEM. As we further elaborate in Chap. 10, competitiveness among peers, especially in a high-achieving college environment, provokes perfectionist tendencies in some students, including hiding signs of struggle so as to make high grades seem effortless.

*Competitive individualism* often manifested as efforts by some students to be recognized as "the smartest" in the class. By publicly asserting themselves to be better than their classmates, usually in terms of higher grades, these students were also tacitly undermining the sense of belonging and commitment to persist of other students. This occurred most commonly among freshmen. For example, two women who switched out of STEM noted that individual competition to be recognized as "the smartest" began during freshmen orientation, when some students overtly flaunted their pre-college test scores and grades:

> During welcome week, when you're with a group of people who are the same majors…we were talking about SAT and ACT scores, and GPAs and stuff. And I was definitely not in their range. [In fact], I was on the wait list for the College of Science and Engineering and then I finally got in…So I got into there, but I just felt not as smart as everybody. Because, [the other students would say], 'Oh yeah. I'm doing this, and I did this, and I have all these AP credits.' And I was like, 'I don't.' So that was one thing that kind of deterred me. (White woman, high-math switcher, astrophysics to media studies)

Competitive individualism among students had some basis in the practice of curve grading previously described. By ranking students according to their relative performance, curve grading contributed to a competitive classroom climate. Its wider effect is to create a market scarcity for high grades, since the ranking of students takes no account of how wide or narrow is the range of relative scores that students receive. Along similar lines, as noted in the original study, curve grading does not distinguish between cohorts of students with greater or less intellectual capacity. Departments sometimes also have policies encouraging faculty adherence to stringent grading practices. For example, one student in our interview sample recalled a professor telling her that he could not give out all A grades because he would "get in trouble" if he did so. Curve grading also had the (presumably unintended) consequence of undermining friendships, academic peer support, and a collaborative ethos, since it actively pitted peers against each other:

> No one wants to study with you because, if you do better on a test grade, that means they're going to do worse. For some classes the curve helps. But I think for other classes it's just kind of destructive. Because then you've got this competition where no one wants to study with each other...It creates a competitive, really destructive environment. But sometimes, in research, you have to be cooperative and work with other people. (Hispanic woman, low-math microbiology persister)

> Especially with chemistry courses, the majority [of grading assessments] are based on curves...It's not like you don't want [other students] to do well, but you want yourself to do as well as you can so you can get ahead of everyone else and be ahead of the curve. (Asian woman, high math, persister in biochemistry)

In addition to curve grading, other teaching and testing practices could also intensify student competitiveness with the result of excluding and isolating some students. For example, by publicly naming top performers in lab assignments a teaching assistant sought to motivate performance, but also succeeded in isolating at least one woman who eventually switched out of STEM:

> I didn't know what I was doing in my labs, but I was pretty good at following the steps. My TA, basically to keep people competitive with each other, would name who was getting the top scores in the lab and I was generally in the top three. It had nothing to do with the fact that I knew [the material]. All the friends that I had in the lab withdrew because why would you keep helping the competition? So that time I was alone. And my Organic Chemistry lab friends were also in my Organic Chemistry [lecture] class, which I was failing. (White woman, high-math switcher, biology to government)

In another example, a man who switched from engineering to art recalled that a competitive and work-intensive atmosphere in engineering undermined peer interaction. Implicit in his description were instructional practices that enabled a lack of engagement with others:

> In my engineering courses, no one talked to each other. If they did talk to each other they were the same group of friends... In my art courses, it's gotten a lot better. I actually talk to people in my class. I'm actually making friends. In engineering courses, it just seemed everyone was in such competition with each other and so busy with their coursework that no one was looking out for each other. It felt like everyone was out to get each other. (White man, high-math switcher, aerospace engineering to studio art)

Instructional practices supporting a competitive classroom climate were also said to affirm the sense of belonging for some "types" of students more so than others. This had the effect of advantaging some groups of students over others:

> In engineering there's a couple different classes of students. Some of them are very anti-social, very smart, just get your homework done, nerdy kids. They're there to be engineers. I think that's great, but I don't have much in common with them. And then there's a separate class of kids who are more easy-going. They party sometimes and they get their homework done. They're maybe not the smartest kids in the class, but they're more social, they interact with you more. I've been in groups with both types, especially in the physics labs. And [sometimes] I'd crack a joke and a kid looks at me like I'm crazy because he's trying to get done first so he can get the 'check plus' on the lab instead of just getting the check. And I'm, like, 'We can have a little fun here, make a joke and laugh, and we'll still get done on time and do our work.' So one reason why I changed [majors] is a serious lack of people who are similar to me…I feel like I fit in better with people that academics and school isn't 100% on their mind all the time. That's definitely a big part of my life and that's why I'm here and I want to get a degree, but [there is more to life]. (White man, high-math switcher, engineering to construction management)

Consistent with findings from the original study, competitive individualism was most frequently mentioned as occurring in courses with a high number of students seeking entry to medical and veterinary schools. The dynamics among students in these courses established de facto criteria of "what it takes" to succeed in these pathways. However, these norms might or might not be consistent with the goals of STEM programs or with socialization into the broader set of skills and dispositions needed for a particular career:

> I'd be like, 'Hey, do you want to study?' And [my classmates would] be like, 'Oh, no, I'm not studying tonight.' Meanwhile, they're locked in their room studying. But when I switched into the life science major, everyone's chill. Everyone's just trying to do well and they all help each other. And that's what I like. (African-American man, low-math relocator, pre-med/biology to life science)

> It's definitely been interesting being pre-vet because there are a ton of people that almost won't help you with exams. It's only in certain classes. But they almost don't want to see you do well or they don't want to see you get that internship because it means that's one more thing that they have [to do]...It's almost that some people will try to tear you down so that they can one-up you because everyone's competing for those 100 spots in vet school. (White woman, high-math relocator, animal science to equine science)

In the extreme, these competitive practices included "smart" students overtly taunting peers who were stigmatized as less competent. Students exposed and labeled the less competent through public comparison of performance:

> It was very competitive, it was very cut-throat. People didn't care about your success. Some students would be like, 'Oh, you didn't get that answer? Well I did!?'…I used to cry all the time because of those science classes. Because people made me feel so dumb and horrible. And it was very, very competitive…[I switched] because I thought, 'These people are so smart. They're going to make it. I'm not as smart as them, so I'm not going to make it.' You know? They're getting the answers right. (Asian woman, low-math switcher, biology to communication)

Engineering and computer science programs were also said to be especially affected by these competitive practices, including stigmatization of the "less smart" and arrogant attitudes expressed by some students that were off-putting to others:

> It was sort of like, 'I must prove that I am so good.' Maybe they felt a lot of pressure to be the smartest one in the room, because all of us coming from high school [were] usually the smartest one in the room or one of the top five...I wasn't really in a great position to jockey because I hadn't already taken the course. So, I was kind of at the bottom. (White woman, high-math switcher, engineering to psychology)

> One of the more social reasons that I switched was because…not everybody in engineering is very sociable. [Some are] high and mighty about their major and their intelligence. [Instead of] 'We're all in this together, we're all here, we're all intelligent,' [they were] very snooty. [They had] a lot of pompous attitudes that I did not appreciate at all, nor was I like that. So, there was [an attitude of], 'I'm better.' But [I thought], we're in the same class. We got the same grades. There should be no stigma there. (White woman, high math, switcher, engineering to finance)

> I definitely see it with the students who develop a sense of arrogance and pretention about their knowledge of the subject, which is very off putting to me because it becomes a competition of knowledge and awareness. I've always had the personality that if I don't know what I'm talking about then I am not going to speak up because I don't want to look uninformed. I've definitely run into that with students who can be kind of harsh. (Multi-racial man, high-math computer science persister)

Although the effects of competitive behavior among students in a STEM program could be widespread, a competitive classroom culture was typically created by just a small proportion of students in a class. This small number, however, exercised considerable informal influence on the climate both of classes and programs:

> I've seen a lot of engineering classes. A lot of the students are trying to be smarter than everybody else. That's not my favorite environment to be in. There's people trying to outdo each other, and I don't want to say that they have a stuck-up personality, but I just feel like everybody's trying to one-up everybody…Not all of the engineering students are that way, but there's enough in every class that it's sometimes difficult to stay focused on what I need to be focusing on, and not be annoyed or upset by them. (Multi-racial man, high-math engineering persister)

Students sometimes resisted being positioned by their more competitive peers as having less competence or being made to feel "stupid" or "less intelligent." However, assertions of superiority by other students could take a toll on confidence and commitment to persist in STEM *even when the students who were undermined were not performing poorly in objective terms*. It was also difficult for students to resist the terms of competition set by their peers, which were singular, defined by performance, and by public displays of commitment that could include arrogance and assertions of superiority:

> I like to think that I'm the same intelligence level as them, but I always feel, especially in classes like Organic Chemistry 2 that [students] kind of put each other down and they're always trying to be better than you. Like, 'Oh you got 61 on your test? I got 63.' As if they're just so much greater than you. (White woman, high-math switcher, biology to human development)

> There were definitely times…with all bio/pre-med majors that did make me feel uncomfortable, because I was like, 'Well, I'm not a pre-med major, but I'm not stupid. I can do this too, you just have to give me a chance.' You know? (White woman, low-math switcher, chemistry to psychology)

Student displays of "being stressed out" also set de facto terms of competition that were hard for other students to resist. In fact, *rejecting stress* was a factor in some students' decisions to leave STEM programs:

> I think everyone in my class was pre-med and very intense about it. I didn't want to go pre-med [but it was still] a high-strung environment. I'm a good student, but I didn't care to be around such stressed out people all the time. (White woman, low-math, switcher, biology to economics)

Resisting the terms of competition set by other students was difficult even if students did not aspire to the same ultimate career goals as their competitive peers, such as professional school. It was also difficult even when a student did not subscribe to the belief that their own success required public displays of arrogance or stress.

## Cliques and their Impact on Access to Peer Help

As described above, belonging was underpinned by students' views of their own and others' competence in STEM. These assessments, in turn, could be exploited to create a competitive classroom climate, where competitive students stigmatized others as less intelligent. The discussion above characterized competition within some STEM classes—for grades and, tacitly, for *high-status belonging*—as occurring at the individual level. However, these dynamics, as well as the more general need for academic support, also propelled students into alliances and "cliques" to access peer help through study groups. For the majority of students, study groups and study partners were a key to success. Many who persisted in STEM programs attributed their sense of belonging to having found and relied on a group of peers with whom to study. Chapter 12 details the importance of study groups and peer support for persistence. Here, we focus on the potential downside of the need for peer allies, specifically, the negative impact on students' sense of belonging created by problems in gaining access to study partners and groups.

Many students described study groups as tending to form along lines of commonality. This could include knowing each other prior to a class—due to coming from the same community, residing in the same dormitory, or previously taking classes together —and also through appraisals of each other's relative "smartness." Students who viewed and asserted themselves as "the smartest" were (as one switcher characterized it) apt to get "really cliquey with other really smart people." They might also exercise *strategies of exclusion* toward peers who were seen as less smart, including exclusion from a study group or demeaning treatment within a study group:

> What you find is a lot of the students are out for [themselves] when it comes to the sciences. No one's willing to sit there and be like, 'Well, this is what I found, let's work together.' [Instead] it's, 'I'm not going to give you my answer because I don't want you to be ahead.' So that was hard in labs for me... You're supposed to collaborate, but everyone was, 'Well... we're not going to get anything out of you, so we're not going to help you either.' (Hispanic woman, low-math switcher, biology to communication)

> Sometimes [other students were] like, 'I don't understand why you don't know this?!' Like, 'Duh!?!' I hated group activities, because it was like, 'Well, I think I should know this, or I think I should know that.'[In fact] we hadn't gone over it in class, but (somehow) other students knew it. (White woman, low-math, switcher, chemistry to psychology)

In addition to levels of perceptions of competence, students sorted themselves into study groups and access to peer help by commonalities of age, gender, and race and ethnicity. Women of color, especially those in relatively male-dominated STEM fields, had particular difficulties connecting with other students:

> With geology it was hard, because it was more male-dominated. Like in my geology class there were about 40 of us and about five or six were girls so it was really hard to find somebody to study with. And they were old; some of them were really old. So, it was hard. (Asian woman, high-math switcher, geology to business)

> It's okay for the guys who formed their groups. They're more tight-knit and they're able to relate to each other more. But since I was alone, I felt ashamed that I couldn't do [the work] on my own and that I didn't have people in that major to help me through it. I feel like if I had friends who were struggling with me then I could go ask for help. That would have been extremely helpful. (Asian woman, high-math switcher, computer science to accounting)

First-generation college students also faced special challenges demonstrating their competence and connecting with study groups. One such student described how she was disadvantaged by her inability to use vocabulary in a way that was expected when she began her biology program:

> Being a first-generation college student, I felt it was harder to assimilate to the college atmosphere. I may not have started using the same vocabulary as some people whose families had multiple generations of college students and I feel like that might be one reason why first-generation students don't go further with STEM programs as second- or third- or fourth-generation college students. (Asian woman, high-math switcher, biology to youth studies)

Among the most perverse outcomes of a competitive climate in STEM were cases in which students gave out incorrect information to their peers in study groups with the aim of undermining the competition and bolstering their own performance standing in a class. Accessing effective peer study group support was therefore contingent on identifying allies, including those who would offer genuine rather than fake help:

> I know someone who purposely gives wrong answers. In [study groups] going through questions, she'll purposely give the wrong answer so her homework can be correct and everyone else is wrong. Those people are few and far between, but when you meet them it's a really bad experience. So, you just try to cling to the other people who don't do that. (White woman, high-math relocator, animal science to equine science)

> There is a biology housing-hallway for people who are really gung-ho serious about biology. I remember someone saying that there were students in that house who would give out fake help. Like, if they got together as a biology study group, these students would give out misinformation so that certain students would do more poorly on the test and their own grades would look that much better by comparison. (White man, high-math switcher, biology to dance)

Where students sorted themselves into study groups according to performance and actively excluded those perceived to be lower performers, this affected the sense of belonging, not only of switchers, but also of STEM persisters, particularly men of color and all women:

> There are the students that did really well and they worked together. I wasn't part of that group because I wasn't doing as well. That made me feel like I didn't really belong because there was no way I was going get help from someone smarter than me unless I was friends with them outside of class...So if you're not part of the group that's doing really well sometimes you get left behind, because people aren't going to reach out and try to help you. (Asian woman, high-math biochemistry persister)

> Everyone's civil but no one will want to study with you. I get good grades so all of a sudden, when they realize not how smart I am, but how hard I study, people started wanting to study with me...Most of the time these people just want to study with me because of the grades. And they won't talk to me for any other reason. (Hispanic woman, low-math microbiology persister)

Whether the faculty or departments concerned were aware of, or sanctioned, these exclusionary peer-to-peer dynamics was unclear. However, those switchers who were negatively affected by competitive and exclusionary peer interactions gave no indication that faculty or teaching assistants interceded to counteract these effects.

## STEM Programs, Curricula, and Belonging

A modest number of students—all of who had high incoming math backgrounds—switched out of STEM, in part, because of problems with the climate of STEM as they experienced it in classrooms and academic programs. This aspect of belonging was characterized both positively and negatively in terms of connections as well as the opportunities that could come through connections. Students who felt connected to their STEM discipline as a desired future, and to their peers and teachers within their STEM program, tended to have a positive sense of belonging and commitment to persist. For example, after changing majors within engineering one young man found that his studies grew and developed alongside his life interests and connections with others:

> I've had opportunities to meet people who are interested in biomedical engineering. And I've had a lot of opportunities open up to me without me even trying. I'd like to work eventually with disabled children. And more and more I've had opportunities to interact with people, [for instance] with cerebral palsy. When I go home I have these opportunities. When I come back to school, I have more opportunities from the classes. It's kind of just been a flow. So, I feel like I'm in the right place for that. (Multi-racial man, high-math relocator, mechanical to biomedical engineering)

In contrast, other students with high-math backgrounds who opted out of STEM did so, in part, because their STEM education lacked opportunities for connection of ideas. These students felt intellectually alienated by being denied the prospect of thinking beyond numerical representations to disciplinary applications in "the real world." When STEM was presented in curricula as pure numbers, detached from more holistic and relatable representations, this turned off some high-achieving students, particularly women:

> The one thing that really made me realize that [STEM] was not what I wanted was that it was so impersonal. I'm good at science and math and I enjoy [it], but everything was just a number. And even myself, in a way, [was just a number] because my classes were so big. I never had a small science class. And my TAs weren't that invested in us because they also had their own schooling to do. There was so little personal connection with my classmates. Part of that might have been because they were guys, but I think more it was because of the atmosphere. I don't feel uncomfortable around men. Maybe [gender] plays more into it than I think it does, but from my perspective that wasn't a main reason. Just that it's impersonal—all the data and examples are just numbers and figures. But they do apply to the real world. Physics is very applicable to the real world, but we never connected it to human life. (White woman, high-math switcher, physics to geography)

Because many switchers had taken courses in both STEM and non-STEM disciplines prior to formally switching out, they were able to compare the climate of departments. Consistently, their programmatic and curricular problems with belonging were related to unwelcoming and emotionally sterile STEM environments. One former chemistry major commented that his program gave a "cold welcome" to freshmen: there was no enthusiasm or sharing of information about the kinds of opportunities that would be forthcoming through the major. A Hispanic woman with high incoming math scores switched out of biology and into journalism, in part, because she related more to the enthusiasm and expansive sense of opportunity that was projected by its students and faculty in contrast to the dullness of the biology program. It was not, she had concluded, that biology per se lacked intellectual excitement. Some of her high school friends studying biology at other universities had encountered opportunities to "innovate, feel excited, and make a difference." Similarly, several high-math switchers out of engineering and computer science described a preponderance of "quiet people," a dearth of opportunities to meet new people, and "discarding the emotional side of things" had contributed to the sense that they could not belong in these majors. There were, however, indications that the climate of some STEM programs improved over time. Some persisters reported that it was easier to make enjoyable intellectual and personal connections in upper-level courses:

> I enjoy the atmosphere 100 times more in the upper-level classes because people are there because they're interested….I enjoy the upper-level classes way more than the Gen Eds… We laugh about science. And I'm around a bunch of super nerdy people who are nerdy about the same things I'm nerdy about. My professors will make jokes about cell biology and it's awesome. (White woman, low-math relocator, biochemistry to biological science)

Progression in STEM degree programs could bring a greater sense of belonging because more advanced students had more opportunities to interact and experience *continuity of relationships*. They had taken more classes with peers, studied together, and bumped into each other in departmental offices. The realization that they shared common troubles and were "in the same boat" fostered camaraderie and an "unspoken solidary." Out of these cumulative interactions over time, de facto cohorts could form. How these experiences contributed to persistence is taken up more fully in Chap. 12.

For some students, positive belonging was also cultivated through authentic STEM research experiences. Consistent with our previous research findings (Hunter, Laursen, & Seymour, 2006; Laursen, Hunter, Seymour, Thiry, & Melton, 2010; Seymour, Hunter, Laursen, & DeAntoni, 2004), one of the key benefits of undergraduate research was development of a sense of belonging to their disciplinary community and to science as a shared enterprise—a process that we labeled, "becoming a scientist." The aspect of these experiences that was consistently cited as most relevant to belonging was connections to those working in the lab, to "smaller subgroups of people who work in this field," as well as associated opportunities to get to know graduate students and professors. The value of undergraduate research experiences for persistence is further described in Chap. 12.

One final point related to STEM programs, curricula, and belonging that was consequential for persistence relates back to our findings about competitive individualism and the negative aspects of student study groups. Some students shared examples of departmental policies that affected everyday classroom practices in ways that moderated competition and enhanced students' sense of belonging. In one case, instructors erased the dividing line between grades of A and B in order to reduce petty competition:

> Something that's nice that they did for the upper-level classes is that they've made them less competitive. My virology teacher at the end of the year erases the A-B line so basically everyone that has an A and a B gets an A…I think that's nice because he wants us to actually understand the material and work together and quiz each other, instead of trying to compete against each other. (Asian woman, high-math biochemistry persister)

In another example, a department followed a policy of study groups being required for homework, including randomly assigning, then reassigning, student group membership in the study groups in order to reduce isolation and mitigate against the exclusionary practices associated with informal study groups. The result, one student recounted, was "more integration," less "insularity," and a discernibly stronger sense of cohort identity among students:

> There was a policy that was really effective during my sophomore and junior years. They started making people work on group homework assignments, so people couldn't work alone. And the groups started being assigned. So, they'd just randomly generate teams of people. And that was a really good thing. Their rationale was that when you work in the real world, you work with whoever you have to work with, and you don't choose your buddies. Before, what was happening, was people who knew each other from the dorms or from high school would hang out and always work with the same people once they found someone that they could work with easily. So, you never met new people, and that's why it felt insular. After they started forcing people to integrate, I met a lot of other people that I hadn't met before [because] it's really hard to meet people in a 300-person classroom. You meet the two people that sit on either side of you. So, I think that was a really effective thing. Since

> then, there's been a lot more unity in our major. This year in particular, we've started to feel a lot of class identity, when, in my sophomore year, we had zero class identity. And people have started socializing a lot more outside of class. Which seems like it would happen just automatically, but when you don't necessarily know a lot of people, that's not the case. So I think that was really effective, and I know the majority of the people in my major now as a result of that. While after the first two years of college, I knew maybe only seven people pretty well, out of 70. (White man, high-math engineering persister)

This relatively simple policy and instructional intervention is in line with research-based strategies that have been shown to improve cross-group interactions (Mendoza-Denton & Page-Gould, 2008; Steele, 1997; Walton & Carr, 2012).

## Connections with Instructors and Belonging

Some of the belonging problems experienced by switchers arose from lack of academic support or encouragement from instructors. As described in Chap. 6, instructors often seemed disinterested in students' understanding or progress, or were distant ("very on task") and, in effect, invisible to their students other than appearing in front of them in lectures. Students who struggled with course content and regularly sought help from instructors reported impatience, irritation, put-downs, and rudeness in response to their requests for help. All such experiences checked development of a sense that they belonged.

The physical design and layout of STEM program faculties, including the size of lecture halls, also discouraged a sense of belonging. Large classes often correlated with an impersonal classroom climate, precluded opportunities to connect with instructors—"to make yourself an individual," as one switcher characterized it. Students also derived a sense of the relative warmth or coldness of their STEM program from the design of buildings, the body language of faculty, and whether there were opportunities for informal interactions.

As will be elaborated in Chap. 12, persisters more so than switchers, not surprisingly, reported more positive connections with instructors as having contributed to their sense of belonging and persistence in STEM. These included informally interacting at events, such as scholarship dinners, bumping into instructors outside of class, being greeted by name, and being responded to positively when requesting help. Just as a single negative interaction with an instructor could compromise a student's sense of belonging, a single positive interaction or moment of recognition by an instructor could also be enough to affirm belonging and propel a student's commitment to persist. In the original study, we pointed to the serendipity of these fork-in-the-road experiences and their power to tip the balance between leaving and staying for students who had begun to think about leaving. Here are two positive experiences and their important consequences:

> There was one moment that [my sense of belonging] really crystalized. I was in office hours for cell biology and our professor asked us how voltage-gated channels work. And I rationalized it in my head and then explained it to him. He asked, 'Did you take a class like this before?' And I said, 'No I just kind of put the pieces together.' I didn't even know if I

> was right or not. And he said, 'You're absolutely right.' So that was a very proud moment for me because I was able to apply my prior knowledge and figure out a problem without really learning about it yet. That's when I realized, 'You know what? I actually do belong in biology.' It was also very nice because he kind of gave me a lot of praise for coming up [with the solution]. So, it was very encouraging to be praised and he kind of respected my rationalization behind how I got there. (Asian woman, high-math biology persister)

> There's one person...he's a Ph.D. candidate in the education school but he graduated with his Master's in electrical engineering. He really hit it hard for me by saying, 'You're a good person at the end of the day and it gets lost in a lot of ways at the university.' And that really helped me. (Asian man, high-math engineering persister)

## Belonging Issues Specific to Students of Color, First-Generation College Students, and Economically Under-Privileged Students

Students of color experienced some of the most challenging belonging problems, as did students who were the first in their families to pursue higher education and students from low-income families. Students who experienced belonging problems also tended to be women, whose specific belonging problems are examined in the next section. The issues faced by these students affected persisters as well as switchers. This raises the question of whether belonging problems matter if students nonetheless persist in their STEM programs? Here, we argue that they do matter. While problems with belonging are important to understand and address for the benefit of all students, they are particularly problematic if they disadvantage some types of students over others. Our findings show that belonging problems were particularly acute for students of color, creating extra stress and the burdens of managing stereotypes and disadvantages that potentially exaggerated the already uneven playing fields of STEM (McGee, 2016; McGee & Martin, 2011).

Roughly 18% of students of color in our interview sample experienced some degree of race and ethnicity-based *isolation* during their STEM studies. These problems were often characterized in terms of students' compromised *ability to relate*, which undermined connections with other students and access to help. A number of African-American students had a particularly difficult time finding peers to relate to and study with. Given the importance of peer-to-peer engagement associated with students' persistence in STEM (e.g., Espinosa, 2011; Ong et al., 2011), this represents a notable disadvantage. As one biology persister explained:

> For students of color, it's hard to find someone to relate to here because [this university] is a predominantly white campus. When you can relate to someone you can go to them for homework help. ...If you don't find anyone you can relate to easily, then it's really hard to find help among students [and] you're not going to stick with it. Because, sure, you probably are smart enough. But what's the point if you don't have any friends and you're miserable? And it's hard working when you're miserable. (African-American woman, low-math biology persister)

Like the student quoted above, students of color who felt isolated in their STEM programs also tended to have low math scores upon entering higher education. This suggests that their high school preparation was not as strong as that of other students, which put them at particular risk of being tacitly stigmatized and stereotyped as lacking competence. One African-American student described how she was often "overlooked" by her peers, who seemed predisposed to expect that she would not know the class material. She persisted despite the need to constantly prove that she legitimately belonged in the biology program:

> I'm usually the only black person or the only female in class, or both… and sometimes it's hard to find people in classes [to ask], 'Oh, can we study together?' Just not looking like everybody else kind of doesn't do the best for self-esteem, especially if you're struggling… Like in my Organic Chemistry class, especially in lab, sometimes I'm overlooked. So it's hard to ask for help, to ask, 'Oh, I don't understand what we're supposed to be doing with this reaction. Can you help me?' There's not a warm welcome or reaction… So it's kind of hard to get help. But it's more being overlooked, like I'm expected not to understand. Or sometimes when I offer up solutions they're not taken as possibilities. I feel that I am constantly having to prove that I'm intelligent. And I am [intelligent]. It's just that sometimes I don't understand the material. (African-American woman, low-math biology persister)

As the student quoted above implies, the experience of exclusion and stigmatization may not have been motivated by overt racism or sexism. However, there was evidence that these tacit assumptions encountered by students of color about "how smart" they were, and how likely they were to make a contribution to study groups were grounded in implicit racial and ethnic stereotypes. The theory of *stereotype threat* is relevant here. It contends that negative stereotypes about a person's group can lead her or him to be concerned about being judged or treated negatively based on the stereotype (Spencer, Logel, & Davies, 2016; Steele, 1997; Steele & Aronson, 1995). Our findings show that stereotype threat was most evident for students of color who were from low-income communities and were aware of having had inferior science and math high school preparation compared with their more privileged peers. At the same time, however, like McGee and Martin (2011), we found that stereotype threat was not deterministic and that some disadvantaged students persisted in STEM despite the special challenges that they faced, as depicted by one African-American persister in biology:

> I grew up in [a county] that is basically 90% African-American. So, I feel like the standards education-wise are somewhat lower. I came in top from my high school. I was the best my high school had to offer. But coming into college, mixing with other people, I feel like actually I was probably mediocre compared to other people and the education they got—the rigor of their education, how they were able to study, and how they were able to perform—so that kind of [made it] more difficult to acclimate. I do feel like your background has a lot to do with how you are able to progress…the rigor of your [high school] education has a lot to do with whether you stay in or not...It kind of affects how easy it is for you to relate to your classmates and form the study groups that are necessary. (African-American woman, low-math biology persister)

The lack of diversity among students of color in STEM programs in terms of personal background, given how few there were numerically, was an additional aspect of the problem.

Stigma based on social class background was experienced by some of the black and Hispanic students in our sample who were also the first in their families to attend higher education. This amounted to a triple intersection of characteristics that made students vulnerable to stigma—being students of color, coming from low-income backgrounds, and being the first in their family to attend university. These students reported grappling with hostile stereotypes and racism in particular at the private research university in our sample. Consistent with previous STEM persistence research (e.g., Chang et al., 2014; Espinosa, 2011), our findings show that high institutional selectively disadvantaged the students of color students in our interview sample. In this selective context, these students were pushed to become aware of the advantages enjoyed by their more economically privileged peers, some of whom used these disparities in status competitions to undermine the competition:

> As a Latina student, you get a lot of, 'You're here on a scholarship.' And, 'You're here because you're from a low-income school where the education isn't that great.' Like we should feel, you know, 'You're lucky to be here. You're not on the same level as us and you're lucky to be here.'…The level of education in lower-income high schools is not the same. When a lot of students from lower-income schools come to a college like this, they're not prepared and they struggle. And I think that a lot of students in this school have no idea about that and they just assume that we are all on the same playing field. And [they assume that] we're all, you know, 'stupid.' But instead it just comes down to not having had the same resources. (Hispanic woman, first-generation college student, low-math switcher, biology to sociology)

> The majority of people of color coming to a university, we don't come from money. We don't come from [an upscale high school]. [We don't have] tutors or parents who've gone to college who know what it's like and are going to provide the resources to get there. Or to be able to help you with your homework. My family hasn't been able to help me since elementary school. So, a lack of resources, definitely, definitely [makes a difference]. (Hispanic woman, first-generation college student, low-math switcher, biology to communication)

> Not having that preparation, because …being [the first in my family] to attend university, and foreign too, my parents weren't able to pass me anything —any skills, or anything like that. My dad can fix cars and stuff, but students whose parents took them to museums when they were younger, parents that took them out to see things, they have more—not culture—but just [awareness]. (African-American woman, first-generation college student, low-math engineering persister)

The preoccupations of students from low-income backgrounds could be very different from peers who enjoyed the advantages of rigorous high school preparation and sufficient family income to support their academic success. Disadvantaged students in STEM majors were aware of the uneven field on which they played and were being judged. These students *struggled against internalizing* the stigma of deficient preparation and competence that they sensed was held by other students and by sometimes faculty, as well. An Asian woman, for example, explained how her and her family's financial struggles had direct bearing on her classroom performance, by compromising her ability to access the books and resources she needed to be successful in class. One means of overcoming the stigma of judgment by peers was, as described by an African-American man studying biology, actively trying to make connections with other students and to "be social" despite minority status. Asserting one's competence and appealing to faculty to enforce equal treatment among students were additional strategies used by two Muslim women who had both faced stigma as a result of wearing head scarves. Chapter 12 provides more elaboration about strategies that students of color used to manage and, ultimately, overcome these challenges to persist in STEM programs.

## Belonging Issues Specific to Women

This section details belonging issues that were described by women in our interview sample ($n$ = 174). While a substantial number of women reported a positive climate and sense of belonging in their STEM programs (23%; 41), belonging-related gender issues were of concern to well over half (62%) of all the women. Among those who switched out of STEM, women overall had more belonging issues that contributed to a decision to switch (56%; 34 of 61) than did their male peers (46%; 16 of 35). Among women who persisted in STEM, belonging was especially problematic for white women with high-math backgrounds (78%; 21 of 27) and women of color with low-math background (88%; 21 of 24).

In some ways, the gendered problems of belonging in STEM described by women were variations on themes previously discussed. For example, for women to feel that they belonged in STEM hinged upon self-assessment of their intellectual competence as STEM majors. This could be undermined by the competitive dynamics among peers as well as their estimates of the risk of "standing out" publicly as incompetent. As explained in greater detail below, the double difficulty of being a woman of color in some STEM programs heightened the self-consciousness and anxiety that many women felt about the possibility of being negatively and publicly judged. Therefore, the programs with the greatest gender disparities (specifically computer science and some engineering programs) generated the highest number of complaints by their female students about the difficulties of belonging.

There were no notable differences between switchers and persisters in the experience of gender-based belonging problems. Rather, the main difference between switchers and persisters *was how belonging issues were handled.* Whether a tacit or overt comment implying that a woman was not competent was internalized as true and abiding, or brushed off as the speaker's hubris, ignorance, or rudeness influenced its consequences for developing a sense of belonging and, thus, decisions about staying or leaving. For example, one student in a materials science program described how her male peers sometimes jokingly ask her to "go make us a sandwich" while they are working on a group project. She explained that these comments did not bother her because, "I'm like, 'That's funny. Woman stereotype. Whatever,' and then we do our work." In contrast, she recalled women in her classes who "were not as strong willed or loud mouthed" and were more easily intimidated and quick to assume that "the guys are so much smarter."

Despite a few references to "cattiness," competition between women, and a preference for working with male peers, it was largely men who were seen as fueling the competitive climate of STEM programs, not other women. Women described developing an aversion to the men who sought to "make them feel inferior" or incompetent, even when this limited their access to peer help and relationships:

> Because it's all male engineers, when you try to find study partners or when you hear that people in your class check their homework together, you [think], 'I don't know if I'd really want to fit into that group.' [Because] they might not be as tolerant of a woman, especially if I don't know the answer to all the questions. (White woman, switcher, engineering to psychology)

For some female persisters, avoiding male peers was a means of protecting their self-confidence in competitive classroom cultures:

> In STEM, I'm a lot less apt to raise my hand in class or to speak out—unless it's something that I'm very sure I know a lot about so I can back it up. Because always in STEM fields, people are trying to catch you on something you don't know, like, 'Oh, that sounds right, but, no, you're wrong.' You know? And that's discouraging. So I don't speak out as much, and I'm more quiet. I'm not as apt to form study groups because I'm afraid to look stupid… And in STEM, I never want to be partnered with a guy. That would be horrible. That sounds sad, but that would be really horrible. (Native American woman, low-math physiology persister)

> The very competitive nature of [biology] is why most of my friends are not biology majors. And whether that's part of all of our personalities that brought us to science, or it's something they encourage, I'm not really sure. But there is a feeling after every test that everyone wants to know that they did better than the person sitting next to them. I think men thrive off of that competition more than women do. For me the competitive nature of it never made me really want to work harder. It just made me want to keep everything more private…It made me have less friends in the sciences and more outside of the sciences. (Hispanic woman, biology persister)

While avoiding male peers in STEM programs may have provided protection for women's confidence, it may have also limited their access to peer support and contributed to the perpetuation of gender segregation in male-dominated disciplines.

Although most belonging problems described by female students concerned interactions with their male peers, a few pertained to faculty. One white woman who switched from biology to European studies described a single but particularly negative "patriarchal" interaction with a male faculty member that led her to feel she could not continue in STEM. A Hispanic woman experienced a similar problem. Although she persisted in STEM, she could not shake her lingering resentment at being stigmatized as an underperformer and continually forced to assert her rightful place in STEM:

> [I had a problem with a] chemistry professor…he's very high up in the department…I sat in the front row every day so he definitely saw me on a morning-to-morning basis. And I had done really well on my first test, which meant the world to me. After the second test he came up to me in a lab and he was like, 'You.' He could never address me by my name. That was my first issue with this. I have a name. And he was like, 'You didn't do well on your last exam.' And, for me, I did fine. I think this was a matter of [him trying to] scare the rest of the lab by picking on the one person that he could find. And then he made me come into

> his office to get extra help, though I was doing fine in the class. And when I went into office hours, it wasn't about help. It felt like an oral exam, like, 'Describe this compound. Describe this structure.' I thought I was coming in for extra help, not to see what I know. He's notorious for believing that women don't belong in science and he was like, 'Well I don't see you in class.' But I've been in the front row every single day. Like, 'You've clearly seen me.' [I had] this feeling that I had to defend myself constantly, that I belonged to be there, that I was doing well in the class...It still bothers me. I passed the class. I did well. It was the end of my chemistry minor, but I have not let go of feeling as though I've done everything to deserve to be here and I shouldn't have to defend it on a daily basis. (Hispanic woman, low-math biology persister)

It is worth underscoring how this overtly sexist (but happily rare) faculty member operationalized his beliefs that women did not belong in science though his interactions with this student. In class, he erased her presence by not acknowledging that he had seen her at the front and when she was ordered to come to office hours for help that she did not seek or need, he contested her content understanding and competence. While she withstood this biased treatment and persisted in biology, other women likely realized his prophecy by switching out.

That said, having women as instructors in a program did not necessarily mean that their female students felt more connected. A woman who switched from physics to anthropology explained that, even though there were many female faculty in her department, they were difficult to relate to:

> The physics department is over half women and still, for some reason, it feels like there's this barrier between where I am and where [the faculty] are. It doesn't feel cooperative. (Hispanic woman, high-math switcher, physics to anthropology)

As this student's experience suggests, and previous research has shown, female faculty in the physical sciences, especially older professors who have had to struggle to survive among hostile male colleagues, can be disinclined to provide targeted support to female students (Etzkowitz, Kemelgor, & Uzzi, 2000).

Consistent with the work of Margolis and Fisher (2003), another category of belonging problems identified by women concerned male peers' having greater apparent preparation and familiarity with relevant content through their informal interests. In engineering and computer science, for example, men could come into programs with experience gained through building things and gaming. One woman found the resulting "jargon" and "masculine language" that was embedded in the examples used in her engineering class put her at a disadvantage:

> I know that motors are important and engines are important and we need to learn how engines work and that's awesome. But there are engines and things that are not cars [which are always used as examples in class]. [A lot of guys] and their fathers have worked on [cars] for their entire lives, but I've never seen under the hood of one because I just wasn't raised that way. Because of that terminology, it makes me feel like the males have this background experience just from life that I don't have. That has set me 20 feet behind the starting line. (White woman, high-math engineering persister)

Similarly, another female persister recounted that the experience-based advantages of many male students were particularly evident in the freshmen year of her engineering program. This was when the disparities in men's *informal* STEM experiences translated into advantages in the classroom:

> My freshman year I wasn't up to par with a lot of the students. Guys are more prepared for the classes, especially the introductory engineering classes [that involve] building things. Guys have been building things since they were young. And we [women] kind of come in there not really knowing what to do or what to expect…It gives the feeling like you're...not inadequate, just not at the same level as others. I feel that makes girls sometimes want to give up, because it's harder for us sometimes to get these skills that the men already have. I mean there are some girls that already have them of course. I'm not saying that all girls are like that, but a lot of the girls. (African-American woman, low-math engineering persister)

Several women who persisted in engineering at one of the sample institutions shared stories of being assigned stereotypically gendered roles, such as sewing and paperwork, by the men in a group project for an entry-level engineering course. As a result, the women's opportunities to learn new skills were preempted by their male peers based on their assertion of greater competence:

> I feel like the guys do think they're smarter than we are and they can do better… [We have one class] where we do a semester-long project…In my group I was the only girl. One girl and all these guys, and basically what I did for that project was I wrote the papers. That's what they thought was my expertise. Honestly, [it's true that] I didn't really know that much coming into the class, but I wasn't allowed to do other things. Like, if I was trying to solder something, rather than letting me learn, they were like, 'Oh don't worry, I've got that. I can take care of that.' Or, 'I can do the programming,' and stuff like that. It's the same way with most of the girls I know. We talk about that class, in general, like, 'Oh, all I did was the skirt for [what we built].' 'All I did was the paper.' 'All I did was the PowerPoint.' Unless you were super smart and you could prove it. There were a few [female students] who are super smart and can prove it. (African-American woman, engineering persister)

A classmate of the student quoted above described how the time period in which she struggled to persist in her major coincided with the time it took for her male peers "to grow up"—an insight that was also reported in the original study. While, as freshmen, many men held received adolescent notions and stereotypes about women, the situation improved over time as young men matured, encountered more and more female peers, and came to *recognize their competence*:

> As you keep taking classes, you start to gain skills and so I guess they respect you more. And they see more women because, again, there are a lot of women that do have these skills. So, they see more women and I guess they grow up and lose those notions…So by the end, I feel like we had a good or a respectable relationship with each other. I didn't really have another group project experience until the next year. And that one went completely fine—no issues at all. So I guess it was just that first semester. (African-American woman, engineering persister)

Stereotypes could, however, be detrimental in both directions, as was the case with one woman who initially assumed that all of the men in her computer science program entered with long-term experience with computing until she realized that this assumption was just not true:

> I remember when I first got into computer science and thinking, 'Oh, man, these are a bunch of boys who've been coding since they were 10! And here I am, I've never coded before.' And this, of course isn't true. You talk to some of the guys and they're just starting, as well.

> But I think that mentality is still there, of, you know, '[Women feel] less qualified to do this. It's going to take forever for me to catch up with these guys who know computers inside and out.' (White woman, low-math computer science persister)

There were, of course, variations evident in the interview data about how male students treated their female peers, with many men (especially in later years) described as very supportive. However, one female switcher drew a distinction between being "socially accepted" by male peers in her STEM program—with some lauding her as "cool" since she was "the only girl"—and being "intellectually respected" with mathematics competency being the litmus test for assessing competence. Again, this observation matches similar observations by women in the original study. Some male students were also vulnerable to the stigma of perceived incompetence and found themselves sidelined in group projects.

As with the disproportionate representation of students of color in STEM majors compared with white students, the skewed gender ratio in many STEM programs worked against women developing a sense of belonging. Challenges to women's competence and the risks of stigmatization were reported to be rarer in programs with near gender parity, such as bio-engineering. Computer science and other engineering programs, notably mechanical engineering, where the gender ratio could be as low as nine men to one woman, women reported it was difficult to belong. The small number of women in these programs experienced "extra pressure" and stress that heightened the consequences of not being taken seriously by male peers. Logically, as one women pursuing a computer science degree pointed out, with a substantially greater number of men in comparison to women in a program, there is a high statistical probability that among the male students someone will know the answer to a particular question. Thus, a presumption of greater competence among male students is partly an artifact of their greater numerical representation. This gives male students control over setting the *informal terms of performance and productivity* in their classes. For example, several women switchers described how men in their major worked faster and with less deliberation than the women—as if all tasks were a race to a finish line and all solutions derived from straightforward analyses that did not benefit from discussion:

> It was difficult to express myself in those groups [of male STEM students] because I don't want to be wrong … and I don't want to sound like an idiot when I speak up. I felt like I never contributed much...But I always out-performed them on tests because when I could sit down and work on things and go see the teacher, I was the one actually figuring it out. Whereas they were speeding [through it], throwing out ideas, going really fast. I could sit back and really think it through. (Multi-racial woman, switcher, mathematics to management)

> A lot of women are more social beings, more critical thinking, and [focus] more on broader aspects as opposed to just problem-solution. Women like to be heard, you know? Not, 'Give me a solution right away.' So, I feel like it's important for a woman to fit in in that way. (Hispanic woman, low-math switcher, biology to communication)

Even when female STEM students felt a secure sense of belonging in their academic program, career-related interactions in internships and job interviews could still challenge their assumption that they belonged in a STEM field. A female computer science student, for example, described a summer internship in which she was assigned to a coding group of "men's men," who were largely hostile to her presence:

> They were extremely manly. And they didn't hire me. Their functional manager hired me. I think that if it had been up to them, they definitely would not have hired a girl to infiltrate their group. It was clear that I was an outsider the whole time. This was exacerbated by the fact that I was an intern, and also because a lot of them were ex-military and a couple of them played football...It was clear that I was not really welcome in their group. At the end of the summer, I felt kind of like a daughter maybe to some of them, like that kind of relationship maybe. Which makes sense because I was at least 20 years younger than most of them. But I think a boy would have fit in much better with their group than me. (White woman, computer science persister)

In a similar instance, a female engineering student recalled being overtly confronted in a job interview with the fact that the employer hires mostly men. She was put off by the exchange with the interviewer, assuming as a result that the company would have a hostile work climate:

> I had an interview recently with a mechanical engineering company. And one of the questions he asked me was, 'Well, we mostly hire lot of guys and predominantly this field is dominated by men. How would you deal with that?' And I said how I would deal with it, but at the same time I realized that I don't know that I want to work for a company that says it predominantly hires men…I know that the field is growing and [there are more] women engineers even in mechanical, so it was kind of off-putting for me. (Multi-racial woman, chemical engineering persister)

## Conclusion: Climate, Belonging, and STEM Persistence

Problems with belonging and the climate of STEM programs affected a variety of students who participated in this study, but in different ways. Students experienced the most problems with belonging and climate in engineering and computer science programs and also in classes that included a high proportion of students who anticipated medical or veterinary school entry. Belonging problems were most prevalent among women and students of color, especially those who were first-generation college students or came from low-income families.

There were also discernible differences in the type of belonging problems that students reported, depending upon their math readiness. Students that entered with poorer math preparation from high school were more likely to have belonging problems that were rooted in low self-assessment of their competence in science and mathematics, competitive STEM classroom climates, and difficulties in connecting to other students, most importantly gaining access to peer academic support. Students with high-math backgrounds were not immune to the negative effects of competitive classroom climates or to being excluded from study groups. However, they were also more likely to have their sense of belonging within STEM disciplines confounded by aspects of curriculum and instruction, especially issues of connectedness. Specifically, these were whether a STEM program: connected students to each other, to instructors, and to opportunities; made intellectual connections; and, applied concepts in ways that were real-world relevant.

Despite the varieties of belonging-related problems that students in our interview sample experienced, they were all problems that were generated *during STEM programs*. Like other researchers (e.g., Walton & Cohen, 2007; Wilson et al., 2015), we found evidence to support the idea that a sense of belonging within a STEM discipline is not something immutable that students bring into classrooms, or that is inherent to STEM education and cannot be changed. Rather, it is malleable and shaped by contexts, interactions, and resources that are largely *under the control* of STEM departments. Concrete, yet simple, instructor and departmental practices, such as assessment policies, program design, group-work policies, and the provision of academic support, can positively influence students' sense of belonging and their commitment to persist in STEM majors. Instructors and STEM program representatives also need to be aware of, and ready to intercede in, *informal peer interactions*—including status competitions, stigmatization, and exclusion from peer support—that occur among students. Negative peer dynamics systematically and unfairly disadvantage some students over others and promote tacit versions of *professional socialization* that might not be consistent with what is intended by STEM faculty and their departments.

## References


Aronson, J., Fried, C. B., & Good, C. (2002). Reducing the effects of stereotype threat on african american college students by shaping theories of intelligence. *Journal of Experimental Social Psychology, 38*(2), 113–125.

Astin, A. W. (1993). *What matters in college? Four critical years revisited.* San Francisco, CA: Jossey-Bass.

Bean, J. P. (1980). Dropouts and turnover: The synthesis and test of a causal model of student attrition. *Research in Higher Education, 12*(2), 155–187.

Benbow, R. J., & Vivyan, E. (2016). *Gender and belonging in undergraduate computer science: A comparative case study of student experiences in gateway courses* (WCER Working Paper No. 2016-2).

Berger, J. B., & Lyon, S. C. (2005). Past to present: A historical look at retention. In A. Seidman (Ed.), *College student retention: Formula for student success* (pp. 1–30). Westport, CT: Praeger Publishers.

Blackwell, L. S., Trzesniewski, K. H., & Dweck, C. S. (2007). Implicit theories of intelligence predict achievement across an adolescent transition: A longitudinal study and an intervention. *Child Development, 78*(1), 246–263.

Bonous-Hammarth, M. (2000). Pathways to success: Affirming opportunities for science, mathematics, and engineering majors. *The Journal of Negro Education, 69*(1/2), 92–111.

Brown, E. R., Thoman, D. B., Smith, J. L., & Diekman, A. B. (2015). Closing the communal gap: The importance of communal affordances in science career motivation. *Journal of Applied Social Psychology, 45*(12), 662–673.

Cech, E., Rubineau, B., Silbey, S., & Seron, C. (2011). Professional role confidence and gendered persistence in engineering. *American Sociological Review, 76*(5), 641–666.

Cech, E. A. (2014). Culture of disengagement in engineering education? *Science, Technology, & Human Values, 39*(1), 42–72.

Chang, M. J., Cerna, O., Han, J., & Saenz, V. (2008). The contradictory roles of institutional status in retaining underrepresented minorities in biomedical and behavioral science majors. *The Review of Higher Education, 31*(4), 433–464.

Chang, M. J., Eagan, M. K., Lin, M. L., & Hurtado, S. (2011). Considering the impact of racial stigmas and science identity: Persistence among biomedical and behavioral science aspirants. *Journal of Higher Education, 82*(5), 564–596.

Chang, M. J., Sharkness, J., Hurtado, S., & Newman, C. B. (2014). What matters in college for retaining aspiring scientists and engineers from underrepresented racial groups. *Journal of Research in Science Teaching, 51*(5), 555–580.

Chen, X. (2015). STEM attrition among high-performing college students: Scope and potential causes. *Journal of Technology and Science Education, 5*(1), 41–59.

Christe, B. (2013). The importance of faculty-student connections in STEM disciplines: A literature review. *Journal of STEM Education: Innovations and Research, 14*(3), 22.

Clark, E. K., Fuesting, M. A., & Diekman, A. B. (2016). Enhancing interest in science: Exemplars as cues to communal affordances of science. *Journal of Applied Social Psychology, 46*(11), 641–654.

Demetriou, C., & Schmitz-Sciborski, A. (2011). Integration, motivation, strengths and optimism: Retention theories past, present and future. In R. Hayes (Ed.), *Proceedings of the 7th national symposium on student retention*, 2011, Charleston (pp. 300–312). Norman, OK: The University of Oklahoma.

Diekman, A. B., Brown, E. R., Johnston, A. M., & Clark, E. K. (2010). Seeking congruity between goals and roles: A new look at why women opt out of science, technology, engineering, and mathematics careers. *Psychological Science, 21*(8), 1051–1057.

Diekman, A. B., Clark, E. K., Johnston, A. M., Brown, E. R., & Steinberg, M. (2011). Malleability in communal goals and beliefs influences attraction to stem careers: Evidence for a goal congruity perspective. *Journal of Personality and Social Psychology, 101*(5), 902–918.

Espinosa, L. L. (2011). Pipelines and pathways: Women of color in undergraduate STEM majors and the college experiences that contribute to persistence. *Harvard Educational Review, 81*(2), 209.

Etzkowitz, H., Kemelgor, C., & Uzzi, B. (2000). *Athena unbound: The advancement of women in science and technology*. Cambridge, England: Cambridge University Press.

Geisinger, B. N., & Raman, D. R. (2013). Why they leave: Understanding student attrition from engineering majors. *International Journal of Engineering Education, 29*(4), 914–925.

Hausmann, L. R. M., Ye, F., Schofield, J. W., & Woods, R. L. (2009). Sense of belonging and persistence in white and African-American first-year students. *Research in Higher Education, 50*(7), 649–669.

Hunter, A.-B., Laursen, S. L., & Seymour, E. (2006). Becoming a scientist: The role of undergraduate research in students' cognitive, personal, and professional development. *Science Education, 91*(1), 36–74.

Hurtado, S., & Carter, D. F. (1997). Effects of college transition and perceptions of the campus racial climate on Latino college students' sense of belonging. *Sociology of Education, 70*(4), 324–345.

Hurtado, S., Han, J. C., Sáenz, V. B., Espinosa, L. L., Cabrera, N. L., & Cerna, O. S. (2007). predicting transition and adjustment to college: Biomedical and behavioral science aspirants' and minority students' first year of college. *Research in Higher Education, 48*(7), 841–887.

Hurtado, S., Eagan, M. K., Tran, M. C., Newman, C. B., Chang, M. J., & Velasco, P. (2011). "We do science here": Underrepresented students' interactions with faculty in different college contexts. *Journal of Social Issues, 67*(3), 553–579.

King, B. (2015). Changing college majors: Does it happen more in STEM and do grades matter? *Journal of College Science Teaching, 44*(3), 44–51.

Koenig, K., Schen, M., Edwards, M., & Bao, L. (2012). Addressing STEM retention through a scientific thought and methods course. *Journal of College Science Teaching, 41*(4), 23–29.

Laursen, S. L., Hunter, A.-B., Seymour, E., Thiry, H., & Melton, G. (2010). *Undergraduate research in the sciences: Engaging students in real science*. San Francisco, CA: Jossey Bass.

Locks, A. M., Hurtado, S., Bowman, N. A., & Oseguera, L. (2008). Extending notions of campus climate and diversity to students' transition to college. *The Review of Higher Education, 31*(3), 257–285.

Margolis, J., & Fisher, A. (2003). *Unlocking the clubhouse: Women in computing*. Cambridge, MA: MIT press.

McGee, E. O. (2016). Devalued black and Latino racial identities: A by-product of STEM college culture? *American Journal of Educational Research, 53*(6), 1626–1662.

McGee, E. O., & Martin, D. B. (2011). "You would not believe what I have to go through to prove my intellectual value!": Stereotype management among academically successful black mathematics and engineering students. *American Education Research Journal, 48*(6), 1347–1389.

Mendoza-Denton, R., & Page-Gould, E. (2008). Can cross-group friendships influence minority students' well-being at historically white universities? *Psychological Science, 19*, 933–939.

Montfort, D., Brown, S., & Pollock, D. (2009). An investigation of students' conceptual understanding in related sophomore to graduate-level engineering and mechanics courses. *Journal of Engineering Education, 98*(2), 111–129.

Museus, S. D., Yi, V., & Saelua, N. (2017). The impact of culturally engaging campus environments on sense of belonging. *The Review of Higher Education, 40*(2), 187–215.

Nora, A., & Cabrera, A. F. (1996). The role of perceptions of prejudice and discrimination on the adjustment of minority students to college. *The Journal of Higher Education, 67*(2), 119–148.

Ong, M., Wright, C., Espinosa, L., & Orfield, G. (2011). Inside the double bind: A synthesis of empirical research on undergraduate and graduate women of color in science, technology, engineering, and mathematics. *Harvard Educational Review, 81*(2), 172–209.

Ost, B. (2010). The role of peers and grades in determining major persistence in the sciences. *Economics of Education Review, 29*(6), 923–934.

Rask, K. (2010). Attrition in STEM fields at a liberal arts college: The importance of grades and pre-collegiate preferences. *Economics of Education Review, 29*(6), 892–900.

Seymour, E., & Hewitt, N. (1997). *Talking about leaving: Why undergraduates leave the sciences*. Boulder, CO: Westview Press.

Seymour, E., Hunter, A.-B., Laursen, S., & DeAntoni, T. (2004). Establishing the benefits of research experiences for undergraduates in the sciences: First findings from a three-year study. *Science Education, 88*(4), 493–534.

Smedley, B. D., Myers, H. F., & Harrell, S. P. (1993). Minority-status stresses and the college adjustment of ethnic minority freshmen. *The Journal of Higher Education, 64*(4), 434–452.

Smith, J. L., Cech, E., Metz, A., Huntoon, M., & Moyer, C. (2014). Giving back or giving up: Native American student experiences in science and engineering. *Cultural Diversity & Ethnic Minority Psychology, 20*(3), 413–429.

Smith, M. K., & Wood, W. B. (2016). Teaching genetics: Past, present, and future. *Genetics, 204*(1), 5–10.

Spencer, S. J., Logel, C., & Davies, P. G. (2016). Stereotype threat. *Annual Review of Psychology, 67*, 415–437.

Steele, C. M. (1997). A threat in the air: How stereotypes shape intellectual identity and performance. *American Psychologist, 52*, 613–629.

Steele, C. M., & Aronson, J. (1995). Stereotype threat and the intellectual test performance of African-Americans. *Journal of Personality and Social Psychology, 69*(5), 797.

Stetzer, M. R., van Kampen, P., Shaffer, P. S., & McDermott, L. C. (2013). New insights into student understanding of complete circuits and the conservation of current. *American Journal of Physics, 81*(2), 134.

Streveler, R. A., Litzinger, T. A., Miller, R. L., & Steif, P. S. (2008). Learning conceptual knowledge in the engineering sciences: Overview and future research directions. *Journal of Engineering Education, 97*(3), 279–294.

Talanquer, V. (2014). DBER and STEM education reform: Are we up to the challenge? *Journal of Research in Science Teaching, 51*(6), 809–819.

Thiry, H. (2017). *The importance of community, belonging and support: Lessons learned from a decade of research on Hispanic retention in STEM*. Computing Alliance of Hispanic Serving Institutions. Draft White Paper.

Tinto, V. (1975). Dropouts from higher education: A theoretical synthesis of recent literature. *Review of Educational Research, 45*, 89–125.

Walton, G. M., & Carr, P. B. (2012). Social belonging and the motivation and intellectual achievement of negatively stereotyped students. In M. Inzlicht & T. Schmader (Eds.), *Stereotype threat: Theory, processes, and application* (pp. 89–106). New York, NY: Oxford University Press.

Walton, G. M., & Cohen, G. L. (2007). A question of belonging: Race, social fit, and achievement. *Journal of Personality and Social Psychology, 92*(1), 82–96.

Wilson, D., Jones, D., Bocell, F., Crawford, J., Kim, M. J., Veilleux, N., … Plett, M. (2015). Belonging and academic engagement among undergraduate STEM students: A multi-institutional study. *Research in Higher Education, 56*(7), 750–776.

# Chapter 10
# The Processes and Consequences of Switching, Including the Loss of High-Performing STEM Majors

**Dana G. Holland, Raquel P. Harper, Anne-Barrie Hunter, Timothy J. Weston, and Elaine Seymour**

## Variations in the Switching Process

Students' departures from STEM programs entailed decisions to leave. Yet prior to making their decisions, typically, there were lengthy processes involving personal reflection, interactions with others, and sometimes the experience of fraught and distressed emotions. The process also often included imagining and appraising future career pathways associated with different degree programs.

An important point of variation among switchers in our interview sample was the degree to which *struggle*—with grades, comprehension, STEM program climate and access to academic help, advice, or other support—figured into the switching

D. G. Holland (✉) · R. P. Harper · A.-B. Hunter
Ethnography and Evaluation Research (E&ER), University of Colorado Boulder, Boulder, CO, USA

Center for STEM Learning, University of Colorado Boulder, Boulder, CO, USA
e-mail: Dana.Holland@Colorado.EDU; Raquel.Harper@Colorado.EDU; abhunter@colorado.edu

T. J. Weston
Ethnography and Evaluation Research (E&ER), University of Colorado Boulder, Boulder, CO, USA

National Center for Women and Information Technology (NCWIT), University of Colorado Boulder, Boulder, CO, USA
e-mail: westont@colorado.edu

E. Seymour
Ethnography and Evaluation Research (E&ER), University of Colorado Boulder, Boulder, CO, USA
e-mail: Elaine.Seymour@colorado.edu



E. Seymour, A.-B. Hunter (eds.), *Talking about Leaving Revisited*,
https://doi.org/10.1007/978-3-030-25304-2_10

experience. Almost always when students struggled in their STEM programs, they also suffered emotionally. Ten percent of all students participating in this study (both persisters and switchers) reported chronic levels of stress, depression, fear, anxiety and/or guilt and self-blame that were severe enough to produce clinical diagnoses or worsen existing health conditions. One-third of switchers described struggling with depression, high levels of stress, chronic anxiety, overwhelm, feelings of guilt, regret, shame, and self-blame for "failing," or being stigmatized as someone "who couldn't hack it." As is illustrated in Table 10.1, the same kinds of negative personal consequences were reported by about 10% of persisters but in smaller proportions than switchers.

Also as indicated in Table 10.1, the personal costs of STEM education experiences are broadly of four types, the most common of which is a sharp drop in students' entering levels of self-confidence in their ability to learn and succeed. For women, this effect was noted in early research on reasons for the disproportionate loss of women from the sciences. (Reviewed in TAL, pp. 11, and pp. 268–274, and discussed as a finding of that study in TAL Chap. 5.) As we described in Chap. 3, loss of confidence and feeling discouraged or demoralized by the largely unaccustomed experience of low grades—or perceived low grades—was one of the top five contributors to switching decisions for all student groups. Experienced by 79% of switchers and 44% of persisters, loss of confidence in face of low grades contributes to 61% of switching decisions. As also reported in Chap. 7, 43% of switchers and 18% of persisters described how losses of confidence were created by weed-out experiences, with greatest losses occurring among women, half of whom were women of color. Our findings echo those of Bressoud and colleagues about the nationwide impact of Calculus 1 courses, namely, "how effective this course is in destroying (incoming) confidence" (Bressoud, Mesa, & Rasmussen, 2015, p. 182). This effect was particularly marked among women: even with final grades of As and Bs, twice as many women as men with the same grades abandoned the idea of continuing to Calculus 2. Descriptions of lost confidence are often coupled with loss of self-esteem ("feeling stupid all the time") and loss of hope:

> There were a couple of things that contributed most to my deciding to move. It was the diminished esteem after getting those terrible grades—things I had never got in my life before. In O-Chem I got a B and in Chem a C-minus. And I thought, 'They're just going to continue to decrease.' Just the whole weed-out process was really so rough on me. (White woman, switcher from chemistry to psychology)

A similar proportion of switchers (viz. 34%) and 8% of persisters reported depression, high levels of stress, chronic anxiety, feeling lost and overwhelmed, or living

**Table 10.1** The personal costs of experiences in STEM majors as percentages of STEM switchers and persisters interviewed

| Negative consequence | Persisters ($N = 96$) (%) | Switchers ($N = 250$) (%) |
|---|---|---|
| Lost confidence | 7 | 34 |
| Stress, depression, anxiety, fearfulness | 8 | 34 |
| Feeling bad about self | 3 | 23 |
| Mental or physical health problems | 7 | 15 |

with intolerable fear. These include: the fear of admitting that one cannot cope with the work, the fear that this will only get worse, and dread of going to class. These outcomes are not reported as direct causes of switching and relocation, but are presented as the emotional context within which such changes of direction can occur:

> I got frightened about whether I was doing all that I needed to do. So I left the class because I didn't want to fail. (African-American man, switcher from life sciences to psychology)

The third group of accounts reference outcomes from which students may or may not eventually recover. They precede decisions to switch or relocate, and also linger beyond them. Reported by 23% of switchers and 3% of persisters, these are: feelings of guilt, failure, or regret; shame, self-blame for what has happened, and feeling stigmatized as someone who failed or "couldn't make it"; being tired of feeling bad about one's self, miserable, or too paralyzed to act:

> You look at the other students who are excelling in the course and you suck compared to them. It hurts so much that you feel like less of a person. And it's not that they are that much smarter than you, but they've been better prepared. I don't appreciate being challenged so much that it makes me feel that I can't succeed no matter what I do. (White woman, environmental science persister)

Finally, chronic levels of stress, depression, fear, anxiety and/or guilt, shame, stigma, and self-blame took their toll on 15% of switchers and 7% of persisters in the form of mental or physical health problems, including exacerbation of existing health conditions and clinical diagnoses of depression or anxiety. One woman described the outcome of struggles rooted in her transition from high school, combined with a lack of sensitivity by her STEM department that led to her switching decision:

> I felt pretty bad about myself. I graduated high school with over a 4.0 and then I got to college and I wasn't doing nearly as well as I thought I would be. So I felt really bad about myself. I thought I was stupid. I thought that I was going to fail. I thought that I was going to disappoint my family. I thought that I wasn't living up to my full potential and I just didn't really understand why it was happening… I talked to my doctor and he actually diagnosed me with anxiety. So that kind of helped after I started some medication for that. And then I talked a little bit to my academic advisor here and he wasn't very helpful about that. He said, 'Well, that's typical.' Not very personal, just kind of shove you off a little bit. I understand that they're very busy, but it just wasn't what I had expected. (White woman, switcher, chemistry to family studies)

Emotions could be especially turbulent when students' long-held career goals were undermined as part of the switching process:

> It was upsetting because my whole life, I anticipated being a doctor. I fell in love with the medical field, [but] I obviously didn't plan it accordingly when I came to the pre-med track. (Hispanic woman, switcher, biology to communication)

Struggling, sometimes with only one course, could cause students to panic and provoke fear of failure that hastened the switching process, regardless of actual performance. For some, emotional distress was partly due to the disconnection between the experience of academic struggle in a STEM higher education program and their previous experience of having been a high-performing student in high school.

For other students, academic struggle or performance problems did not figure so prominently in their switching process. However, stress reactions at varying levels of severity were endemic among students in our interview sample. Switching decisions that emerge entirely rationally through learning, maturation, or purely pragmatic thinking were, as discussed below, the exception more than the rule.

Nationwide, campus student services have reported high and increasing rates of students seeking help to address stress and anxiety. The American Psychological Association reports that, "Since the 1990s, university and college counseling centers have been experiencing a shift in the needs of students seeking counseling services from developmental and informational needs, to psychological problems," and cites two surveys showing the rapidly growing need for college mental health services (APA, 2019). The 2014 National Survey of College Counseling Centers reported that, "52% of their clients had severe psychological problems, an increase from 44% in 2013. A majority of survey respondents noted increases over the past five years of anxiety disorders, crises requiring immediate response, psychiatric medication issues and clinical depression" (APA, 2019). The second report from a 2016 student survey by the American College Health Association, found that, within the previous year, 53% of students surveyed reported feeling that things were hopeless and 39% reported feeling so depressed that it was difficult to function" (APA, 2019). In an age of rapid technological change and globalism, psychologists and other researchers link reports of students' increased anxiety and angst to the constant pressure to achieve to ever-higher standards and the "toxic" outcome of an affluent culture (Cashman & Twaite, 2009; Curran & Hill, 2017; Luthar, 2003; Luthar & Sexton, 2004; Twenge, 2014; Verhaeghe, 2014).

## *Overcoming Stigma in the Switching Process*

Even students whose switches did not involve academic struggle still often grappled with disconcerting stigma and stereotypes about STEM and non-STEM fields during the process. In particular, they had to work through stereotypes about non-STEM fields having poor career pathway opportunities. Students dealt with these stereotypes both personally through reflection about their own beliefs and views about themselves and their futures, and interactionally in terms of others' views about them and their choices. This student, for example, had been interested in writing before entering university, but pursued physics because of its presumptively better employment prospects. Over the course of his first few semesters, he gained insights and became more confident that he could secure a viable career in writing through a switch to journalism:

> I'd been considering going into writing before I even stepped onto [this campus]. I went into physics because, at that time, I was overly concerned about being able to find a job. I figured, 'I know I'm going to find a job and make a living in physics. I don't know about writing.' And it was actually because I got more confident that I was able to decide, 'Alright,

> I'm going to do what I enjoy the most and I'm going to move into writing.' (White man, switcher, physics to journalism)

Another student, who switched from math to environmental design, described herself as having been "super artsy" in high school, but dismissed becoming an art major when she entered university as a "waste of money" on a field that "goes nowhere." In discovering environmental design she realized that it "could be legitimate" as a practical version of the arts program.

*Legitimation* is a key theme and legitimizing non-STEM academic pathways could be challenging for students who switched due to the prestige associated with STEM disciplines and careers. For these students, switching involved working out a conceptualization and pathway plan for a non-STEM field that included validation of its worth. This student, for example, recounted how bias against non-STEM pursuits in his high school propelled him into engineering prior to his eventual switch to psychology:

> Psychology, I love it! That's my passion! [But, before] I thought it was a dead end, you know? It was just, 'Oh, everybody does that. Losers do that.' Like, 'Real people, worthwhile people do engineering.' You know? ... In high school, people who were good at math and science were encouraged. They were [told], 'You're awesome.' But if you were good at English, who cares? Anybody can be good at English. (White man, switcher, engineering to psychology)

As part of his switching process, he had formulated a non-STEM pathway plan that justified his switch out of engineering:

> I'm a very success-oriented person. I'm extremely motivated. I anticipate getting into a top graduate school for psychology, which is saying something in light of the fact that I lost a year, essentially. (White man, switcher, engineering to psychology)

Before switching out of biology and away from a dentistry career pathway, another student similarly had to rework her own latent biases against the "dumb major" of Spanish, despite her abiding interest in the subject and its alignment with a future she desired:

> I always liked Spanish, but I thought, 'No, that's such a dumb major. Why would you major in something that's how people live their life?' You know, like, people just speak Spanish every day. You don't have to study it. And so I [thought] 'That's so dumb. That's embarrassing to go to school for that.' Although I always loved it...So I was on track for dentistry and [thought] 'But why? I'm not happy.' I love working with kids and I wanted to teach. It was because of pride. I was too proud to give it up that easy. (White woman, switcher, biology to Spanish)

Sometimes the stereotypes that students worked through concerned STEM fields or their STEM programs. This woman, for example, initially majored in physics because of its prestige as well as her talent for the field. She ultimately rejected it, however, because she believed it would not support a career pathway with sufficient human interaction:

> One of the basic self-serving reasons I chose physics initially was because I was good at it and the prestige of eventually being an engineer or being good at that sort of thing. And then I just realized, more and more, that that prestige was only going to take me so far and that I

> needed to find something that I could be happy doing. So, in just working through and just realizing how important human interaction and human relations are to me, [then it seemed] science made less and less sense as a choice. … I realized that prestige, although it's nice and it's tempting, it just wasn't going to be enough for me. I needed to find something that I enjoyed more and that allowed me more connection with people. (White woman, switcher, physics to geography)

Switches out of STEM were therefore at times grounded in student rejection of some stereotype about a STEM field or STEM career pathway. This student, for example, grew concerned about the potential military application of his work should he follow through with his initial plan to pursue mechanical engineering. He considered shifting to environmental engineering, where he could see himself working on issues of global sustainability; ultimately, he settled on religious studies because of its compatibility with his ideals.

> [With engineering] I felt I was crushing my inner idealist and I was just going to play it safe, and get a good job, and live a comfortable life, and then just go to Disneyland with my friends and family and then probably die as a mechanical engineer. Work up through the firms, whatever that may be. I didn't like that future. … I remember reading some old texts in Confucian Analects or the Tao Te Ching, and thinking how remarkably applicable to the problems of contemporary society [they were]. I remember thinking, 'Wow, if people actually understood what this means, then maybe we could focus our engineering efforts on our earthly domains instead of our economic domain, and maybe we could realign to a track that is more sustainable. Maybe I could fulfill goals that I want to fulfill about the environment by exposing a new angle of approach. … I sort of aimed my goal in switching to that and maybe looking at our problems in a different light. (Hispanic man, switcher, engineering to religious studies)

## *The Roles of Others in the Switching Process*

Switching out of STEM is something that individual students do, and indeed the switching process as described above involved students in internal reflection and conversations with themselves over their decisions. For some the switching process also involved interactions and negotiations with others, such as parents, friends, peers, faculty, TAs, advisors, or program administrators. These interactions were sometimes overtly about switching and entailed gaining the advice or assent of former high school teachers or parents, for example. The prestige of STEM and the stigma of some non-STEM fields as impractical could underlie these interactions. This was the case, for example, for the student switching from engineering to religious studies quoted above. He also described having to regularly explain and, at times, defend his choice:

> It's got an undertone of failure, switching from the prestige [associated with] the mighty engineering to, [people thinking], 'Oh, religious studies, what is this? Now you're going to be some scholar prancing around in Egypt, looking at ancient texts, who is ignoring the contemporary needs of the world! What are you going to do? Like, things need to get done these days!' I always feel like that's going to be [everyone's] approach. …With my close friends, they will respect my choice. And [for others], if I have an opportunity to provide

> them with [my] reasoning, the rationale behind my choice, then they love the rationale. (Hispanic man, switcher, engineering to religious studies)

This student's high school teachers had proved to be valuable allies, both in the process of making his switch and in his formulation of a rationale that legitimized his move out of engineering:

> I really got a lot of help from some of my high school teachers. I went back and talked to my AP European history teacher. That was my favorite class at the time, and where I put all my effort in those days. And she said, 'Yeah, you need to get out of engineering. Go find something else and that seems like it is probably going to be in the arts and sciences.' [She said], 'I know that it doesn't have the prestige that you're looking for. It doesn't have the credibility or the official recognition that you're going to be able to achieve with a major in engineering. But if you're good, then you will be successful, and if you like it, you will be successful.' [And she meant] 'successful' not in the terms of money, but in the terms of fulfilling what I want to do. And I [thought], 'Okay, well, that pretty much sums it up. I'll switch.' (Hispanic man, switcher, engineering to religious studies)

In another instance, a student worked through the stigma of leaving a chemistry program and pharmacy pathway by developing a legitimizing view of psychology as "not easy":

> Definitely telling people that I was going into pharmacy was pretty cool. They were very proud or whatever. So I was a little embarrassed to tell people that I switched to psychology when I first did it. But now I'm just like, 'Whatever, this is awesome! I love it!' I'm writing an honors thesis, and it's really cool. And I'm going to go to grad school. But it was definitely a little hard in the beginning. ... And now I know that [psychology] is not easy. Because doing research is one of the hardest things ever! And if I have to read another journal article, I'm just, I'm done! (Laughing). But, yeah, that's definitely the perception that people have, and I can tell them that it's wrong now. (White woman, switcher, chemistry to psychology)

As we discuss in more detail in Chap. 11, parents, unsurprisingly, sometimes played a prominent role in their children's switching process, helping them identify options and providing feedback and guidance about their decisions. For example, after a lengthy period of reflection and trying alternatives within STEM this student formulated a solid rationale for switching out of physics that could withstand her parents' intermittent pressure for her to pursue something "safer" than geography:

> [In some ways] I wish I had [switched] sooner because there's a lot of things that I could've taken advantage of had I been able to plan a little more. I essentially only have two years in the major that I'm in now. But, at the same time, I'm glad I took so long to make the decision and thought about it so long, and tried so many other things. [Because] by the time I made [the decision], I was sure and I was ready. ... My parents still pressure me to have the stereotypical well-providing, safe job that pays well. Even in moments [when they bring that up], I think back about it, and it's easy for me to remember everything that I went through to get to the point that I'm at now. So I don't regret [that I took the time]. (White woman, switcher, physics to geography)

In contrast to the overt role that other people played in the switching process, sometimes others played covert or unwitting roles. This specifically occurred when the switching process included some negative interaction with faculty, TAs, advisors, or program administrators, who seemed often unaware of their complicity in students' exodus from a STEM program. For example, this student recounted the instrumental role that an advisor had played in confirming to her that she could not handle her STEM program, and, in effect, that there could be no recovery from receiving poor grades in two gateway courses:

> Freshman year chemistry, I got a D, which is terrible. And then calculus, I got a C or a D. [The grades] pushed me away, telling me I couldn't do it. ... And then I talked to my advisor about it. I think, it wasn't their intentions or anything, but they kind of made me feel like I couldn't do it either. You know what I mean? That's what really pushed me away. (Multiracial woman, switcher, biochemistry to human development)

Another student described a lack of academic support and guidance from her engineering program as playing into her switching process:

> I got a 59.7%, so I passed physics, which was really good. But I [thought] 'I passed physics because my parents were willing to pay an extra $70 or $80 bucks a week for a tutor.' I worked really hard, [but] I didn't have any support from the physics department or from the School of Engineering. ... Both the physics professor and the Dean of Engineering gave me advice that was really bizarre and not helpful. So at that point I realized that engineering wasn't what I wanted to do, and I wasn't going to be successful in it. ... I think it's good that I made that decision, because I love geography. I'm so happy in geography. But at the same time, I think I could have been successful at engineering with more support. Because, at least in applied math, I really used the resources that were given to me [and was successful], whereas there just weren't resources to be used [in physics]. (White woman, switcher, engineering to geography)

Aloof, detached, unhelpful, or even hostile behavior on the part of STEM program representatives was, not surprisingly, alienating to students. This bad behavior signaled a negative climate and indicated to students that they did not belong and should leave a STEM program. It figured very prominently in switching processes when students had had dramatically different educational experiences in high school:

> Anytime I went to a teacher in high school it was, 'How can I help you?' Instead here, it's, 'I've got research to do. You're not my problem.' And that's really bothersome because you think it's an institution of higher education and it really was more of, 'I have favorites. You're not one of them. Get lost.' I actually had a chemistry professor tell me, 'You don't belong here.' (White woman, switcher, biology to sociology)

Similarly, when negative interactions with STEM program representatives contrasted with positive interactions and welcoming climates in non-STEM undergraduate courses and programs, this could clinch students' decisions to switch—a phenomenon noted and labeled in the original study as a "push-pull" process:

> Over the summer [after freshman year] that's when I [thought], 'What am I doing? I'm dreading going back to classes.' I had taken a communication class as an elective. And that's when I [thought], 'Oh, what am I doing in the sciences? I really like this. I'm doing well in this. My professor's helping me. They're engaging.' And so, I was like, 'I'm just going to change my major.' So I did. (Asian woman, switcher, biology to communication)

> [In chemistry] I had diminished esteem after getting all those terrible grades, things that I've never gotten before in my life. ... So, it was kinda just like, 'I really don't like this [chemistry program]. What can I do that I would really like?' And I was already in the

> [psychology] lab, so that really helped. ... The [psychology] teachers have been just phenomenal, because they're interested in what they do. And they can actually tell you what it is in English, instead of going a roundabout way. The teachers have been really influential in my staying there. (White woman, switcher, chemistry to psychology)

Along similar lines, competitive dynamics among students in a STEM program could make a similar contribution to the switching process. This woman, for example, recounted "peer pressure" to appear intelligent and tacit competition in terms of the number of AP credits students had earned prior to entering higher education. The competitive climate ultimately pushed her out of engineering and into finance:

> My class had a lot of AP students and I was one of them. [But] I still felt a stigma [from those who had taken] even more AP classes, or they just had an attitude of prestige. It's more the people within it feeling that they're entitled versus what's actually [the case based on performance]. ... Engineering [has] a lot of [students with] attitudes of, 'I'm smarter than you.' They're not going to blatantly say it, but it is, 'I don't want to ask a question 'cause I don't want to look stupid,' and things like that—peer pressure to be this really intelligent person. (White woman, switcher, engineering to finance)

When the switching process involved rejecting competitive dynamics among peers, students were in effect taking themselves out of aspirational contention in a STEM program. This could include rejecting both the terms of competition, and its toll on students' lifestyles, as well as rejecting the general atmosphere:

> Basically, I had organic chemistry and I loved science, and I still love science. My boyfriend is still in his major and I'm always asking him, 'What are you learning.' So [my switching] wasn't because I was disinterested in science and I was still doing really well in my classes. I had a 3.9 GPA when I switched. [My GPA] actually dropped now that I am in non-STEM classes. But it was just the competitiveness. Everyday all I did is, I woke up, went to class and I went home and I studied until 4:00am in the morning and then I would go to sleep and wake up at 6:00am. And it was really rough. (White woman, switcher, biology to political science)

> Living in that honors dorm with people who were always so pressuring and competitive and we're all in exactly the same classes and we're all trying to be doctors, I just was done. I mean when I was done, I was just done. There was no question. (White woman, switcher, biology to human development)

In addition, the negative climate created by competition among peers in a STEM program could spur the switching process, despite students' ongoing interest in a STEM field:

> There are a lot of topics in computer science that are fascinating to me. [But] at the time [I switched] I thought my interests had dipped. Because while I might have liked the topic, there are very evident people who are more attuned to that kind of thinking. So for a little bit I took that too far and convinced myself that I wasn't as interested in it anymore. But I still find a lot of computer science topics really interesting. I talked to my roommate, who is a computer engineer, all the time. I talked to him about his programming projects. Because it's still definitely of interest to me. But at the time, when I was making the switch, I was like, 'Do I really like this?' I was questioning my own opinion on it. (White man, switcher, computer science to economics)

As these student experiences illustrate, the switching process could involve a distinction that students drew between a STEM discipline, which they were still positive about and valued, and their former STEM programs and departments. This point makes it very clear that some switching is *institutionally or programmatically contingent* and that in a different program context at another higher education institution, the student could well have persisted in STEM. This student, for example, explained that her interest in biology and chemistry is ongoing, and that it was instead how the chemistry department at her university implemented its program that was the problem for her:

> I've always loved biology. I've always loved chemistry. But it's the department at [my university] that just didn't fit me. It was not about what we were learning, because I was fascinated constantly. But when you have professors that don't respect your time, that when you ask a question they either refuse to answer it or don't know how to answer it and won't find out, or even bother to write exams that are legible, it's really difficult to commit. … It's kind of a futile effort at that point. So I found sociology. It's got the science components that I've always loved. … I've always been fascinated by people and finding a department that would allow me to analyze that through social institutions was a way of finding science that would work [in a way] that I couldn't find in biology. (White woman, switcher, biology to sociology)

The role played by negative experiences in a particular disciplinary department, as well as that played by other people who are significant to the student, echoes the framing of switching offered in the original study as a process of pushes and pulls that are often held in an uncomfortable tension for some time until resolved by a decision to move (cf., pp. 392–393).

## *Projecting Future Pathways*

Earlier in this section it was pointed out that a key reason why switching is more of a process than a decision was that students typically engaged in reflection that included imagining and envisioning forward in time, to the immediate future in a degree program and longer-term to an anticipated career pathway. It is important to note, however, that this reflective process was delimited by the scope of information that students had acquired about a STEM discipline, their STEM program, and likely career options, which could be incomplete or inaccurate. This student, for example, assumed that majoring in biology would lead to a career "in a lab" and did not seem aware that a biology degree could support alternative futures:

> [My leaving biology] was a little bit of I didn't feel competent, it was a little bit I don't see myself in a career doing this, and a little bit of I'm just not happy doing this. And whether its college, whether it's a career, I'm not going to put myself through it if I don't love it. [Because] if I don't love it, and don't feel very good at it, then why am I doing it? If I pictured myself doing this sort of work every day, would I be happy? And you know, seeing myself in a lab, I said, 'No.' (White woman, switcher, biology to management)

Along similar lines, others assumed that what they were doing in their first- and second-year STEM classes was what they would be doing throughout their academic pathway and beyond:

> I didn't find the material interesting [in terms of] what I truly cared about and it sort of [undermined] my views of engineering, sort of killed it for me. ... I just knew if I could picture myself doing the same stuff as what I was learning in 10 or 15 years...and I didn't want to be doing that, so I just switched majors. (White man, switcher, engineering to construction management)

> By [the end] of my first year...I had struggled so much just trying to keep up, and learn the biology and the chemistry, that there was no way that I could see myself continuing that. It was a lot of tears shed and a lot of stress to keep up. (Hispanic woman, switcher, biology to communication)

## Conclusions About the Switching Process

Consistent with the evidence presented throughout the book, these findings show that switching out of STEM programs is generally not a single moment in time, but is instead a process represented in series of events, personal reflections, emotions, interactions with others, and reactions to those interactions. Some amount of switching out of STEM programs—as well as any degree program—is arguably a valid response to the discovery and maturation process that is implicit in the higher education experience. This represents, in effect, a *rational choice model of switching*, in which the switching process results from valid factors and is fully informed. Instead, for a great many students in our interview sample, the process of switching was much more *haphazard and situationally contingent* than the rational choice model would predict. By this we mean that in many cases, the switching process, as it played out for individual students, may well have played out for them differently had they been in another institutional or STEM program setting. This was likely the case in particular when students grappled with certain absences during their switching process, such as insufficient academic help, unhelpful advising, inadequate information about academic and career pathway alternatives, lack of intercession in a negative program climate, or insufficient accommodation for mental health or other life crises. Chapter 12 takes up many of these themes in illustrating how the presence of these supports aid students' persistence in STEM.

## The Loss of High-Performing Students from STEM Fields

In the balance of this chapter, we examine the loss of high-performing switchers—students who leave their STEM major with high GPAs. We define as "high-performing" those switchers who left STEM majors with a 3.5 GPA (B+) or greater. In our interview sample of highly capable students, 69% of women, and 37% of men, left their STEM major with a GPA of 3.5 (B+) or better (i.e., 42 of 61 women and three of 35 men). As indicated earlier, we over-sampled students of color to get a clear picture of their experiences and concerns; however, we subsequently found that half of the high-performing women in our sample who switched were women of color ($N = 21$), with a smaller number who were men of color (i.e., 38%, $N = 5$). As these high-performing switchers are precisely the kind of students that institutions of higher education seek to recruit, retain, and prepare for advanced degrees and careers in STEM fields, it was important to learn why they are moving into non-STEM fields.

From our analysis of the interview data, we found that high-performing switchers had distinctive characteristics and reasons for moving to non-STEM majors. We present three patterns that emerged, largely in response to our questions about students' decision-making processes. First, we examine how perfectionism puts the persistence of high-performing STEM majors at risk. We then describe the rationales of two types of high performers who leave their STEM majors—those who move into the arts and humanities, and those who intend a career in a STEM field but without a STEM degree.

## *The Risks of Perfectionism for High-Performers*

In interviews, both high-performing persisters and switchers, described themselves as extremely motivated to achieve in high school. They often took multiple AP classes and intensive International Baccalaureate coursework in order to enter a good college or university. They commonly recounted how their identity had been built upon the achievement of high grades and they had come to see themselves as "an A person," "a high achiever," or "kind of a perfectionist." As also noted in Chap. 9, notwithstanding their record of success, these high performers often experienced a difficult adjustment to college, especially when confronted with their first grades. Having succeeded in high school, they found it difficult to come to terms with the realities of college-grading practices:

> I shoot for A's usually, but…but they curved the class a lot….and it didn't really feel like you were excelling when you were getting low grades on the test freshmen year. You don't understand how college classes curve and when you're getting like a 67 or 70 something on the test, it just doesn't feel good to you. It feels like you're failing almost. (White woman, high math, switcher, biology to sports medicine)

Some high-performing seniors continued to struggle with the adjustments required:

> I have a lot of friends in the sciences and math—people that all through high school got straight A's. They're the poster children of their high schools, just like me—we all came to this school, we're very smart. But they've had to come to terms with the fact that they'll get a C … And for someone who is used to that 4.0 mentality—that you live with through four years of high school—I think that's very challenging to come to terms with. … I don't know if I've fully struggled with that yet….Even though I can do very well with getting a B in a

> class, I do have to have a little mental conversation with myself when I do see that come up on my transcript. 'I worked hard for that. The test grades weren't very high, but I still learned a lot from that class.' (Hispanic woman, high math, mathematics persister)

In the latter example, the influence of talented peers seems more influential than that of parents in building up the pressure to score high grades, although, in other cases parental expectations were significant.

From the accounts of their struggles, women tended to be more perfectionist about their grades than men and had a much harder time adjusting to curve-grading practices in college. This often generated a good deal of confusion for women about whether or not they "belonged" in their STEM major:

> I think boys are a lot more laissez-faire in terms of the grades they get, whereas girls are more oriented to perfectionism and what that grade means to their self-worth and self-esteem....When I took that first test, it was, 'Oh, my gosh! I can't believe I did this poorly! I don't even deserve to be a biology major 'cuz I got a 66 on this exam!' That was, literally, the first thought that I had, whereas I'm sure some people said, 'Okay, I got a 66%. That's not the end of the world; I'm moving onto the next thing' you know? So, I think girls place a lot more pressure on themselves in terms of excelling in their unrealistic expectations. Because you can't get a perfect score on all of your exams, but that's...what we're trying to aim for.... It's just ingrained from a very early age, to the point where when you get to college, it's almost innate—as a girl, you just need to do well and succeed and work hard. It's expected of you, especially in my family. My brother doesn't really study as hard as I do, but it doesn't seem unnatural, because that's just how things are in our society. (Asian woman, high math, biological sciences persister)

Male peers also noticed this behavior and its consequences among the women in their STEM courses:

> Girls I normally work with, they got 4.0s in high school. In my fluid dynamics class, if you get a 70 overall, it's an A, and he makes the tests so that the average is supposed to be like a 50. And so, even though they got a 55 and it's above the average, they still couldn't wrap their heads around it. They would always freak out.... I definitely see that a lot more with girls—that perfectionism that they get from high school, where they need to get a 100 on everything. And it can really bite them in the ass. (White man, high math, chemical engineering persister)

Several psychological studies report that high-achievement and perfectionism are associated traits (Chang, 2006; Chang, Watkins, & Banks, 2004; Hewitt & Flett, 1991, 1993; Stoeber, Hutchfield, & Wood, 2008). Perfectionism in college students has been extensively studied: as a motivator to achieve academically and athletically (Carter & Weissbrod, 2011; Gnilka, Ashby, & Noble, 2012, 2013; Neumeister, 2004), and especially in relation to its maladaptive manifestations, such as procrastination, burn-out, anxiety, depression, eating disorders, and suicidal ideation (Chang et al., 2004; Dean & Range, 1996; Flett & Hewitt, 2012, 2013; Rice & Ashby, 2007; Rice, Lopez, & Richardson, 2013; Zhang, Gan, & Cham, 2007). Building on social comparison theory (Festinger, 1954), Travers, Randall, Bryant, Conley, and Bohnert (2015) recent work focuses on the phenomenon of "effortless perfectionism" among high-achieving college students. In this form of perfectionism, individuals seek to conceal the effort and struggle put into achieving high grades, such that their success appears to be both natural and easy. Thus "the effortlessness of success…rather than the success itself…may be the distinguishing factor that individuals believe sets them apart from their peers in high-achieving communities" (p. 2). As detailed in Chap. 9, the struggle to belong in the negative, at times-hyper competitive, atmosphere pervading some STEM programs (particularly with large proportions of students' set on medical school) often led to students jockeying to be on top in precisely this way. As the following examples show, comparing oneself to others, tended to lead to negative self-evaluation and, again, raised questions of belonging:

> I wasn't happy doing pre-med and everyone was so competitive…I was doing fine in my classes, but the way they were was not what I wanted to be a part of. (African-American man, low-math relocator, pre-med/biology to life science)

> At orientation I was a little bit nervous about being a math major. And I'm in the Honors College so I came to Honors Orientation. We were sitting in a room of very intelligent people. The College of Math and Natural Sciences was one of the smallest groups…And everyone was talking about their standardized test scores…It just seemed like everyone was, 'Oh, I got a 2,400 [on my SAT]. Oh, I got a 36 on my ACT.' 'Oh, I got a 36.' 'Oh darn, I got a 35!' And while I did have good test scores, I wasn't to that caliber of having almost perfect scores. That scared me a little bit. I was already deciding maybe I don't want to do mathematics, and I think that could have been a little extra nudge for me [to switch out]. Because I was like, 'Well, what if I'm not good enough for this...what if I'm going to fail in [comparison] to the rest of my peers?' (Hispanic woman, high-math switcher, math to management)

The student quoted above was, in effect, saying that because she did not have perfect scores on incoming standardized tests she feared not being good enough to succeed in the math program. The implied assumption is that there is no space for program participation between perfection and failure, and that there is only *one very narrow and extremely high standard* of preparation that can lead to success. This student's account also implies that there had been no effective departmental intervention to dissuade such a belief.

As both switcher and persister accounts attest, the competitive ethos in STEM was difficult to endure, with women and students of color experiencing greater negative effects. In addition to Travers and colleagues' research, two recent articles describe the rise of "effortless perfectionism" among women on competitive college campuses and the personal and emotional toll of sustaining such standards (Ruane, 2012; Yee, 2003). Both in the struggle to belong detailed in Chap. 9, and here in our discussion of high-achieving STEM students' experiences, we note that these descriptions match those reported in the original TAL volume where men indicated that gaining high grades with apparent ease was seen as an ultimate proof of one's intelligence and natural ability to succeed in a competitive and prestigious STEM major (pp. 250–251). Yet, as Travers et al. (2015) research finds, one of the consequences of effortless perfectionism (or EP) is that it increases students' sense of separateness:

> EP involves hiding not only imperfections, but also the time and effort spent in achieving high-level performance. The self-concealment associated with EP may require greater social isolation than the former, given the difficulty of hiding how hard one works when in

> the company of peers. Furthermore, because struggle is unacceptable among effortless perfectionists, this further intensifies their social isolation because they are unable to commiserate with or solicit support from peers (p. 8).

Although not, as yet, formally studied with respect to STEM undergraduate education, we documented earlier in this chapter some of the negative personal consequences of switching, including those of high performers. On learning about the high suicide rate among veterinarians, particularly for women, this student took her experiences in her major and the status of her mental health into account in reaching a decision to change fields:

> Getting perfect grades was very, very important 'cause, you know, you hear how vet school is more competitive than medical school right now. Like everyone will have the 4.0, so you need that and all these *other* things. So it felt like the end of the world to get a D.... The third-year vet students I was living with were having to take stress management classes because being a vet is so stressful. They have very high suicide rates among vets, which I never knew. And I thought given my temperament and my reaction to high-stress situations, maybe I should not head in that direction.... I got a B and an A- the semester that I decided to change majors. It's probably the healthiest thing that's ever happened to me. (LAUGHS) (White woman, switcher, high math, biology to European studies)

That perfectionist students are at particular risk is attested by a recent study that found the suicide rate among female veterinary students is three and a half times that of the general US population, and for male students it is over twice the rate. The authors speculate that, "one potential factor associated with an increased risk of suicide among veterinarians is the selection of veterinary students with certain personality traits. The veterinary school application process commonly selects for perfectionism to meet the rigorous veterinary school academic requirements. However, perfectionism has been associated with higher risk for developing mental illnesses, including anxiety and depression" (Tomasi et al., 2019, p. 110). In discussing the issue of perfectionism and rising rates of college student mental health issues with a senior administrator at one of the institutions participating in this study, we learned that four engineering students had committed suicide that semester. According to the Suicide Prevention Resource Center, suicide is now the second leading cause of death among college and university students (SPRC, 2019).

Results from Curran and Hill's (2017) recent meta-analysis of studies on college students and perfectionism show that, in this population, perfectionism has significantly risen over the past 25–30 years. The authors theorize that the increased cultural drive for competitiveness, individualism, and meritocracy stems from neoliberal policies that have shaped American and other world economies over this period. Most recently, these economic trends have encouraged perfectionism in students:

> Recent generations of college students are demanding higher expectations of themselves and attaching more importance to perfection than previous generations. As to why self-oriented perfectionism is rising, we speculated earlier on several cultural shifts that include competitiveness, individualism, meritocracy, and anxious and controlling parental practices that may be promoting perfectionism generally (p. 10).

Travers et al. (2015) agree, hypothesizing, "effortless perfectionism is particularly apparent within the Westernized culture of affluence (one that strongly emphasizes achievement and success)" (p. 10).

The cultural shifts seen to drive perfectionism—competitiveness, individualism, meritocracy—are also evident in student accounts of faculty, departmental and institutional attitudes and practices that are dominant in STEM fields. As we also reported in the original study, men respond more positively to the competitive ethos in STEM (and in other spheres of life) than do women. Their sense of identity is also less extrinsically dependent, so competition does not have such a damaging effect on men's self-images. It simply determines who wins:

> I think that men are inherently way more competitive than women, just in general. And my classes are all about competition and comparing yourself to others. I'm constantly thinking, 'Where am I falling in this class? Am I in the top 10%? Am I in the top 20?' What do I have to do to get up to that next level? What do I have to do to compete with my peers?' And I think I'm much more comfortable thinking that way than women are normally. (White man, high math, chemical engineering persister)

Similar to Festinger's (1954) social comparison theory, Fletcher and colleagues' work describe, "social perspective taking"—measuring oneself against others—and is "a skill that adolescent perfectionists use to check how others view their behavior and performance" (Fletcher, Speirs Neumeister, & Flett, 2014, p. 897). It flourishes where cultural and educational systems define achievement solely by high grades and students internalize the perspectives and opinions of others about themselves to construct their identity. We might add that, the power of social media enhances the focus, particularly of young women, on constant comparison of their physical attributes and achievements with peers—both known and in the wider youth culture. This further reinforces the trend to extrinsic measurements of identity with results that include depression, anxiety, eating disorders, isolating behaviors, and the risk of suicide. As we described in Chap. 4, a student's choice of a STEM degree often derives from praise expressed by parents and teachers and peer recognition for their mathematics and science abilities, as evidenced, largely by their grades and other markers of academic success, such as awards and scholarships. Bressoud and colleagues also found that both the choice of a mathematics major and students' sense of identity were strongly influenced by extrinsic sources, such as parents and high school teachers (Bressoud, 2015; Bressoud et al., 2015). In the following example, we see this process at work. A high-performing senior woman demonstrates a sense of herself that is defined in relation to others. She thinks her performance is "unacceptable," but is frustrated by her inability to accurately assess it. She also regrets not having switched earlier to a major where she might have achieved a GPA that she could feel good about, and, thus, feel better about herself:

> Maybe we just have the bar set higher for ourselves. Because I know that I've been reconsidering my major even though I have no choice but to graduate with it. I probably would have changed a few times by now just because of my GPA, just because I feel like it's too low. I don't know how I compare to my peers, but I feel like it's unacceptable. And so, to me it's horrible and I wanna switch, but I can't. But I don't actually know if it's actually

> horrible or not 'cause I don't have anyone to compare it to. Really, I just don't feel comfortable asking people. I have very high expectations and it's hard to know... So I would want to switch to something where I think I could meet those expectations. (White woman, high math, neuroscience persister)

As we have illustrated, the push-and-pull decision-making process around switching majors includes deep introspection and questioning by students about themselves and their future. It comes at a time when they are maturing as young adults and developing an understanding of who they are as well as deciding what they want for themselves and for their future. As Baxter-Magolda (2014), also clarifies, deciding for themselves what does or does not make them happy, what career will best suit their talents, interests and future sense of self, involves a break with dependence on external sources to define their identity and requires them to make their own choices. Wrestling with whether or not to switch majors, and coming to terms with that decision, pushes students to rely on their own authority. Thus, for high-achieving STEM majors, acceptance that they have become a "small fish in a big pond" comes at a critical juncture in their development of independence and autonomy. Either they manage to re-calibrate the meaning of their grades and set their expectations within a range that is normal in STEM courses, or they remain at high risk of leaving their major. These two seniors explain their difficulties of making this adjustment, but both had done so:

> [My grades] defined me. Until very recently, I could not accept an A-minis and when I came to [college] and started getting A-minuses for working my butt off in classes, it broke my heart. In electronics, my junior year, I got my first B, and I've never worked so hard for a class. ... I essentially slept in the electronics lab. And that was the first time that I was like, 'It's okay. You gave it your all.' Now that I'm applying for jobs, I'm realizing how little GPA matters. It matters a lot for grad school for sure, but now that I'm not looking into grad school, suddenly, I can hardly care [about] my GPA. I'm really trying to get myself to keep trying to get good grades, but I'm having a hard time. ... My perfectionism has quickly faded. But for a very long time I would cry if I got an A-minus instead of an A. I got over it though. (White woman, high math, neuroscience persister)

> My current computer science GPA is lower than my regular GPA. I don't like to talk about my GPA. I'm kinda average, but my expectations since high school have dropped dramatically in terms of the grades I've expected. Which is, yeah, it's my fault because I know friends who are getting straight As in computer science classes. But I guess I just wasn't as willing to put out the same amount of work. So, there were times where I'd be like, 'Yes! I passed!' ... For me, my grades dramatically dropped, but rather than quit, my standards dropped instead. (Asian male, high math, computer science persister)

All of these accounts demonstrate the perfectionist tendencies of many high-achieving students and their difficulty in adjusting to STEM grading norms and practices and to greater competition from a larger pool of highly talented peers than they were accustomed to in high school. Conditioned to high-performance expectations for themselves, high-achieving STEM majors, particularly women, who are unwilling or unable to dissociate their grades from their identity are at particular risk of switching out of STEM majors.

## Do High-Performing Men and Women Switch Out of STEM Majors for Different Reasons?

While reviewing our data to explore issues of perfectionism, we also found distinctive patterns in the reasons that prompted high-performing men and women to leave STEM majors that distinguished them from each other and, in some cases, from other switchers. These differences reflect our overall findings that there are patterned variations in the ways that students respond to the negative consequences of their STEM major experiences.

Two significant gendered variations among high-performing switchers were that 60% of women compared with 38% of men cited loss of confidence due to low grades in early courses as a contributory cause of their decision to leave (see Appendix F for the "Problem Iceberg" breakout for high-performing students). Loss of incoming interest and motivation was also cited as reasons to leave by 67% of high-performing women, but only 46% of high-achieving men:

> It was frustrating. It was different from high school. … It was hard. On many tests in high school and on standardized tests I'd always performed really well and I kind of thought of myself as a top student...so it is demotivating to find yourself kind of moving down. … I think that the depression played into the dropping grades and the dropping grades played into the depression. I lost confidence. I did. I felt like I wasn't succeeding in the sciences and I wasn't sure that I could succeed. (Multi-racial woman, switcher, microbiology to psychology)

> I like getting As and Bs and when you get Fs and Ds that's like really discouraging and it makes you think, 'Can I really do this? Is my brain able to handle it? Is my mind capable of doing this?' I think it makes you feel dumb. (Asian woman, switcher, geology to business)

Low tolerance for less than perfect results is perhaps a more accurate explanation for the depression and doubts about their own ability expressed by these and other high-performing women switchers than the loss of confidence described by women of lower ability who also struggled with the difficulty of the work itself, as well as the norm of lower-than-accustomed STEM grades.

High-achieving female switchers also reported greater difficulty than did high-achieving men in transitioning to college (48% vs. 31%) and were concerned that their high school STEM courses had not adequately prepared them for college-level work (24% vs. 8%). Some of this gender difference may reflect the high representation of high-performing women of color in our sample, who are more likely than high-performing white women to have experienced poorer quality high school preparation:

> I excelled well in high school. My GPA—I think I graduated with like a 4.16, so I thought I was really ahead of the game when it came to coming to college. But I wasn't, and was nowhere near prepared for the sciences, and the math…. I performed well in high school. I got As in everything. And then when I got to college, everyone was coming in with (a better) knowledge base from their high school classes. They had taken AP Calc, AP biology, all these classes, so they already knew things. Like, the beginning classes were just so easy for them; it was review. For us, it was learning something entirely new. (Hispanic woman, switcher, life sciences to communications)

High-performing male switchers reported more problems than high-performing female switchers with curricular design, including fast pace and overload (46% vs. 26%), and with conceptual difficulties in one or more of their STEM courses (23% vs. 12%). However, although high-performing men and women who switched were equally critical of faculty teaching practices (46% vs. 50%) and of the competitive ethos of STEM majors (54% vs. 57%), as discussed in Chap. 7, high-performing women cited the negative effects of weed-out courses as a contributory cause of their switching more than did their male counterparts (21% vs. 8%).

Career-related concerns also affected high-performing students' switching decisions. A slightly larger percentage of high-achieving female than male switchers recounted leaving their STEM major because, either they rejected the STEM careers and associated lifestyles that they projected for themselves (64% vs. 54%), or saw the shift to a non-STEM field as offering more appealing career options (64% vs. 46%):

> Coming into school here, I was studying the sciences for about a semester and a half. I switched over actually because I decided that I like talking to people about how the world works better than staying in a lab and working on discovering how the world works. (Hispanic woman, switcher, life sciences to journalism)

> I think the biggest thing that worried me is, 'What am I gonna do when I graduate? I love mathematics, and it's something that I enjoy doing, but where am I going after I spend these four years here?' And I think that's something that I could have researched, but instead I just kind of shied away from it and said, 'It's not what I wanna do for the rest of my life. It's just something that I enjoy doing kind of on the side.' (Hispanic woman, switcher, mathematics to business)

Discovering an aptitude for a non-STEM subject as a reason for their switching decision was far more common among high-achieving male switchers (85%) than high-achieving women switchers (69%) However, 26% of high-achieving female switchers, compared to 8% of high-achieving male switchers had also come to the conclusion that their STEM career options and rewards were not worth the effort and stress required.

> I was competing with kids that would get, like, 99s on exams where most people were failing. And I was, like, 'Oh, okay, well, I don't know if I can compete with this. And I don't know if I want to, either! Like, I could compete, but is it worth it to be so stressed out? Be studying so much and then failing exams and ending up with Bs somehow? And I didn't know if it was worth it. So, I don't think I felt like I belonged. Because there's a lot of people that, like, knew they wanted to do that. And that made me question, 'Well, I'm not sure I want to do this like they do. So maybe I should find something I do want like that.' (White woman, switcher, biology to dietetics and nutrition management)

An issue that we discuss in the following chapter—the impact of difficulties in financing college and the stress created for students who work long hours to pay for their college education—also surfaces in this analysis: 70% of switchers compared with 48% of persisters struggled with financial problems in completing a degree, and women struggled more than men with paid working hours that competed with their academic work. Financial problems influenced 10% of all switchers to move out of STEM majors and, among these, 14% of high-performing women (compared with 0% of high-performing men) cited financial problems as a contributory cause of their switching. The implication of this finding is that STEM majors are losing some highly talented women, in part, because they cannot support themselves financially through to graduation.

In the descriptions of perfectionist tendencies in the accounts of both high-performing persisters and switchers, we discern that the mental and emotional distress they describe arises from low tolerance for less than perfect performances. Moderate grades misinterpreted as failure are interpreted as an intolerable threat to identities that are extrinsically derived from high grades and the status they confer. Switching enables high performers to regain self-esteem and a valued sense of self without actually changing the reputational criteria upon which these are built. It also stems the psychological damage created by trying and failing to preserve a historic reputation and self-image as a high flyer. High-performing women (including high-performing women of color) appear to have greater difficulty than other students in easing the stress of unrealistically high expectations for their academic performance. Many persisters also spoke about the importance of letting go of these expectations and of finding self-acceptance independent of their grades. As one senior in the original study described this journey, it was akin to learning to see grades as being "like weather—sometimes it rains and sometimes the sun shines—it affects your mood, but you can't control it or take it personally." Explanations for why women leave STEM majors that are common in the research literature—such as loss of confidence or the negative effects of the competitive ethos—are still relevant, but they may more accurately apply to that larger proportion of women entrants to STEM majors who are not the highest performers. If STEM departments and colleges seek to keep more of these highly talented students, it would seem imperative to make systematic efforts to normalize for incoming students—whether high-performing or not—their struggles with norm-referenced grading systems. As one persisting senior suggested:

> It would be nice to have them, or just to know where you stand in terms of readiness in the field. Because I know that the psych department like spells it out. A C is adequate preparedness to continue in the field, and then like a B is above average kind of thing. But I don't quite get that in the other departments. I don't think that their grades are based on whether or not they think that you can continue, I think it's just, 'Show me what you know,' kind of thing. And so I think that would be a good direction to go in for the other departments is to really make the grades reflect, 'How you should do in the future,' kind of thing, 'cause otherwise you really have no idea. 'Is this a good grade? I don't know! How do I stand?' (White woman, high-math, neuroscicence persister)

## High-Performing Students Who Leave STEM for the Arts and Humanities

Among the possible types of moves that students made, a switch out of STEM into arts and humanities programs might seem the most counterintuitive—perhaps because they are thought to draw on very different talents and interests, or because

STEM fields are viewed as leading to careers in more secure and lucrative professional fields. In our interview sample roughly 10% of all switchers followed this pathway. Among these, the loss to the arts and humanities of 18% of high-performing former STEM majors makes it particularly important to address the open question of what prompts these decisions. Why, when talented students assess whether to continue with their STEM major do the arts and humanities sometimes win out over STEM?

Among other researchers who have addressed this issue, Chen (2015), using a nationally representative sample, also found that 18% of high performers—defined as having a 3.5 GPA or higher—shifted out of STEM and into humanities programs. Chen's analysis showed that these high performing students took a lower percentage of STEM courses in their first year compared to high performing students that persisted in STEM, leading him to contend that a weak initial focus on STEM coursework contributed to these STEM losses. However, this pattern of course-taking also suggests that students could have *other subject interests* that they are pursuing concurrently with STEM—interests that might or might not be motivated by future career plans. Along these lines, Heilbronner (2011) found that interest in another field was the most important reason that talented students did not pursue STEM majors. Indeed, interest in the subject matter has been found to play a decisive role in students' decisions to enter and complete a STEM degree (Maltese & Tai, 2011). As a factor influencing persistence, *interest* connotes cognitive and emotional attributes, content specificity, and personal disposition and significance (Krapp & Prenzel, 2011). Choosing STEM and persisting in a STEM program can therefore be seen as a process of assessing one's "fit" in relation to perceptions about a subject, its associated community, and the future life course foreseen through studying the subject (Taconis & Kessels, 2009). Previous research has consistently found that high-performing students readily distinguish between their interest in a subject of knowledge and how it manifests as "school science" in classrooms and degree programs (Boe, Henriksen, Lyons, & Schreiner, 2011; Holmegaard, Madsen, & Ulriksen, 2014; Maltese & Tai, 2011). As a result, students' interest in STEM disciplines may be greater than their interest in the same disciplines as experienced in the classroom.

## *Characteristics of Students that Switched from STEM to the Arts and Humanities*

Over half ($n = 55$) of the 96 switchers in our interview sample were high performers, defined as having a GPA of 3.5 (B+) or higher. Like Chen (2015), we also found that 10 of these 55 students (18%) switched from STEM into arts and humanities programs. These students were also high performing in that they had not experienced actual or perceived performance problems that contributed to their shift out of STEM.

The students had several notable characteristics. The group was nearly evenly split between men ($n = 6$) and women ($n = 4$), and while predominately white, one-third were black or Hispanic. The students' disciplinary moves were highly diverse, representing five different STEM fields and six different non-STEM fields. The students all attended one of the four large, public research universities included in the study. No student attending either of the two small, private institutions followed this STEM to arts and humanities switching pathway, suggesting a possible underlying institutional effect. Most of these students described themselves as extroverts. Three of these students had attempted to pursue double majors before shifting out of STEM altogether. These included physics with theater, dietetics with theater, and chemistry with film studies. They found that double majoring was too challenging to pursue without extending their time to graduation. All but two retained an interest in the sciences while not pursuing a STEM major. None had significant performance issues in STEM courses, and several contended that they could have succeeded in completing a STEM degree had they wished to. One former chemistry student, for example, objected to what he perceived as the likelihood that he would be labeled a statistic of failure in our larger study:

> I know somewhere in this study I'm a statistic of dropping STEM to go to the arts. I know that I'm a statistic and it bugs me, because the stereotype of, 'Oh, he dropped STEM because it was too hard, poor kid.' That's not me. If another [me] went back and decided he didn't want to do theatre, he would have graduated with a chemistry degree. In fact, I'm sure he could have graduated a semester early because he came in with credit...It was never a question of how difficult it was. And I hate that somewhere in this study I'm a statistic about it. (White man, switcher, chemistry to theater)

These students expressed their interest in both STEM and non-STEM subjects in terms of "love," "passion," and "excitement." One student, for example, who switched out of physics, recalled "falling in love" with theater after taking his first acting class in high school. "Passion" for theater drew another away from chemistry, despite his enduring "love" of it. More generally, the students described these sentiments as applying beyond "school STEM" and continuing after they switched:

> I just love the challenge of the sciences. I think I naturally think that way analytically and I love the challenge of piecing it together. (White woman, switcher, chemistry to film studies)

Several students also described themselves as all-rounders, that is, "interested in everything."

## Why High-Performing Students Switched Out of STEM and into the Arts and Humanities

***Multiplicity of Interests*** Because of their multiple subject matter interests, these students chafed at the restrictiveness and lack of flexibility that they had experienced in their STEM degree programs. They alternatively sought a more "liberal arts" education. One woman who switched from biology to journalism made the decision because her biology program afforded little opportunity to accommodate her wide-ranging interests in the world:

> My skills are in the humanities. Naturally, that's where my talent lies. But I cultivated such a deep interest in the sciences that I became very well rounded in my interests. So, I can go anywhere from quantum physics and turn around and talk about history or the politics of war or something like that, and at a moment's notice. Because of my wide, diverse, very broad range of interests, I felt that the sciences were very restrictive for me. (Hispanic woman, switcher, biology to journalism)

Another female student also found that the dietetics program in which she initially majored was too rigid to accommodate her interest in exploring different subjects, "to try bits and pieces of everything." After dropping her STEM major and entering the theater program she felt more able to pursue a broad liberal arts education:

> Once I dropped the science major I felt freer to just do whatever I want in college and use it as a liberal arts degree, a general higher education, rather than focusing specifically on something. (Black woman, switcher, dietetics to theater)

***Narrowness of STEM Curricula*** In epistemological terms, these students characterized STEM as having pros and cons, with some of the cons influencing their diversion to the arts and humanities. Questions and problems for which there are no definitive right and wrong answers were interesting to them and they had a high tolerance for ambiguity and complex problems. They sought creativity, open-ended inquiry, opportunity for interpretation, and intellectual space for "play," as one woman characterized it. For a male student who switched from physics to journalism a large part of the appeal of physics was its "paradoxes." This had contributed to his parallel interest in science fiction and belief that "physics is mysteries, possibilities," and not "just all math."

***Problems Fitting Oneself to "School STEM"*** Although these students recognized that if they had stayed in STEM their experiences in upper-division STEM courses would likely have been different, they expressed disappointment that lower-division STEM courses lacked theory and opportunities for open-ended inquiry. The teaching and learning experiences that they encountered in STEM were delimiting and the questions driving instruction tended to have "yes/no" answers. Interest in the theoretical was important for them, yet most switched—out of physics and biology programs, in particular—before those programs had progressed into theoretical content. One student, who switched out of physics to cultural studies, explained:

> I noticed that right before I quit physics was when it was probably about to get more theoretical and less just 'yes-no' or at least 'yes' and 'no' is very unclear. But I never got to experience that. (Black man, switcher, from physics to cultural studies)

More than particular instructional techniques, these students valued an instructional environment full of excitement and enthusiasm, both for the subject and for the students. A female switcher from chemistry to film studies, for example, contended that professors' enthusiasm serves the goal of stimulating and challenging students, rather than appeasing or entertaining them, and can overcome some of the negative aspects of large classes. Like themselves, these students expected that their instructors should both have, and express, "love" for their discipline, and should lead students towards that type of relationship. Unfortunately, the opposite was often true in their STEM programs. For example, in her biology program, a female student experienced a lack of enthusiasm among the biology professors. This attitude spread to the students, whom she described as treating their studies like a chore. The lack of enthusiasm among biology faculty and students contrasted with the enthusiasm that she found in the journalism department and was a primary reason why she switched.

***Problems with Climate and Belonging*** For this group of students, passion and enthusiasm on the part of instructors should, they thought, quite naturally extend to "caring" for students and their learning. Like enthusiasm, "care" (as discussed in Chaps. 6 and 8) had relational and interactional qualities that could contribute either to positive or negative feedback loops. As one male student who switched from physics to theater illustrated from his experience:

> I need the teacher to be very passionate, which seems like a given. But I've had so many teachers where it's just, this is the 10th year that they've taught [a particular course], and [during lectures] words just kind of flow out. They don't care what they're teaching about, so I don't care what they're teaching about. (White man, switcher, physics to theater)

In terms of classroom climate, this group of students sought a "personal" environment and rejected the "impersonal." Continuing to explain, the student quoted above attributed his switch from physics to theater to the "very personal environment" of the theater department. That the faculty there took a "personal interest" in every single student appealed greatly to him.

Along similar lines, "being recognized" was important to these students. In some cases, when recognition did not happen within STEM this was taken as a sign that switching out was warranted. A female switcher from biology to film recounted an occasion in which she made a chemistry joke to a lab partner. In it, she likened herself to "the type of hydrogen atom that is lazier than the others" as a playful excuse not to get up and retrieve a piece of paper from someone across the room. In response, her lab partner gave her, "the meanest look ever" and then ignored her. This interaction acted as the final clue that confirmed her suspicion that she would never fit in with the chemistry community. Commenting on another form of recognition, a male student objected to chemistry faculty not knowing his name, and, implicitly, not recognizing him as a person with interests in a subject and the subject community. He contended that, had he felt more connected to the chemistry program, he could have found his way to declaring a double major in chemistry and theater.

## Persistence, Attrition, or Something in Between

The STEM persistence literature tends to analyze students' choices and degree completion in the dualistic terms of retention in or attrition from STEM. However, these alternatives do not align well with the self-perceptions or future orientation of most of this group of students. All but two of the students retained their interest in STEM. Despite switching out of "school STEM," they continued to value and use aspects of STEM-thinking and subject knowledge. Several of the group cited ways in which they envisioned combining STEM and non-STEM interests in the future, including keeping STEM as a hobby. As one student explained:

> [STEM] is not going to go away completely just because it's not my major. Your major is not your entire life. It may seem like it, but it's not. (White woman, switchers, chemistry to film studies)

More substantially, several of these students were considering ways to combine STEM and non-STEM in their professional pursuits. One male student who switched from biology to history, for example, was thinking about getting teaching credentials in both fields. Another former engineering student, now pursuing religious studies, still hoped to contribute to solving some of the problems involved in environmental sustainability, but from a humanities viewpoint. A female student was inspired to follow the example of a woman she met on a flight who had combined her medical degree with film studies. While another woman who left dietetics to pursue theater hoped to create theatrical works that address heath care issues and nutritional awareness. Finally, two students sought to combine their interests in physics with creative and scientific writing.

The students featured in this section were not pushed out of STEM by poor performances; nor did they make the shift for instrumental reasons related to future job prospects. Rather, their decisions to switch were largely due to *other disciplinary interests* and because they found their STEM majors *too intellectually and emotionally restrictive*. Neither were problems with classroom instruction per se particularly influential with these students. What did matter were their instructors' affect, and finding elsewhere greater emotional and relational possibilities than was afforded by their STEM programs. These students were disappointed with the dearth of passion and enthusiasm that they encountered among instructors in STEM programs. They found the relationships with both instructors and other students in their STEM majors to be impersonal and lacking in humor and excitement. They also believed that their former majors were too narrow to accommodate their multiple interests and allowed too little opportunity to incorporate non-STEM subjects into their undergraduate education. Despite leaving what they described as "school STEM," most of these students retained interests in the sciences and hoped to combine their STEM and non-STEM interests in the future.

These findings suggest that some high-performing students are likely both to benefit from and contribute to the sciences even without earning a STEM degree. However, the potential contributions of these students may have been greater had their STEM programs recognized and supported their dual interests. The ease with which these students *connected* their STEM knowledge and skills to those gained in the arts and humanities represents the kind of innovative thinking that is argued to be necessary in order to solve broad societal problems in the future (Wolfson, Cuba, & Day, 2015).

## Pursuit of a STEM Career Without a STEM Degree

We discovered that many STEM switchers do not leave the university-defined STEM disciplines completely. Some students switch to majors such as biopsychology and nutrition that are not traditionally classified as STEM subjects but that require substantial math and/or science knowledge. Others who have switched to non-STEM majors still plan to pursue graduate education in a STEM discipline, thus they continue to complete STEM requirements for their desired graduate programs. These include both professional health-related degrees (e.g., medicine, dentistry, and nursing) and other applied degrees, for example, statistical sociology, biostatistics, and applied mathematics. We discovered that 21 (22%) of our switcher sample of 96 students were intending to pursue a STEM career and/or graduate program although they had switched to a non-STEM bachelor's degree and would graduate without a STEM degree.

### *Characteristics of Switchers Who Still Pursue STEM-Like Career Goals*

Seventy-six percent of these 21 STEM switchers were women and 28% were students of color. Their second-choice majors of this group were wide-ranging. One-third of those taking alternative STEM pathways (seven students) had changed to a social science major (either psychology or sociology); another seven had switched into a non-STEM health-related major, such as community health and sports medicine; three students had switched to a non-STEM, math-related, major, such as accounting and finance; and the remaining four students chose arts- and humanities-related majors, such as film studies, marketing, and landscape architecture. What their choices had in common was their potential to serve STEM-like career goals.

Many of the switchers who chose alternative STEM pathways were also high performers in their STEM subjects. Ninety percent of these students (19/21) entered college with high SAT or ACT math scores, and 62% (13/21) had GPAs of 3.0 or greater in their STEM majors prior to switching out of them. Although 76% (16/21) of the switchers raised their overall GPA following their move, their average GPA while within STEM majors was 3.10, with only a slight increase to an average cumulative GPA of 3.28 upon graduation.

Most (71%) of these 21 students reported that they planned to pursue health-related careers after graduation, and, of these, 14 intended to pursue a professional graduate program in medicine, dentistry, physical therapy, or nursing. Their career plans include descriptions of how they saw their new majors helping to accommodate their long-term goals:

> I'm [still] hoping to go into nursing. It appeals to me because it has a lot of the same [kinds of] medical aspects of medical school [that] being a doctor would have. And I like that it will help me apply what I learned in psychology about whole-person care and the mental

> well-being of a person — maybe affecting their outcome as a patient. (Multi-racial woman, switcher, microbiology to psychology)

They also outlined what benefits are added to their career plans by their new majors:

> I feel like I understand everything more [in my new major]. And I like going to class. And I actually understand what I'm learning while I'm in class. It's really interesting and practical and gives me a lot of different aspects of health care that I can go into if I decide I don't want to go to dental school anymore, or don't want to go right away. (White man, switcher, biological sciences to community health)

Some students were keeping their options open by taking prerequisites for intended careers or choosing majors that offered an array of career possibilities:

> I'm thinking I'll go into naturopathic medicine. So, while doing my four-year plan here, I'll look into what are the pre-requisites and put that into my schedule too. (White woman, switcher, mathematics to community health)

> I came in thinking I was going to be a bio major, then I was going to be a neuroscience major, then I was going to be a psychology major, and now I'm a sociology major, but I still wanna be a nurse practitioner, so I'm taking all the prereqs I need, both here and at community college. (White woman, switcher, neuroscience to sociology)

> I think I want to still do something where I'm dealing with the body and with the science, and I can always be expanding my knowledge with that … so I want to either go to physiology grad school or med school. I just want to set myself up for everything and then just try for med school. I just wanted to have a major where I'm set up for a lot of possibilities and I felt like by doing psychology it was on track. (White woman, switcher, chemistry to psychology)

Four students reported their intentions to pursue math-related graduate programs after finishing their non-STEM degrees. Two envisioned business careers that would make use of their technical skills:

> I wanna start a software company [after finishing my degree in economics], and work on analytic software used for HR departments. (White man, switcher, chemical engineering to economics)

> I really want to work for an aerospace company. Even though my degree is now in business, I want to do aerospace or medical devices. Maybe I'll do marketing for an aerospace company or work for a medical device company. (Asian Pacific Islander woman, switcher, geology to finance)

Two students were already combining their scientific knowledge with other fields to help prepare them for interdisciplinary careers:

> I would like to do a master's in applied mathematics [after finishing with my landscape architecture degree]. Kind of molding the applications and the theoretical and then possibly move into an engineering field. I am trying to ramp up the mathematics classes again this next semester. (White woman, switcher, mathematics to landscape architecture)

> My goal has always been to help people and I found a way to do that using sociology. I've actually got a plan to set up a treatment center for victims and families that have been abused and getting my Ph.D. in sociology with an emphasis in statistical sociology will allow me to develop a research and treatment center that allows me to put all of the resources in the same place. (White woman, switcher, biology to sociology)

Most of these alternative STEM pathways were health-related, but about one-third of the students intended to stay in math- or technology-related fields. What they shared, however, was a conviction that switching out of STEM would not prevent them from achieving their goals.

## *Why Students Choose Alternative Pathways to STEM Goals*

### Switching to Increase the Chance of Graduate School Entry

A sub-set of students chose alternative pathways for a variety of career-focused reasons. One of the most common of these is to make an instrumental or tactical move that is calculated to increase their chances of acceptance into competitive STEM graduate programs. This often takes the form of choosing a major that will enable them to increase their GPA:

> I knew I was gonna do premed all along. I knew my GPA probably wouldn't be higher than a 3.6 when I graduated … I wanted to make sure that I could have a career in medicine, so that's why I decided to switch. (White man, switcher, biological sciences to community health)

> Getting into the Master's program is going to be difficult with my grades from here because my overall is a 2.86 because of my biology minor. But it's more about making sure that I talk about what I've been through on my application and about how I've improved [in my new major]. (White woman, biological sciences to sociology)

> Physical therapy isn't as tough as med school to get into but it's still tough … My grades aren't very high [from first major] … I need to bring my grades up in the next few semesters. (White man, switcher, mechanical engineering to sports medicine)

Improving one's grades for competitive-entry graduate programs could also be combined with a calculated means to deal with tough future courses:

> My career path hasn't changed—I'm still doing the same thing. I want to be a physician assistant. I was scared—honestly, I was scared by the needs of the classes I was going to take, and I knew I needed really high grades to get into PA school, so I was like, I should probably do something that's a little different … that was my first thought, that I was scared of the classes. And then the second thought was, 'What if I get burnt out?' and, 'I want to get a different experience,' because I know I'm going to get all those classes again in grad school. (White woman, switcher, biomedical sciences to psychology)

Other instrumental rationales prompted a move to degree programs that took less time to complete or were less expensive:

> I came in with 25 college credits. But I realized when I looked through [the program] that I would end up having to spend five years here, even with the 25 credits that I already had coming in. I didn't want to spend that much time and that much money, that much extra money on college when I already know it's gonna be hard to pay it back in the future for both me and my parents. (African-American man, switcher, physiology to psychology)

These students were among the 26% of all switchers in the sample (described in Chap. 3) who switched for pragmatic reasons. In this sub-set, students switched majors for several reasons: to increase their chances of success in competitive-entry graduate programs; to avoid courses that—by their level of difficulty, length, or costs—jeopardized their career goals; or to broaden their range of skills for careers that required cross-disciplinary experience. For all of these students, their end goals either remained intact or were broadened to make use of knowledge gained in a second major.

### Seeking Particular Skill-Sets

Some students switched in order to follow specific or multi-disciplinary interests or to gain particular combinations of STEM and non-STEM knowledge and skills.

> I'm looking into law school for post-graduation to hopefully do some sort of policy degree with a background in science. Especially 'cause medical ethics and bioethics are large up-and-coming fields and they are looking for people with science backgrounds. (White woman, switcher, independent STEM major to bioethics and public policy)

> I switched into psychology, with a track in neuroscience ... I really didn't want to stick just within science ... I wanted to have...a better understanding of different things and of science incorporated into other aspects of life. (African-American man, switcher, physiology to psychology)

> I wanted a more global view of health and a broader application of biology concepts. (White man, switcher, biological sciences to community health)

Overall, we noted that blending science with other disciplines, such as art, business, communication, or social sciences, is a growing trend that was barely discernable in the original study. In the current study, the desire for interdisciplinary learning was evident, for example, in the numbers of students who opted for double (or more) majors—some within the sciences, and some that crossed disciplinary lines—and or to earn minors in a wide variety of fields that accompanied STEM majors.

### Rejecting the Laboratory Environment: Stereotypes About Working Scientists

Some students believed that the pure sciences only prepared them for research or laboratory-intensive careers and rejected their original majors in line with that belief. These were students for whom social interaction at work was important and imagined that all scientists worked in solitary situations. We also encountered similar stereotypes about the working life of scientists in the original study. These assumptions are illustrated in the following example:

> I guess just thinking about like jobs afterwards—I think I wanted to interact with people, and, with chemistry, it seemed like most of the jobs... didn't have as much of that, or it wasn't guaranteed that the career that I had afterwards would be with that.... So, the jobs that you could do with chemistry didn't sound appealing to me. ...I knew that nutrition had some science factors in it and it was related to chemistry, but I feel like there was more people interaction [in nutrition], and it just seemed more applicable right away. (Asian woman, switcher, chemistry to nutrition)

This example highlights the practical significance of student perceptions about what work in STEM-based careers would be like, and how well or badly these ideas "fit" with students' own sense of what kinds of workplaces would suit them best. Their perceptions could be mistaken but, unfortunately, their departments did not present them with an array of possible career settings to address their stereotypes about a working life in the sciences. As in *Talking about Leaving*, we highlight the need for STEM departments to be more proactive early in their programs in discussing the variety of STEM, near-STEM, and non-STEM careers (and types of work within them) that are open to graduates in their disciplines.

### STEM Majors Are Too Stressful

As described earlier in this chapter, the experience of stress, anxiety, and other forms of distress in STEM majors provided the conditions in which many switching decisions were made. This was the case for some of the students in this group who reported high stress or feeling overwhelmed by their experiences in STEM majors. They switched to majors that were less stressful, but that still allowed them to pursue their STEM-based career goals.

> My grades suffered. I just felt like I wasn't being successful and I felt like, even though I love biology … being a biology major was making me hate biology. And I didn't want to hate biology so I knew I had to switch … I felt like I went from being a really good student in high school to basically a terrible student in college. And it just made me feel like I—like I didn't belong in that major. … Community health is just more practical for me. And I understand it more and I actually like it. I actually like the health aspect of things. It's really interesting, and I also get good grades. (African-American woman, switcher, biological sciences to community health)

> My original intention was to go to medical school or to apply to medical school after undergrad. And eventually I realized that would be better if I switched into the College of Biological Sciences. So I switched over to that with a degree in microbiology. I kind of had a bout of depression my sophomore year. Struggled a lot and came to the conclusion that I really didn't want to go to medical school. That was kind of my parent's idea of my life. So, I switched back into liberal arts, into psychology, with the aim of getting a Master's degree in nursing later. (Multi-racial woman, switcher, microbiology to psychology)

The combined effect on their mental and physical health of the kinds of struggles described at the start of this chapter led this group of students to feel pushed out of STEM and drawn into majors where they still pursued STEM-linked career goals but without experiencing severe stress.

Over 20% of the students who switched out of STEM still intended to pursue a STEM-related career, with women accounting for over three-quarters of those who switched into alternative STEM pathways. This finding suggests that a significant number of women who switch may not be leaving because they have come to dislike the sciences; nor do they leave them altogether. Rather, they are finding more appealing routes to the STEM career goals that they are still pursuing.

Among men and women with comparable high math aptitudes, several studies report that women outperform men in verbal ability (Lubinski & Benbow, 2006,

Riegle-Crumb, King, Grodsky, & Muller, 2012; Wang, 2013; Wang & Degol, 2013). This wider range gives women more flexibility than their male peers in their choice of majors and careers, including consideration of both STEM and non-STEM fields. Wang and Degol (2013) found that mathematically capable women are more likely than men of matched ability to pursue careers in "people fields" (such as the humanities and social sciences). Mathematically capable men are more likely to pursue STEM fields that involve work with objects, machines, and tools, such as engineering and the physical sciences. Mann and DiPrete (2013) also posit that women are more likely than men to value both the humanistic and vocational aspects of education. They found that female STEM majors take a more diverse set of courses than male STEM majors, and are more likely to add humanities and social science courses to STEM requisites. Consistent with these findings, many women in our sample described their non-STEM majors as more enjoyable than their former STEM majors because they offered more multi- and interdisciplinary experiences, more interaction with instructors and peers, and more opportunity to pursue altruistic career goals.

Most of the STEM switchers who took alternative routes to STEM career goals were high performers, especially in math. About 90% had entered with high SAT or ACT math scores, and 62% had a GPA of 3.0 or more in their STEM courses before switching. Thus, the majority of these students did not switch because of academic difficulties in their STEM courses. However, as also noted by Wang and Degol (2013), high math-performers who also have high verbal skills are less likely to pursue STEM careers than students with high math skills but moderate-to-low verbal skills. These well-rounded students do not have to choose between multiple interests, but are able to pursue alternate interests while preserving their ultimate STEM career goals.

As we have stressed, 70% of the alternative pathway switchers intended to enter health-related graduate programs, and were, thus, still taking STEM requisites for graduate school applications. Many also realized that majoring in a STEM discipline was not obligatory for medical, nursing, or dental school entry; and some had made deliberate strategic moves into non-STEM disciplines in order to increase their overall GPA and, thus, their chances of acceptance into highly competitive programs.

Overall, our findings suggest the need to rethink what constitutes "a STEM major," and how STEM persistence should be defined in future research. However, their broader implications include reconceptualization of STEM majors as part of a liberal arts education with applications to, and relevance for, many career pathways and lifetime interests beyond those more narrowly defined by each discipline. Operationalizing this might include institutional promotion of inter- and multidisciplinary studies as pathways toward STEM careers, and departmental dissemination to its students of information about the breadth of career options open to graduates with particular STEM degrees—a strategy that, in itself, could reduce the flow of high-performing STEM majors (notably women) into non-STEM majors.

## Patterns of Switching and Persistence Among High-Academic Performers

Many of the findings from interview data discussed in this chapter highlight moves out of STEM majors by high-performing students. Thus, we now turn to the institutional records data to learn more about patterns of switching and persistence among high-academic performers across the whole institutional sample. As explained in Chap. 3, for the purposes of analysis, we divided the total population of STEM majors at our six institutions into four quartiles by GPA and considered students in the quartile with GPAs of 3.5 or higher as *high-academic performers*. In this group, we found 5231 current and former STEM majors of whom 12% (513 students) had switched out of their original STEM major. High-academic performers also accounted for roughly one-quarter (25.7%) of all STEM switchers across six institutions participating in this study.

We also took into account the ACT or SAT math scores of students in the highest GPA quartile. This identified a doubly high-performing group (i.e., by both GPAs and ACT or SAT scores), of whom 9% had switched. However, when comparing switchers and non-switchers on both performance measures, we found that STEM switchers in the high GPA group had lower SAT and ACT math scores than high-performing STEM persisters. Switchers averaged 28 ACT math scores and 665 SAT math scores, compared with averages of 30 ACT and 694 SAT math scores for persisters. This probably means that even though 12% of these high-performing switchers had high GPAs, they came in less well prepared than did high GPA persisters.

Using the high GPA criterion for the balance of our analysis, we found that switching patterns among high-performing students varied by discipline. High-achieving biology and life sciences majors were more likely to switch than would be expected given their representation at the six institutions. As is indicated in Fig. 10.1, 46% of all high-performing switchers came from biology or life science majors, while 54% came from all other STEM majors combined. At 36%, biology and the life science majors were also over-represented among persisters (with 64% from all other disciples). However, the preponderance of this disciplinary group was more marked among switchers than persisters. The opposite pattern is evident for engineering, where 26% of all switchers are from engineering majors, compared to 43% of all engineering STEM persisters and with 74% of switches coming from other disciplines. These results suggest that academically high-performing students in the life sciences are more likely to switch than their overall representation in our sample would suggest, but the opposite is true for engineering students.

We also noted that high-performing switchers were distinctive in their choices of subsequent majors. Sixteen percent of all high-performing switchers chose undeclared majors in their first term after leaving a STEM major. Six percent of all high-performing switchers—compared with 12% for other switchers stayed in undeclared majors for longer than one semester after switching. High-performing switchers also moved to different types of majors than did switchers with lower GPAs. Sixteen percent of high-performing switchers (primarily in engineering and

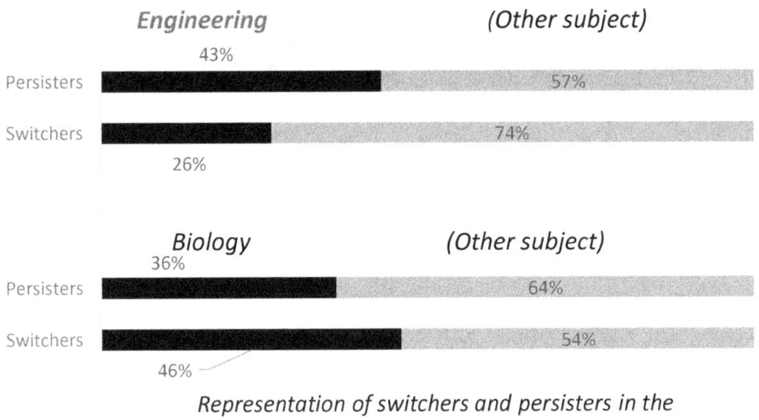


**Fig. 10.1** Representation of switchers and persisters in the high-performing quartile by discipline. *Note*: These are high academic performers with GPAs greater than 3.5

biology) switched into business majors, such as accounting, marketing, or finance. Only 6% of switchers in the lower GPA quartile choose these majors. With the exception of psychology, high-performing switchers were also less likely to choose social science majors, such as economics and sociology, than were switches in lower GPA quartiles. Thirteen percent of all high-performing switchers took the path out of STEM into social sciences (other than psychology), compared with 21% of lower-performing switchers.

Given our concern throughout this book to understand which groups of students are at greatest switching risk, the results of our analysis of high academic performers by gender are perhaps the most significant. As discussed in Chaps. 1, 3, and 7, we and other researchers have already established that, regardless of GPA, more women switch than would be expected given their representation in STEM majors. In the high GPA quartile, 12% of women switched versus 6.5% of men. These losses of high-performing women also occur in a female population that is over-represented in the high-performing group. As is indicated in Fig. 10.2, across the six institutions, of all switchers in the high GPA quartile, 62% are women, while only 48% are men. Losses from the three lower GPA quartiles combined are almost evenly divided between women and men.

By definition, students with higher cumulative GPAs do not receive many poor grades. However, even these students experience difficulties at points in their academic careers. They also seem at high risk of switching in face of a single poor result. During their first two years of college, high-performing STEM switchers were almost twice as likely as high-performing STEM persisters to have received one poor grade or withdrawn from one class. As shown in Fig. 10.3, for both groups, the commonest reason for withdrawal was receiving "incompletes" rather than D or F grades (which would have also removed them from the high GPA category). Thus, high GPA switchers are at risk of leaving STEM majors partly in response to receiving a single incomplete or course withdrawal.

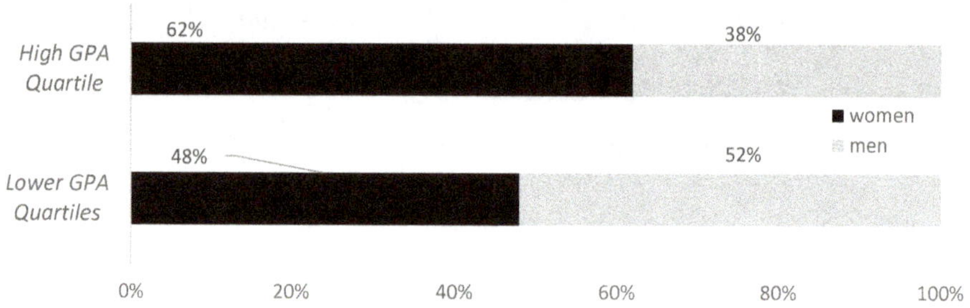


**Fig. 10.2** Percentage of women and of men in the highest GPA quartile and in the three lower GPA quartiles. *Note:* High academic group has GPA greater than 3.5

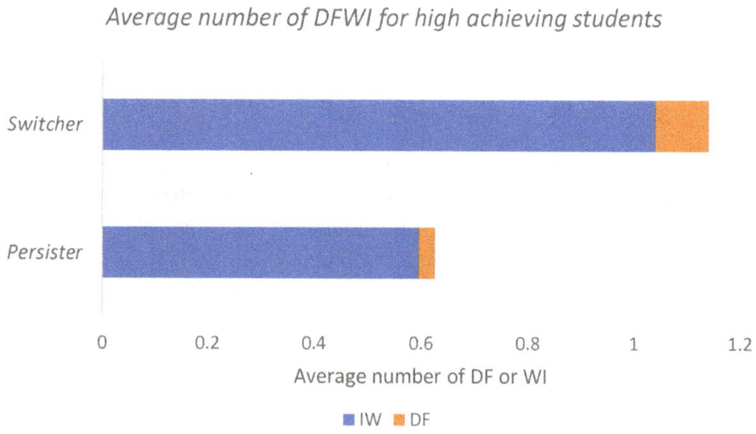


**Fig. 10.3** Average number of DFWIs given to STEM majors in the high GPA Quartile

This finding, coupled with the over-representation of high-performing women among STEM switchers, reinforces the high risk of switching among high-performing STEM majors. These are significant findings because they highlight high loss rates among the most intellectually able women entering these majors. These findings also align with those from our interview data (as described earlier in this chapter), in reflecting the low tolerance of high-performing students (often, but not exclusively, women) for set-backs caused by less-than-perfect performances in particular courses.

## Conclusions on the Loss of High-Performing Students from STEM Fields

In this chapter, we have paid particular attention to switching and relocation among high-performing switchers as well as to the processes by which both they and other students make these moves. We established from the institutional records analysis that 12% of STEM majors with GPAs of 3.5 switched into non-STEM majors and that high-performing switchers were 25% of all switchers. This amounted to the loss of 543 students across the six sample institutions.

The interview data estimate of 18% of high-performing students who switched into arts and humanities majors aligns with that of Chen's, 2015 study. What we learned from the interviews about these particular high-performing switchers was that they:

- Were intrinsically interested in the sciences, but were disappointed in the "school science" that they experienced in the first two years of STEM majors—especially the lack of theory and open-ended inquiry;
- Enjoyed creativity, inquiry, opportunities for interpretation and intellectual play;
- Had high tolerance for ambiguity and complexity;
- Had multiple academic interests that some sought to pursue via double majors;
- Found STEM too intellectually restrictive and unable to provide a liberal arts education of which the sciences were a part;
- Valued the exciting, enthusiastic learning environments they had found in non-STEM programs;
- Valued teacher–learner relationships where they were recognized as individuals;
- Were self-defined extroverts who found STEM education emotionally constraining;
- Had creative ideas about combining their STEM and non-STEM interests and careers.

We argued that these students may be seen as a loss to the sciences because their innovative and synthetic ability to connect knowledge and skills from both STEM and non-STEM disciplines is valuable in addressing complex societal problems.

A second group of high-performing students that accounted for 22% of all switchers were pursuing a STEM career without a STEM degree. Ninety percent of these students had entered college with high SAT or ACT math scores, and 62% had GPAs of 3.0 or greater in their STEM majors prior to leaving them. Most (71%) were continuing to follow academic pathways into the same health careers that they originally planned or had broadened their end goals to use the knowledge gained in their alternative majors. Others in this group planned to stay in math- and technology-related fields, and many in both career groups continued to take STEM requisites for graduate school application.

Unlike the high performers who moved into arts and humanities, these students were among the 26% of all switchers who (as discussed in Chap. 3) often switched for instrumental or tactical reasons. Among these were:

- Increasing their chances of acceptance into competitive STEM graduate programs by improving their GPAs;
- Moving into degree programs that took less time to complete or were less expensive;

- Gaining particular combinations of STEM and non-STEM knowledge and skills. (As with the arts and humanities switchers, some also sought to blend science with other disciplines by undertaking double majors or adding minors.)

Other shared group characteristics were:

- Women comprised 76% of the group, and students of color of both sexes 28%;
- Many had high verbal as well as high math skills. They, thus, had more choice of alternatives to STEM pathways to achieve their career goals;
- They did not switch because of academic difficulties in their STEM majors, but were unable to maintain their grades without high stress and feelings of overwhelm;
- They rejected the careers and lifestyle for which STEM degrees appeared to prepare them, based on stereotypes about working scientists reinforced by STEM experiences;
- They moved to majors that were less stressful but still allowed them to achieve their career goals.

Some high-performing switchers might subsequently contribute to STEM-based fields. However, their contributions could be greater if STEM programs recognized and supported students who have broad interests, addressed aspects of the design and pedagogy of STEM classes outlined in Chaps. 6, 7, and 8, and broadened and promoted the career foci of STEM majors.

We have also stressed in this chapter the loss to STEM disciplines of high-performing women. From the institutional records analyses, we reported that, in the high GPA quartile, 12% of women switched versus 6.5% of men—losses that also occur in the context of the over-representation of women among high-performing STEM students. As is indicated in Fig. 10.2, of all switchers in the high GPA quartile, 62% are women, while only 48% are men. (Losses from the three lower GPA quartiles combined are almost evenly divided between women and men.)

Our interview data on high-performing STEM majors reveal that high-performing women, most especially women of color, and high-performing men of color are all at higher risk of switching than high-performing white men. From our analysis of why high-performing students switched, we found that, compared with high-performing men, women of all races and ethnicities experienced a higher incidence of:

- Loss of confidence.
- Loss of interest and incoming motivation.
- Poor high school preparation.
- Difficult transitions to college.
- Financial difficulties in completing college.

Women slightly more often than men also described as contributory causes of their leaving a STEM major their rejection of STEM careers and associated future lifestyles or being drawn into a non-STEM field that they saw as offering more attractive career opportunities.

High-performing male switchers cited higher rates than high-performing women switchers of:

- Problems with curricular design, including fast pace and overload.
- Conceptual difficulties in one or more of their STEM courses.
- Discovering an aptitude for a non-STEM subject as a reason for switching was also far more common among high-achieving male switchers than high-achieving women switchers.

Both male and female high-performing switchers were equally critical of the aspects of STEM learning experiences and of the competitive ethos of STEM classes, with nearly half of high-performing men and women mentioning this as a reason for leaving STEM.

We have drawn attention to women who find other ways to meet their career goals than persisting in STEM courses that induce high levels of stress. Another group of high-performing switchers that we have labeled "perfectionists" (who are often, but not exclusively, women) have very low tolerance for set-backs caused by less-than-perfect performances in particular courses. In our institutional data analysis, we found that high GPA STEM majors are at risk of switching in response to receiving a single incomplete or course withdrawal during their first two college years: high-performing STEM switchers are almost twice as likely as high-performing STEM persisters to have received one poor grade or have withdrawn from one class. Descriptions of their own perfectionist tendencies by both high-performing persisters and switchers included mental and emotional distress caused by their low tolerance for less than perfect performances. However, high-performing women of all races and ethnicities with perfectionist tendencies had greater difficulties in adjusting unrealistically high self-expectations to STEM curve-grading practices than their white male peers. By men's and women's accounts, high-performing women struggled more to adjust to the weed-out practices that are a trademark of gateway STEM courses. Moderate grades misinterpreted as failure are interpreted as an intolerable threat to identities that are extrinsically derived from high grades and the status they confer. As we have argued, traditional explanations for the loss of women from STEM—such as loss of confidence or the negative effects of the competitive ethos—are still relevant, but we conclude that they may more accurately apply to that larger proportion of women in STEM who are not the highest performers. Efforts to help highly talented students better understand and adjust to STEM grading practices may lessen their loss. The losses of high-performing students from STEM majors for the distinctive reasons we discuss in this chapter will be, we expect, matters of concern to institutions and STEM departments. Having actively recruited men and women of all races and ethnicities with demonstrated talent and intrinsic interest in STEM disciplines, they may be unaware of and alarmed to learn how many such students they are losing and why.

Finally, findings from our interview data refute the myth of switching as an outcome of largely rational and maturational decisions. Rather, most switchers struggle to reach their decisions, and many persisters also struggle as they consider the idea of leaving or relocating. Switching and relocation are predominantly not the outcomes of decisions made in single moments in time, but are processes represented in series of events, personal reflections, emotions, interactions with others, and rejections of perceived STEM career scenarios compared with alternative projections—both of which may be poorly or well-informed. Reaching a decision to move, and dealing with its aftermath, are also costly in personal and emotional terms that include dealing with stigma and needing to legitimate alternative choices to self and others.

The processes of switching are also far more *haphazard* than the rational choice model would predict. This is especially evident in the decisions of perfectionist students for whom experience of one setback (often a moderate grade misinterpreted as failure) were often the basis for a move intended to restore self-esteem. We propose that low tolerance for less than perfect performances is a more accurate explanation for switching by many high-performing women than is loss of confidence—which is an explanation of long-standing in the research literature. There is still strong evidence in both our current and former study, and in ongoing research by others, that lost confidence is a major contributor to many women's switching decision. However, it would seem to apply more accurately to that larger proportion of women entrants to STEM majors who are not the highest performers.

Reaching a decision to move was also *situationally contingent* in that it might have turned out differently had those students been in another institution or STEM program. This was very evident when students grappled with the *lack of resources or appropriate interventions* that would have enabled their persistence. At the individual level, unmet needs that could be pro-actively met by STEM departments and their faculty include insufficient academic help and mentoring, inappropriate advising, inadequate information about academic and career path options, timely intercession to check the bad behavior that fuels negative climates, and insufficient accommodation for health or other life crises.

We also note that blending science with other disciplines, such as art, business, communication, or social sciences, is a growing trend that was barely discernable in the original study. In addition to students who made pragmatic moves out of STEM majors the better to pursue a career that included science and math skills and knowledge, it was also evident in the numbers of students who opted for double (or more) majors—some within the sciences, and some that crossed disciplinary lines—and in minors in a wide variety of fields that accompanied STEM majors. These findings suggest broader remedies that imply reconceptualization of STEM majors as part of a liberal arts education with applications and relevance for many career pathways and lifetime interests beyond those more narrowly defined by each discipline. Operationalizing this might include institutional promotion of inter- and multi-disciplinary studies as pathways toward STEM careers, and departmental dissemination to its students of information about the breadth of career options open to graduates with particular STEM degrees—a strategy that, in itself, could reduce the flow of high-performing STEM majors (notably women) into non-STEM majors.

In Chaps. 6, 7, and 8 we focused on situational needs and interventions in course design and quality of teaching that, when poorly met, contribute to losses among students with a wide range of abilities. In this chapter, we have highlighted the risks of loss among the top quartile of STEM entrants—a sub-set that includes a high proportion of very able women, multi-talented, intellectually agile and creative students, and career-goal-oriented students who abandon STEM majors but not their interest in the sciences. All of these may well be seen as students that the STEM disciplines might not wish to lose. Happily, many of the remedies that might prevent such losses lie within the imaginative capacity of STEM departments and their institutions.

## References

American Psychiataric Association. (2019). Retrieved April 4, 2019, from https://www.apa.org/advocacy/higher-education/mental-health/index

Boe, M. V., Henriksen, E. K., Lyons, T., & Schreiner, C. (2011). Participation in science and technology: Young people's achievement-related choices in late-modern society. *Studies in Science Education, 47*(1), 37–72.

Bressoud, D. (2015). Insights from the MAA national study of college calculus. *The Mathematics Teacher, 109*, 178–185.

Bressoud, D., Mesa, V., & Rasmussen, C. (2015). *Insights and recommendations from the MAA national study of college calculus* (MAA notes). Washington, DC: The Mathematical Association of America.

Carter, M. M., & Weissbrod, C. S. (2011). Gender differences in the relationship between competitiveness and adjustment among athletically identified college students. *Psychology, 2*(02), 85.

Cashman, O., & Twaite, J. A. (2009). *Toxic wealth: How the culture of affluence can harm us and our children*. Santa Barbara, CA: Praeger/ABC-CLIO.

Chang, E. C. (2006). Conceptualization and measurement of adaptive and maladaptive aspects of performance perfectionism: Relations to personality, psychological functioning, and academic achievement. *Cognitive Therapy and Research, 30*, 677–697.

Chang, E. C., Watkins, A., & Banks, K. H. (2004). How adaptive and maladaptive perfectionism relate to positive and negative psychological functioning: Testing a stress-mediation model in black and white female college students. *Journal of Counseling Psychology, 51*(1), 93.

Chen, X. (2015). STEM attrition among high-performing college students in the United States: Scope and potential causes. *JOTSE: Journal of Technology and Science Education, 5*(1), 41.

Curran, T., & Hill, A. P. (2017). Perfectionism is increasing over time: A MetaAnalysis of birth cohort differences from 1989 to 2016. *Psychological Bulletin, 145*(4), 410–429. https://doi.org/10.1037/bul0000138

Dean, P. J., & Range, L. M. (1996). The escape theory of suicide and perfectionism in college students. *Death Studies, 20*(4), 415–424.

Festinger, L. (1954). A theory of social comparison processes. *Human Relations, 7*, 117–140. https://doi.org/10.1177/001872675400700202

Fletcher, K. L., Speirs Neumeister, K. L., & Flett, G. L. (2014). Introduction to the special issue: Perfectionism in the school context. *Psychology in the Schools, 51*(9), 897–898.

Flett, G. L., & Hewitt, P. L. (2012). Perfectionism and cognitive factors in distress and dysfunction in children and adolescents: Introduction to the special issue. *Journal of Rational-Emotive and Cognitive-Behavior Therapy, 30*, 53–61.

Flett, G. L., & Hewitt, P. L. (2013). Disguised distress in children and adolescents "flying under the radar": Why psychological problems are underestimated and how schools must respond. *Canadian Journal of School Psychology, 28*, 12–27.

Gnilka, P. B., Ashby, J. S., & Noble, C. M. (2012). Multidimensional perfectionism and anxiety: Differences among individuals with perfectionism and tests of a coping-mediation model. *Journal of Counseling & Development, 90*, 427–436. https://doi.org/10.1002/j.1556-6676.2012.00054.x

Gnilka, P. B., Ashby, J. S., & Noble, C. M. (2013). Adaptive and maladaptive perfectionism as mediators of adult attachment styles and depression, hopelessness, and life satisfaction. *Journal of Counseling & Development, 91*, 78–86. https://doi.org/10.1002/j.1556-6676.2013.00074.x

Heilbronner, N. (2011). Stepping onto the STEM pathway: Factors affecting talented students' declaration of STEM majors in college. *Journal for the Education of the Gifted, 34*(6), 876–899.

Hewitt, P. L., & Flett, G. L. (1991). Perfectionism in the self and social contexts: Conceptualization, assessment, and association with psychopathology. *Journal of Personality and Social Psychology, 60*(3), 456.

Hewitt, P. L., & Flett, G. L. (1993). Dimensions of perfectionism, daily stress, and depression: A test of the specific vulnerability hypothesis. *Journal of Abnormal Psychology, 102*, 58–65.

Holmegaard, H. T., Madsen, L. M., & Ulriksen, L. (2014). To choose or not to choose science: Constructions of desirable identities among young people considering a STEM higher education programme. *Journal of Science Education, 36*(2), 186–215.

Krapp, A., & Prenzel, M. (2011). Research on interest in science: Theories, methods and findings. *International Journal of Science Education, 33*(1), 27–50.

Lubinski, D., & Benbow, C. P. (2006). Study of mathematically precocious youth after 35 years. *Perspectives on Psychological Science, 1*, 316–345.

Luthar, S. S. (2003). The culture of affluence: Psychological costs of material wealth. *Child Development, 74*, 1581–1593. https://doi.org/10.1046/j.1467-8624.2003.00625.x

Luthar, S. S., & Sexton, C. C. (2004). The high price of affluence. *Advances in Child Development and Behavior, 32*, 125–162. https://doi.org/10.1016/S0065-2407(04)80006-5

Magolda, M. B. B. (2014). Self-authorship. *New Directions for Higher Education, 166*(2014), 25–33.

Maltese, A., & Tai, R. T. (2011). Pipeline persistence: Examining the association of educational experiences with earned degrees in STEM among US students. *Science Education Policy, 95*, 877–907.

Mann, A., & Diprete, T. A. (2013). Trends in gender segregation in the choice of science and engineering majors. *Social Science Research, 42*(6), 1519–1541.

Neumeister, K. L. S. (2004). Understanding the relationship between perfectionism and achievement motivation in gifted college students. *Gifted Child Quarterly, 48*(3), 219–231.

Rice, K. G., & Ashby, J. S. (2007). An efficient method for classifying perfectionists. *Journal of Counseling Psychology, 54*, 72–85. https://doi.org/10.1037/0022-0167.54.1.72

Rice, K. G., Lopez, F. G., & Richardson, C. M. E. (2013). Perfectionism and performance among STEM students. *Journal of Vocational Behavior, 82*, 124–134. https://doi.org/10.1016/j.jvb.2012.12.002

Riegle-Crumb, C., King, B., Grodsky, E., & Muller, C. (2012). The more things change, the more they stay the same? Prior acheivement fails to explain gender inequality in entry into STEM college majors over time. *American Educational Research Journal, 49*, 1048–1073.

Ruane, I. W. (2012, July–August). Effortless perfection. The Undergraduate. *Harvard Magazine*, 55–56. Retrieved from http://harvardmagazine.com/2012/07/effortless-perfection

Stoeber, J., Hutchfield, J., & Wood, K. V. (2008). Perfectionism, self-efficacy, and aspiration level: Differential effects of perfectionistic striving and self-criticism after success and failure. *Personality and Individual Differences, 45*, 323 327. https://doi.org/10.1016/j.paid.2008.04.021

Suicide Prevention Resource Center. (2019). Retrieved 4/1/2019 from https://www.sprc.org/settings/colleges-universities

Taconis, R., & Kessels, U. (2009). How choosing science depends on students' individual fit to 'science culture'. *International Journal of Science Education, 31*(8), 1115–1132.

Tomasi, S. E., Fechter-Leggett, E. D., Edwards, N. T., Reddish, A. D., Crosby, A. E., & Nett, R. J. (2019). Suicide among veterinarians in the United States from 1979 through 2015. *Journal of the American Veterinary Medical Association, 254*(1), 104–112.

Travers, L. V., Randall, E. T., Bryant, F. B., Conley, C. S., & Bohnert, A. M. (2015). The cost of perfection with apparent ease: Theoretical foundations and development of the Effortless Perfectionism Scale. *Psychological Assessment, 27*(4), 1147–1159. https://doi.org/10.1037/pas0000109

Twenge, J. M. (2014). *Generation me: Why today's young Americans are more confident, assertive, entitled and more miserable than ever before* (2nd ed.). New York, NY: Atria.

Verhaeghe, P. (2014). *What about me? The struggle for identity in a market-based society*. London, England: Scribe Publications.

Wang, M.-T., & Degol, J. (2013). Motivational pathways to STEM career choices: Using expectancy-value perspective to understand individual and gender differences in STEM fields. *Developmental Review, 33*, 304–340.

Wolfson, A. J., Cuba, L., & Day, A. (2015). The liberal education of STEM majors. *Change: The Magazine of Higher Learning, 47*(2), 44–51.

Yee, C. (2003, September 23). Committee unveils women's Initiative report. *The Chronicle*. Retrieved from http://dukechronicle.com/article/committee-unveils-womens-initiative-report

Zhang, Y., Gan, Y., & Cham, H. (2007). Perfectionism, academic burnout and engagement among Chinese college students: A structural equation modeling analysis. *Personality and Individual Differences, 43*(6), 1529–1540.

# Chapter 11
# Influences Beyond College that Shape Revised Choices

**Dana G. Holland, Raquel P. Harper, and Elaine Seymour**

## Working While in School

As reported in Chap. 3, 10% of switchers reported that the problems created by working "a lot of hours" were a direct contributor to their decision to switch out of their STEM major. "Working too much" was also cited as a problem by 70% of switchers compared with 48% of persisters. Only 1% of persisters recounted problems that were severe enough to make them consider switching. Thus, an important distinction between switchers and persisters is the degree to which the difficulties of balancing academic work with employment commitments place them at risk of switching. However, the struggle to manage both study and paid employment was also an ongoing source of stress for working students who persisted. These sources of stress were reported by 54% of students overall and was the fourth most cited of students' concerns.

As indicated in Table 11.1, half of STEM switchers and over one-third of STEM persisters worked while pursuing their studies. Similar proportions of male switchers and persisters worked (37% and 36%, respectively), but a higher percentage of female switchers than female persisters worked (viz., 57% and 37%, respectively).

D. G. Holland (✉) · R. P. Harper
Ethnography and Evaluation Research (E&ER), University of Colorado Boulder, Boulder, CO, USA

Center for STEM Learning, University of Colorado Boulder, Boulder, CO, USA
e-mail: Dana.Holland@Colorado.EDU; Raquel.Harper@Colorado.EDU

E. Seymour
Ethnography and Evaluation Research (E&ER), University of Colorado Boulder, Boulder, CO, USA
e-mail: Elaine.Seymour@colorado.edu



E. Seymour, A.-B. Hunter (eds.), *Talking about Leaving Revisited*,
https://doi.org/10.1007/978-3-030-25304-2_11

**Table 11.1** Percent and number of STEM switchers and persisters who worked while in college

| Type of student | Persisters ($N = 250$) | Switchers ($N = 96$) |
|---|---|---|
| All students | 37% (92) | 50% (49) |
| Men | 36% (37) | 37% (13) |
| Women | 37% (55) | 57% (35) |
| Students of color | 47% (47) | 54% (14) |
| Men of color | 49% (19) | 60% (3) |
| Women of color | 46% (28) | 52% (11) |
| White students | 30% (45) | 49% (34) |
| White men | 29% (18) | 33% (10) |
| White women | 30% (26) | 60% (24) |

Among persisters, fewer white STEM students than students of color worked (30% compared with 47%) with only slight variations by gender. Among women of all races and ethnicities, more switching than persisting women worked (viz., 57% compared with 37%). The two groups with the highest proportion of working students were both switchers, namely white women and men of color (each 60%).

## *How Much Do Students Work?*

Both switchers and persisters worked for a wide range of hours, from four hours a week up to full time. However, the majority of students who were employed during the school year worked between 10 and 25 h per week. The most rewarding and manageable number of hours reported was in the 10–16-h range. Both switchers and persisters assessed this level of work commitment as providing enough time for their studies while also providing enough income to be worthwhile:

> I'm probably working like 15 hours a week. I mean still a lot, but enough less that I have more time and I feel like I can focus on schoolwork and stuff more, which is good. (White woman, relocator, biology to environmental health)

> I am working and that has helped me feel more secure for myself I guess just have an income rather than just expenses during the year. I put in about 12 hours per week. (White man, materials science persister).

> I work 16 hours a week. … I'm a nanny for an infant. … It fits well. I mean, school is my priority so I always try to find a family who's flexible just in case something comes up -- so I can switch my schedule. So, once I get my school schedule, then I schedule my nannying. That always works out pretty well. And since he's a baby he sleeps most of the time so I can do homework. … What I have to cover is bills, gas, groceries, leisure. I'm barely making that much money a month but I don't really have pressure financially…I'm, you know, grateful. (White woman, switcher, biology to human development and family studies)

Although students who worked in the 10–16-h range seemed relatively happy to work while in school, this group was also more likely than those who worked longer hours to have significant financial help from parents, loans, or scholarships. Thus, their earnings were a top-up to more significant contributions to their education

A few students worked less than 10 h per week. These students mostly worked for "extra spending money" and some described their jobs as "fun":

> It's only like 7 to 10 hours, so it's not very many, but it's enough for extra spending money. (White woman, mathematics persister).

> Right now, I'm working about fifteen hours a week. ... Everything I earn pretty much just went to having fun. (Asian woman, switcher, chemistry to Spanish)

> It's not much, it's probably like six [hours] or so. ...The work study job that I have is I do communications for a non-profit in town... It's really fun. (Hispanic woman, switcher, physics to anthropology)

> I'm working four hours a week....Right now I teach dance, which is more of a hobby than anything. (Multi-racial woman, chemical engineering persister)

Students who worked less than 10 h reported little time conflict between their college work and meeting their employment commitments. They mostly worked by choice, either for spending money or because the work involved something they enjoyed and did not experience the kinds of disruption in their school work reported by students who worked longer hours.

Although we cannot offer a precise number, it was clear from students' accounts that many of them worked in the 20–25-h range, and that, at this level of employment, STEM majors and their college work both start to suffer. STEM persisters reported difficulties in balancing work and school and some reported a drop in their grades:

> There's absolutely no way to get a STEM degree while working full time. ... I work 20 hours a week and it's a push. And I know my grades suffer somewhat because of it but it's necessary. ...It goes to gas, groceries, that kind of thing. If I had to pay for my own rent I would need to be working more. If I had to pay for college and my own rent I probably wouldn't be going to college. (White woman, biomedical engineering persister)

Working in the 20–25-h range created a switching risk when it proved too hard to balance both work and school, and some students simply had to work this much. One woman reported that switching from chemistry to language studies had made it possible for her to work:

> I work at a yogurt store. It varies between like 20 and 30 hours... ... But I can manage it now with my current schedule and class load [since switching out of STEM]. (White woman, switcher, chemistry to French)

Only a few students reported working full time or almost full time. And, surprisingly, some of them were persisters. Two persisters really struggled, but were motivated enough to keep powering through. Another, who worked full time found it more manageable to be a part-time STEM major:

> I try to reach 40 hours every week. But with six credits, two classes, it's easy to do. (White man, wildlife biology persister, full-time work, part-time student)

Another persister viewed herself as lucky that her tuition was paid for by her parents, and that she only had to work to pay living expenses:

> I usually work about 25-30 hours a week. And that usually takes up most of my time outside of class. ... And I usually don't have much free time ... I have an agreement with my Dad

> that he pays my base tuition and I'm responsible for everything else. So, that includes books, food, rent, gas, insurance--everything like that I pay for myself. I took one loan out for my study abroad but everything else is paid for. I got very lucky in that regard. (Hispanic woman, equine science persister)

Despite the heavy demands of their STEM programs, some STEM persisters found ways to juggle full-time employment alongside their college work.

## *Why Students in STEM Majors Work*

We asked students why they worked and what benefits they gained from their jobs. The most common reasons were that they were contributing to their college education and/or to their own maintenance. A common theme in their rationales was an awareness that their family could not pay for any, or only part, of their college costs. Students seemed very ready to make a good contribution and expressed their concern not to place more of a financial burden on their parents than they could manage:

> I don't want to put a burden to pay for my housing on my parents. I can't just allow them to do that. They have their own house to take care of and my two brothers … I've had to work and I found a pretty good job. (White man, relocator, mathematics to actuarial science)

> I have a job working for Kaplan Test Prep. It's a company that does things for the MCAT, the LSAT. … I'm making my own money, which is nice, 'cause I don't want to be a financial burden [on my parents]. I'm ready to be independent. (White man, physiology persister)

As with the last speaker, some STEM persisters expressed pride in their ability to contribute to their educational and living costs and experienced the beginnings of financial independence. However, some also alluded to family tensions around shared responsibility for college costs:

> I need the mental gratification of putting in my part. … It's definitely a tense subject to talk about with family because it is not the most affordable school. (Multi-racial man, relocator, computer engineering to environmental science)

One switcher's account of her need to work full time is a reminder that some students (as we found in the original study) carry the responsibility of caring for a family member as well as for themselves:

> I work full-time. I live with my mother because she has a mental illness. I actually have to be the mother to her because she's not mentally able to be … We were actually in an abusive household; she married a man that sexually, physically and mentally abused both of us … She really doesn't have a resource besides me to take care of her … I'm the only support system that's holding my family together. … I pay for just about everything but food and rent… I cover doctor's bills, I cover medicine, I cover everything else. (White woman, switcher, biology to sociology)

We also noted a willingness to take responsibility for their own upkeep, a strong work ethic, and an appreciation of the value of work itself:

> I thought it was important to take on responsibility … I am at college to learn, but I decided it's important to also get some work experience. (Hispanic woman, switcher, biology to journalism)

> And I took out some student loans because I'm paying for this year of school myself, which is fine (Multi-racial woman, chemical engineering persister)

> It helps me with time management, because I know when I have to work I can't do school, but if I didn't have to work I would have too much time and I wouldn't be able to focus I would feel lazy I think if I wasn't [working]. (White woman, switcher, biology to psychology)

Being financially responsible for some part of their education was also recognized as making them more appreciative of the cost of their education and better at handling their own money:

> I pay for my own rent, I have a job that I can work 20 hours a week, 40 hours a week during breaks and the summer. And because I pay my own rent and have to buy my own food, that's really shaped where I live; it shapes what I choose to do recreationally, and it's helped me learn to find ways to be as financially responsible as I possibly can. I see a lot of students who go and live in apartments that are small and horrifically expensive … and they go eat out several times a week, and meanwhile I've learned how to build a diet that satisfies my caloric and nutritional needs while costing a couple dollars a week. (White man, astrophysics persister)

> My parents help me out as much as they can. When they went to college, they didn't get their college paid for. They had to take out loans and pay for it. And, I'm the youngest of three kids. Both my sisters, they had to pay for their college. It teaches accountability and responsibility. It's challenging. I mean, if I didn't have to work, yeah; maybe my grades would be a little bit higher than what they are now. (White man, wildlife biology persister)

The last speaker also illustrates the tradition of paying as much of your own way through school as you can as a moral responsibility. Both speakers also refer to the personal costs of shouldering this responsibility—one in terms of the enforced frugality of his diet; the other in accepting lower grades than he might otherwise have achieved.

Working students also explained some of the benefits of working. One benefit of having your own income is greater financial independence from parents, including, in the following instance, the pleasure of independent living:

> When I got my job at [the national laboratory], that enabled me to do a whole lot of things, 'cause I suddenly had my own income. And that allowed me to move out of my mom's house … I lived at home my sophomore year, and then I got a job at LASP and I was like 'Oh, now I can pay for rent, and I can now support myself outside.' So that allowed me to move out. (African-American man, aerospace engineering persister)

Other working students cited greater appreciation of their education through their own investment in it. This was, for some, a powerful source of motivation to gain as much as possible from their degree course:

> [Other students] are going through college without any debt and they're just like, 'Well, if I learn it, I learn it. If not, I don't.' And I'm like, 'No, this is it. This is make or break for me.

I'm putting the money into it, so, I'm really going to get as much as I can out of it.' And I feel like I have. (White man, mechanical engineering persister)

I work in the summers waiting tables and then I work on campus tutoring Italian.

… It's a great job. I think that the fact that I know how much is going into my education and what I'm gonna end up with debt-wise when I get out has me like, 'I need to do as much as humanly possible in the four years that I'm here.' (Hispanic woman, biology persister)

Another practical benefit of work experience was to enhance their future graduate school or job applications:

I work on campus. I'm a student coordinator, so that's not related to my major, but it looks good on your resume. (Asian woman, switcher, geology to finance)

Whatever the benefits of paid employment while in college, students were very clear that it is also difficult and often stressful. Some persisters who had to work described their struggle to make it through their STEM programs:

There's absolutely no way to get a STEM degree while working full time. … I work 20 hours a week and it's a push. I know my grades suffer somewhat because of it, but it's necessary … It goes to gas, groceries, that kind of thing. If I had to pay for my own rent, I would need to be working more. If I had to pay for college and my own rent, I probably wouldn't be going to college. (White woman, biomedical engineering persister).

My life pretty much consists of work and school. Like I don't get to talk to my mom very often, and when I do, it's usually quick conversations. My sister just had a baby, which I probably won't be able to see until I'm done with school. And that's partially because of working while in school. I think normally, I'd be able to go there for Thanksgiving, but Thanksgiving I'm going to work. (White man, relocator, chemistry to chemical engineering).

We tried to tease out what enabled some students to persist while maintaining significant paid work commitments, while other students resolved these stresses by switching into a major that more easily accommodated their need to work their way through college.

## *Why Do More Working Persisters than Working Switchers Survive?*

The most obvious answer to this question is that, as noted at the outset, more switchers than persisters worked, and more of them worked long hours. References by our interviewees—particularly by students of color and women—to families that cannot afford to help them financially point to the theme of structured disadvantage that we have tracked throughout this book. The decreased availability of financial aid and increasing resort to student loans have made going to college more difficult for working class and first-generation students—including many students of color. Many such students opt for community college as a more affordable alternative. For those who enter STEM majors and work long hours to self-fund their education, switching may be the only way to stay in college if managing the dual load becomes unsustainable.

About half of switchers who worked described their experiences as a serious, constant struggle to balance work and school. Although 17% of persisters reported similar difficulties, almost three times as many working switchers (50%) experienced financial strains and feeling "overloaded," "overwhelmed," or "stressed all the time":

> It's challenging, trying to balance work and school. You know, I need to work, but I also need to do well in school. I need to have the time in school. It gets really challenging: My weeks are really crazy … I'm working about 12 [hours per week]. I can't really go any lower than that with my budget. There's been times already this semester where I just don't have time. …I know if I didn't have to work at all, I would do really well. There would be a difference in my grades if I didn't have to work at all. (White man, switcher, physiology to architecture)

> I work 16 hours a week … and that can be tough 'cause usually if I have a couple tests the next week. So. I'm in the library all weekend studying. When I had my mid-terms I had to work the weekend before and that kind of threw me up a bit. (White woman, switcher, biomedical sciences to sports medicine)

These switchers were frustrated because they knew that they were performing less well than they would have done had they not needed to work.

The next speaker introduces an additional source of obligation—the requirement that she regularly contribute to the life of her community. In the original study, we also found this obligation to be a distinctive feature of the lives of Hispanic and Native American students:

> I don't come from money. I never had money, so having to work two jobs, maintain a full course load, and then be involved in my community was something that was really hard. (Hispanic woman, switcher, biology to communications)

Another possible explanation for differences in the extent to which persisters seemed more able than switchers to tolerate dual college-work stresses and to continue in their STEM majors may involve their deployment of time management skills. Several of the working persisters and switchers cited above commented on their lack of good time management abilities, and others credited working as having helped them improve these skills. This STEM relocator points to the critical role in survival under stress played by efficient time management:

> I'm taking 16 credits in math and science and on top of that I'm working 20 hours a week. So sometimes that doesn't reflect well on my GPA, when I'm trying to balance it. It's definitely hard to manage and I have poor time management anyway, so I'm still trying to figure it out. (Hispanic woman, relocator, biology to chemistry)

A persister who credited his job with forcing him to use his time well cited as one positive outcome that "I feel like I can focus on school work more." We hypothesize that, already possessing the ability to manage one's time efficiently, or learning how to do this through necessity while juggling demanding academic and employment roles, may contribute to differences in the survival rates among STEM majors who work.

Other explanatory clues arise in persisters' accounts of what they gain by working despite having less time for study and other aspects of life. Indeed, taking a positive view of employment may, in itself, enable persistence. We found that more STEM persisters than switchers described the positive values of working over the negative consequences. For example, one male persister studying materials science described feeling "more confident and secure" because of having a job while in college; a mechanical engineering persister reported that working in a job on campus had "helped him get to know people" better and feel more connected at the university; a mathematics persister who tutored middle- and elementary-school girls explained that, despite having to stay up late after work to finish homework, she was "very passionate" about her job; and an African-American student studying mathematics told us that, although "it's quite stressful at times," he "thrives under pressure." Thus, working while in school actually seemed to help these students feel more confident, connect to others at school and work, and, as illustrated earlier, motivated them to get the most out of degree courses.

These lines of explanation are tentative, but they suggest the need to learn more about the positive and negative consequences of working while in college, the work load limits on positive (or neutral) effects, the role of time management in persistence, and, perhaps most significantly, the role of financial hardship in limiting the success of even talented students. It is to this issue that we now turn.

### *Has the Impact of Paid Work on STEM Majors Changed since the Original Study?*

In 1997, we reported on the effects on STEM students of a trend that began in the late 1970s, of decreasing public contributions to higher education, including student tuition and fees. Students discussed their difficulties in getting student aid, and we noted that the competition for shrinking financial aid had become racially divisive. White students often suspected that students of color were receiving financial assistance that they found it increasingly difficult to get. Approximately two-thirds (63%) of the interviewees had taken out loans and 56% were meeting some proportion of their educational and personal expenses by working, the average being 18 h per week. In the current study, what students then regarded as a change away from more generous policy of student financial support is now regarded as simply the norm. Thus, student expectation of public financial aid is now far less discussed. Although some students reported scholarships based on performance, most students who worked because of limited family support or financial aid simply expected to do so.

That said, the stress that significant hours of work creates for STEM majors appears to have increased. As indicated in Table 11.2, 7% fewer interviewees in the current than the original study reported that difficulties in balancing the work of their STEM major with that of paid employment had contributed to their decision to switch. However, the work-school stresses reported by switchers, persisters, and students overall have doubled.

**Table 11.2** Comparison of the original and current study in contributions to STEM switching and student stress of working while in college

| Book | % factor in switching decisions | % all switchers' concerns | % all persisters' concerns | % all students' concerns |
|---|---|---|---|---|
| *Talking about Leaving* (1997) | 17 | 30 | 23 | 27 |
| *Talking about Leaving Revisited* (2019) | 10 | 70 | 48 | 54 |

A number of recent studies illuminate some of the causes and consequences of increases in students who work while attempting STEM degrees. For a family earning a yearly average of $20,000, Broton and Goldrick-Rab (2015) calculated that, since 2008, the average net price of a STEM degree at a public university (i.e., after all Pell, state and institutional grants are subtracted) has grown to over $12,000 a year. (The comparable price for a year at a community college began at $8000.) For students at this family income level Kirshstein (2013a) cites the students' net price range as $7800 for a public 4-year university to $30,000 at a private university. The College Board (2014) cites a recent rise in percentage of students receiving Pell grants (which are need-based) from 25% in 2007–2008 to 36% in 2012–2013. However, Feeney and Heroff (2017) found that students in most need may not get a Pell grant for reasons that include limited awareness by students and their families of the process and deadlines involved in applications. They may also lose grants because they exceed the allowed time to degree completion—a contingency that can arise from working long hours and opting for fewer credit hours per semester.

The College Board also reports that, over the last 10 years, even with more students receiving financial aid, student debt from public 4-year universities increased by 12% between 2001 and 2007 and by an additional 20% over the subsequent five years. Thus, by 2012, about 60% of majors graduated with an average debt of $26,500. This rises to 65% and more than $30,000 for STEM majors. To which Kirshstein (2013b) adds, that a larger proportion of STEM students from underrepresented minorities graduate with debts of over $30,000 than do students from all other groups.

We have indicated from our interview data some of the stresses that STEM majors who work longer hours face. Broton and Goldrick-Rab (2015), Chaparro, Zaghloul, Holck, and Dobbs (2009), and Freudenberg et al. (2011) document the "material hardships" of students from low-income ($2000,00 p.a. or less) families, including students who work 20 or more hours a week. In their surveys of first semester college students, Broton and Goldrick-Rab found a high degree of student anxiety about having insufficient money to buy what they needed to attend college. Their responses to this situation included: spending more time working (38%); cutting back on food (71%) and utility usage (23%); reducing or stopping driving (48%); postponing medical and dental care (24%), paying off bills (23%), and buying required books or supplies (15%); managing without a computer (19%); and borrowing money or using credit cards more (39%). Chaparro and colleagues found that 21% of students had reduced their food intake because of limited resources and a further 24% were anxious about how they would manage to eat given their lack of money. The risk of food insecurity was unevenly distributed, with students at greatest risk being those with low incomes, poorer health, Hispanic, Black, and financially independent students, those working 20 or more hours per week but with annual incomes under $20,000. The risk for housing insecurity is also high because of insufficient savings for a deposit and lack of a rental history or a guarantor (Joint Center for Housing Studies of Harvard University, 2011; Dworsky, Dillman, Dion, Coffee-Borden, & Rosenau, 2012; Wilder Research, 2008). Broton, Frank, and Goldrick-Rab (2014) assert that insecurity in food and housing has now become greater among college students than it is in the general population. Given this evidence of unmet basic needs, the Committee on Barriers and Opportunities in Completing Two-year and Four-year STEM degrees (Malcom & Feder, 2016) concluded "Students with the greatest financial need have lower rates of degree completion than other students" (p. 124). They add that we lack research into the factors that create this risk and how they interrelate. However, we would argue that the efforts of working students to overcome insufficient income and the stresses and deprivations that this entails are a likely culprit.

In our original study, we cited (then contemporary) research of Porter (1990) and Rotberg (1990) that shared our conclusion that the constant strain of juggling time and energy between the demands of employment and work for a STEM major created a major risk of switching and also dropping out of college altogether. We concurred with the authors of both studies that the risks of loss were especially great for students of color. One further observation from the original study is perhaps worth repeating. Some of our student commentators on the weed-out system described it as, in effect, a means test that was biased in favor of students with sufficient independent funding for college. Among those with inadequate funds—especially those who worked longer hours—weed-out courses selected only those with the stamina to simultaneously meet both the heavy economic and academic demands made on them.

## Parental Influence on Students' Change of Major and Career Intentions

In Chap. 4, we reported that parental attitudes toward their education and potential career paths were students' most influential considerations in selecting a STEM major. Reported by a greater number of switchers (54%) than persisters (40%), parental influence over choice was highly effective when offered as support or encouragement, but it could be counter-productive when perceived by the students as undue pressure. The influence of parents on student choices seems to have increased since the original study, while the advisory role of K-12 teachers has declined—a trend also noted by Hall et al. (2011), Sjaastad (2012), and VanMeter Adams et al. (2014).

The majority of students described parental encouragement and support for their choice of a STEM degree. However, switchers were more likely than persisters to have a parent who had pressured them into a STEM field that they viewed as the most desirable, and a few parents made their financial support for college contingent on pursuit of a STEM degree. Parental pressure was more common in immigrant communities that highly valued particular STEM-based professions. We found that switchers who had selected a STEM major largely in response to parental pressure entered the discipline with less interest than did persisters. They were also more likely to discover that they had chosen a major for which they were ill-suited. Parents of either sex who were scientific or technical professionals were often followed by their children into the same or related fields. In the wider parental population, as Sjaastad (op cit) and Simpson (2003) also noted, fathers exert more influence than mothers in the choice of STEM majors.

Here we take up the story of how parents—both together and individually—responded to their children's decisions to switch out of STEM majors and/or career pathways. Sometimes, as discussed in Chap. 4, parents played an active role in the process by which students changed majors; sometimes they expressed their views but left the decision to their children. Students' descriptions of their parents' reactions to their decision to leave a STEM major provide us with information about how parents prioritize particular aspects of their children's higher education and expected career pathways. These include the criteria, assumptions, and past experiences that they draw on in judging the validity of their children's proposed change of direction.

As also noted in Chap. 4, we found that the gender of the student mattered little in how parents reacted to announcement of an impending move to a non-STEM major. This was a significant change from our finding in the original study that daughters received more indulgent treatment than sons, especially from fathers, when proposing a move out of a STEM major. As we reported in our 1997 study, sons were commonly admonished to resolve their problems by toughing it out, working harder, and sticking to what they had begun. However, both parents commonly expressed concern about high levels of stress in their daughters and more readily gave them permission or support to move into fields where they would perhaps be happier. Some parents also expressed relief that their daughter was moving to a major or career path that they saw as "more appropriate" for a young woman. Some women who persisted, but who had expressed frustration with their STEM major to their parents along the way, reported having to resist parental encouragement to switch to something less stressful and "more normal."

We found no trace of these parental attitudes in the current study. Indeed, fathers, in particular, now regard the STEM disciplines as a sound basis for greater certainty of employment in well-paid, prestigious careers for their daughters. In focus group discussions with female STEM seniors, some daughters expressed concern that their fathers' expectations might be under-informed or inflated, but none of them had been obliged to argue with their parents in order to choose or persist in a STEM career path. We also found no vestige of a former parental view of college as a place to find a suitable marriage partner. Indeed, our female interviewees found it incredible that such views had been common only twenty years before.

Where gender divergence was evident in the current study was in the gender of the parents rather than that of their children. As we shall illustrate, mothers and fathers often reacted in distinctively different ways to a child's proposal to switch into a non-STEM major and career path.

## *Supportive Parental Responses to Switching*

Overall, almost twice as many students described their parents as supportive than unsupportive of their switch from a STEM to a non-STEM major, namely 60% compared with 33%.

For some parents, recognition of their child's move out of a STEM program was interpreted in the larger framework of a positive assessment of his or her level of maturity. These parents trusted students to make good judgments about what degree and career pathway was best for them. They had also anticipated that the experience of higher education would involve self-discovery and that their children would quite possibly rethink some of their initial ideas about their future:

> I don't think [my switch] bothered my parents at all. They understand that the freshman year of college is figuring out what you want to do a little bit and they understood. (White man, switcher, biology to film studies)

> My parents were really supportive, [and took the attitude that], 'I don't want you to be doing a major that you're not enjoying.'…So, it was nice that they trusted me and let me make my own decisions. (White woman, switcher, computer science to health and wellness)

> My parents have always been supportive of whatever I wanted to go for. And [despite] the struggles that I had freshman year and throughout my college experience, they have always been like, 'I don't understand why you don't see that you're smarter than you're giving yourself credit for?' They've been very reserved and … mostly they step back and want me to really make those decisions for myself and not feel like I was pressured by something they wanted me to be or wanted me to do. So, while my parents have been very supportive and always there for any help that I needed, they've never pushed any particular route, because they really wanted me to have the ultimate say. And they thought if they shared their opinions they would influence how I felt about it. (Multi-racial woman, switcher, mathematics to hospitality management)

Parents often qualified their support for their child's decision to move out of their STEM major by applying particular criteria in judging the situation. Common among these was whether the new direction would make their child happy. Many parents simply wanted their daughter or son to be happier and less stressed regardless of the degree they pursued. As we also found in the original study, mothers were more likely than fathers to view the happiness criterion as an appropriate rationale for moving from a STEM to a non-STEM major and career path—whether for a daughter or a son. Mothers were more apt than fathers to be aware of the history of stressful experiences that had led to the student's proposed switch. They were often the student's primary confident and sounding board in recounting their troubles, seeking advice, and talking through the alternatives:

> I was just miserable freshman year. The whole chemistry thing just broke me down. I cried all the time. I called [my mom] crying and said how I'm not cut out for this, 'I'm gonna fail.' And then, I took some writing and communication classes, and I loved those. And [my mom] just saw how much of a difference it was. All summer long, I [asked], 'Mom, should I switch my major?' And she's [said], 'You need to do what you love. It will be okay. The money will come.' (Asian woman, switcher, biology to strategic communication)

Mothers were cited as advocates of "choosing something that you really care about"—as a way to find satisfaction in a field of study or a career:

> My mom was very supportive. She was like, 'I don't care if you are in college for ten years until you find what you are passionate about. I want you to do what you want to do.' It was that sociology class; I couldn't wait to get the reading assignments, I loved it. And my mom says that you will have that kind of moment, when [you realize], 'I love this.' I had it and I was just like, 'Mom, I love this!' And it's been true ever since and I still enjoy my classes. (White woman, switcher, computer science to sociology)

Mothers were also close observers of their child's state of mind (and of health) before and after a change of direction:

> [My parents] support me either way…I remember one phone call a couple months after I switched, and my mom just kinda ended the phone call [by saying], 'Kyle, you know, you sound a lot happier now than you did a year ago. You just sound happier on the phone.' (White man, switcher, environmental science to economics)

And some mothers legitimated or proposed the change, based on their observations over time of their child's ongoing distress and their appraisal of how best to resolve it in the longer term:

> They just wanted me to do something that I enjoy. I was very stressed and didn't want to drop the course but my mom actually convinced me to do it. She was like, 'I can tell you're really stressed out about this. I don't think it's worth it to stick with it if you're going to be like this for all four years.' So, yeah. They were okay with it. (White woman, mathematics to hearing and speech therapy)

One mother was aware of her daughter's stress in her biology program, and blamed her decision to leave on her mistreatment by one particular professor. Because mothers tend to have more frequent "check-in" conversations with their children on behalf of both parents, they are often more sensitive to their child's emotional states as they move toward a change of direction.

Some parents qualified their support for a son or daughter's switch out of a STEM program based on their views of the proposed alternative major. These parents valued what they viewed as the career-related practicality of STEM fields and the same criterion was deployed in judging the new major. They particularly wanted their children to pursue majors that would ensure viable employment. Although both parents were often concerned about the career implications of the change of direction, it was more often fathers than mothers who voiced this concern. Where parents differed over their child's decision, typically it was the mother whose primary concern was for her child's happiness and the father who emphasized the practical implications of the change of pathway in terms of career options and likelihood of employment:

> I talked to my mom. She is kind of a, 'Do what you want to do'-kind of person…I knew when I [selected] the film program that I would want to be teacher afterwards. And since my mom lectures here and also teaches at the art institute, she's cool with that. She thinks it's a good idea. My dad, he's more worried about money… [My mom has] told me things to look into, [like] volunteering with film programs around here, or helping to videotape football games…And she always sends me little emails with links. (White woman, switcher, astrophysics to cinema and media studies)

> It was difficult for me to tell my father that I wanted to change out of the pre-med track because he had been really proud of me for that. But I knew I wasn't happy and I didn't want to do that anymore. So, I told him and he was okay with it. But my mom was really supportive with me. She wanted me to be happy and enjoy myself and not have to be miserable at school. So, she was helpful. (White woman, switcher, biology to psychology)

This divergent pattern in mothers' and fathers' primary concerns was also a common finding in the original study.

Also related to concerns about practicality, some parents qualified their support of a switch out of STEM with the caveat that a switch was preferable to dropping out of higher education altogether—a move that carried even more worrisome future implications:

> They were really supportive. They had seen me struggle, and really understood that it probably wasn't right for me. My dropping engineering was fine. [But] if I had said, 'I want to drop out of college'…they would have struggled with that. (White woman, switcher, civil engineering to geography)

> They were totally all for it. I said this is how I feel, this is what I like, this is what's available, and they were like, 'Yeah, go for it, do it.' My mom helped me a lot. My dad helped me a lot with applications and supporting me… Saying, 'If it doesn't go exactly how you want it, we're still your support. Just do it and figure it out.' 'As long as you don't drop out.' 'Just don't come live at home.' (White man, switcher, mechanical engineering to supply chain management)

Family and parent-specific occupational backgrounds also conditioned the nature of parental support for leaving a STEM major. For example, a father who is an architect was more supportive than his wife of their daughter's switch from engineering to film studies because, like his own professional field, he saw this as a "good marrying of art and sciences." His daughter saw the character of his own profession as predisposing him to appreciate her creative interests. In another case, a mother with an undergraduate degree in Spanish was excited that her daughter left chemistry also to study Spanish. Yet another mother who is a teacher was "really excited" when her daughter who switched from engineering was considering a career in elementary education. Parents who had themselves followed an indirect path to their current field of work were more comfortable with the idea that there is not always a direct correspondence between a student's field of study and the type of work that they end up pursuing. This was the case for a father who had studied finance but now ran a handyman business. Two other fathers endorsed their sons' decisions to switch out of engineering because it echoed their own past challenges and negative experiences with that discipline:

> [My parents] didn't care [that I switched]. My dad hated engineering anyways. He's just, 'Do what you want to do.' But he hated [it]. He has a mechanical engineering degree, and

> he's got a Master's in Construction Management, and an MBA. And he hated his mechanical engineering degree. He liked construction management, but he hated the mechanical (program) and work as a mechanical engineer, so he didn't care at all. (White man, switcher, civil engineering to economics)

> A lot of my family did [engineering] so I thought that was kind of a path for me to fight through and get through. I told my parents that I was not doing very well in my classes and I was not extremely interested in it. Luckily for me, my dad went through the same thing, where he really struggled in engineering classes before transferring to business. So, he said, 'Yeah, it is tough. [And] you'll get to a point where you'll have to decide, okay this is what I really want to go into or not.' In hindsight, that point should have come earlier, but it actually came at the end of my sophomore year. (White man, switcher, biomedical engineering to psychology)

As illustrated, a student's switch out of a STEM program could provoke not only support, but active parental enthusiasm. Switches into non-STEM programs sometimes validated a parent's long-held view about where the student most appropriately belonged. This was true, for example, for one woman who switched from biology to community health, thereby realizing her mother's hope for her. A father supported his daughter's switch from mathematics to economics as opening a pathway to "greater things" than the math teaching career she had originally envisioned. Some students' choices of an alternative major were welcomed because they were in the same career field or area of interest as one of their parents, thereby creating commonalities and opportunities to share resources. For example, a father whose daughter had switched from chemical engineering to his field of accounting was "so excited" by her decision and regularly sends her links to accounting-related websites. A young woman who switched from civil engineering to geography inspired her father to consider a new career path for himself in geographical information systems. Other examples similarly revealed positive parental approval for moves that connected the students' revised interests with those of a parent:

> My dad's in business and I definitely think that influenced my decision to do business because I'm definitely exposed to it. And he is more excited about it [than biology] because he can help me with it and we can talk about business now. (White woman, switcher, biology to economics)

> My dad was very supportive of [biology], but he was just as supportive of my looking into other things, as long as it was a career that I could potentially make money in. He was happy about it…He had tried my entire life to get me interested in history, and I wasn't interested in learning about dates and memorizing names. And now I'm in that area, so (Multi-racial woman, switcher, biology to language and literature)

Thus, the rationales that prompt parental support for a student's decision to leave STEM for a non-STEM major reflect a spectrum of criteria that range from a primary concern for their child's happiness at one end to an equally dominant concern for their future employability in well-paid jobs at the other. We found a common pattern of divergence between mothers and fathers at these extremes, but also observe that these parental priorities are not necessarily mutually exclusive in that they are both motivated by concern for a child's well-being. Where they differ is in the timing of "happy" outcomes—either immediate release from present stress and pleasure in a more congenial education experience and career pathway or deferral of current gratification for future happiness in a career with a good income and prospects. The latter view taken by many fathers (but also by mothers based on their own personal and work experiences) is that happiness in the longer term depends on economic independence and security. Indeed, the tougher line that we now perceive fathers taking with their daughters than that we documented two decades ago registers a shift toward the expectation traditionally deployed with sons—that, to be happy in a world where marriage no longer ensures financial security, young women must achieve this by their own efforts.

Another rationale for supporting, encouraging, or promoting an alternative career pathway arises when a parent relates to their child's choice from their own disciplinary interests, aspirations, or work experiences. Some parents also view a switch of majors as an entirely predictable and desirable outcome of higher education that their children will discover who they are and what they want out of life. Contrary to the "deferred gratification" route to a happy life, motivation by passion for a discipline and openness to career options were seen as the best routes to both present and future happiness. For a student to rethink their initial academic and career preferences in light of what they now see as a good fit for themselves is viewed by these parents as normal, desirable, and worthy of their support.

We now turn to what motivates parental disapproval of their children's proposed changes of direction.

## *Negative Parental Responses to Switching*

As indicated, 33% of switchers described their parents as disappointed, concerned, and unaccepting of their decision to switch out of their STEM major, and some parents withdrew, or threatened to withdraw, their financial support. The central question for these parents was what their children could do with a non-STEM degree. They were concerned that non-STEM majors would not provide the knowledge and skills that would secure future employment, and were especially worried that, in an uncertain job market, their children's decisions would lead them into insecure, poorly paid work:

> My mom was like: 'What are you gonna do with it? I just don't know what you're gonna do with that.' (White man, switcher, molecular biology to international studies)

> My father is an attorney and he's like: 'Oh, my gosh! You idiot! The job market is horrendous. What are you doing?' (White man, switcher, microbiology to political science)

> They are kinda worried about me being able to get a job since international relations is so vague. (White woman, switcher, biology to government and politics)

A particularly hard line was taken by parents whose view of the main purpose of higher education is to ensure high-paying jobs. This was a source of contention with children who thought that education served wider purposes:

> My parents were very against it. They didn't like that I might go into something that would not produce a lot of revenue for me in the future. (Asian woman, switcher, biology to youth studies).

> My mom called me and she was like, 'I was looking up salaries for dieticians, and they're not that high.' ... She was getting at the meaning that it's such a waste ... She thinks I'm really smart, and she's like, 'Oh... you should be doing more.' (Asian woman, switcher, computer science to nutrition).

Some students with parents who focused on expectations of high financial returns from investment on a STEM undergraduate degree questioned the factual basis of this view—defining it as inflated or ill-informed about the post-graduate requirements needed to achieve such outcomes:

> My dad was really mad [about the switch]. I think it was a money thing because he thought the only way I was going to be successful or make money is to be a doctor, and that was my only source. He thought, 'Oh, the humanities. You're not gonna make any money with that degree.' So that was a big thing for him because all he wants is for me to be rich and famous one day. (Asian woman, switcher, biology to communications).

> I think my parents would prefer that I'd stuck with chemistry, just for the profitability. They think I'm gonna end up a homeless person with my degree. (White woman, switcher, chemistry to French).

As discussed in Chap. 4, some students had experienced strong parental pressure to choose STEM majors. This was especially true for students from immigrant communities that highly valued STEM-based professions that they saw both as prestigious and as commanding good salaries. When these students decided to switch to a non-STEM major, their parents often found it difficult to accept. For example, this multiracial woman explains how hard the decision was on her Chinese father:

> My father is Chinese and was raised in a very traditional Chinese family. There was a lot of emphasis in his life on having access to the prestige that comes with academia. And then being able to get a wealthy, good job. I guess he and the family don't really think of the arts as worthwhile ... So, he was concerned about my ability to really succeed and find a comfortable career later on. (Multiracial woman, switcher, microbiology to psychology)

Among parents who disapproved of their children switching into non-STEM majors, we found no gender differences in how they treated their children's treatment: that is, they were equally hard on their daughters as they were on their sons.

> My parents are not thrilled. They just want me to have a solid job where I can support myself and be independent. ... It's just harder for them to see that with the major I have now. (White woman, switcher, physics to geography)

That said, fathers were distinctly more unaccepting of switching decisions than mothers, though made no distinction between daughters and sons in this regard. Their motivation for taking a hard stance against proposed moves out of the sciences were the same as fathers who were motivated by concern for their children's future happiness: they wanted both sons and daughters to follow studies that would lead to financial security and believed that this was best secured with a STEM degree:

> My dad, he saw on my schedule that I had entered a film class and he said, 'Tara, I swear to God if you enter this film class, I'm not talking to you for the next six months. This is ridiculous. Why would you ever take this class?' (White woman, switcher, physiology to film studies).

> My dad is super economically motivated. Like he has been so stressed out by money his entire life, and he thinks it's ridiculous that I wanna go from being a science person to being a wilderness therapist where I will get paid like basically minimum wage. … I've already said I'm not in it for the money. But he doesn't really seem to understand that. (White woman, switcher, biology to psychology).

Concern for their child's financial future could also prompt parents to push them into career paths for which the student saw themselves as unfitted. This father reluctantly accepted his daughter's move but his concern for her career options and job stability remained unchanged:

> My father wants me to be financially set … He looked up all the highest paying jobs and gave the list to me. I was like, I'm not going to do this, Dad, because I don't want to be a doctor. I can dissect things that are already dead, but alive people it just freaks me out. He just wants me to be financially okay. He is supportive of me doing what I want to do, but now and again he interjects like, 'Well, you should think about how stable that is or where you can go with that kind of job'. (White woman, switcher, biology to psychology)

Although more students reported that their fathers had the most difficulty in accepting their move onto non-STEM majors, both women and men reported that their mothers also struggled with the proposed change. One man who switched from biology to dance described his mother's concerns about his future financial implications of his choice:

> I called [my mom] and said that classes were not going very well, that I didn't think that I wanted to be a biology major anymore. The first thing my mom said was, 'Well, how about chemistry? Chemistry is easier.' … And I said, 'Well, I was thinking that maybe I'll be a dance major.' And there was just dead silence over the line. And mom said, 'You don't make a lot of money with that. You can't be very successful if you're doing that. Are you sure you can't be a biology or chemistry, or… you know, you can drop the double major. You don't have to double major.' (White man, switcher, biology to dance)

His mother's response affected him so strongly that he planned an additional major to assuage her concerns for his future financial security:

> I think what finally convinced her in the end is I told her that I would find a second major. I would double major with something else that would actually be—I can't remember the word I used—but basically along the lines of, 'It would be intelligent for me to major in something else so that I could make money'—so that I could be marketable, I guess. (White man, switcher, biology to dance)

In the following case, the mother is distressed that her daughter does not share her own disciplinary interests. It reflects the same rationale encountered in support of an approved change to a discipline favored by a parent:

> My mom was mad because she thought it was that one professor's fault. … Because she wanted me to be a science major. She loves science. She loved sharing that with me and she was sad I wasn't going to take any more chemistry classes. (White woman, switcher, biology to European studies)

Where parents' financial support was contingent on graduating with a STEM degree, these switchers lost both parental approval and support:

> To my parents, getting an English degree or a history degree, or something similar, is a huge joke to them. … It was a constant battle with them 'cause they had in their head that, 'This

> is what she's gonna do.' … They don't pay my tuition anymore. (White woman, switcher, physiology to film studies).

> I've always been telling my parents I sort of wanted to go into music, and it was like, 'No. No. No.' … I remember my mom specifically saying once, 'I'm not paying for you to play with instruments for four years in college.' …It made me feel like I was restricted specifically to science … I had to do something like engineering. (White man, switcher, aerospace to studio art)

One persister explained that she stayed in her STEM major because she didn't want to lose her parents' financial support:

> My parents seemed to be discouraged that I was trying to switch into teaching, and said they would not pay for out-of-state tuition for teaching unless I was concurrently getting my engineering degree. They had an influence on my going back to chemical and biological engineering. (Native American woman, persister, chemical engineering)

Some students reported that their parents influenced which majors they subsequently switched into. These guided choices were usually those that were heavy in either math or science and with obvious "practical applications" beyond graduation. For example, this Asian woman explains that her mom suggested she switch into accounting:

> I definitely wanted them to approve so that played a part in [in my decision]. … My mom suggested I try accounting … I guess she just thought that there was always a need for accountants and that I would have job security. (Asian woman, switcher, computer science to accounting)

Some students who faced an initial negative response to their change of direction had found that their parents' opposition could be eroded by their own persistence or, as below, by a traditional children's strategy—manipulation of family dynamics:

> So, there's two sides: There's my mother and my stepdad, and my father and my stepmom. My father and my stepmom were very much, 'Do science. Do something that makes money.' … I was finally able to convince my father to let me drop it and just do acting because that's what I wanted to do. (While man, switcher, physics to theater)

However, parental concessions did not remove their over-riding concern for a secure future for their children:

> My mom was a little let down. Dad was like, 'What are you doing with your life? Oh my God. That's an awful decision; you're gonna regret this later.' But he's sort of coming around more slowly. He's still very much trying to make sure that I'm on the top of the game and very competitively placed. (White man, switcher, engineering to psychology)

In summary, when parents disapproved of a student's intention to switch out of a STEM major, their dominant concern was for their child's future employability and financial security. They worried equally for daughters and sons about the long-term consequences of moves into non-STEM fields. However, more fathers than mothers opposed STEM switching for these reasons. Parents of both sexes told their children that non-STEM degrees would be useless or inferior to those based on STEM degrees, particularly in an uncertain job market. However, some students questioned how much their parents actually knew about the career paths they promoted.

Students also clashed with parents over estimations of how much money they could earn as the primary criteria for a parent-approved career path. Earning potential was not the only reason for disapproval. Some parents were disappointed at what they saw as a move away from a prestigious field; notably, medicine or engineering that would have reflected their own aspirations to be parents of professionals. As with some approving parents, there was some evidence of parental interests and aspirations as the basis for a reaction to a switch out of a preferred STEM field. Finally, some parents were sufficiently adamant in their disapproval to threaten withdrawal of financial aid as a way to leverage compliance or to end financial support by way of punishment for disobedience to their wishes.

As noted, some parents included in their reasons for disapproving a move into non-STEM fields their concern that STEM degrees offered more certainty of employment in an economy and job market that they perceived as problematic. We included in our interview protocols some questions intended to discern how students perceived the economic world and their own chances of employment. We also wanted to know whether such considerations played a part in shaping their decisions. In the concluding part of this chapter we discuss what we learned.

## Students' Appraisals of Economic Conditions and Job Opportunities as Factors in Switching Decisions

As outlined in Chap. 3, we found four career-related factors that informed switching decisions. Two factors that were reported by more than half of switchers reflect the push–pull nature of switching decisions: rejection of the careers and projected lifestyles toward which their STEM degrees appear to be taking them (58%) and choice of career options that seem more appealing (54%). Two other career-related factors, that also embody push–pull processes, reflect more instrumental considerations. As a push factor, 17% of switchers decided that the potential benefits of a STEM degree were not worth its financial costs, effort, and stress. (A comparable number of switchers in the original study referred to this calculation as "weighing the profit-to-grief ratio" and also cited it as prompting their move out of STEM.) As a pull factor, one-quarter (26%) of switchers in the current study made pragmatic career-focused moves into majors that they saw as better ways to pursue particular career aspirations. As also described in Chap. 10, switching to a non-STEM major was sometimes undertaken because it seemed a more achievable way to enter a desired career (often in a health field).

We wished to know to what extent students' considerations about their revised choices of majors and career paths were informed by their perceptions of current economic conditions and work opportunities in various fields. Thus, we asked all students whether, in reaching decisions about particular STEM and non-STEM degree programs, they took into account what they saw as the employment opportunities available in particular fields and in current economic conditions. One-third of switchers (33%) and one-fifth of persisters (19%) reported that they did weigh these odds in reaching career-related decisions. Concerns about economic problems—whether nationally, regionally, or in particular fields—led these students to doubt their prospect of securing a job with a high salary. They also sought career pathways that seemed to offer recession-resistant job security. Persisters sometimes viewed STEM fields as a safer bet in that they offered more employment and financial security than they imagined were available to graduates in non-STEM fields. Like other STEM majors, this computer science student was simultaneously encouraged by his own employment prospects while concerned for his friends who were majoring in liberal arts degrees:

> Personally, the computer science field is immune to [downturns] because of how rapidly the market is growing for jobs. I read statistics every day that tell me within the next however many years, there are going to be a million different computer scientist positions that need to be filled, and there's only going to be 800,000 people looking for those positions…That's good news…If anything, it's made me feel sorry for all the people who get liberal arts degrees. No offense to them, but with that kind of a degree…you just have no options. And in an environment where the market has gotten so unbelievably competitive, a liberal arts degree won't help you. You might as well not even have a degree, because it's not going to help you. (White woman, persister in biology)

Across the entire student sample, persisters majoring in engineering or computer science were the most confident that current economic conditions would favor their ability to secure well-paid employment. Engineering was characterized by one student as "one of the few things that everyone needs" because "making things [and] consumption are not going to go away." The related field of construction management was also viewed as recession-proof because, as one engineering switcher into this field explained, "there are always going to be buildings being built somewhere." The allied field of physics was, by contrast, seen as less tied to tangible needs, and therefore less secure and lucrative than engineering. Other STEM fields, such as environmental science and biology, were viewed as offering less predictable employment. Students majoring in biology who hoped to pursue careers in healthcare held varied and complicated views about their future employment prospects, which are described below.

Another STEM option—teaching science and mathematics—was seen as less desirable largely because of perceived negative societal attitudes toward teaching as a career:

> Even though I'm in the educational track myself, I…question whether or not I want to be a teacher. And it's not because I don't enjoy teaching or because I don't want to deal with kids. In fact, those are the things that I enjoy most about it. What detracts me the most is that as a teacher in America, you cannot get ahead. The system is designed where teachers do not get paid adequately, and they do not get enough benefits for what they do. I have personal experience with that because not only am I friends with some of my high school teachers…I also am the child of a teacher. (White man, persister in engineering)

Only a handful of students from across our total sample intended to teach—a marked decline from our *Talking about Leaving* findings where one-fifth of the total sample had considered a teaching career and 8% intended to pursue this for reasons that included both altruism and pragmatism. However, in students' accounts of discussing their teaching aspirations with parents, peers, and faculty, there were signs that teaching has lost its respected place in U.S. society. Teaching other than in college was discouraged by some parents for reasons similar to those voiced above. STEM seniors avoided mention of their interest in teaching to some STEM faculty because they decried this as an aspiration unworthy of any serious science student and were apt to withdraw interest in those who professed it (cf., Seymour & Hewitt, 1997, pp. 197–201).

In addition to assessments of the potential security and lucrativeness of particular career paths, a degree's perceived flexibility and versatility were valued by both switchers and persisters. Examples include a switcher into economics who viewed this choice as offering greater flexible than a specialized degree such as accounting. Another switcher chose economics as an "up-and-coming" field with many business applications and, thus, employment opportunities. Persisters also cited the energy sector as a flourishing area of employment. An Asian woman moved from biomedical engineering to material science because this field appeared more flexible and more likely to provide future opportunities. A related planning strategy was to seek employment with a growing new company that could offer new hires career growth opportunities. Several students mentioned the then recent U.S. federal government shutdown and were, as a result, wary of pursuing jobs in government service, except in the defense sector that was seen as more resistant to loss of funding.

Some switchers who anticipated a generally tough job market found solace in having skills and experience that would set them apart from others. For example, one psychology major, who was bilingual in Spanish and English and had overseas experience, thought that these attributes would give her an edge with employers. Another switcher, who described the competitive job market as a "squeeze on millennials," had undertaken a number of internships to set himself apart from competing job applicants. However, a persister argued that the new norm of unpaid internships was, in itself, an indicator of a highly competitive job market. Somewhat resentfully, she complained that, in contrast to previous generations, current undergraduates have been obliged to pursue "tons of unpaid internships," involving "extremely, extremely competitive application processes," that typically do not lead to employment. As she elaborated:

> [The internship] is likely going to unpaid, under-paid, or below minimum wage work. And maybe it leads to a job, but most likely it won't. Most likely they'll say, 'We've paid you an experience. Now you know more about the field, you can go on. We can't offer you a job here.' (Multi-racial woman, persister in biology)

In contrast to these pessimistic views, other students expressed optimism about current employment conditions, and viewed the present era as a time of economic upswing and recovery. A geology major, for example, took the view that although mining was at a lull it would, inevitably, rebound. A female mathematics major was confident she would find well-paid employment because of her disciplinary proficiency and the additional advantage of being a woman in a predominately male discipline. Optimism was also discipline-based with some STEM fields seen as less competitive than others—notably, chemistry (because it has fewer graduates and therefore more job opportunities) compared, for example, to biology.

For many students, assessments of economic conditions were viewed through the lens of their financial obligations beyond graduation. Repayment of student loans was the predominant financial concern of both switchers and persisters. It was a universal source of fear and worry, even among engineering persisters, and could strongly affect a student's choice of a professional pathway. Concern about loan repayments caused some students to reject or delay graduate school, or to consider it out of reach unless a paid graduate assistantship could be secured. Loan obligations also caused some students to question whether they could afford to follow their ideals or whether their preferred career was financially viable. For example, a persister in mechanical engineering who had hoped to find a job in the renewable energy sector accepted that it was more realistic to enter the oil and gas industry because he could earn more and be better able to pay off his student loans.

For some students—often students of color and/or from first-generation, low-income or immigrant families—career decisions were grounded in their obligations to family. An African-American persister in civil engineering who grew up in a household in which his parents struggled financially made finding a well-paid job a priority so that he could "give back" to his parents. Also largely absent in working class families was the social capital of socio-economic networks through which knowledge of and access to jobs can be secured. Students who had access to these family-based networks were aware of their advantages. A computer science major was confident about his future job prospects because, he explained, his father worked for a large international technology company. A biology persister acknowledged his privilege in coming from a family with substantial financial means that freed him from worry about adverse economic conditions and the competitive job market. Some students who became aware of the significance of social capital to building a career, but lacked these familial advantages, were actively developing academic and professional networks. They viewed these efforts as being as important to their future career prospects as the skills and expertise that they were gaining through their STEM program.

The single most notable aspect of economic conditions to which students reacted concerned professional fields that were formerly considered secure and lucrative but that have recently been subjected to market and reformist pressures—namely medical care and veterinary medicine. Some students who had switched out of biology did so because they had calculated that becoming a doctor required more money than they would reliably be able to recover in a realistic timeframe, especially given their substantial student loan obligations:

> I think there's [an awareness] now that it's harder for doctors to pay off their student loans. I mean, in the media, I definitely get [the view] that doctors are struggling. And then you see TV shows that follow [medical] interns around and they're living like crap. And, you know, I think that definitely makes a difference. (Asian woman, switcher, biology to strategic communication)

Students who switched out of the medical school pathway could also become disenchanted after gaining a clearer understanding of the length of time that it takes to become a doctor—from undergraduate and professional studies, to residency, fellowships, etc.—all before you can earn a paycheck and begin to pay back student loans. The passage of the Affordable Care Act also contributed to some switchers' wariness about their initial medical school intentions. They were nervous that pursuit of a medical degree would require taking on substantial debt in return for what they now saw as a financially unstable profession:

> With the Affordable Care Act being in place, and with the government shutdown right now, the way that doctors are paid is going to change...If I'm going to spend a long time in school I want to make enough to support myself without having hundreds of thousands of dollars in debt. (White woman, switcher, chemistry to family studies)

Even applying to medical school presented financial challenges for some persisters who still planned to pursue it, with each application expected to cost roughly $50 and additional expenses associated with forwarding MCAT results.

There was also indication that some persisters in biology were nervous about what they would do if their medical school plans did not come through, given how competitive the application process for admission had become. In contrast to other STEM fields, biology was viewed as having limited future options for graduates with only a bachelor's degree. In contrast with engineering and computer science peers who obtained jobs immediately after graduation, some biology majors felt caught between the vagaries of securing a place in medical school and the challenges of obtaining a position in a biology lab—where it seemed, "you can't get experience without [first having] experience."

Veterinary school was also seen as too risky by some switchers in that it represented a pathway with one of the highest "debt-to-salary ratios." The veterinary pathway was characterized as extremely competitive, requiring many years of study and substantial financial investment with increasingly little assurance of a well-paid career:

> Like I said it's very, very competitive to get into vet school. And it's very, very expensive to get to vet school. So, it'd be a huge investment in money to get this education with no guarantee of a solid job after I got it. (White woman, switcher, biology to European studies)

For some switchers, although medical school was deemed overly expensive, time-intensive and less secure than in the past, other professional careers in healthcare and therapeutic fields were appealing and (as discussed in Chap. 10) remained as career goals that were undisturbed (and sometimes enabled) by switching. These students switched into caring professional pathways such as psychology and marriage and family counseling, and a wide array of health-related specialties (e.g., dentistry, physical therapy, nursing, community health and sports medicine) that required specialized post-graduate studies. Among the benefits of these choices were career trajectories with expanding opportunities and versatility in an era of fluctuation and uncertainty in medicine per se:

> It's a little bit easier to get into a graduate school for Marriage and Family Therapy rather than medical school. Not that I had any doubts that I would have gotten it into medical school. I think that I would have been a pretty good applicant for it, but I also feel there's a lot more options with a Marriage and Family Therapy degree. And I feel like it would be a little smaller and tight-knit [program]. I actually did a tour of a graduate school last week and fell in love with the campus and the smaller class sizes. (White woman, switcher, chemistry to family studies)

> I don't really understand what's going on with healthcare, but psychology's becoming a really great field to be in. I guess because they need people to give a social [perspective]. A lot of engineering companies need psychologists to give them an idea of what society wants or needs so it's a flexible field to be in. (Asian woman, switcher, biology to psychology)

Nursing and physical therapy were appealing because they were seen as growing professions that required fewer years of education than those required to become a medical doctor, and, thus, less aggregation of debt. In addition, these pathways were seen as accessible to switchers without a formal STEM degree.

Many persisters in STEM majors were also skeptical about a medical school pathway for reasons similar to those described by switchers. More appealing alternatives for many persisters interested in healthcare were becoming physician assistants (PAs), nurse practitioners, nurse anesthetists, or physical therapists. Becoming a PA, as one persister characterized it, was like becoming "almost a doctor," but "in way less time, for way less money." Similarly, physical therapy was viewed as a growing profession that was lucrative, yet not as competitive, lengthy, or costly as pursuing medical or veterinary schools. As touched on previously, persisters couched their career plans in terms of their impressions of trends in the healthcare sector that signaled advantages to pathways other than medicine:

> I've talked to a lot of my parents' friends, and they believe that the direction that the health field is going [encourages becoming] nurses and physician assistants, as opposed to only doctors. Nurse practitioners and physician assistants are very well educated and have the ability to see patients... So, to spread the patient base across a bigger field of professionals is really nice. It seems like there would be a little more job security in that field. (Multiracial woman, persister in biology)

> I've thought about going either to get my Master's in public health or to PA school. At least for monetary reasons, it seems like PA is a safer bet. So, I definitely thought about that...I keep hearing that nurse practitioners and PAs are going to be in the majority [in the healthcare sector] rather than doctors. [So] they're in pretty high demand right now. (White woman, persister in biology)

> [Medical school] is starting to become less worth it... so that's basically shut down that path for me...economically, with the job payoff and everything. Because med schools are majorly expensive...You don't make a lot of money until seven years out and then you're just indebted. But from what everyone has been saying, the PAs and the PTs, they're kinda flying under the radar. (White woman, persister in biology)

## Summary and Conclusions

As we indicated at the outset, these three external sources of influence on students' rethinking of their educational and career objectives are often interrelated and can modify college experiences in contributing to such changes. The most obvious of these connections are evident among students who work more than 20 h per week while struggling to do justice to their college work and maintain adequate grades. Although we found evidence that working to contribute to college costs can be motivational and satisfying, these benefits rapidly fade as the number of working hours increases. Persisters who worked often credited their survival to efficient time management and persisters may be more skilled in this than switchers. However, more switchers than persisters worked; switchers worked longer hours than persisters; and three times as many switchers as persisters reported financial strains, work overload, and feeling "stressed all the time." Thus, working 20 or more hours a week while undertaking a STEM major is another characteristic that distinguishes switchers from persisters. Working while in college can (as cited) also lead to failure to complete a degree. However, among the highly qualified sub-set of students that enter STEM majors, the strains of working more than 20 h a week also prompt switches into programs that enable students who must work to do so while completing a degree.

We also noted demographic patterns among students who face persistence risks that are created by their need to work. More students of color and more women of all races and ethnicities worked than did white men, and the groups with the higher proportions of working students were both switchers, namely white women and men of color. The most common reason for students to work was that their families were unable to contribute anything, or only a little, to their college fees and maintenance costs. Thus, social class, gender, and race/ethnicity combine as a source of persistence risk that amounts to structured disadvantage. Indeed, some students described how the weed-out system acts, in effect, as a means test that is biased against those who have to work their way through college.

Compared with original study findings twenty years prior, work-school stress has doubled for switchers, persisters, and students overall—a finding that aligns with (cited) studies of the growth in costs for STEM degrees. Also cited were studies indicating that food and housing insecurity that are now greater among college students than among the general population, and that students with the greatest financial need have lower rates of degree completion than other students.

Paying a substantial portion of one's own college and maintenance costs does offer one advantage—greater freedom to choose one's own educational and career pathway. Among students whose parents are able and willing to contribute to their undergraduate education, financial concerns often shape parental attitudes toward their children's original and revised degree choices. One-third of switchers reported that one or both parents had pressured them to choose a STEM major because it was perceived as leading to a financially secure career. Parental pressure on initial choices was more often exerted by fathers than mothers, by scientific or technical professionals who advocated related fields to their children, and by families who held particular professions in high esteem. A few parents made financial support for college contingent on pursuit of a STEM degree. Switchers who undertook STEM majors under such pressures entered with less interest than persisters and more often saw themselves as ill-suited to their original major and/or projected career.

That said, 60% of switchers described their parents as supportive of their decision to move to a non-STEM major and career path. The dominant concern of these parents was that their children would find a good fit for their talents and interests. They regarded this as a better criterion for a sustainable future than instrumental choices focused on career prestige or likely earnings. However, as in the original study, mothers and fathers differed somewhat in the criteria by which they judged a proposed switch of majors and careers. Mothers, who were often the student's primary confident and sounding board in their education and career and rethinking process, took into account the student's enjoyment, interest, and investment in the new discipline and projected career when assessing whether a revised choice would reduce stress, increase engagement, and secure future career satisfaction. Some parents, particularly fathers, qualified their support of a switch out of concern that the proposed alternative would ensure, not only greater enjoyment of the discipline, but also viable future employment—a distinction between happiness now and happiness later.

Divergence between mothers and fathers was more pronounced where one or both parents opposed the moved to a non-STEM major. The hardest line was taken by parents (often fathers) who viewed higher education (and especially choice of a STEM degree) as an investment made in expectation of high future financial returns. Where parents differed, and fathers took a tough stance against a move out of STEM, we noted a distinctive change from the original study findings. In the 1990s fathers were less enthusiastic about their daughters' choice of a STEM major—preferring something more "gender-suitable"—and also took a more indulgent attitude toward a switch into a non-STEM degree by their daughters than by their sons. In the present study, daughters describe their fathers as strongly favoring STEM degrees as a sound way for young women to secure financial security, and as equally unsympathetic to moves out of STEM pathways for both daughters and sons. (We also noted the complete disappearance of marriage as a parent-supported plan for a daughter's financial security.) Other parents who disapproved the proposed switch were worried that, in an uncertain job market, all or most non-STEM degrees would lead to poorly paid, insecure work. Some students doubted, however, that their parents' expectations of good financial payoff from STEM degrees were well grounded. Concern for loss of an entrée to a prestigious profession was most strongly expressed by parents in Asian-American and immigrant communities.

Regardless of whether students or their parents were contributing to the costs of undergraduate education, we found far greater awareness of, and concern about, what prevailing economic conditions implied for job availability and financial opportunities than was evident in the original study. When students were asked whether they took into account what they saw as the employment prospects and limitations that they saw in particular fields in current economic conditions, one-third of switchers and one-fifth of persisters replied that their decisions to switch or relocate were influenced by these assessments. A second dominant concern was the long-term cost of undergraduate and graduate education that are increasingly dependent upon student loans. This was reflected in a marked shift toward instrumental choices of career fields at the expense of choices based on altruism or pure interest. Indeed, repayment of student loans was the predominant financial concern of both switchers and persisters. It often shaped rethinking of majors (especially biology) and career pathways away from those with a high "debt-to-salary ratio." Thus, students' concerns for the future now reflect those of many parents. This trend is evident in the continuing decline of aspirations to teach science and mathematics in K-12 settings, and re-assessment of post-graduate education for medical and veterinary careers as too long, costly, and risky. Shorter undergraduate and post-graduate programs with more certainty of employment in medical, nursing, and other health-related fields were favored instead. As also discussed in Chap. 10, these pragmatic re-appraisals were a contributory cause of switching in and of themselves.

## References


Broton, K., Frank, V., & Goldrick-Rab, S. (2014). *Safety, security, and college attainment: An investigation of undergraduates' basic needs and institutional response*. Madison, WI: Wisconsin HOPE Lab.

Broton, K., & Goldrick-Rab, S. (2015). *Public testimony on hunger in higher education*, Submitted to the National Huger Commission. Madison, WI: Wisconsin HOPE Lab.

Chaparro, M. P., Zaghloul, S. S., Holck, P., & Dobbs, J. (2009). Food insecurity prevalence among college students at the University of Hawaii at Manoa. *Public Health Nutrition, 12*(11), 2097–2103.

College Board. (2014). *Trends in College Pricing 2013*. Retrieved April, 2015, from http://trends.collegeboard.org/sites/default/files/college pricing-2013-full-report.pdf

Dworsky, A., Dillman, K., Dion, M. R., Coffee-Borden, B., & Rosenau, M. (2012). *Housing for youth growing out of foster care: A review of literature and programs typology*. US Department of Housing and Urban Development Office of Policy, Developments and Research.

Feeney, M., & Heroff, J. (2017). Barriers to need-based financial aid: Predictors of timely FAFSA completion among low-income students. *Journal of Student Financial Aid., 4*(2), 65–85.

Freudenberg, N., Manzo, L., Jones, H., Kwan, A., Tsui, E., & Gagnon, M. (2011). *Food insecurity at CUNY: Results from a survey of CUNY undergraduate students*. New York, NY: City University of New York.

Hall, C., Dickerson, J., Batts, D., Kauffmann, P., & Bosse, M. (2011). Are we missing opportunities to encourage interest in STEM fields? *Journal of Technology Education, 23*(1), 32–46.

Joint Center for Housing Studies of Harvard University. (2011). *America's rental housing: Meeting the challenges, building on opportunities*. Cambridge, MA: President and Fellows of Harvard College.

Kirshstein, R. (2013a). *The price and cost of a STEM degree*. Presentation at the second meeting of the Committee on Barriers and Opportunities in Completing 2-Year and 4-Year STEM Degrees, National Academy, Washington, DC.

Kirshstein, R. (2013b). *How much debt do science and bachelor degree recipients accrue?* Delta Cost Project. Washington DC: American Institutes of Research.

Malcom, S., & Feder, M. (Eds.). (2016). *Barrier and opportunities for 2-year and 4-year STEM degrees*. Washington, DC: The National Academies Press.

Porter, O. (1990). *Undergraduate completion and persistence at four-year colleges and universities: Completers, persisters, stop-outs, and drop-outs*. Washington, DC: National Institute of Independent Colleges and Universities.

Rotberg, I. (1990). Sources and reality: The participation of minorities in science and engineering education. *Action in Teacher Education, 15*(4), 36–46.

Simpson, J.C. (2003). Mom Matters: Maternal Influence on the Choice of Academic Major. *Sex Roles, 48*(9/10), 447–460.

Sjaastad, J. (2012). Sources of Inspiration: The role of significant persons in young people's choice of science in higher education. *International Journal of Science Education, 34*(10), 1615–1636.

Seymour, E., & Hewitt, N. (1997). *Talking about leaving: Why undergraduates leave the sciences*. Boulder, CO: Westview Press.

VanMeter-Adams, A., Frankenfeld, C. L., Bases, J., Espina, V., & Liotta, L. A. (2014). Students who demonstrate strong talent and interest in STEM are initially attracted to STEM through extracurricular experiences. *CBE—Life Sciences Education, 13*(4), 687–697.

Wilder Research. (2008). *Overview of young and young adult homelessness in Minnesota: Facts and analysis of data from the 2006 statewide study*. St. Paul, MN: Author.

# Chapter 12
# What Enables Persistence?

Heather Thiry

## Introduction

So far, we have largely focused on the factors that contribute to students' decisions to leave STEM majors, yet many different aspects of students' backgrounds and college experiences also contribute to their persistence in STEM majors. Pre-college experiences matter as early interest in STEM fields and high school achievement have been shown to predict retention in STEM majors (Chang, Sharkness, Hurtado, & Newman, 2014; Crisp, Nora, & Taggart, 2009; LeBeau et al., 2012; Maltese & Tai, 2011; Riegle-crumb & King, 2010; Tai, Liu, Maltese, & Fan, 2006; Tyson, Lee, Borman, & Hanson, 2007). Yet there are many points along students' pathways in STEM where their interest and commitment may be strengthened or weakened. Once students enroll in STEM degree programs, peer support and positive departmental climates enhance their commitment to STEM majors and retention (Callahan, 2009; Espinosa, 2011; Garcia & Hurtado, 2011; Liztler & Young, 2012). As we have already noted in Chap. 9, social and cultural factors within STEM classrooms, departments, and other disciplinary settings contribute to the retention of students in STEM majors, especially for students from underrepresented populations in STEM (Gayles & Ampaw, 2014; Hurtado, Eagan, & Chang, 2010; Museus & Liverman, 2010; Ong, Wright, Espinosa, & Orfield, 2011). As we highlighted in Chaps. 6, 7, and 8, pedagogy and curriculum can play an outsized role in encouraging or discouraging students' persistence. Other researchers have also demonstrated the

H. Thiry (✉)
Ethnography and Evaluation Research (E&ER), University of Colorado Boulder, Boulder, CO, USA
e-mail: Heather.Thiry@Colorado.edu



E. Seymour, A.-B. Hunter (eds.), *Talking about Leaving Revisited*,
https://doi.org/10.1007/978-3-030-25304-2_12

importance of classroom practices; for example, coursework that is related to students' lives has a significant impact on underrepresented minority students' academic and social adjustment in STEM (Hurtado et al., 2010). Active learning strategies, such as peer-led team learning or inquiry-based learning, also increase student achievement and retention (Drane, Smith, Light, Pinto, & Swarat, 2005; Freeman et al., 2014; Thiry, Hug, & Weston, 2011). Thus, students' academic background, their interest and commitment to their major, and their classroom and departmental experiences all affect their decisions to persist or to leave STEM majors.

Yet other institutional factors and supports also influence STEM persistence. Out-of-class academic experiences, especially sponsored undergraduate research, have been shown to promote retention and achievement in STEM. In particular, apprentice-style undergraduate research experiences introduce students to the technical and collaborative nature of research, promote student identification as a scientist, influence students' career aspirations, and increase graduation rates in STEM majors (Chang et al., 2014; Chang, Cerna, Han, & Saenz, 2008; Clewell, de Cohen, Tsui, & Deterding, 2006; Eagan et al., 2013; Espinosa, 2011; Graham, Frederick, Byars-Winston, Hunter, & Handelsman, 2013; Herrara & Hurtado, 2011; Hunter, Laursen, & Seymour, 2007; Hurtado et al., 2010; Jones, Barlow, & Villarejo, 2010; Laursen, Hunter, Seymour, Thiry, & Melton, 2010; Perna et al., 2009; Seymour, Hunter, Laursen, & DeAntoni, 2004; Thiry, Laursen & Hunter, 2011). Academically oriented peer support also makes a difference in STEM retention, especially for underrepresented minority students who are more likely to complete a STEM degree if they join a STEM-related club during their undergraduate studies (Chang et al., 2008; Chang et al., 2014; Espinosa, 2011; Herrara & Hurtado, 2011; Hurtado et al., 2010; Palmer, Maramba, & Dancy, 2011). Thus, prior research has illustrated many academic, pedagogical, social, and cultural factors that contribute to students' STEM persistence.

## *Findings from TAL1*

TAL1 largely focused on the reasons that students leave STEM majors, yet the original study focused some attention on the factors that help women and students of color persist in disciplines that are often seen as hostile climates for students from underrepresented groups (Seymour & Hewitt, 1997). Individual coping skills, including strong interest in the field, self-efficacy, assertiveness, and a willingness to accept critique and to let go of the fear of being wrong, helped women to persist in TAL1. Peer support from other women, including professional societies for women, study groups, and other formal and informal opportunities to bond with other women, also fostered persistence in women. Moreover, women undergraduates benefited from women faculty who served as role models and mentors. Although not widely available at the time, campus support programs contributed to the retention of students of color by offering academic assistance, advising, counseling, and a structured orientation to college.

## Factors that Contribute to Persistence

In the current study, students also discussed a range of individual, social, and institutional factors that enabled their persistence in STEM. Most students credited their persistence to a complex mix of all of these factors. Students rarely cited a single factor in supporting their persistence, but often described an interaction among individual factors, such as self-efficacy or determination; behavioral adjustments, such as refining their study habits; practical behaviors, such as navigating the college system and STEM courses in a way that will best ensure their success; and social and institutional factors, such as peer support or university services. Table 12.1 provides an overview of this complex mix of persistence factors. No one factor stands out as essential for students' success, yet most students drew on a number of these factors to enable their persistence in STEM majors.

When we examined students' persistence based on math readiness, students with low math indicators used slightly different persistence strategies than students with high math readiness. For instance, low-math students were more likely to cite sheer determination and "grit" in their persistence (see Table 12.2). Perhaps more importantly, low-math students were more likely to find a support system and more likely to figure out how to shrewdly navigate college STEM courses and coursework. Perhaps most importantly, low-math students relied on almost all persistence strategies to a greater extent than high-math students. Therefore, low-math students expended an enormous amount of effort to enable their persistence, drawing on individual, social, and institutional resources to support their success in their STEM major.

**Table 12.1** Factors cited by persisters as promoting their retention in STEM

| *Individual characteristics* | % of all persisters |
|---|---|
| Maintaining determination; will to persist | 49 |
| Sustaining interest | 44 |
| Believing in self; self-efficacy | 22 |
| *Adjustments to behaviors or identities* | |
| Adjusting to lower grades | 39 |
| Developing effective work habits | 39 |
| Leading a balanced life | 32 |
| *Instrumental moves (i.e., practical behaviors)* | |
| Figuring out how to succeed within the system | 46 |
| Switching to more appropriate career or major within STEM | 27 |
| *Institutional and social support* | |
| Finding a support system | 50 |
| Seeking appropriate help | 38 |
| Engaging in out-of-class experiences | 29 |

**Table 12.2** Factors cited by persisters as promoting their persistence in STEM based on math readiness

| *Individual characteristics* | % of all HIGH math | % of all LOW math |
|---|---|---|
| Maintaining determination; will to persist | 36 | 78 |
| Sustaining interest | 17 | 34 |
| Believing in self; self-efficacy | 37 | 50 |
| *Adjustments to behaviors or identities* | | |
| Adjusting to lower grades | 40 | 56 |
| Developing effective work habits | 27 | 54 |
| Leading a balanced life | 25 | 43 |
| *Instrumental moves* | | |
| Figuring out how to succeed within the system | 51 | 74 |
| Switching to more appropriate major or career within STEM | 28 | 29 |
| *Institutional and social support* | | |
| Finding a support system | 43 | 66 |
| Seeking appropriate help | 25 | 61 |
| Engaging in out-of-class experiences | 23 | 27 |

## *Individual Characteristics*

Students often credited their success to individual characteristics, such as determination, resilience, interest, and self-efficacy. These individual explanations sometimes, but not always, reflected a meritocratic narrative that success is dependent upon the individual only, and that those with the most "grit," talent, and intelligence will succeed. Likewise, our interviews with faculty and focus groups with students enrolled in gateway courses revealed that this was one of the dominant explanations for persistence in STEM majors (Ferrare & Miller, 2019). Thus, belief in the meritocratic narrative is widespread in STEM disciplines which belies the ways in which courses and institutions may be structured to facilitate or impede student success. However, there were slight differences in the ways that faculty and students viewed individual ability. Faculty were more likely to perceive ability as a "given," in that students were either capable or not capable of college-level STEM coursework and there was not much that faculty could do to change this (Ferrare & Miller, in press). In this sense, some faculty approached STEM learning with a "fixed ability" mindset. In contrast, students were more likely to perceive ability as the result of determination, persistence, and hard work, thus exhibiting more of a "growth" mindset to STEM learning and success.

In keeping with this dominant narrative of meritocracy in STEM, our interviews with persisters near the end of their studies also illustrated that many students ascribed their success to talent, intelligence, or self-efficacy. In addition, low-math students (78%, compared to 36% of high-math students) and students of color (60%, compared to 41% of white students) were more likely to comment that maintaining determination, especially in the face of obstacles, helped them to persist. There were no gender differences in persisters' references to individual characteristics, such as determination and interest, as factors in their retention in STEM. For example, 24% of men and 21% of women mentioned self-efficacy, and 40% of men and 38% of women noted that strong interest had helped them to persist.

In contrast to switchers who were often devastated by unexpectedly low grades, strong academic achievement and high grades boosted persisters' confidence and their belief that they could succeed in the major. Still, persistence was never related to a single factor such as self-efficacy or grades. For example, a biochemistry major mentioned that high achievement bolstered his confidence and intentions to persist, yet he also cited connections with faculty as the most important factor contributing to his persistence.

> First point [in my persistence], most certainly is encouraging professors that really invested in me during the semester and after the semester. A second one would be actually achieving success in highly rigorous classes. (Asian-American man, biochemistry persister)

Low-math students, in particular, credited their grit, drive, and determination in aiding their persistence in STEM majors that they often perceived to be challenging. For some, family expectations fostered a strong will and determination to succeed in their STEM degree program, as mentioned in the following comment from a first-generation college student. Notably, first-generation college students or children of immigrant parents were likely to credit strong parental expectations in their drive to succeed in STEM.

> I know a lot of [my persistence] is very internal. Because I'm the first child in my family. And I'm first generation. When I set my mind to something, I have to complete it, otherwise I feel kind of like... for lack of better terms, inadequate. (Native American woman, physiology, low math persister)

Although many students felt that their determination or internal drive had helped them to persist in STEM, many of these same students also acknowledged that they had received support from other sources. For instance, a life sciences major attributed her success to her internal drive, interest, and motivation, yet, like the previous biochemistry student, she also recognized that faculty had played a role in her persistence.

> I think it was my self-motivation, honestly. Because there's only so much that external factors can really do to keep you in the major. If you don't like it yourself, then you're not going to want to do it. So, I think definitely me wanting it that much and really also enjoying it kept me in it. But I think professors also do have a big role. Like they have a lot of influence. (Asian-American woman, biological sciences persister)

Thus, many students felt that sheer determination and "grit" had contributed to their persistence, yet these factors alone were not the only reasons that students persisted. Peers, faculty, and other supports also contributed to students' persistence and bolstered their motivation and determination. For example, an electrical engineering student described how sheer will had kept her going through a difficult first year, yet at a certain point in her studies, she developed a network of friends and peers in the department that, in turn, sustained her motivation to persist.

> So I'm very stubborn, and when it comes to a challenge I'm always, like I don't care if I just fail. I'm gonna keep going, cause I didn't want to quit. I've never quit anything in my life. And so I was like, 'I don't like this. It's a rocky start, it should be better.' Second semester was actually even worse. I did fail my first class and I was just so shocked by it, but at the same time I just don't know why I didn't drop it then, I don't know why I didn't. But I guess, I just kept pushing and making new friends and just creating a bigger network and I got more comfortable once I had my kind of network because I was like we're all in this together, we got this. (Asian-American woman, high math, electrical engineering persister)

Maintaining and sustaining interest also played a significant role in students' persistence. Interest was closely connected to student's identity in their major and many persisters maintained high interest in either their field of study or their future career. Without a strong interest in their discipline, it was difficult for students to develop an identity as an aspiring scientist or engineer and, thus, challenging to remain in their major. In contrast, most students who switched out of STEM overwhelmingly lost interest at some point in their undergraduate studies. Yet persisters often described their feelings about their major as "love," "passion," or "enchantment" that they sustained even if they were not highly interested in every course in their major. Although individual courses did not always pique students' interests, persisters managed to maintain their interest and commitment to the larger discipline or their future career. Similarly, in Chaps. 9 and 10 we also discussed students' differentiation of science knowledge or the "doing" of science from "school" science. The adoption of this differentiation was an important strategy for persisters when their interest in "school" science may have ebbed. For instance, an Asian woman chemical engineering major stated: "I don't love the minutiae of what I do [in courses], but the overall idea of chemistry still enchants me, and I really love that." These types of statements demonstrated students' commitment to the major and the fact that they were able to successfully navigate challenges to develop a STEM identity.

While sheer determination and sustained interest often enabled students to overcome early struggles, low grades, or other difficulties, some students encountered more significant obstacles during their STEM studies. These students, notably underrepresented minority students and women, to a lesser extent, encountered social, cultural, or structural barriers that required strength, stamina, and determination to overcome. For example, an African-American woman described the loneliness and isolation she felt in her science major and reflected on the factors that motivated her to persist in the face of obstacles, namely a desire to make a difference in the world and a determination to "push through" it. Moreover, her father acted as a role model because he was a professional in a primarily white field which boosted her confidence that she could indeed succeed. Beside having a role model and determination, her altruistic desire to make a difference in the world also sustained her commitment to the major. Altruism was a somewhat common theme for students of color, especially for women of color in sustaining their STEM interest. In particular, altruism motivated some students when they felt a lack of belonging in their STEM major.

> There are times when I get very blue when I think about it because I sit there and I'm like yeah…there's a possibility where I might not interact with another Black person for an extended period of time. I need to hold on to the ones now so that I can email and text them and be close, but whenever I start to feel that way I just remember how much I can possibly do for endangered species or basically it comes back to I want this so much that I'll push through it. And my father understands that because there aren't very many Black men in architecture, and his parents kind of gave him the… 'There's no Black people in that. Why are you doing it?' kind of routine, which my parents have never given me. I really appreciate that. He understands that if you really want something and if you're really driven you'll do whatever you have to do. (African-American woman, environmental science persister)

Therefore, individual characteristics, such as determination, self-efficacy, and interest, played a role in many persisters' retention in their STEM major, yet these factors alone were rarely enough to ensure students' persistence. Most students attributed their success to these individual characteristics as well as to supportive faculty and peers or other institutional factors. Moreover, the nature of students' will and determination to persist varied based on whether they held a privileged or marginalized position in the field of STEM. Students of color and women were more likely to draw on determination and "grit" to overcome isolation or hostile climates, while advantaged students were more likely to draw on these characteristics to overcome low grades or other early difficulties related to coursework.

## *Adjustments to Behaviors or Identities*

Another strategy that helped students to persist was to adjust certain behaviors, strategies, or identities to better enable their persistence. Some persisters learned that they needed to adjust how they balanced their life and spent their free time, how they studied and learned, and how they interpreted their grades, especially low grades. Many of these issues were stumbling blocks for switchers, yet persisters developed strategies for overcoming the challenges or obstacles posed by these factors.

### Leading a Balanced Life

Despite stereotypes that STEM majors are one-dimensional "geeks" in their discipline, many students persisted *because* of their interest and talents in other subjects, not because they were wholly absorbed in their STEM major. Some groups of students were more likely to use this strategy than others, notably students who may have been more marginalized in their majors or who may have faced a more challenging pathway through STEM. For instance, women were slightly more likely than men to rely on this tactic to persevere in STEM (36% of women and 24% of men described the importance of varied interests to their persistence in their STEM major). Students of color (38%) were also somewhat more likely than white students (27%) to seek outside interests to balance their time spent on STEM coursework. While one might assume that students with lower math readiness scores may have needed to devote more time to their studies, low-math readiness students were much more likely than their high-math peers to seek a balanced life. In fact, nearly half of low-math students (43%) utilized this tactic, while only a quarter of high-math students did. Still, many students found diversion and solace in non-STEM interests, such as sports, art, literature, or languages, as noted by a biology major.

> The way I have coped mainly is…I still draw. I still sketch and do those things, and I find that's my coping mechanism because I transitioned away from art. Having another thing that's completely not science related. I also do martial arts and physical activity. Those are the things I've done mainly to cope, but, recently it's been more of just taking naps (LAUGHS). After a while, I can only stare at those pages for so long. (Non-switcher, Asian/Pacific Islander man, biological sciences persister)

Some students added a double major or pursued a minor in a non-STEM field as a way of developing their interest in another subject. Though not common, 12 persisters (out of 250) had a non-STEM double major, most often psychology or music. Balancing STEM and non-STEM interests offered students a way to pursue their various talents and interests, especially when the student found more joy in the non-STEM interest, as suggested by an Asian/Pacific Islander chemistry major who was also minoring in Japanese. In contrast to her STEM major which she viewed as "work," the Japanese minor served as a counterbalance because it was "play" and "fun."

> For Japanese, for me at least, it's really fun, it's not work, it's kind of play, a little bit. Except I'm in 4th year Japanese this year, I had to skip it last year, and 4th year Japanese is a lot harder, a lot more work. It's kind of like reading a newspaper versus reading a children's book. It's that kind of a level difference, but it's still fun. (Asian/Pacific Islander woman, high math, Chemistry persister)

In contrast to the widespread belief that non-STEM interests detract from STEM studies, some persisters discovered that nurturing their non-STEM interests enabled their STEM persistence. In contrast to the switchers described in Chap. 10 who had a difficult time navigating and integrating STEM and non-STEM interests, these persisters felt energized and reinvigorated from their pursuit of outside interests and managed to find the time to balance their STEM major with outside interests and non-STEM coursework.

## Developing Effective Learning Habits

As seen in students' experiences in transitioning to college in Chap. 5, some students were not prepared for postsecondary education, whether because they were not adequately equipped for the conceptual focus of STEM coursework or, more often, because they lacked time management or study skills. Persisters were often able to adjust and develop their capacities more quickly than switchers, and were better able to tolerate low grades as they developed appropriate study strategies.

There were no gender differences or differences between students of color and majority students in developing effective study and learning habits. Additionally, there were no disciplinary differences as students from all STEM majors often struggled in adjusting to the workload and content of higher education STEM courses. However, low-math students were much more likely than high-math students (54% of low-math students compared to 27% of high-math students) to adjust their study habits in order to succeed in STEM coursework.

In Chap. 5, we described the struggles of switchers who were underprepared for the rigors of college-level STEM coursework, often because they had attended under-resourced high schools. Similarly, some persisters struggled with a lack of prior preparation for the same reasons as switchers, whether because their high schools had not been demanding or rigorous enough or because they had attended schools in underserved communities. However, persisters gradually adapted to the new reality of undergraduate STEM coursework by adjusting their studying habits and strategies to accommodate the different expectations they faced in college. In the following comment, a first-generation college student whose parents were immigrants described her struggles in transitioning to college, especially in the realization that the rigorous coursework in her high school had not adequately prepared her for STEM coursework in college. As a result, she adjusted her approach to studying and classes as she realized that she was not prepared conceptually for the STEM material she encountered in college. She also described how the rote memorization offered in her high school STEM courses contrasted with the emphasis on conceptual learning and application of learning in college coursework, an adjustment that she noted she was still grappling with.

> Being in IB, I felt like the rigor of the classes and the pace of having constant homework, I felt like I was prepared in that area, but material-wise, I wasn't, and I don't feel like I still am. Like I struggle a lot in school. I'm just trying to keep up, just at the regular, the average level, I struggle a lot. And the idea of how to take tests and kind of actually learning instead of just memorizing is something I struggle a lot with, so.... it's really hard to grasp concepts the first time they come around. I feel like in high school, it was really drilled into us. We had a certain set of concepts, and it was just constantly drilled in different aspects, but now it's just a constant learning of new concepts and ideas, and it's really difficult 'cause these are new things that I've never heard of, and the vocabulary and my reading is not that strong. It takes me a really long time to read my homework... So it's just tiring. (Asian/Pacific Islander woman, biological sciences persister)

As persisters advanced in their coursework, they began to better understand the nuances of the workload, expectations, and work habits required in college STEM courses. They also learned that they needed to adjust their study strategies based on the course, the professor, or the material. For instance, a computer engineering major described how he learned to develop effective study habits in college because he did not have these habits when he entered the university. He also described it as an ongoing learning process that he still has not mastered.

> Studying in college is a lot different from high school so I would say that no, [I was not prepared for college]. Honestly, I would say that I'm still learning how to study effectively because it almost seems like different classes require different methods of studying because there's different resources available, the professor teaches differently, the book presents the

material differently. So, while I will definitely say I'm better at studying now than I was when I came in, I still don't, I'm still kind of learning the best way to study. But coming in, I definitely didn't know how to study, I found out. (White man, computer engineering persister)

Students, persisters and switchers alike, often spoke of the first semester of college as a shock or as a challenging transition compared to their high school experience. A civil engineering major who had been a high-achieving student in high school described how her first semester in college had been a wake-up call that she needed to work harder and expend more effort on her classes to be successful. As a consequence of her experience, she had begun to mentor incoming students so that she could share her experience and help other students in their transition to college.

I think it really takes that first semester, it's funny cause I mentor first semester kids now, in their first semester of college, just because of my experience and stuff and they always think, until you get your grades back, you don't really realize like, 'Wow I should have [studied more], this wasn't as easy as I thought.' So I think after you get your first semester grades back and you're like, 'Okay I need to buckle down and try harder.' So after first semester I think I realized this is different from high school. (African-American woman, civil engineering persister)

Low grades often served as a catalyst to prompt students to reassess and revise their approaches to studying. Poor grades on a test or in a course helped students to realize they needed to adjust their study strategies. In Chap. 9, we described the demoralizing effect that low grades had for many women switchers, often contributing substantially to their decision to leave their STEM major. In contrast, persisters, notably men, were not as traumatized by poor grades and often viewed a low or failing grade on a test or course as a prompt to work harder and revise their approach to studying. For example, a biological sciences student described receiving his first low grade in college in Organic Chemistry which motivated him to change his approach to studying for tests. He also described the poor grade as an "awakening" to adapt his learning strategies. By trial and error, he devised an effective study strategy that helped to reinforce his learning.

I was obviously, I was pretty upset about it. But it kind of like in a way was like my awakening to, 'Hey, what I'm doing is not working. I need to actually figure out however and whatever I need to do to study to do better on tests.' And then I took a class where we actually were able to use cheat sheets on the test. So, I would write up the cheat sheet and then I'd realize that I never had to look at it because all that stuff that I wrote down I already knew. So that's what I started doing before every test is going through and writing out the important stuff, going all the way back through my notes. And that has really helped since then, on exams and whatnot. So, I think it was honestly kind of better that I did that, just because it was kind of my awakening that I needed to do something different with how I was studying. (White man, biological sciences persister)

On the other hand, some students entered university with strong study skills and learning habits yet found themselves intimidated by the new environment. Students were not always comfortable relying on their old learning habits, such as asking questions, because of the large size of introductory courses or the perceived aloof nature of university faculty. Yet many persisters gradually adjusted as they became more comfortable in the university environment and as classes became smaller in subsequent years, as mentioned in the following comment from an animal sciences major:

> So I went to recitations more, asked questions more. In high school I had my hand in the air all the time. I raised my hand at everything, had like a billion questions. And then coming into college, freshman year, I wasn't as like bold, you know. In the lecture halls like 350 people, you're just like—no one wants to get behind, and I don't want to make anybody get behind, so I'm not going to raise my hand. And so it also helped when the class sizes starting getting smaller. But asking questions like I normally do, because I ask a lot of questions. But freshman year, I just remember I was really quiet. (White woman, animal sciences persister)

Over time, some persisters began to realize that the benefits of practicing effective learning habits extended beyond simply improved academic performance in a single class. An astronomy student described how she began to attend office hours during her sophomore year and made a positive impression on the faculty member which resulted in a research position in the department.

> My freshman year, I was actually, you know, I had this pride thing going of like, 'I don't need extra help. I can do this myself.' And then, after that wore off, I realized that was really stupid. So I went to, I actually went to office hours for a class that I really didn't need to go to office hours for. My sophomore year, it was Modern Cosmology, and I was just acing the course. My professor told me at the end of it that I would have to fail the midterm to get an A-minus instead of an A. Or fail the final to get an A-minus instead of an A. But I would go to office hours to help out other kids. And, and it was also a convenient slot of time to just knock my homework out. So I'd go there and I'd help explain it to the students, and that made a huge impression on my professor I think. He still sees me in the hallway and I've only ever had him for that one class, even though he's in the department. But, he still recognizes me in the hallway. And he's the person who asked me, like, 'Hey, do you wanna do research?' And got me into my research thing. (Native American woman, astronomy persister)

Some persisters learned the importance of adjusting their approach to their education in general, such as the need to take initiative in their own education and to stay engaged in STEM coursework. A civil engineering student described the connection between interest, engagement, and learning that contributed to his persistence, noting that his interest increased after his first semester, and only then did he begin to truly learn and retain the material.

> I guess my first year was, 'Okay, lemme just read all this and spend a lot of time in school and just try to figure out.' That was my first semester. But I just never understood it because the stuff, the stuff wasn't as interesting for me as it has been my second semester. So really I wasn't really learning anything and I wasn't retaining anything. But when I came to my second semester, when I actually studied and talked to other students and actually getting to know my professors and, and I was getting more engaged in the class because I knew the professors and we can talk in the class. And then I started, you know, started to retain more stuff because the class was getting more interesting to me because everything started getting more interesting to me. And that's the only way you can learn in these classes. And actually, that is the only way to succeed in these courses, is being engaged in the class. Because there's a lot of information they throw at you, and it's hard to keep all that information. But as soon as you start being engaged, you decide to get all the important stuff. And that's what I was doing my second semester and so I was still studying a lot, but I knew what to study

> as well as understanding what the professors wanted me to learn. (African-American man, civil engineering persister)

Therefore, persisters made certain crucial adjustments to their study habits and to their approach to their college education to support their persistence. Many persisters described a catalyst, such as a poor grade on a test or in an important course, that motivated them to change their approach to their education. Some persisters made these adjustments in their first year in college in response to a grade or another issue, yet many persisters did not make these adjustments until later in their undergraduate careers. Indeed, many persisters noted that they still struggled with effective study and learning habits even at the end of their studies. Overall, some persisters also reflected on the importance of taking initiative and becoming engaged in their education because this approach enhanced their learning and sustained their interest and commitment to their major.

## Adjusting to Lower Grades

Gaining an ability to realistically assess their own academic performance and grades was a major factor in many students' persistence in STEM. As described in Chap. 9, students, notably switchers, had difficulty in interpreting their grades, especially when grades were based on curved grading practices. As a result, some students, particularly high-achieving students or those with perfectionistic tendencies (usually women), were unable to tolerate grades and grading practices that did not seem to match their effort or that conflicted with their cherished academic identities as smart, high-achieving students. In contrast, some persisters were able to adapt to curved grading practices, especially when they adjusted their own expectations of themselves or began to understand their performance relative to that of the rest of the class. For instance, a biochemistry major described her adjustment to a slide in grades in her sophomore year as she realized that she needed to let go of her perfectionistic tendencies related to academic achievement. In turn, this shift in identity bolstered her confidence that she could manage STEM coursework and gave her a sense that she did indeed belong in her major. Nevertheless, receiving lower grades than expected shook her belief that she belonged in the major, just as it did for many switchers.

> [I needed] an attitude adjustment, I think. I sort of felt overwhelmed, by all the things I was forcing myself to do. I'm a very perfectionistic person and, so when I got to college the first year, I was performing very, very well in all of my classes and on all of my exams. Classes were not difficult at all, and then when things sort of started sliding downhill my sophomore year, I was like, 'Wow, what the heck is happening? I don't know how to deal with this,' and I sort of went into a depressive spiral because I was like, 'I'm not good enough for this, and what am I doing?' And then, after that, I was like, 'Okay, if you just focus and do the work, you're capable of this,' so that's basically what happened and after that, I haven't had a problem. (Asian/Pacific Islander woman, biochemistry persister)

Persisters also learned to adjust their perceptions and expectations as they gained a better understanding of how their grades related to those of the rest of the class, particularly in relation to curved grading. For example, students learned to recognize that everyone in the class had received a "failing" grade on a certain test and thus adopted a different perspective on their own ostensibly failing grade. Students gained a sense that they were not alone and gained comfort in the realization that their performance had been average, rather than failing, compared to the rest of the class, as noted in the comment below from an aerospace engineering major. He described how he re-oriented his perceptions of his own academic performance as he realized how his academic performance compared to his peers in curved grading situations. Initially, he had thought about switching because of difficulties in interpreting his grades and its effect on his sense of belonging in the major, yet as he advanced in his coursework, he adjusted his own expectations for grades and began to better understand his own academic performance in light of his peers' performance. This comment is from an African-American and, indeed, students of color were more likely to feel that misinterpretation of grades contributed to self-doubt or a lack of belonging in the major. The realization that they were not alone in their "poor" performance offered solace and motivation to persist in the major.

> I think initially a lot of [my thoughts about switching to a non-STEM major] came from, just all the tests I've failed. 'Cause they're like 'Oh, you have to get a C.' And your sophomore year, you don't realize that everybody's failing. So you just kinda see your grades and you're like, 'Oh, I got a 40%. Like obviously I can't do this.' But then your entire class gets a 40%. And I think that's by design, 'cause you fail and you're like, 'Well, how did I fail?' And then you learn it, 'cause that's that extra step thing again. And that happened so much my sophomore year- the first lab report I got, I got like probably a 40 on it, 'cause... You get so many fails when they're telling you you have to get a C. It's extremely stressful. But you survive it, if you do, you're just as good as everybody else. (African-American man, aerospace engineering persister)

Most importantly, some persisters began to disentangle their academic identity from their academic performance as measured by grades. They acknowledged that grades did not always reflect their learning or understanding of the content, and that their grades did not dictate their future success in a STEM profession. This realization was often the result of a long process of adjustment and reflection, but it was a necessary process for students who were used to receiving high grades and defining their sense of worth or identity through their grades. A multi-racial chemistry major described this type of process of identity adjustment that was sparked by an interaction with a STEM employer which provided her with a more accurate understanding of the tenuous relationship between STEM coursework and career success.

> So when I was in high school I was like 4.0 GPA, I think it was actually higher than 4.0 because I took a lot of honors classes, so then I came here and my GPA was not a 4.0, and I kind of realized that I don't, GPA is not an accurate measure of a person's intelligence because it's just how well you do on that test or if you have a good day or a bad day, you're trying to cram all this information in fifteen weeks….I thought GPA was, 4.0 I'm a perfect student, and now I don't think it is. I've heard from an employer that, he said, 'When I see a 4.0 I get a little suspicious.' I think they just look for well-rounded students, and if they see a 4.0 they wonder like, is this person not social? Or do they have a life beyond their books maybe. (Multi-racial woman, chemistry persister)

Many persisters adjusted to lower grades in their major because other students or even faculty members affirmed that many students struggle in gateway courses. A persister who participated in a minority STEM scholars program described how he was comforted by the fact that his peer mentor from that program also had similar struggles in engineering classes.

> Persister: So I got Cs and Bs throughout my [first two years]. I never thought of quitting ever or like, OK, this is too hard, I'm going to switch into something else. It was more like, OK, I understand why I did bad. Then people keep telling me, even like my professor or other students or upperclassmen, and they will always say, 'Yeah, they're hard but you just got to push through and just keep working. Try your best.' And so, I was like, 'OK.' And now, I got a C but I'm still going forward and I'm still passing my classes. It's not the grades I want but I have to do better next time.
>
> Interviewer: How important is it to have people say things like that to you?
>
> Persister: I think it's very important especially in stuff or something that's very hard. Because if you're like…if you've become too sad, thinking more likely to give up and then someone tells you, 'OK, it was tough for me too.' There was someone my age that was mentoring me [through the Scholars of Color program]. And they're like, 'Yeah, there are some classes like that. It was really tough.' So I was like, 'OK, I see that it's not just me that's struggling in this classes, a lot of people do, too.' (Hispanic man, civil engineering persister)

Adjusting to grades and understanding how to properly interpret them was an essential persistence strategy for many students. While some switchers were demoralized and deterred by lower than expected grades, persisters learned to understand their grades in relation to those of their peers. Students came to realize that many of their peers also struggled with grades and that the entire class may have received "failing" grades on a curved grading test. The realization that others struggled and also received poor grades was an especially important source of solace and comfort for students of color and women who were more likely to doubt their skills and belonging in the major because of perceived poor academic performance. The ability to correctly interpret grades was vital to students' persistence in STEM.

## *Instrumental Moves*

In keeping with their adjustments to behaviors, attitudes, and identities, many persisters also became savvy navigators of the STEM higher education system. Persisters figured out how to navigate their coursework and STEM pathway in such a way to maximize their chances for success. In this way, students learned how to pick the right courses and professors, enroll in other institutions, or re-take classes as necessary to improve their success in their major. Some persisters even referred to these strategies as a "game" or "gaming the system." Another persistence strategy was to switch majors within STEM to find a better fit or more appropriate career path for their interests and talents.

## Figuring Out How to Succeed Within the System

Rather than simply adjusting to lower grades or disentangling their learning from grades, some persisters figured out how to navigate the system to increase their chances of achieving high grades. These students put great effort into researching potential instructors, course sequences, and other aspects of their degree program that were in their control in order to maximize their chances of receiving high grades. Some students connected these behaviors to the emphasis on curve grading within many introductory STEM courses which enabled them to research which sections and/or instructors offered the greatest number of passing grades, as suggested by a biological engineering major.

> I comfort myself with the fact that I don't do worse than at least the average, yeah. But I think the problem with the curving is that it lets people game the system really easily, so like when I have to pick classes for my major I'll go to the ranking site and I'll look at which class has given the highest number of As, and so I'll pick that class because I have a better chance of doing well in it, whereas not necessarily because I like the material. So when I pick my classes for the following semester I always ask what my friends are taking, what the people in the level above me are taking, and I always like check out you know what the grade point average breakdown was for the last class that they taught the year before. (Asian/Pacific Islander woman, biological engineering persister)

Students often used online resources to investigate professors and review student ratings of faculty. Many students also relied on their social network, such as friends or siblings in their major, to provide recommendations about courses or certain instructors. For instance, a biochemistry major described how she relied on her brother's advice about faculty members and also used an online course evaluation site to review student ratings of courses. This student also investigated average grades in specific sections of courses. Thus, many students drew on multiple resources to investigate options and navigate their STEM degree programs to maximize their chances for passing grades.

> My brother went here so I ask him, 'If this class is hard, is this teacher good?' I use the course eval site where you can like look at teachers, look at classes and their ratings and stuff. And another one that is similar, like you can see what others say about the classes, and what percent grades they are. I've taken some advice from my parents about what to take. (Multi-racial woman, biochemistry persister)

Students also carefully researched specific professors because they had learned that the quality and clarity of instruction had a direct influence on their success. A zoology major described how he withdrew from a particular faculty member's section and continued with a different professor so that he could be successful in the course. As noted in Chaps. 6 and 8, students often learned to differentiate between "good" and "bad" teaching early in the semester so they would have a chance to move to a different section, if necessary.

> I've noticed even the same exact class is—has a different outcome, depending on which professor I've had. So I've had to, you know, withdraw from a class and continue with a different professor. And then I succeeded. (Asian/Pacific Islander man, zoology persister)

Once enrolled in courses, students also learned how to navigate the course to increase their chances of success. For instance, students learned the implicit rules of particular faculty members, such as recognizing in which courses they could miss assignments or turn homework in late, as indicated in the comment below from a mechanical engineering major.

> And so there's this level of, you have to bite the bullet eventually, and you learn which classes it's okay to fail homework in, which classes you can get away with turning stuff in late, which professors you can schmooze into letting you turn stuff in late. (White man, mechanical engineering persister)

Students were also selective about whether they attended recitations or tutorials and which students they chose to study with. Many persisters developed strategies to selectively use the available resources within a course, including their peers, to increase their chances for success in the course. Yet, students' active management of peer study groups had drawbacks as demonstrated in Chap. 9. Students who were excluded or isolated from these peer interactions (more often switchers in our study) felt a sense of isolation and lack of belonging as a result of the exclusion. Students who were able to successfully navigate the uncharted path of informal peer study groups found belonging and academic support from their peers. Many of these persisters learned to successfully navigate peer study groups through trial and error. For instance, an environmental engineering major described how she changed her approach to the course and her selection of peers to study with after a poor experience in Physics 1. Subsequently, in her Physics 2 course, she navigated the course quite differently and was more selective in the peers that she sat with and studied with in the course.

> You almost have to like scope out groups, and you have to be really selective about who you sit with. 'Cause I remember during Physics 1 I chose an awful group for my recitation tutorial, and it was really difficult. Like, I hated it. And then for Physics 2 I was very selective about which group I sat with, and I ended up sitting with two or three junior-level, aerospace engineering majors, 'cause that's when they take Physics 2. Just the fact that they were older, they took academics much more seriously, and they knew a little bit more physics than I did- 'cause I was never very good at physics. But they were really helpful. And we all just kind of approached it with the same mindset of like, 'Okay, we're gonna try, but we'll see what happens.' And that was really helpful. (White woman, environmental engineering persister)

Students also actively managed their course schedules, such as delaying classes that were perceived to be hard until later in their degree programs, taking hard classes during the summer, picking professors with reputations as good teachers, or dropping or withdrawing from a course in which they were seriously struggling. Some students also chose to take difficult or "weed-out" courses at other institutions, such as community colleges, where they might benefit from smaller classes or more individual attention from faculty. Students with low math readiness were much more likely to actively manage their schedules in a way that best facilitated their success. Low math students were also more likely to take classes at community colleges, although students with high math readiness also used many of these same strategies as seen in the following comment from a high-math chemistry major:

> This university is notorious for bad physics teachers. So, the next step was going to the community college [to take the class] which ended up being a great decision because I had an amazing teacher there. He was for one the best teachers I've had throughout my college career. So, it's good to know there are good teachers out there. (Multi-racial woman, Chemistry persister)

When students used these types of persistence strategies, they often talked about managing their grades and coursework as a "game" to "play." This discussion revolved around courses that students perceived as weed-out courses in which beating the class average became more important to retention and advancement than actual learning or interest, as noted by a chemical engineering major.

> Statistically, if you look at the averages on the exams, because the averages were so low, it didn't really matter if you got things right. I started having to think about where am I in proportion to the average, not do I know things? It's like, do I know more than everyone else? So that's the way that I would distinguish between a course where it feels like the goal is to make sure that I'm learning everything and a course where the goal is to play the game really well. (White male, chemical engineering persister)

Some students simply played the "game" to improve their grades, while others offered insight that the "game" is not necessarily designed for optimal student learning and success.

> It took me a while to figure out that going through college is a system and you have to play the system in order to succeed. And so that is a huge part in why people leave science programs, because it is a system. You have to learn how to play each class, manipulate each class to get the best grade you can. This system, which we've gotten good at playing, is flawed. And it doesn't necessarily have the student's best interest in mind. (Asian female, physiology persister)

Thus, persisters employed a variety of strategies to navigate the higher education system and to increase their chances of achieving high grades in courses. While many of these students had instrumental motivations in increasing their grades or their chances of passing the course, some students used these strategies to find the appropriate faculty member or learning environment that will enhance their learning and retention in the course. Whatever the motivation, many students went to great lengths to ensure that they selected a course, section, or professor that would increase their likelihood of success.

## Switching to a More Appropriate STEM Career or Major

Nearly one quarter of persisters in our interview sample switched majors within STEM at some point during their degree program. Unlike switchers, who shifted to a non-STEM major for a variety of interconnected reasons, relocators were often motivated by fit or interest (often related to career interests as noted in Chap. 4). Students may not have known about the pathways within the major they first selected or the career options within that field. As they became more familiar with STEM disciplines, they realized that their interest lay in another field or that they were more attracted to the careers that might be available to them in a different, but related, major. Students' relocations usually represented a move to a highly related major, for instance, chemical engineering to chemistry, or computer engineering to electrical engineering. Yet some relocation was prompted by a lack of fit with the initial STEM discipline, either temperamentally, socially, or intellectually. Students' realization that their initial major was not a good fit often resulted from their burgeoning understanding of the field and the career options within that field. For instance, some relocators were attracted to the variety of career options available in a different, yet similar, field within STEM, such as described by a chemical engineering major who switched to materials science after learning more about the nature of her initial nature and the available career options within it.

> So I came in thinking I wanted to do chemical engineering as a freshman and then I changed right away, I found out what chemical engineers do and that just didn't interest me, and then material science engineering was in the same department and so I just talked to a whole bunch of professors in material science and what they seemed to be doing is something that I could see myself doing so I switched over like right away in the second semester of college. So material science you like work with all different types of materials and what I liked about it the most was that I could work in a whole variety of industries, I could go into biomed or like aero-space or like packaging is what I did this summer or like a polymers company so I could go into almost anything, any like engineering company needs a materials engineer, so that's why I really like it. With chemical I was just very limited to where I could go but material science I could go anywhere. (Hispanic woman, chemical engineering to materials science relocator)

Thus, the majority of relocators moved to a new major because they perceived it to be a better fit for their interests and career goals. Many of these students had shifted career aspirations during their studies or had clarified the field within STEM that they hoped to pursue. While most relocators moved to similar majors within STEM, a few students undertook more unexpected moves to entirely different STEM disciplines. For example, a computer science student described how she stumbled upon her major during her college studies, a field in which she had not previously envisioned herself.

> So when I was applying to colleges, I was kinda looking at architecture, and then, when I was talking with some different schools about architecture they were like, 'Oh, you really should do architectural engineering, because it opens up so many doors for you!' So I went into engineering because I still didn't know too much about architectural engineering and I wanted to keep my options open for when I started college. Then in one of my classes, we had a speaker come in, and she's actually the director of a master's program in Technology and Developing Communities. And then after, I kinda went up to her and I was like, 'Listen, I really wanna get involved in this, it seems like something I could get really passionate and involved in, what do you think would be a good step for me?' And she's like, 'Let's have a meeting, like, and really talk a little more so I can get to know you.' We met and she pretty much was like, 'You seem like you could go into computer science.' And I was like, 'Computers! That's like the last thing I would ever imagine myself doing.' but I was like, 'Okay, I'll try it..' So I took an intro to computer science class, and I was like, 'Alright. I could swing this.' So, yeah, it kinda just fell into place from there. (Asian/Pacific Islander woman, Engineering to Computer Science relocator)

Certain STEM fields seemed to prompt more relocation than others. For example, engineering majors were more likely to move around as they settled on the sub-field of engineering that provided the greatest fit and interest for them. Likewise, students often moved among the array of life sciences choices available to them, such as moving from cell biology to physiology, or biology to animal sciences. Students may have known that they wanted an engineering or biology major when they first entered college, but they did not always know enough about the sub-disciplines to pick the right field for them. As they learned more about their discipline in college, they were able to select the field that best matched their interests and career goals. These students were similar to the under-informed switchers; however, they managed to find a new "home" within STEM, as described by the following relocator:

> I decided [initially] to major in chemical engineering because of my AP chemistry class. I actually didn't like science in high school, but I was good at chemistry and I went into AP chemistry and realized I really loved chemistry. And my AP chemistry teacher kind of I just talked to him, and he was saying I should do chemical engineering. So I looked up chemical engineering and I saw that they have a really good starting salary and they do interesting things so I just came in as a chemical engineer. I had no idea what it was….. [And once I got here] most of the time I just didn't understand what I was doing and why. That was my biggest problem, I didn't know why I was learning this. Cause I wasn't sure what my goal was in the beginning. But I got into energy and I realized chemical engineering isn't going to get me there. 'Cause the way our curriculum is it kinda prepares you to work in like a chemical plant and like work with reactors and stuff and I had like no interest in that whatsoever. And civil [engineering] has tracks. You can do a structural track to build things, you can do a transportation track to deal with roads and stuff, the environmental track to deal with environmental things. And the chemical is just one lane, I didn't like that. And then so I saw the environmental track and so I decided why don't I just switch to civil. And that's what I did and I like it better because it's going to help me get into this field. (African-American woman, chemical engineering to civil engineering relocator)

In conclusion, relocation was another persistence strategy employed by STEM students. Moves within STEM were often motivated by a process of self-discovery as students narrowed their interests and career aspirations within a certain field. Thus, relocation was almost always inspired by the pursuit of a "better fit" for students' temperaments, interests, or career paths. Relocation was yet another way that students shrewdly navigated STEM coursework and majors to enable their persistence and success.

## *Institutional and Social Support*

Despite the individual characteristics and savvy navigation of pathways displayed by many persisters, the most frequently utilized persistence strategies were social and institutional in nature. Some students found support in formal institutional programs, such as student clubs or organizations, tutoring, or research opportunities, while many others drew upon informal sources of support, such as family, friends, and most importantly, peer networks in their major.

## Finding a Support System

Half of all persisters relied on some type of external support for their persistence in STEM. Likewise, in the gateway course study, instructors often identified social networks as essential to students' persistence (Ferrare & Miller, in press). In student interviews, support was often described as informal, consisting of family, friends, or academic peers, but students also cited institutional support systems as beneficial, such as formal study groups, honors programs, or academic clubs. Out-of-class experiences (e.g., undergraduate research experiences, etc.), another form of institutional support, are discussed separately because they bridged institutional and non-institutional support (e.g., internships were often not sponsored by the institution) (Table 12.3).

Given that institutional or formal support structures were vital for some students' persistence, it is not surprising that there were institutional differences in students' reliance on support. For instance, students at PB3R1 were more likely than students on other campuses to comment that support had helped their persistence. Students at PV1E2 and PV1R3 were least likely to mention these types of support as essential to their persistence. There were few disciplinary differences, although students in math (47%) were much more likely to mention informal peer study groups, particularly compared to computer science (27%) or physical science (27%) majors. In contrast, engineering faculty were the most likely to mention the importance of social ties to students' success (Ferrare & Miller, in press). Disciplinary expectations and culture most likely contribute to these differences. There were no gender differences in students' reliance on support systems, these were important for both men and women. There were also no differences according to math readiness (e.g., 33% of high-math and 38% of low-math persisters cited informal or formal sup-

**Table 12.3** Sources of informal and institutional support for persisters

| Source of support | % of persisters |
|---|---|
| *Informal support* | |
| Informal peer support in major | 34 |
| Parental support or encouragement | 25 |
| Friends outside department | 10 |
| Non-faculty mentors in discipline | 3 |
| *Institutional support* | |
| Minority support programs | 18% of minority persisters |
| Faculty support/mentoring | 10% of all persisters |
| Honors or leadership programs | 5 |
| STEM-related academic clubs, students chapters | 5 |
| Non-academic organizations (church, fraternity/sorority, counseling center, disability services, etc.) | 6 |
| Departmental advisors | 2 |

port). However, students of color were more likely than White students (43% and 29%, respectively) to note that informal peer study groups had been important to their success.

## Informal Sources of Support

Students drew on many sources of informal support during their undergraduate studies, including parents, other family members, friends, and academic peers. These informal relationships helped students to maintain their motivation in STEM, build resilience in the face of low grades or setbacks, and, in the case of academic peers, contributed to their learning and mastery of course material. Parents often served as a source of motivation, encouragement, and comfort for students, as noted in the following comment from a chemistry major. In particular, for first-generation college students or students of color, such as this student, parents also served as a source of motivation and inspiration to remain in college despite challenges and struggles.

> Well, my parents have sacrificed a lot for me to be here, both financially, emotionally. They've done a lot for me and I love and respect them for it. I feel like if I had gone to school for three years and suddenly dropped out that I would have been just a huge disappointment 'cause they would want to see me put in effort and motivation and succeed. Disappointing 'cause they put a lot of money towards it, tens if not hundreds of thousands of dollars, towards my education, and so that was always kinda motivating me to just keep going. (Hispanic man, chemistry persister)

Parents were often integral to students' persistence, especially when students were considering changing majors or leaving the university entirely. Parents supported and encouraged their children and reminded them of their passion for their field. For example, a first-generation college student biology major recounted how she began to struggle with grades during her sophomore year and considered transferring to a community college, but her parents encouraged her to remain at the university and motivated her to stay because a 4-year degree would be better for her future.

> Student: Sophomore year is when [my motivation] really started to decrease 'cause I started struggling, like my exams weren't that good, and then it's discouraging to feel like you understand something, but the test shows you that you don't. So that's when it started to decrease. It really is a struggle.
>
> Interviewer: Did you ever in that time wonder if you were in the right major and think about going to another major?
>
> Student: Yeah, definitely. Well, it's not even just another major, it was just another school. I want to do biology. 'Cause I know like I love it. It's something I care about but I was just like, 'I don't know if [this university] is right for me.' I was gonna transfer to a community college 'cause it's just so hard, and it's like, I don't feel comfortable or I feel like almost inadequate to be here. Like I don't understand the information. I really thought about just leaving schools, but my parents were supporting me and like, 'You know, you should stay here. Like, it's good for your future, you know,' instead of going to a community college (Asian/Pacific Islander woman, biological sciences persister)

As much as parents supported students' persistence, the most important source of informal support was overwhelmingly from peers in the discipline. Students supported and motivated each other through struggles or challenges, particularly in gateway courses. A mechanical engineering major described how late-night study sessions with friends in his major supported his persistence because he found it more motivating and fun to work with friends than alone.

> I remember every single Sunday night my sophomore year, I'd be doing homework with two of my friends for dynamics. Cuz every single Sunday night, I would be in their lounge, doing homework with them. Laughing, joking around, and eventually doing homework, finishing at like 3:00am. But it'd also be fun, just doing homework, but you're also with a bunch of friends--it's not even work then. (Asian/Pacific Islander man, mechanical engineering persister)

Friends and peers in the major also served as a source of support and provided a sense of community that students were not alone in their struggles. It was very important for persisters to know and recognize that others also struggled in coursework and perceived it to be "hard" and challenging. In contrast, many switchers felt that they were alone in their difficulties in the major. Other students persisted, in part, because they knew that others shared their struggles. Friends offered support and a sense of community and belonging that motivated students to remain in the major in the face of difficulties, as mentioned by a low-math microbiology major. A community of peer support not only promoted belonging in the major but also alleviated some of the competitive culture that can be inherent in the life sciences from the pressure of medical school admissions. Additionally, peer support was particularly important for students of color, as this African-American woman demonstrated.

> I really think it's a support system of other science majors because it sucks to know you're struggling, first of all. And it sucks to know you're not doing well. But knowing that other people are not doing well with you, you know that you're not to blame, you know. And then you all get together and you can study together, and you can figure things out together and things like that. I think that's generally what has gotten me do it because there are some days we all play this game, like, 'Man, if I wasn't a science major, what would I be?' And I was like, 'I'll be psych major. And I wouldn't study as much. And I'll go out way more.' That's our game. But I really think the community of science students and it kind of sucks because like you have like let's say if there was a medical school here and I was a pre-med student, I would see people in my class as competition. I really…I really think that hurts. I think you need at least like one person, one or two people that you're like, 'Oh man.' You sit down. You talk about how much that physics test got you. And then you study together the next day, you know. I really think friends and even my TAs are like, 'Don't worry about it. No one did well.' I'm like, 'All right, it's not my fault.' Because science is hard. And I don't think they emphasize that enough, science is hard. (African-American woman, microbiology persister)

Academic peer networks also enhanced students' learning and mastery of concepts. Informal peer study groups allowed persisters to learn from their peers, as students were able to explain challenging concepts to each other, and ask clarifying questions in settings that were comfortable and non-intimidating. Peer study groups were particularly important for low-math students who were more likely to struggle with difficult concepts and were often less well prepared in STEM by their K-12 schooling. For example, a low-math biology major mentioned that studying with peers has been the "most helpful resource" to his persistence.

> I feel like you have to learn, you can't go into a group not knowing what you're doing because then other people will just basically dominate the group and you'd just be in the side, just not knowing what's going on. So I feel like as long as you do your studying by yourself or learn your material by yourself. And then you can go into the group and I feel like that's when it's best to be in a group because the best way to know if you know something, is if you're able to tell or teach it to somebody else. That way, it means that you have to know the material very well to be able to explain it to somebody else, to have them understand it. And when they don't understand it and they keep asking you questions and you're able to answer their questions until they understand it, that's the best way to judge whether or not you know something. So I feel like studying in groups is important. At first I thought I could just stay by myself, just study by myself. That's pretty much what I did my whole first semester. I just did everything by myself or maybe a group of two, just me and one other person. But I've definitely incorporated larger groups since my studying has gone on. I think it's necessary. You can't stay in your own little bubble and try to succeed. You definitely need to branch out because, like I said, my peers have been my most helpful resource. So the more you know, the more people you know or the more people you interact with in your, that's the best, I think the better you'll do. (African-American man, biological sciences persister, low-math)

Thus, informal sources of support, such as family and friends, were vital to students' STEM persistence. In particular, STEM peers were essential to the persistence of many students, especially for students of color and first-generation college students. Some switchers moved to non-STEM majors because of a lack of peer support and a subsequent lack of belonging in STEM (as described in Chap. 9); in contrast, peer support enhanced persisters' learning and also bolstered their feeling of belonging in the major. Sophomore year was often a critical juncture in which persisters needed to find peers to study with in order to better learn the material and advance in the major.

## Institutional Sources of Support

Many persisters benefited from institutional supports, such as tutoring, peer mentoring, study sessions, or student clubs and professional societies. Interestingly, instructors noted the importance of informal social connections yet did not comment on institutional mechanisms for fostering or enhancing these connections (Ferrare & Miller, in press). In this sense, peer support becomes the students' individual responsibility and absolves the institution of any responsibility for encouraging peer relationships. However, in student interviews, many noted the importance of formal, institutional programs in helping them to connect with academic peers and access institutional resources. In particular, underrepresented students benefited from targeted programs, whether formal institution-wide support programs or disciplinary-specific clubs or organizations geared toward women or minorities. Formal programs for underrepresented minority or first-generation college students in STEM were particularly beneficial for students' persistence. These programs often offered mentoring, tutoring, advising, and provided a cohort of peers in STEM fields. Such programs also served as institutional brokers and helped students to navigate the institutional structure and locate the resources and support that they needed to persist. For example, an African-American woman with low math readiness described how the Minority Scholars in STEM program at her institution connected her with important resources and a community of support so that she would not suffer from the isolation in STEM so often experienced by women and minorities.

> Being a part of the [Minority Scholars program], I definitely got a lot of assistance. They have tutors that help you with your classes, like math classes, chemistry classes, physics classes. Sometimes they pair you up with students that help you, like, a student that has taken the class before. So if you're having difficulties in that class, you can always go and talk to that student. And they also have study groups as well….Going to the teachers and TAs, that's something I had to learn, like going alone. But, talking to someone in the program, like these events that we go to and I meet all these people at this event and they're like, 'If you need help in this class, come and talk to me.' And stuff like that. Like, I knew that these were people I could always talk to. But sometimes you have to go to the teachers or the TAs. Like, that took me a while to understand that you have to go to the teachers like if you don't understand something. But it took a while. (African-American woman, civil engineering persister, low-math)

Institutional programs also provided financial support to low-income and underrepresented minority students which contributed to their ability to persist in STEM. Institutional programs offered financial, academic, and social support as described in the following comment from an aerospace engineering major. The Minority Engineering program at his institution also offered a summer bridge program that introduced students to the college experience. Four years later, the cohort had still maintained a strong sense of community.

> I think the number one factor of why I made it through my first year was the [Minority Engineering Program]. 'Cause I did that summer thing and they gave me a scholarship, and they actually gave me like four of my books which are like 400 dollars right there, which is awesome. But to get your scholarship, you have to fulfill certain requirements. So you have to do calculus workshop classes, and it's once a week for like 2 or 3 hours. So it was probably a group of 20 of you, you'd all come and then there'd be… I guess you'd call them a TA, but he would have math problems, and put 'em on the board or just give 'em to you. You'd work in little groups and you'd all work on them, and then again as a big group you'd go through them, and work it. And that's pretty much how I learned my calculus, I'd say. My engineering friends are all from the [Minority Engineering Program]. We were all in the same summer program, so most of my roommates are actually from that program. We met then and we've never left each other. (African-American man, aerospace engineering persister).

Faculty also offered mentoring and support for some students, though faculty were constrained in the number of students they could help individually because many gateway courses had large enrollments with high faculty–student ratios. Also, as discussed in Chap. 6, many students continued to perceive that faculty were more interested in scholarship and research than in teaching or mentoring students. Nevertheless, a few persisters benefited from faculty mentoring and it had a substantial impact on their persistence and commitment to their field. Faculty mentoring appeared to be particularly influential for women in fields where they are severely underrepresented, as demonstrated in the following comment from a woman computer science major. She described how a woman professor motivated her to stay in the major at a time that the student felt lost and was considering switching.

> One of my next computer science classes after my intro class, I had a really motivating professor, and, it was over the summer but she didn't single me out, but she kinda gave me, what I would call special attention. Kinda when I reached out for help, she really just made it clear, again I don't know if this is just a characteristic of her being kinda motherly as a person, but kinda having that and talking with her and being like 'Y'know, this is kind of hard, I don't know anything about this.' If anything, that was my one time I had almost a drop-out time, just because I didn't really know how I was doing, but she was really motivating and I did really well in her class. Having that individual attention definitely helped. (Asian/Pacific Islander woman, computer science persister)

Departments also provided formal peer mentoring and student organizations, clubs, or professional chapters that enabled students' persistence, especially for women and students of color. Occasionally, students themselves founded and organized these opportunities to create a community of support for themselves and their peers to fill a gap or meet a need in their education. For instance, a student described a peer mentoring group that she and her fellow students founded because they were dissatisfied with the formal advising offered in the department. Subsequently, they created a structure where students could advise and learn from each other.

> It would be hard to get through this school without having a group like this. Primarily because you need a support system in order to figure out what you're doing. And like I said, the advising system at this school does not provide that for you. And that's why I personally spearheaded this event two weeks ago, which is like a student-to-student mentorship program which had a pre-health focus. So me and one of my friends who just think that like you cannot survive without the help of other people. Especially if you're going into the pre-medical field. (Asian/Pacific Islander woman, physiology persister)

Student chapters of professional societies also contributed to students' persistence. Some persisters took advantage of student organizations such as the National Society of Black Engineers, National Organization of Black Chemists and Chemical Engineers, or Society of Hispanic Professional Engineers. The Society of Women Engineers was beneficial, too, as it provided peer mentoring and support for women engineering students, as noted in the following comment from a woman biomedical engineering major. She found the advice that she received from peers in the Society of Women Engineers program to be invaluable, especially the advice to go to faculty office hours which she would not have ordinarily done on her own.

> So a lot of things was like, Society of Women Engineers, so we have a like a mentoring program where we're paired up with somebody, so upperclassman are paired up with somebody from lower class of the same major, or they try to do it within the same major, that way, you know, upperclassman can tell you where they struggled and how they got past it or what they wished they would have done differently. So I don't think I ever would have gone to office hours because I didn't want to be that one kid who always went to go see the professor, but like, that's something that upperclassman have always like really encouraged us to do was go to the professor, like go get help, you know, just because you're the kid that goes to get help, that doesn't mean you're stupid, that means that you're trying to succeed. So a lot of like being encouraged by people that have already succeeded or by people that have been there and can tell me like what they would have done differently, like to avoid the

> mistakes that they've made, was really helpful. (Asian/Pacific Islander woman, biomedical engineering persister)

In sum, institutional support programs such as minority STEM scholars programs, peer mentoring or peer instruction, and student clubs and organizations were integral to the success of the students who took advantage of them. These programs offer guidance, mentoring, and role modeling and created a community of learners who supported one another's success. However, many persisters did not take advantage of these resources and opportunities. Nevertheless, they were vital to the success and persistence for students who utilized these institutional supports.

## *Seeking Appropriate Help*

A significant part of students' persistence rested on their realization that most, if not all, students struggle in STEM coursework at some point during their degree program. Students who felt isolated or that they were the only ones struggling with the material were most likely to leave STEM entirely for a non-STEM major. In contrast, students who found peer support or recognized that many others were having similar problems were more likely to persist. This help was not only provided through informal peer study groups, as was previously discussed in this chapter, but was often provided through formal institutional sources, such as tutoring, organized study sessions, or faculty office hours. There were slight institutional differences in students' use of institutional "help" resources; for instance, 25% of students at most of the sites mentioned use of appropriate help services, while 40% of students at PB2R1 reported the same. Online searches of services and STEM support at the universities revealed that PB2R1 offered more formal sources of support than many of the other sites, and many of these support services were STEM-specific, such as tutoring, guided study sessions, and other resources. Across all sites, there were no gender differences in students' use of help (39% of persister women and 34% of persister men). However, students with lower math readiness were more likely to take advantage of institutional resources (61% of low-math persisters and 25% of high-math persisters). Additionally, students of color were more likely to access these resources than white students (51% of students of color and 29% of white students). There were few disciplinary differences although life sciences persisters were more likely to use these resources than engineering persisters (41% and 27%, respectively).

More than half of students who credited their persistence to institutional support services had used organized study groups, tutoring sessions, math labs, or peer-led supplemental instruction. Teaching assistants, tutors, and peer instructors helped students to understand and apply complicated, conceptual material. Some students were intimidated by faculty members or felt faceless in large classes, and students found recitation sessions or peer-led supplemental sessions to be more comfortable learning environments than faculty office hours. Persisters often felt more comfortable with peers or peer instructors than with faculty. Thus, organized study sessions offered an opportunity for students to seek help that did not involve going to faculty office hours which some students found to be uncomfortable or intimidating. Students also did not have to navigate the challenging terrain of finding and sustaining informal study groups, an ability that some students were more successful at than others. In the following comment, an aerospace engineering student noted that he learned to seek out resources through the minority summer bridge program that he attended prior to his freshman year:

> I never really interacted much with the professors because I didn't feel like it was really, appropriate for me. Because I was one of 500 students, and so I didn't want to depend completely on that because there would be 499 other students like that. So what I did is, I interacted more with the TA or Peer instructors that were there, or I tried to find like other resources like tutoring and stuff. And that was kind of what got me through some of the more complicated subjects I had. Like the first few semesters, I felt like I had a good grasp on just because it wasn't as complex, but the summer bridge program kinda taught me what to do if you run into this problem. (Asian/Pacific Islander man, aerospace engineering persister)

Students often did not know or even realize when they entered the university that seeking help when needed is an essential facet to student success in STEM majors. Students needed to learn over time how to ask for help and where to receive the academic help that they needed. Students often didn't realize at the beginning of their studies that resources were available to them. Again, this is an example of the process of persistence in which it takes time for students to adjust to the nature and culture of their institutions and STEM degree programs. Just as students needed time to learn to adjust to grades or to identify friends and study partners in their degree programs, they also needed time to understand the institutional landscape and the resources available to them.

As noted in Chap. 5, some students switched because they could not access or did not know about appropriate help during their transition to college. Likewise, persisters needed to learn where to find help when they needed it. Minority STEM programs often served the function of educating students about available academic resources. Other students often learned about resources through peers or word-of-mouth. Perhaps most importantly, students needed to recognize that they needed help and that formal academic help and support could enhance their learning and success, as mentioned in the following comment from a chemistry major:

> Something that I didn't really do but everyone emphasized and that I eventually picked up on was learning to ask for help when you need it, like go to tutors, go to office hours, go to the Learning Commons that we have here. There's a lot of resources at the U that are available to students. When I was a freshman, I was ignorant and I didn't take advantage of, but having progressed through my curriculum I realized I do need to get some extra help on this. (Hispanic man, chemistry persister)

Learning to seek help was difficult for students who had been high achievers in high school and whose academic identity was grounded in grades and academic performance. Persisters often realized after a single poor performance or grade that they would have to seek academic support which had never been necessary for them before. For example, a chemistry major described how she realized that she needed to seek help after a poor grade on her first midterm in Calculus I, a gateway course that was necessary to advance in her major.

> Calc 1, I remember my first midterm I got a D and I cried because I had never gotten anything below like a B-plus in high school. So that was tough. And then I started going to the learning center, it's in the library so then you can go and they have a tutor there and people just kind of congregate there and most of the people are taking the same classes, a bunch of freshman level classes, calculus and physics, then you can get tutoring help so that was a big help. Cause they went over problems but then it wasn't just here's the problem, it's here's how you do it and how we work through it. So that was really helpful. (Multi-racial woman, chemistry persister)

Other students commented that faculty had helped them during office hours and that they had learned that seeking help during office hours was critical to their success in gateway courses. A biochemistry major credited her success in organic chemistry to her use of Teaching Assistant office hours and faculty office hours. Like many students, she was reticent to go to faculty office hours at first, but she discovered that it was helpful and "not so bad."

> I just kinda like started going to, well like starting with the TAs was nice to go to their office hours and figure it out. But like it was really helpful I went for my Organic Chemistry 1 and Organic Chemistry 2 labs, cause they're taught by TAs, and I figured out that if I went to office hours that my grade would be higher so I was like okay. So then I started going to professor office hours and I think really that just going and forcing myself to go and then figuring out that it wasn't so bad. And then like everything kinda fell into place after that. (Asian/Pacific Islander woman, biochemistry persister)

While some students, as described in Chaps. 5 and 6, felt that professors were inaccessible and intimidating, some persisters found faculty to be very accommodating during office hours. These persisters recounted how faculty explained concepts, answered questions, and talked through problems with them during office hours which greatly facilitated their understanding of key concepts, as noted in the following comment from a math major:

> Just going to the professors has been probably the most helpful thing for me. And so the willingness to be able to do that and just have the humility and say, 'I don't get this.' Five, six, seven times. As many times as it takes, that's what it took for me to get over the hump. And really feel like I had got it, and was able to progress. And so yeah, that's the number one thing I would say [that helped my persistence]. (Hispanic man, mathematics persister)

In sum, institutional support services and academic help were integral to the persistence of many students. However, students often did not know about these services at the beginning of their studies or realize that they needed to use them to be successful. Over time, students learned about academic services through organized campus programs, such as minority scholars programs, or through word-of-mouth or trial and error. Seeking appropriate help from organized study sessions, peer mentors, TAs, and faculty office hours was vital to many students' learning and enhanced their success in key gateway courses that they needed to advance in their majors.

## *Engaging in Out-of-Class Experiences*

Students' experiences and opportunities out of the classroom often had as much, or even more, influence on their persistence than their classroom experiences. Nearly one-third of students credited their persistence to participating in professional, out-of-class experiences in their field. In particular, undergraduate research and internships offered students a glimpse into the professional practice of STEM and helped students to confirm or clarify that they had made the correct choice of college major or career. There were no differences in the impact of out-of-class experiences for high-math or low-math students, both groups of students were equally likely to benefit from these experiences. There were differences in access to research or internships by institution; for instance, PB3R1 and PB2R1 had participation rates of 35%, while PV2R3 had a participation rate below 20%. Moreover, out-of-class professional experiences were more common in the sciences (e.g., 35% of physical science persisters and 33% of life sciences persisters) than math (e.g., 17% of persisters). Despite these differences, the group that benefited the most from professional, out-of-class experiences was women of color (e.g., 46% of women of color compared to 30% of white women reported that these experiences had helped their retention in STEM); however, this was not the case for men of color who were equally likely as white men to report that out-of-class experiences had contributed to their persistence.

Out-of-class experiences affirmed students' commitment to their major as they gained understanding of the nature of work in their chosen profession and gained confidence that they could succeed as a STEM professional. Internships gave students a new-found perspective on school and their coursework as they learned about life after college in STEM professions. In this sense, students were motivated to persist in their major as they saw how a career in STEM may differ from their experiences in college or in STEM coursework. Internships were especially important for engineering majors, as shown in this comment from an electrical engineering major.

> I did my internship in that summer though. So I kinda solidified, it was a great like look into this is what happens after school and it's definitely not as stressful. I don't use anything I learned in school so I was kinda just like I just need to pass! And like so I had a different mindset, I was more motivated by my experience, and it definitely helped a lot to see what the rest of life will look like after school, after all these stressful exams. (Asian/Pacific Islander woman, electrical engineering persister)

Similarly, research experiences offered students the opportunity to work with graduate students and to learn more about life in graduate school. This insight into graduate school allowed students to decide whether graduate school or a research career were the right paths for them. Undergraduate research experiences also allowed students to deepen their learning and mastery of their field. In out-of-class research experiences, students progressed in their knowledge and understanding of the project over time and some even began to mentor other undergraduates once they had advanced to a certain level of expertise. Students also appreciated that they could apply their classroom learning in a research project, as noted in the following comment. For some persisters, the opportunity to apply their learning outside of class reinforced their learning and also influenced their persistence.

> I know how to do everything now. The PhD student that I work under, I used to go ask her questions like every ten minutes. Like 'How do I do this? What do I do? Why am I doing this?' But you know, now we kind of like have a rhythm where, I ask her questions for sure at times, but I know what I'm doing, and then now I'm training other undergraduates. It's really nice, and I get along really well with my research professor, and then also all the other students that are working in the lab, like the PhD students. And they've been really helpful. I think that's part of the reason, going back to kind of the overarching question of why I'm really sticking to the sciences, I think undergraduate research, it gives you an application of what you're learning in class. And without the application, like, why study the material? So that was really, really, helpful for me. Just to sort of understand why I'm learning what I'm learning in class. (Asian/Pacific Islander woman, chemistry persister)

Perhaps most importantly, professional out-of-class experiences strengthened persisters' commitment to their majors. For some, research and internships provided motivation to stay in STEM when they may have been contemplating moves to non-STEM majors, as described in the following comment:

> I was wondering, 'Is this really the way I wanna go?' And for a time I was even contemplating business (MAKES CHOKING NOISE). And I was thinking business so that maybe I could start some kind of organization so, privatize the space industry or something. If NASA can't do it, then maybe some corporations could. And so I had times where I considered, you know, 'How can I get around this direct route that doesn't seem to guarantee me anything? How can I get like a good work around to where I wanna go?' But then I got a research job off campus, and once I started playing around with actual numbers comin' from a hunk of metal spinning around Jupiter 1,000 miles an hour, I was like, 'S∗∗∗, I'm sold.' (White man, astronomy persister)

Multiple internships allowed students to investigate several different career directions, while undergraduate research students often stayed with the same research group throughout their research experience. Thus, many internship students had the opportunity to "try out" different industries and positions to assess the right fit for their future career direction. Some students even found new career opportunities through internships, as described in the following comment from a materials science major. In fact, this student's second internship position was such a good fit for her interest and skills that she received a full-time job offer from the company.

> So, when I got my like first internship position, like I was working on roofing granules, which like isn't like cutting-edge or anything, it's painting rocks and boiling rocks and stuff like that. And I actually didn't really enjoy it, but you know, it was some money and it was some experience that I can put on my resume. But [in my second position], I actually went into packaging, which is like, I totally didn't even know that packaging engineering existed, but apparently it does. And I'm sitting here like, I'm materials science, coming into this packaging engineering role and kind of like getting thrown into it, like not really caring about like the itty bitty details of materials, but rather how they're applied. And so I wasn't really prepared so much for that, but I had a really good time like learning about all those things and I had a really good time like, at the company I had a really good time, like having my own project, doing the things I needed to do for that. They gave me a full-time offer, so I assume that I was pretty good at it. So I feel like because I found success in industry work and industry experience, that's, you know, if other people tell me that I'm good at it and I

> enjoy doing it, then that that should be the path that I should go. (Asian/Pacific Islander woman, materials science persister)

Volunteer opportunities in healthcare settings were very important to the motivation and persistence of pre-med or pre-healthcare students. Thus, volunteer opportunities offered students a glimpse into their future healthcare career and inspired them to persist in their studies. For some, volunteering was more motivating and inspiring than their coursework, as mentioned in the following comment from a biology major:

> I didn't get an internship last summer, I just decided to volunteer at the hospital. It was nice. I liked it, it kind of motivated me to keep going. So I have like equal parts motivation to succeed but also like I don't care anymore as well. Like it's just, I guess junior year just sucks. I'd rather do that than take classes. I feel like I still haven't got anything from my classes. I feel like I'm just getting through them to get through them. I know this is just a foundation that I have to do in order to do things I want to do. [At the hospital], I got to see the things I'm learning in action and how they actually like, I got to see firsthand how people are helped so it was nice to see that. It was less theory and more practical so that was cool. (African-American woman, biological sciences persister)

On the other hand, out-of-class professional experiences enabled students to rule out certain career paths and to narrow their interests to a certain sector or occupation. Sometimes students realized a particular sub-field or career path was not a good fit for them after experiencing it in the "real world" through internships, undergraduate research, or volunteer opportunities. For instance, in the following comment, a Microbiology persister explained how she ruled out being a doctor after a volunteer experience at a hospital. She had never pictured herself in a research lab, yet an interesting and engaging lab section in one of her biology courses inspired her to pursue an undergraduate research experience. To her surprise, she discovered a penchant for research and a desire to pursue a research career. She also found a research project that connected to her initial motivation for being a doctor, to engage in development work in Africa. Thus, she changed her career direction from medical school to graduate school.

> The only lab I had ever had was chemistry lab and I was like, there's no way in hell to grow up and like working in a lab like this is not what I want to do in my life. So that's something awful. I don't want to do any of that, and then honestly I think the turning point was when I actually took classes for my major, because you can't take a biology class here until you're in organic chemistry. So, I never really knew my major. I took the class and I thought it was really cool. I took the lab and I was like, 'I'm in love with this.' I really enjoyed lab, it's the weirdest thing. And so, I really enjoyed it. And so it's like I want to keep working on undergrad research labs. But now, I was like, let me see if I can get into a lab. So maybe when I go to grad school, because I'm not going to go to med school so I want to go to grad school, I'll be prepared. And I got into this lab, and everyone is just so nice and the work we're doing is so interesting and it's so applicable. I guess since the beginning of this I really wanted to be a doctor, so I can help Africa and now, we're doing mosquitoes and malaria in Africa. And it's just applicable. I feel like I'm learning more in lab sometimes than I actually learn in class. I found something I wanted to do. I don't know exactly what area I wanted to do. But now I'm going to grow up and I'm going to work in a lab. (African-American woman, microbiology persister)

As mentioned in the previous comment, altruism in research or internships was an important motivator and factor in the persistence of some students, notably women of color. Another persister who had attended conferences and made presentations related to her research findings discussed how the opportunity to conduct research in childhood language development had cemented her interest in her field and a future research career. When asked if she had ever considered leaving STEM, she answered that she had not because of her research position.

> Not really, just because, um…I started getting involved in research my- summer of my sophomore year I feel like, yeah, and so- my research is on how kids learn language actually, and so I do- yeah, I do computational models of how kids learn language. I use it every day. My initial reason why I came into computer science is knowing that technology can be used as an enabler, like that was a big thing. And just having that reinforced with all my experiences and all the news articles I read on how like technology is the future I don't think I would stray from it. (Asian/Pacific Islander woman, computer science persister)

Thus, professional, out-of-class experiences, such as undergraduate research, internships, or volunteer opportunities, were important for affirming students' commitment to their major and encouraging their persistence. From these experiences students were often able to confirm or clarify that they had chosen the right correct field or career path. Most importantly, out-of-class experiences offered students a glimpse of professional life and provided an opportunity to apply their learning from courses.

## *Doubts Along the Way*

Despite the breadth of persistence strategies employed by students, nearly half of persisters (44%) had doubts at some point during their undergraduate studies that STEM was the correct path of them. In fact, over a quarter (28%) of persisters actively considered switching to a non-STEM major during their degree program. Similar to switchers (as described in Chap. 9), these moments of doubt were often prompted by a poor grade or unexpectedly poor academic performance on a test or in a course. These academic crises sparked moments of doubt and thoughts of switching. To a much lesser extent, and aligned with our findings about a sub-set of switchers, a small group of persisters thought about leaving STEM to pursue an alternative interest, such as Japanese or theater, more fully. However, the vast majority of persisters who had thoughts about leaving STEM were motivated by grades that they found unacceptable. Unlike switchers, persisters pulled back from this tipping point and remained in the major, often at the behest of a friend, advisor, or faculty member who normalized low grades for the student and placed them in context.

While many persisters considering switching out of STEM, some groups of students were more susceptible to thoughts of leaving. As with other issues related to grades, women persisters were more likely to consider switching out of STEM than men (32% of women compared to 21% of men persisters). Asian and African-

American persisters were also more likely to consider switching out of STEM (49% and 42%, respectively), compared to White persisters (22%). Low-math readiness students were also more likely to consider switching out of STEM during their undergraduate career (35% of low-math and 25% of high-math persisters). There were no disciplinary differences in persisters' considerations of switching. Although certain groups of students, such as women, low-math, or African-American or Asian students, appeared to be at greater risk of thinking about switching out of STEM, all of these persisters ended up remaining in the major. Often, the intervention of a more advanced peer or faculty member was enough to make a difference in a students' trajectory. In the following comment, a biology major had actively considered switching because of her GPA, but her undergraduate research advisor counseled her to remain in the major because her interest in the subject still remained strong. The research advisor also helped to redirect her focus away from grades and toward learning and interest.

> I talked to the pre-med advisor here and I thought about switching. Because I'd been told a lot of times that you don't have to major in the science you just have to do the pre-reqs and I was done with the pre-reqs at the end of my second year. Because after my second year I was done with all the pre-reqs so I could have switched my major if I wanted to and I knew a lot of people that did that just to cushion their GPAs because they took easier, like psychology is an easier major. But I also talked to my research advisor, I just thought about different things and he just asked me like, 'What are you genuinely interested in?' And as hard as biology is for me to grasp, I still think it's really interesting and that's what he said that you should do whatever you're interested in, the grades will come, just don't worry about that. (Asian/Pacific Islander, molecular biology persister)

Thoughts of switching among persisters were often sparked by crises of confidence, usually related to grades or performance. In switchers, these crises often fostered a lack of belonging and a feeling of isolation if the student felt that they were the only one struggling or performing poorly. In contrast, persisters often had a peer or faculty member who normalized the struggle for them. In this way, persisters were able to overcome these temporary crises of confidence and renew their commitment to their STEM major, as described in the following comment from a computer science major. A more advanced peer provided perspective on a low midterm grade in a gateway course and helped her to overcome her doubts about her ability to succeed in her major.

> I'm remembering one point in particular my freshman year. It wasn't the first computer science class that you take, but the one where you actually learn a language that you're going to use and where you're building. It's like your first real computer science class, that's what people say. It was like halfway through the semester and I had just gotten a C on my midterm and I was just really frustrated because, I understand the material and then actually implementing it is the hard part. So I was really frustrated, really considered switching majors for a couple weeks. I decided to stay mostly because I talked to a few people older than me, a girl who was a junior, she's two years older than me who I'd just met kind of randomly and we had become friends and then another guy that I'd met who was in my major. So I talked to them and it was just helpful I think, like I think they helped me see things more in perspective in terms of, this is one class in a big major, and kind of helped me understand my strengths more of like you are gifted in this and just because you're struggling doesn't mean it's necessarily gonna be the end. And I think it just helped me put

> it into more perspective and this isn't the world's ending. You're gonna be okay. (Asian/Pacific Islander woman, computer science persister)

Therefore, thoughts about switching out of STEM were relatively common among persisters. The thoughts were often motivated by poor performance or grades that led persisters to doubt their ability and belonging in the major. However, persisters were almost always pulled back from the tipping point by peers, faculty, or an advisor who insisted that an occasional poor grade, even in a required course, is normal and that the persister does indeed belong in the major. This encouragement and support was essential in keeping persisters in the major and was often missing for switchers.

## Conclusion

Just as switchers left STEM for a variety of interrelated reasons, a complex and interconnected mix of factors supported students' persistence in STEM. Students cultivated an array of supports because STEM pathways were often difficult to navigate. There was not a single strategy, asset, or resource that alone fostered students' persistence, rather, a constellation of factors helped students to maintain interest and shaped their identity in their major. While determination, passion, and "grit" propelled many students to persist in their STEM majors, these individual characteristics alone were never enough to ensure students' success. Thus, it was necessary, but certainly not sufficient, for students to remain determined and committed even in the face of challenges or difficulties. Still, the nature of students' determination and persistence varied for different groups of students based on whether they were in a privileged or marginalized position within STEM. Students of color and women were more likely to have to push through isolation, loneliness, and self-doubt in addition to grades and identity issues, while privileged students were more likely to push through poor grades. While individual characteristics certainly supported students in their pursuit of a STEM degree, almost all persisters also highlighted an array of social, institutional, or other factors that supported them during their studies. In rebuke of the dominant meritocratic narrative in STEM education, talent, interest, and "grit" were not enough to retain students in STEM fields. The popular meritocratic narrative of individual effort and intelligence absolves faculty and institutions of any responsibility for student learning and success. Yet, almost all persisters relied on some sort of external support to succeed, whether informal peer study groups, organized study sessions, or faculty office hours. This finding alone has implications for the design of STEM courses and education because it illustrates the role that institutions and systems play in students' success. Thus, determination, passion, and a strong will to persist were important components of persistence, but hardly sufficient.

Switchers often struggled with grades and academic identity, as did a fair number of persisters. On the one hand, academic achievement and high grades bolstered some persisters' self-efficacy and increased their commitment to their major. However, more often, persisters needed to adjust their academic identity to include lower than expected grades. Many high-achieving students were stunned to fail their first midterm or receive their first ever C in an important gateway course. While switchers (most notably women) were often demoralized and thrown off course by these obstacles, persisters (most notably men) viewed them as a catalyst that sparked adjustments to their approaches to studying and learning. Some persisters realized that they needed to spend more time studying, while others developed more effective learning habits. Perhaps more importantly, many high-achieving persisters began to decouple their academic identity from GPA and grades as they came to realize that grades did not always reflect their effort or actual learning in a course. This identity work was a difficult and long transition for persisters, but nonetheless important for their retention in the major. Two main factors helped students to let go of the importance of grades. For one, students came to understand curved grading practices and recognized that an ostensibly failing grade might not represent failure depending on the performance of the rest of the class. Second, many switchers changed their orientation toward grades after more advanced peers or a faculty member normalized poor grades for them and assured them that many students struggle and perform poorly in gateway courses. In contrast, switchers typically did not come to this astute understanding of STEM grading practices, and most notably, they lacked the intervention of caring peers or faculty members who normalized the experience for them. Since this type of intervention was so important for the trajectories of many persisters, it could be more systematically structured into STEM coursework to reduce isolation and to enhance students' sense of self-efficacy and belonging.

Another difference from students' STEM experiences in the original *Talking about Leaving* Study rests in the creative and improvisational ways that persisters are now navigating their STEM pathways. In recent decades, the discourse around higher education has becoming increasingly business-focused, emphasizing the role that universities play in providing a workforce to a capitalist economy. Students are increasingly cast as "consumers" of education, and students in this study proved to be very savvy "consumers" of STEM education. Students carefully and painstakingly researched courses, sections, and faculty members to make the right selection and to optimize their chances of receiving high grades. They reviewed online resources and crowd-sourced comments to identify the "right" faculty, course sections, or even college campuses where they could succeed in gateway courses. Occasionally, students were motivated to undertake this work because they realized that the quality of instruction influenced their learning and retention, but more often students took these steps to try to ensure high grades and persistence in their major. Students described this process as a "game" in which they intended to maximize their chances of winning or, in this case, achieving a STEM degree. Students' enrollment mobility between institutions, movement among majors, and circuitous course-taking patterns defied the traditional conception of a straightforward, linear degree path undertaken at a single college or university. Clearly, these persistence strategies have profound implications for teaching and advising in higher education particularly to ensure continuity and consistency in students' foundational learning and undergraduate experiences.

Students also utilized a number of supports and resources, seeking both informal and formal support to enable their persistence. One of the most common strategies was the use of informal peer study groups. Not only did these groups enhance students' learning and conceptual understanding, but peers offered social and emotional support, a sense of belonging and community, and important guidance and advice about STEM pathways. While some institutions offered organized study sessions and supplemental peer instruction, the majority of students still connected with peer study groups informally. Although peer support was absolutely vital to the persistence of many students, it is troubling that students so often needed to informally navigate and negotiate these relationships. As a result, some students may have difficulty or may be unable to connect with a peer support system, such as students who are marginalized or underrepresented in STEM, students with health or emotional disabilities, underprepared students, or students with outside responsibilities, such as children or full-time jobs. Despite institutions' efforts to create systematic and structured study groups, not all students took advantage of these opportunities. Institutions would be well served to identify why students are not accessing these resources in greater numbers and experiment with ways to more strongly encourage students to utilize formal peer support structures.

In sum, students accessed institutional, formal, and informal resources to persist in STEM. Persisters made many adjustments along the way, improving their study and learning habits, adjusting their academic identities and behaviors, and re-orienting their approach to education as they learned over time how to succeed in STEM. They often used creative ways of navigating their coursework and degree program to ensure their success. Like switchers, they encountered obstacles on their STEM pathway, such as challenging material, poor instruction, and low grades. Unlike switchers, most persisters found a community of peers and cultivated a sense of belonging by recognizing that they were not alone in the struggle. Institutions provided many supports and services to promote student learning, including peer instruction, study sessions, faculty office hours, tutoring, undergraduate research, and comprehensive support programs for underrepresented students. While these services were highly beneficial for students who accessed them, many persisters relied more heavily on informal persistence strategies, such as peer study groups or actively managing their STEM pathway. Thus, although a significant minority of students received beneficial institutional and academic assistance, the majority of persisters still perceived institutional and classroom environments in STEM as challenges that needed to be creatively overcome, rather than as supports for their learning or success.

## References

Callahan, K. M. (2009). Academic-Centered peer interactions and retention in undergraduate mathematics programs. *Journal of College Student Retention, 10*(3), 361–389.

Chang, M. J., Cerna, O., Han, J., & Saenz, V. (2008). The contradictory roles of institutional status in retaining underrepresented minorities in biomedical and behavioral science majors. *The Review of Higher Education, 31*(4), 433–464.

Chang, M. J., Sharkness, J., Hurtado, S., & Newman, C. (2014). What matters in college for retaining aspiring scientists and engineers from underrepresented racial groups. *Journal of Research in Science Teaching., 51*(5), 555–580.

Clewell, B. C., de Cohen, C. C., Tsui, L, & Deterding, N. (2006). *Revitalizing the nation's talent pool in science, technology, engineering & mathematics*. White paper published by the Urban Institute and the National Science Foundation.

Crisp, G., Nora, A., & Taggart, A. (2009). Student characteristics, pre-college, college, and environmental factors as predictors of majoring in and earning a STEM degree: An analysis of students attending a Hispanic-serving institution. *American Educational Research Journal, 46*(4), 924–942.

Drane, D., Smith, H. D., Light, G., Pinto, L., & Swarat, S. (2005). The gateway science workshop program: Enhancing performance and retention in the sciences through peer-facilitated discussion. *Journal of Science Education & Technology., 14*(3), 337–352.

Eagan, M. K., Hurtado, S., Chang, M. J., Garcia, G. A., Herrara, F. A., & Garibay, J. C. (2013). Making a difference in science education: The impact of undergraduate research programs. *American Educational Research Journal., 50*(4), 683–713.

Espinosa, L. L. (2011). Pipelines and pathways: Women of color in undergraduate STEM majors and the college experiences that contribute to persistence. *Harvard Educational Review, 81*(2), 209–240.

Ferrare, J. J. & Miller, J. M. (2019). Making sense of persistence in scientific purgatory: A multi-institutional analysis of instructors in introductory science, technology, engineering, and mathematics (STEM) courses, *Journal of Higher Education*. https://doi.org/10.1080/00221546.2019.1602392

Freeman, S., Eddy, S. L., McDonough, M., Smith, M. K., Okoroafor, N., Jordt, H., & Wenderoth, M. P. (2014). Active learning increases student performance in science, engineering and math. *Proceedings of the National Academies of Science., 111*(23), 8410–8415.

Garcia, G. A., & Hurtado, S. (2011). *Predicting Latina/o STEM persistence at HSIs and non-HSIs*. Paper presented at the American Educational Research Association annual meeting, New Orleans, LA.

Gayles, J. G., & Ampaw, F. (2014). The impact of college experiences on degree completion in STEM fields at four-year institutions: Does gender matter? *The Journal of Higher Education, 85*(4), 439–468.

Graham, M. J., Frederick, J., Byars-Winston, A., Hunter, A.-B., & Handelsman, J. (2013). Increasing persistence of college STEM majors. *Science, 341*, 1455–1456.

Herrara, F. A. & Hurtado, S. (2011). *Maintaining initial interests: Developing Science, Technology, Engineering, and Mathematics (STEM) career interests among underrepresented racial minority students*. Paper presented at the American Educational Research Association Annual Meeting, New Orleans, LA.

Hunter, A. B., Laursen, S. L., & Seymour, E. (2007). Becoming a scientist: The role of undergraduate research in students' cognitive, personal, and professional development. *Science Education, 91*(1), 36–74.

Hurtado, S., Eagan, K., & Chang, M. (2010). *Degrees of success: Bachelor's degree completion rates among initial STEM majors*. Los Angeles, CA: Higher Education Research Institute, University of California.

Jones, M. T., Barlow, A. E., & Villarejo, M. (2010). Importance of undergraduate research for minority persistence and achievement in biology. *The Journal of Higher Education, 81*(1), 82–115.

Laursen, S., Hunter, A.-B., Seymour, E., Thiry, H., & Melton, G. (2010). *Undergraduate research in the sciences: Engaging students in real science*. San Francisco, CA: Jossey-Bass.

LeBeau, B., Harwell, M., Monson, D., Dupuis, D., Medhanie, A., & PostLeBeau, T. R. (2012). Student and high school characteristics related to completing a science, technology, engineering or mathematics (STEM) major in college. *Research in Science & Technological Education, 30*(1), 17–28.

Liztler, E., & Young, J. (2012). Understanding the risk of attrition in undergraduate engineering: Results from the project to assess climate in engineering. *Journal of Engineering Education, 102*(2), 319–345.

Maltese, A. V., & Tai, R. H. (2011). Pipeline persistence: Examining the association of educational experiences with earned degrees in STEM among U.S. students. *Science Education, 95*(5), 877–907.

Museus, S. D., & Liverman, D. (2010). High-performing institutions and their implications for studying underrepresented minority students in STEM. *New Directions for Institutional Research, 148*, 17–27.

Ong, M., Wright, C., Espinosa, L., & Orfield, G. (2011). Inside the double bind: A synthesis of empirical research on undergraduate and graduate women of color in science, technology, engineering, and mathematics. *Harvard Educational Review, 81*(2), 172–208.

Palmer, R. T., Maramba, D. C., & Dancy, T. E. (2011). A qualitative investigation of factors promoting the retention and persistence of students of color in STEM. *The Journal of Negro Education, 81*(4), 491–504.

Perna, L., Lundy-Wagner, V., Drezner, N. D., Gasman, M., Yoon, S., Bose, E., & Gary, S. (2009). The contribution of HBCUs to the preparation of African-American women for STEM careers: A case study. *Research in Higher Education, 50*, 1–23.

Riegle-Crumb, C., & King, B. (2010). Questioning a white male advantage in STEM: Examining disparities in college major by gender and race/ethnicity. *Educational Researcher., 39*(9), 656–664.

Seymour, E., & Hewitt, N. (1997). *Talking about leaving: Why undergraduates leave the sciences*. Boulder, CO: Westview Press.

Seymour, E., Hunter, A.-B., Laursen, S. L., & DeAntoni, T. (2004). Establishing the benefits of research experiences for undergraduates: First findings from a three-year study. *Science Education, 88*, 493–534.

Tai, R. H., Liu, C. Q., Maltese, A. V., & Fan, X. (2006). Planning early for careers in science. *Science, 312*(5777), 1143–1144.

Thiry, H., Hug, S., & Weston, T. J. (2011). The computing Alliance of Hispanic-serving institutions: Enhancing the success of Hispanic undergraduates in computing disciplines. *Journal of Enrollment Management, 5*(1), 32–56.

Thiry, H., Laursen, S. L., & Hunter, A.-B. (2011). What experiences help students become scientists? A comparative study of research and other sources of personal and professional gains for STEM undergraduates. *Journal of Higher Education, 82*(4), 357–388.

Tyson, W., Lee, R., Borman, K. M., & Hanson, M. A. (2007). Science, technology, engineering, and mathematics (STEM) pathways: High school science and math coursework and postsecondary degree attainment. *Journal of Education for Students Placed at Risk (JESPAR), 12*(3), 243–270.

# Chapter 13
# Then and Now: Summary and Implications

**Elaine Seymour**

## How Have Rates of Persistence, Loss, and Relocation Changed Since the Publication of *Talking About Leaving*?

Analyses from two national data sources (NCES and CIRP) discussed in Chap. 2 converge on the conclusion that the loss rate from STEM majors caused by switching into non-STEM majors has substantially dropped. In 1997, our *Talking about Leaving (TAL)* study reported the then most recent (1991) CIRP switching estimate as 44%, averaged across all STEM disciplines. By contrast, the overall STEM rate of switching reported in analyses, both of CIRP data for 2017 (Eagan, Hurtado, Figueroa, & Hughes, 2014) and of NCES data for 2013, was 28%. This considerable improvement occurs, however, alongside a second source of loss that is highlighted in Chen's (2013) NCES study: in addition to the 28% of STEM majors who switch into non-STEM majors, a further 20% of STEM majors leave their college or university without a degree in any major. Thus, the total loss of STEM entrants is 48%. Expressed another way, only 52% of students who enter a major in a STEM field complete a STEM degree.

Information about STEM majors who leave college rather than switch majors was not available from any national data source at the time of the original study, so we have no way to determine whether these losses have changed over time. We also lack research evidence about which students are lost. However, in Chap. 7, our Gardner Institute collaborators, Koch and Drake, contribute to our understanding of these losses from their study of DFWI rates in STEM "severe" gateway courses. As

E. Seymour (✉)
Ethnography and Evaluation Research (E&ER), University of Colorado Boulder, Boulder, CO, USA
e-mail: Elaine.Seymour@colorado.edu



E. Seymour, A.-B. Hunter (eds.), *Talking about Leaving Revisited*,
https://doi.org/10.1007/978-3-030-25304-2_13

we discuss in the context of weed-out course effects, students with socio-economic disadvantages are at risk of leaving their institution following just one DFWI grade in a severe STEM gateway course even when their grades in other courses place them in good academic standing. As discussed in Chap. 1, our colleagues, Lee and Ferrare, also add to our understanding of switching as a form of permanent wastage. They report that: STEM switchers are far less likely than STEM persisters to attain a degree within 6 years, take significantly longer than persisters to obtain a bachelor's degree, and are at higher risk of dropping out of college altogether (Lee & Ferrare, 2019).

As we discuss in Chap. 2, there is consistency between the CIRP and NCES analyses in the switching patterns that they report, and most of these patterns are reflected also in our representative sample of six institutions. Some (CIRP) patterns reported in 1997 have continued: switching rates are still higher for women than for men, although the gap has narrowed—from 52% for women in the 1991 data to 30% in 2011. Relative differences among STEM disciplines continue. However, persistence has improved much more in some disciplines than in others. There are notable increases in persistence in both mathematics and biology. Engineering retains its position as the STEM discipline with the highest persistence rate, but there was a decrease in persistence of 5% in computer science. Relocation within STEM majors has increased slightly, except in mathematics where fewer students moved to another STEM major than in 1997. Switching rates also varied substantially among institutions. Thus, which institutions and what STEM disciplines students enter have important consequences for their chances of graduation with a STEM degree.

In Chap. 2, and throughout the book, we comment on changes in STEM persistence by men of color and women of all races and ethnicities that were evident in our sample institutions. At all six institutions, and across the range of students' standardized math scores, disciplines, and GPAs, women switched at a 7% higher rate than men (viz., 18% compared with 11%). Thus, the factors contributing to higher switching rates for women were present at all institutions and in all STEM majors. However, as noted, women's overall switching rates have decreased since 1997, although with marked variations by discipline: the rate fell sharply in mathematics (from 72% to 30%) and in computer science (from 69% to 31%). There were also marked improvements in the physical sciences and biology. The switching rates for men also decreased (from 41% to 26% overall) but less so than for women, with the lowest rates in engineering, the physical sciences, and the biological sciences. However, the switching rate for men in computer science rose to 55% from 46% in 1997.

Switching rates for students of color were not reported in the original CIRP analysis but are reported in the most recent CIRP and NCES analyses as 42% for African American, 41% for Hispanic students, and 28% for white students. However, as we discovered, traditional designations by race/ethnicity—often expressed in terms of underrepresented minority (URM) status in institutional and other records—beg questions about what these designations actually mean. When students' math scores and URM status were assessed for their contribution to switching, math scores accounted for switching better than did URM status. Indeed, in our logistic regression model, URM status, by itself, did not predict switching at all. As our interview data findings also confirm, an important part of the greater switching risk experienced by students of color is created by poor high school preparation. When gender, URM status, and math scores are combined, underrepresented minority women who enter university with lower standardized math scores switch at much higher rates than any other student group: one-third of URM women switched from STEM majors compared with 14% of all students. In Chen's (2013) analysis of NCES data, academic performance also explained rates of switching better than race/ethnicity by itself. In this study, much higher proportions of lower performing than higher performing URM students switched out of STEM (19% vs. 6%).

The larger implications of these findings are, first, that focusing on race/ethnicity as if it were a significant independent variable appears to be inherently, if unintentionally, racist. Second, the characteristics that create what appear to be issues related to race/ethnicity, are more accurately, issues of socio-economic and educational disadvantage. Thus, raising the quality of math and science preparation in high schools that serve large numbers of students of color has the potential to significantly increase their STEM persistence rates. Indeed, all of our study sources triangulate on the conclusion that major improvement in persistence rates would be achieved by raising the level of science and math preparation in the K-12 system for all students.

Mapping student switching patterns for all students allowed us improved insights into who is most at risk for switching and when. All demographic and academic factors (student GPA, standardized math scores, incomplete grades, and the average difficulty of courses experienced by students) significantly predicted student switching. Of special note, "being a woman" remained a significant predictor of switching with all other variables held constant. Academic duress and incoming level of preparation both play a major role in these patterns. As we discussed in Chap. 7, students who received poor or incomplete grades in gateway courses in their first and second years were particularly prone to switching. Throughout this book we also highlight the loss of high-performing women from STEM majors. Women not only switch more than men but (as we also reported in the original study) are also over-represented among switchers with high academic performance levels. Much higher proportions of women in the higher versus the lower performing group (59% vs. 41%) switched from their STEM majors. As to when students switched, 50% did so by the end of the first year, and 80% by the end of the second year. Students with higher standardized math scores, URM students, and Pell grant recipients all switched later than their comparison groups.

We also found patterning in the destination majors of STEM switchers. Some pathways were based on affinity, similarity between subject matter, and pursuit of a career related to an original aspiration. Frequent pathways were: from biology to psychology and other social sciences, engineering to social sciences (mostly economics), and from all disciplines into undeclared majors. Pathways from engineering and computer science to undeclared majors were higher than expected for students with lower academic performances. A larger than expected number of lower-achieving than higher-achieving students (15% vs. 6%) left STEM for undecided majors and remained in these for more than one term.

In what follows, we draw on findings from the student interviews, SALG survey, and other sources to explain the patterns found in these statistical analyses and to weigh the contributions of a wider array of factors that contribute to switching, loss from college, and relocation within STEM.

## What Contributes to Decisions to Switch and to Problems for Those Who Elect to Stay?

As detailed in Chap. 3, all of the contributory causes of switching decisions that students identified in the first study were also identified in the present study. While no new concerns emerged, there were, as we outline below, changes in their relative ranking. We also found increased complexity in the array of reported concerns that were being simultaneously handled by STEM undergraduates. These are reflected in increases in the sheer number of issues that prompted STEM switching decisions (from averages of 4.2 in TAL to 12 in TALR), in the numbers of concerns reported by all participants, and in students' accounts of their difficulties described throughout this volume.

Although, as discussed in Chap. 2, the extent of STEM switching has reduced since the original study, what has not changed are the contributory causes of switching. Important among these are the negative effects on persistence of students' learning experiences in STEM classrooms:

- *Problems with STEM instructor pedagogy* were found to be slightly greater than was reported in the TAL study: 48% of switchers mentioned poor teaching in their STEM courses as prompting their decisions to leave, and issues with instructor pedagogy were described by 96% of all switchers and 72% of persisting seniors.
- *Problems with STEM curricular design*, notably, content overload, over-fast delivery pace, and poor alignment between course elements, contributed to leaving decisions for a comparable proportion (31%) of switchers to those in TAL. However, in TALR, it affected far more (86%) of switchers overall, and 56% of STEM persisters.
- *Conceptual difficulties with one or more STEM subjects was found to* play only a small role in students' decisions to leave them STEM majors, then or now, but it was of concern to 80% of TALR switchers, overall. This finding (as we discussed in Chap. 5) is related to the high reported incidence of under-preparation in high school.
- As discussed in Chap. 7, *issues of under-preparation*, which create serious difficulties in surviving "severe" (i.e., weed-out) STEM foundation courses, were found in similar proportions in both studies. However, a higher number of TALR switchers overall cited under-preparation as an important aspect of their difficul-

ties (viz., 64% compared with 40% in TAL). Under-preparation also continued to create survival issues for about one-third of persisters.

- *Loss of interest*, which was often a consequence of poor learning experiences in foundation courses, still ranked highly (3rd) in its contribution to switching decisions, and (similar to TAL) was 61% of all switchers' concerns. As noted below, losing interest is commonly paired with finding alternative interests in other majors—in non-STEM disciplines for switchers and within STEM for relocators. Together, they reflect the push–pull nature of the decision-making process.
- *Finding and accessing timely appropriate help*—which was often critical to persistence—continues to be as serious a problem as it was 20 years ago: 80% of STEM switchers overall and 31% of TALR persisters struggled to find the academic resources and the support they need to survive.

We found marked changes since the original study in other "iceberg" items:

- *Discovery of an aptitude for a non-STEM subject now* ranks first among all factors prompting switching. It was cited by three-quarters of switchers as directly influencing their decision to leave and as a consideration by all switchers, compared with 10% and 12%, respectively, in the original study. The large jump in citation of this concern may reflect the large percentage of our interview sample who were high performers. Their representation reflects our institutional records count of high performers as 26% of STEM switchers across the six participating institutions. As we discussed in Chap. 10, high-achieving students often pursued multiple majors and minors in both STEM and non-STEM disciplines and moved to non-STEM majors for reasons that reflect their cross-disciplinary interests and options.
- For 61% of TALR switchers (compared with 23% in TAL) *loss of confidence* was a factor in their decisions and was also a concern for 79% of switchers overall. Losing confidence was also a problem for 44% of persisters. The increased ranking of losses of confidence from ninth to second place may, again, reflect the high proportion of high-performing switchers, two-thirds of whom were women and half of whom were also women of color.
- There was a large upward shift in students' *negative reactions to the competitive climate experienced in STEM classes*: 52% of TALR switchers (compared with 14% in TAL) cited negative class climate as a reason for switching. This experience also created problems for 81% of all switchers and was an issue for 42% of persisting seniors. Competitive class climate issues not only continued but also appeared to be growing as major deterrents to persistence. Intense status competitions among peers, encouraged by steeply curved grading practices, created isolation and failure to develop a sense of belonging that we found to be greatest among women of all races and ethnicities, and men of color.
- *Problems in financing college* emerged as a far more widespread concern in the present than the original study: 30% of TAL switchers cited financial problems as a factor in their switching decision. This rose to 70% of TALR switchers overall, and 48% of persisting seniors also registered financial problems as a serious

concern. In Chap. 11, we discuss how students in this study were paying for college and note the increase (since TAL) in both student working hours and their worry about large loans, which affected the career-related decisions of both switchers and persisters.

- *Choosing STEM majors for reasons that prove inappropriate* was a concern mentioned as a contributor to switching decisions by 48% of the TALR switchers compared with 14% of students in the first study. It also continued to be a concern for switchers overall and for persisters.

Career-related concerns were also found to be a far more pressing influence on students' decisions in the current than in the original study:

- About half of switchers overall, both now and then, *rejected the future careers and lifestyles* to which they projected STEM majors would lead. However, in the current study, twice the number of switchers (58%) than in the first study (29%) identified this a reason for their decision to switch.
- Similarly, nearly twice the number of switchers in this study (54%) than the prior one (27%) explained that they changed to a non-STEM major partly because it *offered more appealing career opportunities.*
- Making *system-playing moves into other majors as a means to further career goals* was a far more prominent strategy among switchers in the current than the original study: 26% of all switchers either sought or had considered non-STEM majors in which they could both achieve their career goals and graduate with higher GPAs. Their motivation was to gain a competitive edge in professional, graduate school, or job applications. We discuss these strategies in Chap. 10 and the rise in system-playing as a persistence strategy in Chap. 12.

As noted above, women's overall switching rates have decreased since the original study although with marked variations by discipline. Interview study results that clarify which factors keep women's persistence rates lower than those of men are summarized in the balance of this chapter. They also offer clues about what has contributed to the upward shift in women's persistence rates. The iceberg tables, broken out by gender, reveal that differences between male and female students in many categories of concerns are less than in the original study. The relative importance that women and men assigned to concerns that prompted switching or relocation was then seen to reflect broad differences in the ways in which men and women approach their college and careers. An overall finding of the present study is that the gender gap in attitudes toward STEM-related education and career goals has narrowed. For example, 20 years ago, male STEM students (and their parents) took a more instrumental approach than women to their education and career choices. Thus, men were found to be more willing than women to place career goals above intrinsic interest and personal satisfaction. In the current study, roughly one-quarter of both men and women reported readiness to switch to non-STEM majors to improve their GPAs and thereby their career prospects. This instrumental, consumerist trend, which is also reflected in parents' attitudes toward their daughters' education, is reported in Chap. 11 and summarized later in this chapter.

In our overall findings about the difficulties of persistence for students of color, the issues were not significantly different from those reported in the original study. As discussed in several chapters (notably, 2, 6, 7, and 9), several of these problems arose from the same source. Inadequate preparation (often in under-resourced high schools) in academic readiness, study skills, and how to navigate the college system was reported by 73% of all switchers of color; it was a contributor to switching for 35% and continued as a problem for 41% of persisting seniors. As we discuss in Chap. 5, consequential difficulties arose in transition to college that contributed to switching decisions for 73% of students of color compared with 31% for white switchers. Discouragement and loss of confidence because of low grades in early courses was a common concern for all students and was reported by 74% of white switchers. However, among switchers of color the figure rose to 92%. Among persisting students of color 78% of seniors of color described how difficult they had found it to adjust to college and 59% reported loss of confidence related to course grades as part of their struggle to survive. Difficulties in seeking and getting appropriate timely help was a problem for almost all (92%) students of color, compared with 76% of all white students, and the competitive, unsupportive STEM culture which (as described in Chap. 9) made it difficult to belong contributed to 62% of switching decisions for students of color compared with 49% of white switchers. For all switchers and seniors of color, this rose to 88% and 60%, respectively.

Because, in the original study, we found no variation between the participating institutions in the top-ranked problems contributing to switching, we checked to see if this had changed. In Chap. 2, we reported variations in the extent of switching by institutions with two of the research universities registering the highest rates. Given the findings of the observation study (cf., Chaps. 6 and 8) of research-based instructional strategies in use in some foundational courses in some institutions, we speculated that the likely cause of institutional variations was the extent to which improvements in teaching methods were in place. Our institutional analysis in Chap. 3 indicates that problems with STEM learning experiences were the highest ranked concerns for all switchers and persisters regardless of the extent of switching in any institution or disciplines, and the appeal of alternative majors for switchers was grounded in these issues at all six institutions.

In the following sections, we expand on these overall findings, noting which issues continue to prompt the continuing gap by sex and race/ethnicity in persistence rates.

## *The Centrality of Curriculum Design, Teaching and Assessment Methods*

In 1997, criticisms of faculty pedagogy contributed to one-third of all switching decisions. Complaints about what students referred to as "poor teaching" were cited as a near-universal concern by switchers overall and (at 74%) were the most commonly cited problems of persisting seniors. Concerns about curriculum structure, assessment practices, and pedagogical effectiveness pervaded all but 7 of the 23 factors driving switching decisions.

In this study, our examination of the consequences of students' learning experiences is wider in scope. In Chaps. 6, 7, 8, and 9, we examine both negative and positive student learning experiences in STEM courses, drawing on the interview study; our institutional data analyses across all four academic years; and the SALG surveys deployed in STEM foundation courses. We also draw on findings from a collaborative observational study led by Ferrare (2019) of teaching methods in the same courses as the SALG survey, and from a multi-methods study of "severe" foundation courses (often referred to as weed-out courses). This is augmented by a study of the consequences of DFWI rates in these courses conducted by our colleagues, A.K. Koch and B.M. Drake, at the Gardner Institute.

Our findings about the negative consequences of STEM pedagogy, course design, and assessment methods for both switchers and persisters are comparable to those of the original study. However, a higher proportion of switchers (48% compared with 36%) reported that problems with their learning experiences were key reasons for their decision to switch out of a STEM major, and 96% of all switchers (compared with 90% in TAL) registered problems with the quality of STEM teaching. Similar proportions of persisting seniors (72%, compared with 74% in TAL) expressed frustrations with STEM teaching methods, and 78% of all students (compared with 83% in TAL) described how particular aspects of STEM course design and educational practice had negatively affected them. In the original study, we reported little variation across the sample institutions in these findings. In this study, problems with poor quality teaching were ranked first by persisters in all schools (91%–96%) and by switchers overall in all but one school (91%–100%). However, the highest negative scores for teaching were given by students at three (of the four) large universities in the sample. Over half of the switchers at these three schools also cited poor teaching as a major contributor to their decision to leave STEM. As suggested by the foundation course observational study, these variations may reflect institutional differences in the extent of efforts to improve students' learning experiences in STEM majors as part of a nationwide effort that has been ongoing since publication of the original study.

We have far more information in this study about what kinds of teaching methods students encountered. There is a high degree of concurrence across our studies in their portrayals of the teaching methods used. In the interview study, all but one of the 95 switchers and 57% of the 143 persisters reported that non-interactive lectures were the dominant modes of STEM teaching, especially in introductory courses. The SALG survey results from foundation courses echo those of the interview study: the most frequently reported teaching methods and student class activities were the most conventional. In almost all classes, students report that they were taught by lecturing. Interaction was predominantly via problem sets, practice tests, in-class discussions and reviews. Ferrare and his team of class observers identified two different forms of the lecture method that, taken together, accounted for 75% of teaching styles in foundation courses. "Chalk talks" (lecturing while writing on a board) was observed in 41% of courses, and "slide shows" (lecturing aided by pre-made slides) in 34%.

From his analysis of semi-structured interviews with the instructors in these observed courses, Ferrare found that the teaching strategies deployed in these two types of lectures are informed by distinct, coherent, and tacitly understood beliefs about how students learn science. Chalk talk lecturers emphasized what students should do to facilitate their own learning. Thus, an underlying belief that informs their approach as teachers is that posing problems facilitates practice through individual "perseverance": students should "grind away" at conceptual problems until mastery is achieved. Chalk talk lecturers explained that their use of Q&A (whereby students pose and respond to questions through dialog with the instructor) reflects the importance that they give to intellectual risk-taking. Slide show lecturers believe it is important to introduce students to the theory and mathematics of new concepts, then model applications through repetition and variability until students can solve the same type of problem—a process enabled by clicker questions.

Unlike the original study which contained few reports of teaching methods other than "straight lecturing," in our current studies we find evidence of more active and interactive teaching and learning methods. SALG write-comments recorded some· group work in 71% of the foundation courses surveyed and the use of clickers in 53% of classes. In an intensive inquiry into a sub-sample of 28 foundation classes, students described participation in group work in 20 classes, the use of clickers in 15 classes and group projects in 7 classes. SALG respondents' written descriptions included methods whose common characteristic was their incorporation of learning technologies. In interviews, small group work was reported by about one-third of all students, and interactive forms of lecturing by one-quarter of switchers and one-third of persisters. Other methods mentioned were clickers, demonstrations, some online instruction, and a scattering of other classroom activities. Ferrare's observation team also reports interactive forms of teaching in 26% of foundation courses. These were also reported in interviews by 26% of switchers and 33% of persisters. Small group work was the most common interactive method recorded in all three studies. In these classes, the boundary between instructors and students that is sharply preserved in lectures was replaced with more open interaction.

As in the original study, students explained that what makes STEM learning "hard" is both the nature of the subject matter—intrinsic hardness—and hardness that is created or enhanced by prevailing instructional strategies. From experience, students found that the same material, taught better or worse, could be made more or less "hard." Paramount among instructional methods that made learning artificially hard for both switchers and persisters are: failure to present topics in a logical sequence, incoherence and inconsistent pacing in presentation of material, leaving out important information, poor management of class time, and reading Power Point slide content without inviting discussion, taking questions, or engaging in two-way exchanges. Conceptual understanding was further compromised where instructors fail to offer applications, examples, and sample problems or to provide context for theoretical material via conceptual connection to other bodies of knowledge and real-world phenomena. These omissions make it harder for students to apply what they are learning. Used in combination, these methods limit student comprehension and generated a sense of overwhelm. They are commonly reported in foundational courses but are still described in senior-level courses. The relationship between flaws in curriculum design and problems with conceptual grasp and application is similar to that reported in the original study. In sum, poor learning outcomes were achieved by delivering too much course material at a level that was inappropriate for the course designation at a pace that was too fast for digestion, and by misalignment between class content, labs, assessment, and homework. Content challenges can be motivating for students when thoughtfully devised *as part of* the curriculum. However, they can also create unnecessary struggle, confusion, and low levels of comprehension when they occur *because of flaws in* the curriculum structure.

Also frequently cited as deterrents to engagement and motivation were dull, spiritless presentations where instructors showed little engagement with either the course material or the learners. A commonly cited indicator of instructor indifference to student learning was failure to pause and check the degree to which students were following the instructor's line of thought and understanding the concepts being laid down. It is notable that we recorded substantially more descriptions of indifferent, lack-luster teaching than of engagement and passion from persisters than from switchers. As in TAL, students insisted that there is no inherently dull material—only dull teaching. They wanted to be intellectually stimulated by their teachers' passion for their discipline and encouraged by their enthusiasm to share it. Over one-quarter (29%) of switchers migrated to non-STEM majors, in part, because they offered more engaging and interactive learning experiences. The students' claim that disengaged lecturing induces "passive learning" and prevents them from building conceptual understanding, and engaging with subject material in depth is validated from multiple sources. As reviewed in Chaps. 8 and 9, many research studies and reports by public and private foundations, and by disciplinary and professional societies, support students' appraisals that interactive teaching, incorporation of active, hands-on activities, small group work, and use of authentic problems all outperform "straight lecturing" in enabling a solid understanding of, and ability to apply, core concepts and engage students in their own in-depth learning.

Students stressed that they valued teachers who demonstrate by their attitudes and behaviors that they want them to learn. Their descriptions of "how they learned best" included instructors who were approachable and available, encouraged questions, took an interest in students' progress, understood why some things were difficult and were willing to help students surmount them. The characteristics of "good" teaching most frequently identified by students included not only improvements in pedagogy that they enumerated but also improvements in the attitudes of instructors toward learners. Students clearly understood the connection between learning theories and learning practice, and the changes they wanted required shifts in both. As noted above for the lecturing formats that accounted for 75% of observed gateway course teaching, the beliefs of students and instructors about how these courses "should" be taught are entirely divergent. In the minority of multi-modal and group work-based courses observed, instructors' and students' ideas of the best ways to learn science were better aligned.

As discussing the learning process in class was uncommon, students could only guess at why their instructors taught as they did and thus they offered their own theories about their instructors' motivations and rationales. Students commonly explained the prevalence of lecturing, despite its dysfunctionalities for learners, in terms of the research priorities of STEM faculty and lack of departmental rewards for good teaching. They explained instructors distancing behavior and refusal to take an interest in them as individual learners as evidence of an assumption that students are lazy, stupid, unmotivated and, thus, unworthy of their instructors' attention. From modes of teaching that neither stimulated nor sustained their interest, they deduced that instructors took little responsibility for enabling learning and that they were expected to learn on their own. Where lectures were disorganized and incoherent, they assumed that instructors lacked the professional skills to deliver their content to best effect. Failure to provide and require education in teaching methods for instructors and graduate student teachers prompted two assumptions, notably among seniors, that departments and the university itself did not value effective teaching skills, or that they were aware of poor teaching quality but saw intervention as over-riding academic freedom. We imagine that it would have surprised many students to learn that instructors' rationales, particularly for their lecturing methods, were informed by a belief that this was how science had to be learned.

It is important to make clear that, as also found in TAL, most students' problems with their classroom learning experiences were not laid at the door of graduate teaching assistants. There were exceptions where graduate TAs, untrained in pedagogical methods, were assigned to teach introductory courses—as was the case of the whole calculus sequence in one sample institution. Students largely experienced TAs in recitations or labs, and over half of switchers described the value of clarifying their understanding in interactive sessions with a teaching or lab assistant. The other half registered negative experiences with TAs but only 3% of switchers included them as a factor in their switching decisions. Persisters rarely complained about their teaching assistants and one-third reported good experiences with them.

We believe that the modest moves into research-based instructional strategies (RBIS) evident in these studies will be encouraging to the STEM education improvement effort that has been ongoing between the original and present studies. Ferrare's findings, however, underscore conclusions from educational change research that wider uptake of RBIS has to begin with acknowledging how instructors conceptualize the student learning process, then persuading them to consider the research-based learning theories that underpin research-grounded teaching.

As discussed in Chap. 6, Ferrare's (2019) findings that instructors' teaching methods reflect their beliefs about learning align with research indicating a common belief among STEM instructors that the ability to "do science" is innate and fixed rather than something that grows with interest and effort. Through their modes of teaching, assessment, and contact with students, instructors who believe in "fixed intelligence" convey the message that only "innately gifted" students are likely to succeed—a message that many of our interviewees encountered and that some had internalized as "true." STEM instructors who believe that it is part of their job to identify students with natural ability and to encourage others to do something more suited to their presumed abilities were clearly doing just that. Instructors who believe that intelligence can be developed were more likely to show students how to become better learners and motivate them do their best. That such beliefs have important consequences is supported by Canning, Muenks, Green, and Murphy's (2019) finding that instructors' beliefs about the nature of intelligence (whether fixed or capable of growth) predict student motivation and achievement better than other aspects of their teaching. Ferrare and Miller (2019) further report that the patterned ways in which introductory STEM course instructors explain students' success can inhibit their taking steps to ameliorate factors that contribute to failure to persist even where the role of social inequalities is acknowledged. In contrast to our report in Chap. 12 that seniors augmented determination and ability with an array of survival strategies and resources to survive, 22% of instructors' "interpretive frames" explained persistence solely in terms of students' individual ability. A further 23% posited that anyone can succeed if they develop relationships in which students learn by struggling together. Poor preparation was acknowledged by 10% but was not seen as something that instructors could address, and 25% conceded that instructional factors were important but that, given the "great strides" made in instructional improvement, "STEM success is no longer predicated on these constraints." (Ferrare & Miller, 2019, p. 10).

Taken together, instructors' beliefs about the nature of students' intellectual capacity, how science must be learned, and what determines persistence are powerful influences on student outcomes of teaching and on student–instructor encounters grounded in these beliefs. As Ferrare makes clear, the dominant teaching methods used by instructors in foundation courses are entirely consistent with their beliefs about how students learn (or should learn) science and, as such, legitimate their use. However, it is also clear from student accounts of how they learn best that students' learning theories and those of their instructors sharply diverged.

## *The Significance of Weed-Out Courses*

As in the original study, about one-third (35%) of switchers cited weed-out class experiences as major contributors to their decision to leave. It is in these (largely) foundation courses that we found problems with aspects of course design, pedagogy, and assessment methods to be the most extreme. Flaws in course design and delivery occur in other courses, but constructed forms of hardness were consistently reported as features of teaching methods in courses identified as "weed-out" by their nature and consequences. They form, in effect, the tip of the iceberg.

Switchers and persisters described, in the same rank order, the characteristics that distinguished weed-out classes from other foundation courses: assignments are misaligned with content and grading is steeply curved; overloaded content is pitched at too high a level for an introductory class and is delivered at too fast a pace for absorption; teacher behavior conveys indifference about whether or not students learn; curriculum organization is incoherent and its delivery misses steps and explanations; and a competitive class culture is created by curved grading that also has the effect of disconnecting grades from students' own sense of their content mastery. By this combination of methods that students describe as confusing, intimidating, and discouraging, instructors effectively convey the message that the major is "too hard for them." Our discussion of concordance between the distinctive ways in which foundation course instructors teach and behave toward students and what they believe about the nature of intelligence and how science is best learned is particularly relevant to our understanding of why weed-out courses are taught in these distinctive ways and with such consistency of form over time.

## How Grades Contribute to Switching

From our logistic regression model of variables drawn from institutional records and transcript data, we found that receiving DFWIs in "severe" (SF) courses is a good predictor of switching even when other variables are held constant. Students also have a higher risk of leaving a STEM major if they: receive one or more poor grades in, or do not complete, an SF course during their first year; have a lower overall GPA (even with no DFWIs), enter college with a low SAT/ACT mathematics score; come from a family with lower socio-economic status (indicated by a PELL award); are a first-generation college student; or are a high-performing student whose leaving is associated with receiving one poor grade. Combinations of these characteristics substantially increase the switching risk. However, failure to complete, or receiving poor grades in an SF course, by itself, increased a student's chances of leaving a STEM major by 5%. Thus, it is SF courses that grades-related contributors to STEM switching are most evident.

An important contributor to these patterns of risk in SF courses is their distinctive use of curved grading systems. Although used in many other courses, in SF courses, steeply curved grading creates quotas of students with D and F grades to an extent that is large enough to depress STEM department grade averages (Rask, 2010). Although 12% of students who did not receive a DFWI switched after an SF course, the rate almost doubled (to 23%) for students who received one DFWI and jumped to 33% for those receiving two DFWIs. While students with higher standardized math scores switched less than those with lower scores, the difference in switching rates for both high-performing men and women almost doubled for those who received one DFWI in an SF course. As other researchers have also observed, if one aim of these courses is to reduce student numbers to a manageable size, they do so effectively.

We found, as we did in TAL, that grades are a complex, multi-faceted variable that have predictive value because of the significance that students assign to their grades. Low grades put students at high risk of switching even where they are sufficient for them to continue in a STEM program because students respond to grades as significant for their self-assessment and identity. It was common for switchers to describe STEM grade shock as part of their transition from high school to higher education. Typically, this affected students with high incoming math SAT/ACT scores who described themselves as "top students" in high school where they had seldom, if ever, received grades below a B. C or even B grades were defined as "failure" and undermined their prior identities as good students. Disquieting projections of an imagined unsuccessful future prompted thoughts of switching both among those who switched and those who did not. Students who were unable to recover their sense of identity as competent students were at risk of switching into programs where they could regain a sense of successful selfhood. In the original study, 23% of switchers cited discouragement and loss of confidence created by low grades in early classes as factors in their switching decisions. By contrast, in this study, the rate rose to 61% of switchers. Men of color and women of all races and ethnicities were even more likely to report that issues with low grades contributed to their switching decisions (69% and 67%, respectively).

The importance of the distinction between the objective and subjective meanings of grades and their consequences for switching is particularly clear in the reactions of many women to low grades. As discussed in Chap. 7, our collaborators, Koch and Drake, also found that women did better on average than men in SF courses and had lower DFWI rates than course averages. Despite this, women had higher rates of switching than men overall and switched at higher rates than men across the math score distribution. Women of color with low math scores switched at significantly higher rates than other students. Curve grading systems in SF courses played a major role in these departures. Among switchers, half of the women, but less than a third of men, were prompted to switch because of low weed-out course grades. Women had less tolerance than men, whether for receiving low grades or for failing classes. Regardless of actual performance scores, women accustomed to getting good grades who received a single C grade, an incomplete or a withdrawal in a weed-out course were at high switching risk. The gendered effect of low grades was also evident beyond SF classes among persisters: although senior men were over twice as likely as their female counterparts to fail and/or retake courses, persisting women expressed less tolerance for low grades and failing classes than did men and were over twice as likely to report their demoralizing and psychologically traumatic effects.

The apparent indifference of SF course instructors toward novice learners seeking academic help and encouragement reinforced self-doubt that is independent of actual performance. The search for validation from significant others that we encountered among women in the original study had not disappeared. Many young women were still less able than their male peers to diminish the significance of reversals, take them in stride, and refuse to allow low grades or distancing behavior by instructors to throw them off track. By projecting a poor overall future performance from poor grades earned in one class or even on one exam, some students switched pre-emptively. What seems to have increased since the TAL study are parental and societal expectations for young women (see below) which they internalize into self-demands to such a high degree that many senior women reported that they still had difficulties in letting them go. Some such women left to find a major where they could once more feel good about themselves and graduate with a high GPA. As some seniors described the choice, it may be "better to bail than fail." Fragility of self-confidence was particularly marked among women of color. The risk of switching due to weed-out course experiences is particularly high for those women of color who enter with below-average ACT/SAT math scores.

Failure to adjust to low grades, particularly in weed-out courses, distinguished switchers from persisters. Similar proportions of switchers and persisters failed, or had to repeat, classes, including some students with high incoming math scores. However, fewer persisters considered failing a course to be a psychological crisis or described it as threatening their continuance in the major. As a matter of survival, most persisters had found ways to adapt to what they described as the STEM tradition of lower grades. Coming to understand the nature of curve grading was also important in the grade adjustment process. Being assessed by the logic of norm referencing was novel and alarming to most incoming students. Some continued to assume that their instructors used criterion-referenced grading to indicate the extent to which they understood course content regardless of how many others did also. Students who continued to view grades in this light had trouble accepting a C grade as other than as evidence of low conceptual competence even when they performed well relative to the curve. As seniors explained, survival requires normalizing single poor grades as a setback, not a deal-breaker.

The advice of peers, advisors, and instructors helped many STEM majors to make this adjustment so that their academic struggles did not come to signify lack of ability or discount future success. As in the original study, we heard many "fork in the road" stories in which a decision to stay or leave turned on a serendipitous intervention by an instructor or advisor who persuaded a capable student that they should stay. More experienced students often play an important role in explaining to younger students how to put low grades in weed-out courses into perspective. We observed how valuable such "translation work" is to survival, especially for women with high self-demands whose confidence is undermined by the consequences of curve grading.

We also note that this is clearly a place where timely interpretation and encouragement can be brought into play to divert talented but self-doubting students from ill-founded departure. Designated and faculty advisors might make use of this finding by organizing a unified, intentional practice of enabling students to make better appraisals of their own competence and thus avoid precipitate decisions. However, STEM departments might also review whether and when the use of curved grading is appropriate and effective. Is there good correspondence between what is taught and tested and what students are able to demonstrate about their learning? Does it enable the loss of students whom it would be worth retaining?

## Which Students Are Lost from Weed-Out Courses?

Our combined findings clearly demonstrate that a student's chances of passing "severe" STEM gateway courses and of remaining in college to successfully graduate are greatly diminished by belonging to low-income and first-generation families who may also be of color. In Chaps. 2 and 7, we reported several apparent differences between students of color and white students in math and science preparation and STEM course performance. However, in the logistic models, these differences disappeared because other variables explain switching rates better than race/ethnicity per se. We therefore explored what variables intersect with race/ethnicity to explain the switching risks of students of color in weed-out classes. As we further describe in a following section, we found a strong relationship between race/ethnicity, first-generation college status, and the enhanced risk of arriving in college under-prepared for introductory STEM courses. From student accounts, we also learned that working-class parents of all races and ethnicities were also less likely to know how higher education works, what career pathways exist, and how to get the most out of college. They were also less likely to be able to provide funds for college, so their children more often had to work to pay for their education and support themselves while in school. Thus, students of color, along with other working-class, first-generation, and immigrant students face a set of structured socio-economic and educational disadvantages in STEM majors that derive primarily from the limitations of their circumstances with additional problems experienced by women and students of color from disadvantages groups.

The non-random, nature of losses from STEM majors arising from weed-out foundation course experiences is further corroborated by Koch and Drake who also identify high weed-out courses as responsible for losses among students who are first-generation and Pell grant-eligible. This group includes many students of color who consistently had DFWI rates that exceeded both the averages for their course overall and those of their white peers. Students whose families had less financial and social capital have clear disadvantages in STEM SF courses. Koch and Drake's findings also help to explain what contributes to the 20% national rate of college drop-outs from STEM majors: for first-generation and Pell grant-eligible students, an unsuccessful outcome in just one weed-out course is related to the decision to leave the institution altogether, even when the student is otherwise in good academic standing. Thus, not only does earning a DFWI grade serve as a predictor for attrition, it is also a predictor of who ultimately graduates.

Thus, across all our studies, switching as a result of weed-out courses was found to disproportionately occur among students who enter with a constellation of socio-economic disadvantages. These risks are evident when examined singly, but, they greatly increase when they occur in combination.

STEM courses that appear to be designed and taught so as to discourage students presumed to be the least capable of continuing paradoxically produce consequences that are dysfunctional to such aims. Students who leave STEM majors because of weed-out course experiences include high-performing students—some in the highest math scores quartile—whose interest is dissipated by insufficient intellectual challenge, engagement with authentic science, and exploration of theory in the limited "school science" presented to them. Among these are multi-talented students with viable interests both in the arts and humanities and in the sciences some of whom are undertaking multiple majors and minors in a wide variety of fields.

Weed-out experiences can force such students to choose whether to drop or keep a STEM major.

We observe that substantial numbers of STEM persisters are now combining their STEM degree studies with pursuit of non-STEM credentials—a trend that was barely discernable in the original study. As a matter for further research, it is largely unknown to what extent which non-STEM pursuits make a positive contribution to STEM degree persistence. Often, students' pursuit of a non-STEM credential along with their STEM major has the effect of creating a liberal arts education with application and relevance for many careers and lifetime interests beyond the narrower confines of particular disciplines. Addressing this trend might be an important focus for collaborative planning discussions across departments. STEM degree programs might recruit more students into STEM programs and ensure their retention were they to offer degree programs that accommodate students' interests in other (including non-STEM) disciplines.

Another sub-set of high performers make pragmatic moves from STEM disciplines into majors that enable them to improve their GPAs and, thus, increase their chances of acceptance into competitive graduate and professional programs. Even though students with high ACT/SAT scores are less likely to switch, we found them to be surprisingly vulnerable to the effects of DFWI scores. Losses from this talented group following weed-out courses are high, even among both women and men with the highest math scores on entry. As we have clarified throughout this book, the processes of switching are also far more haphazard than the rational choice model would predict. This is especially evident in the decisions of perfectionist students for whom experience of one setback (often a moderate grade misinterpreted as failure) was often the basis for a move intended to restore self-esteem.

In light of these findings we propose that low tolerance for less than perfect performances is a more accurate explanation for switching by many high-performing women than is loss of confidence—which is an explanation of long-standing in the research literature. There is still strong evidence in both our current and former studies, and in ongoing research by others, that lost confidence is a major contributor to many women's switching decision. However, this explanation may more accurately apply to that larger proportion of women entrants to STEM majors who are not the highest performers.

The longevity of the weed-out "tradition" appears to reflect its perceived functionality. However, our evidence from this combination of sources contradicts any presumption that weed-out courses are necessary in STEM majors because they select for those who are best fitted to continue and discard only those who are not. It may be sustained as a system by good intentions such as, being cruel to be kind in diverting poorly equipped or ill-suited students elsewhere—a rationale that may also be seen as ensuring the future high quality of STEM disciplines. However, STEM departments that sustain weed-out courses appear to be mistaken about which students they are discarding. Because the class, race, and gender biases in these losses as well as the loss of very talented students appear to be unknown, whether to instructors or their departments, we may presume that they are unintentional. The weeding out of majors from other disciplines, provided as a kind of "service," by math and the physical sciences, is more overt. But, as attested by some angry switchers from engineering and health-bound professions, it is blind to the interest and potential of students with applied science career trajectories.

Of all of our findings, the patterned dysfunctional outcomes of the STEM weed-out system prompt, perhaps, the greatest need for departmental disciplinary and institutional review and reconstruction of traditional teaching and student assessment practices.

## *Developing a Sense of Belonging and Other Climate Issues*

Problems with aspects of the cultural climate experienced in STEM majors continued (as they had in the original study) to undermine students' sense that they belonged in STEM majors. They contributed to half (52%) of all switching decisions, were an issue for 81% of all switchers, and 42% of persisters continued to struggle with them. Feeling that they did not belong was most often expressed by white women and students of color of both sexes, especially those from low-income families and "first-generation" families who entered STEM programs with poorer high school math preparation. Such students were more likely to have problems with belonging that were grounded in low assessment of their own competence. Their concerns were exacerbated in competitive classroom climates, and by difficulties in connecting to other students. This, most importantly, undermined their access to peer academic support.

For women, difficulty in developing a feeling that they belonged was rooted in the numeric dominance of men in particular STEM majors where male peers, and sometimes instructors, acted out their presumptions that women did not belong in their major. In contrast to the TAL study, we did not hear widespread accounts of male instructors who behaved badly toward women in class or allowed male students to be rude, hostile, or make sexually inappropriate remarks. Such behaviors were rarely reported in this study. However, we still documented instances where male faculty operationalized their beliefs that women did not belong by ignoring women's questions and contributions in class, tolerating male peer behavior that excluded women from participation, and, in office hour encounters, contesting their content knowledge and competence. Another remnant of our earlier findings was difficulty in relating to some of the women faculty in the physical sciences, especially older professors who had struggled to survive among hostile male colleagues and were disinclined to provide individual support to female students. This continued but was much rarer than hitherto.

Presumption of greater male competence, however, continued to be expressed and was a significant contributor to women's sense of isolation and exclusion. We continued to hear stories of male peers who assigned stereotypically gendered roles to women in group projects. Women's opportunities to learn new skills were, thus, preempted by male assertions of greater competence. Women often assumed that men had greater familiarity with course content gained through their informal interests. In engineering and computer science especially, disparities in informal STEM experiences translated into classroom status advantages. Some senior women described how they protected their self-confidence in competitive class cultures by avoiding male peers. However, this strategy also cut them off from sources of mutual peer help and support.

Women with high competence in math were not immune to the negative effects of competitive classroom climates or of being excluded from study groups. Their sense of belonging was confounded by limited connection to instructors and peers in the learning process, and less access to opportunities such as undergraduate research and internships. Challenges to women's competence and the risks of stigmatization were rarer in programs with near gender parity. In computer science and some engineering programs where the gender ratio could be as low as nine men to one woman, women reported particular difficulties in developing a sense of belonging.

The programs in which both women and men most often experienced unwelcoming, even hostile, class climates were engineering and computer science, and majors in which a high proportion of students aimed to enter medical or veterinary schools. In these contexts, climate issues often manifested as status competitions whereby some students who asserted a superior right to belong stigmatized and excluded others. Such artificial competitions were heightened by sharply curved grading.

Partly as an artifact of men's greater numerical representation in some majors, the processes that women describe limit their access to peer support, perpetuate gender segregation, and give male students control over the informal terms of performance and productivity in their classes. Left unchecked, negative peer dynamics systematically and unfairly disadvantage some students over others and promote outcomes that might not be what STEM faculty and their departments intend. Recognition that peer dynamics are a critical aspect of program climates and lie within their purview, a shared readiness among instructors to intercede in competitive peer dynamics and to rethink course design, assessment methods, group work, and academic support systems could all increase students' development of a sense of belonging and, thus, their commitment to persist.

## *External Influences*

Patterns of STEM switching and relocation are shaped not only by aspects of students' within-college experiences but also by how these experiences intersect with variables in the outside world. Important among these are: why students chose particular STEM majors; how well-prepared they are to undertake them; how they finance their education; the appraisals they make of prevailing economic conditions and job opportunities; and the influence of parents and family circumstances in all of these.

## Under-Preparation

In our discussion of weed-out effects, we pointed to the role that adequate high school preparation plays in the degree to which students can engage successfully with STEM foundation courses. Nearly one-third of students discovered on entry that they were under-prepared for these courses, of which General Chemistry and Calculus I and II were the most troublesome. The difficulty of trying to remedy missing understandings while simultaneously tackling new concepts directly contributed to 20% of switching decisions—a slightly larger proportion than the 15% reported in the original study. Under-preparation also prompted some relocation to other STEM majors where catching up could be managed. Students with preparation problems in multiple disciplinary areas were those more likely to cite poor preparation as a major influence in their decision to switch.

As in the original study, under-prepared students from families in working-class communities commonly described their schools as under-resourced. Teachers, though supportive of their talented students, were often under-qualified for the subjects they taught, calculus, advanced science coursework was not offered, and science laboratories were poorly housed and stocked. Our most significant finding is that students of color were over-represented among under-prepared students who switched. Students of color were more likely to come from working-class families, attend under-resourced schools, and to report poor preparation: 36% of under-prepared students were African American, 22% were Hispanic, and 16% were white. Women of color were the most likely students to attribute their switching decisions to insufficient preparation in under-resourced schools. This finding helps to explain the apparent connection between race/ethnicity and STEM switching.

Some under-prepared students both white and of color were aware of these deficiencies while they were in high school; others discovered, in retrospect, that the quality of teaching and intellectual challenge in their high schools (particularly in rural or low-income urban areas) was lower than that experienced by peers from better schools. They also described their families as having limited experience of higher education and financial resources to contribute to it. Early tracking also contributed to inadequate preparation. Women of color and first-generation students were most likely to report that they had been placed into low-ability math tracks in middle or elementary schools where they experienced little encouragement in math and science and found restricted access to more advanced, college preparatory curricula. Thus, the intersection of class, race, and gender is clearly significant in explaining patterns of under-preparation for early STEM courses.

Interview analysis revealed other ways in which students arrive ill-prepared for STEM foundational work. Although 61% of switchers had taken at least one AP or IB science course and over two-thirds of switchers had taken high school calculus, these advanced courses had not necessarily provided adequate preparation. Students described poor teaching, lack of challenge, superficial coverage of important concepts, and a focus on memorization without conceptual understanding. In addition to inadequate disciplinary knowledge and skills, many under-prepared students had experienced learning largely via worksheets and rote memorization and had little experience of abstract or conceptual thinking. Some students were unprepared for the workload, organization, and time management skills that undergraduate STEM courses required. These aspects of under-preparation were more common in poorer high schools but were by no means limited to them. Many students from affluent families and well-resourced schools who were adequately prepared in math or science had entered college with little idea of how to manage their work or study effectively for tests. Students who had earned As in high school with minimal effort often did not understand that they must now prepare for classes. Those who adjusted their study practices recovered relatively quickly, but slowness to adjust learning habits created persistence risks for otherwise prepared and able students. However, it was the constellation of inadequate disciplinary and learning skills preparation together with limited knowledge about how to navigate the college environment that most often demoralized able students from disadvantaged backgrounds. Under-preparation in all its dimensions created difficult transitions to college that often prompted an early decision to leave.

In sum, preparation issues and subsequent difficulties in college transition continue to play an important role in prompting able students to switch from STEM to non-STEM majors. As we have illustrated, problems of preparation and transition both reflect and exacerbate inequities of income, race/ethnicity, and gender that underlie so many of the contributors to loss from STEM majors that we encountered in this research.

## Motivation and Influence in Initial and Subsequent Choices

How questions were posed to students about their choice of particular STEM majors was important. When asked in the SALG survey to rate a set of (research-grounded) reasons for choosing their STEM major, the highest rated reason, "I wanted a career in this field," was one of four answers that focused on ultimate careers. Three other career-related ratings expressed gaining a particular STEM education as a means to a good income, job security, or as a stepping stone to a higher degree. However, as was emphasized in students' written comments, apparently instrumental choices often reflected interest in, and the appeal of particular careers. Indeed, the second-highest rated reason was that, "A career in this major allows me to help others." Choices prompted by altruism, including the desire to make a difference, were common, particularly among women and students of color. Lifestyle goals were also well-represented and often shaped particular career aspirations.

When invited to offer a primary, open-ended reason for their choice, the dominant themes in all students' answers were affective rather than instrumental. Paramount were interest in and enjoyment of the field and a sense of a good fit between their ability and temperament and the kinds of careers to which it might lead. These responses were similar to those given by students in the original study. However, as the interviews revealed, the considerations weighed by students are now more complex, and both switchers and persisters chose between multiple, often competing, interests. While switchers generally had broader disciplinary interests than persisters, including both STEM and non-STEM fields, there were no great differences between switchers and persisters in terms of the primary reasons offered for their choices.

Switchers and persisters alike often changed majors as a result of exploring and honing their career interests during their undergraduate studies. Those who were more career-focused in their initial selection of major explored and refined career options as their understanding and awareness of careers matured. Fifty-eight percent of switchers who made career-related choices switched because they became dissatisfied with their initial choice and found more appealing career paths in non-STEM fields. As we also found in the original study, students who were very likely to switch were those who entered with a narrow career focus based on a long-held but under-informed aspiration or an altruistic but unrealistic career goal. Persisters were more apt to enter STEM majors with a general desire to "do" or "be" in a certain field, and then gradually refine their interests as their studies progressed and their field knowledge grew. Switching because of recognition of a mis-fit between their own interests, temperament and goals, and their experiences and career expectations in their STEM major perhaps comes closer to traditional explanations for switching than many of our other findings about switcher–persister differences.

Where the two groups differed most was in their incoming level of knowledge about their major or their chosen field. It was this variable that most influenced whether they stayed in their original STEM major. Under-informed students were more likely to switch, and lack of incoming knowledge was also a leading contributor to relocation into a different STEM field. More than half of switchers (56%) moved to a non-STEM major, in part, because they were under-informed upon entry about the nature of the STEM degree program and its related career options. This factor affected more than four times the proportion of students than in the original study especially engineering and computer science students.

As the most prominent switching factor related to students' choice of major, this has clear implications for remedial action: policymakers and state departments of education could increase efforts to integrate engineering and computer science more robustly into the K-12 STEM curriculum; educators and STEM-based industries could collaborate to create mentoring and internship programs for K-12 students that provide a more realistic and nuanced understanding of the work of STEM professionals; K-12 school counselors, educators, STEM industries, and disciplinary societies could all do more to inform students about the vast array of STEM career options and help them to reflect on which career may best match their interests, aptitudes, and temperament; colleges and universities could also offer greater access to pre-entry advising to help students select an appropriate major and to inform them about pathways within STEM and other disciplines; STEM departments could create mandatory one-credit courses for incoming majors to educate them about the sub-fields within the discipline and the nature of career options with those fields. Were policymakers, K-12 and university educators, and STEM industries to collaborate to inform, mentor, and provide professional opportunities for students at all educational stages, students will be better prepared to succeed in their STEM disciplines and enabled to make more informed choices for their future careers.

## Parental Influences

When asked who or what had influenced their choice of majors, both switchers and persisters described the influence of parents as paramount in encouraging their entry to STEM majors, followed to a lesser extent than in TAL by high school teachers. Fathers exerted more influence than mothers in the choice of STEM majors and fathers clearly favored STEM-based careers for both their daughters and sons—a considerable shift from 20 years ago. A sub-set of both switchers and persisters chose to follow a parent's career in a scientific or technical field. However, what distinguished switchers from persisters was the *type* of influence exerted by parents. Persisters more often described their parents as encouraging an inclination toward the sciences and helping them identify fields of study that suited their talents, temperament, or interests. Switchers more often experienced parental pressure to choose STEM-based careers perceived to be secure, prestigious, or well-paid. They were also more likely to have family financial support for college contingent on following parental preferences. Selecting a STEM major in response to parental pressure rather than intrinsic interest resulted in choices that were highly unstable and prone to early switching.

Parents had influence in decisions to leave a STEM major as well as to enter it. The dominant concern of parents who disapproved of a student's intention to switch from a STEM major was for their child's future employability and financial security. They worried equally for daughters and sons about the long-term consequences of moves into non-STEM fields. Although more fathers than mothers opposed STEM switching for these reasons, parents of both sexes saw non-STEM degrees as inferior to STEM degrees in an uncertain job market. The tougher line that fathers now take with their daughters reflects their recognition that, in a world where marriage no longer ensures financial security, young women must achieve this for themselves. For fathers especially, viewing STEM degrees as good ways for their daughters to prosper in the world marks a huge change from the original study where young women often had to fight their parents, both to enter and stay in STEM majors. Some students questioned how much their parents knew about the career paths that they promoted, and about post-graduate requirements and the costs needed to achieve them. They clashed over parental estimations of how much money they could earn in parent-approved career paths and questioned the factual basis of expectations of high financial returns from a STEM undergraduate degree. Some parents who were disappointed at moves away from prestigious career fields—notably, medicine and engineering—sought to leverage compliance by withdrawal of financial support.

As in the original study, mothers and fathers differed somewhat in the criteria by which they judged a proposed switch of majors and careers. Mothers, who were often the student's primary confident and sounding board in their education and career and rethinking process, took into account the student's enjoyment, interest, and investment in the new discipline and projected career when assessing whether a revised choice would reduce stress, increase engagement, and secure future career satisfaction. Some parents, particularly fathers, qualified their support of a switch out of concern whether the proposed alternative would ensure not only greater enjoyment of the discipline but also viable future employment—a distinction between happiness now and happiness later. The hardest line was taken by those fathers who viewed higher education (and especially choice of a STEM degree) as an investment made in expectation of high future financial returns.

Where parents differed, and fathers took a tough stance against a move out of STEM, we noted a distinctive change from the original study findings. In the 1990s, fathers were less enthusiastic about their daughters' choice of a STEM major—preferring something more "gender-suitable"—and also took a more indulgent attitude toward a switch into a non-STEM degree by their daughters than by their sons. In the present study, daughters described their fathers as strongly favoring STEM degrees as a sound way for young women to secure financial security. We also learned that fathers were equally unsympathetic to moves out of STEM pathways for both daughters and sons. Other parents who disapproved the proposed switch were worried that, in an uncertain job market, all or most non-STEM degrees would lead to poorly paid, insecure work. Concern for loss of an entrée to a prestigious profession was most strongly expressed by parents in Asian American and immigrant communities.

Sixty percent of switchers described their parents as supportive of their decision to move to a non-STEM major and career path. The dominant concern of these parents was that their children would find a good fit for their talents and interests. This was widely regarded as a better criterion for a sustainable future than instrumental choices focused on career prestige or likely earnings. However, the rationales behind parental support for switching decisions were broadly of two kinds. Although their child's happiness was a central concern, some parents (particularly father and mothers based on their own work experience) urged consideration of happiness now versus future happiness via economic independence and security. Other parents supported switching and relocation moves because they saw passion for a discipline and its career options as the best routes to both present and future happiness. Rethinking their initial academic and career preferences as students discovered who they are and what they want out of life was seen as a normal and desirable outcome of higher education and worthy of their support.

## *Paying for College*

An important persistence variable is the degree to which the difficulties of balancing academic work with employment places a student at risk of switching or of not completing any degree. In the original study, we reported student difficulties created by decreasing public funding for student tuition and fees, and that competition for shrinking financial aid had become racially divisive. Approximately two-thirds of interviewees had taken out loans and half were meeting some proportion of their educational and personal expenses by working, the average being 18 h per week.

Since the original study, average student working hours have increased and many students now work in the 20- to 25-h per week range. At this level of employment, college work starts to suffer.

Working 20 or more hours a week also distinguished switchers from persisters: it was cited as a problem by 70% of switchers and 48% of persisters and was a direct contributor to decisions to leave for 10% of switchers. While it prompted moves into majors where completing a degree while working is more possible, it sometimes led to failure to complete a degree. More switchers than persisters worked and switchers worked longer hours than persisters; and three times as many switchers as persisters reported work overload and stress. Persisters who worked longer hours reported difficulties in balancing work and school and saw a drop in their grades. We also noted demographic patterns in persistence risks created by the need to work: more students of color, and more women of all races and ethnicities worked than did white men. The groups with the highest proportion of working students were both switchers, namely white women and men of color (60% of each group). Students who worked less than 10 h per week worked mostly by choice and did not experience significant disruption in their school work.

The most common reason for students to work was that their families could contribute little or nothing to their college costs. Some students reported performance-based scholarships, and 27% of switchers and 25% of persisters received Pell awards. As in the original study, approximately 60% of STEM majors did not get financial aid. However, they expected it far less than hitherto. Those with limited or no family support simply expected to work and take out loans.

Repayment of student loans, a universal source of worry for both switchers and persisters, could strongly influence career choices. Concern about loan repayments caused some students to reject or delay graduate school and question whether their preferred career was financially viable. We noted a clear move away from careers that were once considered secure and lucrative. Given substantial loan obligations, the time and additional funds required to enter medical and veterinary fields were seen as greater than could reliably be recovered in a realistic timeframe. Such calculations prompted both relocation and switching into majors leading to careers in health and therapeutic fields that have shorter, less expensive trainings.

Overall, we estimate that the work-school stress that significant hours of work creates for both switchers and persisters has doubled since the 1997 study. In Chap. 10, we cited the work of scholars who have documented the reasons for this change—growth in the average net price of a STEM degree and a correspondingly large increase in student debt that is greatest among students whose family income lies just beyond the qualification limits for Pell awards. They also document an alarming rise in food and housing insecurity that is now greater among college students than in the general population, and lower rates of degree completion among students with the greatest financial need. In the original study, some of our student commentators on the weed-out system described it as a means test that is biased against those who have to work their way through college.

## *Student Appraisals of Economic Conditions and Job Opportunities*

We found far greater concern than was evident in the original study about what prevailing economic conditions implied for job availability and their financial prospects. One-third of switchers and one-fifth of persisters described how their decisions to switch or relocate were influenced by the employment prospects and limitations that they saw in particular fields. Repayment of student loans was the predominant financial concern of both switchers and persisters. Consideration of the costs of undergraduate and graduate education that would have to be funded by student loans prompted rethinking of majors and career pathways away from those with a "high debt-to-salary ratio." Thus, initial choices based on interest or altruism were often replaced by instrumental career choices. Most students were doubtful about their chances of securing job with a high salary and focused on career pathways that seemed most likely to offer recession-resistant job security. More than the potential pay level of particular careers, both persisters and switchers saw a career's flexibility and versatility as their best chance of a secure future. Persisters looked for emerging fields with many applications, and relocators moved in order to position themselves in expanding and flourishing sectors. More than in the original study, students were wary about entering government service which (except the defense sector) was perceived as insecure. An indicator of a highly competitive job market that was widely resented was the new norm of unpaid internships that do not necessarily lead to employment.

Thus, students' concerns for their future increasingly reflected those of their parents. The single, most notable, outcome of these concerns was re-assessment of medical and veterinary careers, formerly considered secure and well-paid, but now judged as taking too long and costing too much. Favored instead for their shorter post-graduate programs and greater certainty of employment were other healthcare, therapeutic, and caring professions. Popular career choices that met these criteria were physicians' assistant, anesthetist, nurse practitioner, physiotherapist, and other growing health-related specialties. As discussed in Chap. 11, pragmatic re-appraisals were a contributory cause of switching in and of themselves. Perceptions of poor pay and low status trumped job security, however, in the marked decline of interest in teaching science and mathematics in K-12 settings. Only a handful of students expressed an interest in K-12 teaching, whereas, in the original study, 8% of both switchers and persisters intended to teach, and 20% were considering it.

Optimism was discipline-based with some STEM fields seen as more competitive than others. Across the entire student sample, persisters majoring in engineering or computer science were the most confident that current economic conditions would secure them well-paid employment. Chemists were also more optimistic because their discipline has fewer graduates. With fewer direct applications, physics was seen as less lucrative or secure. For first-generation students, including many students of color and students from immigrant families, career decisions were grounded in a primary obligation to "give back" to family. Largely absent in working-class families were socio-economic networks through which knowledge of, and access, to jobs might be secured. Students with access to family-based networks were very aware of their advantages; those who lacked them were actively developing academic and professional networks.

Overall, students' critical appraisals of the job market in the current economy and their concerns about long-term loan debt prompted widespread rethinking of original career choices. To better position themselves to enter a changing and uncertain job market, some students switched or relocated to other STEM majors. This marks a profound shift from the original study where, notwithstanding growing dependence on loans and working while in school, students expressed more optimism about their job prospects, and were more disposed to follow their interests and to choose careers that they saw as "making a difference."

## *The Personal Costs of a STEM Education*

The processes that result in persistence (including relocation) for 52% of STEM entrants and in loss by switching or leaving college by 48% include struggles with negative personal consequences that affect both switchers and persisters (though in smaller proportions). As discussed in Chap. 10, the most common group of negative consequences (reported by 79% of switchers and 44% of persisters) was loss of confidence in face of low grades that contributed to 61% of switching decisions. Loss of confidence was greatest among women, half of whom were women of color, and most often occurred as a consequence of weed-out course experiences. Our finding echoes that of Ellis, Fosdick, and Rasmussen (2016) about what they describe as the effectiveness of Calculus 1 courses in destroying incoming confidence. Even with final grades of As and Bs, twice as many women as men with the same grades abandoned the idea of continuing to Calculus 2.

Also in Chap. 10, we discussed the role of "perfectionism" in the switching decisions of high-performing students that included men, but more often women, and in the ongoing struggles of persisting seniors. Conditioned to high-performance expectations for themselves, high-achieving STEM majors who are unwilling or unable to dissociate their grades from their identity are at high risk of switching. The mental and emotional distress they describe arises from low tolerance for less than perfect performances. Moderate grades interpreted as failure pose an intolerable threat to identities that are extrinsically derived from high grades and the status that they confer. Rather than making adjustments to stay in their major, switching enables high performers to regain self-esteem and a valued sense of self without having to change the reputational criteria upon which these are built.

Thirty-four percent of switchers and 8% of persisters reported depression, high levels of stress, chronic anxiety, feeling lost and overwhelmed, or living with intolerable fear. These included: the fear of admitting that one cannot cope with the work; the fear that this will only get worse, and dread of going to class. Another group of students described states of emotional and mental distress that not only precede decisions to switch or relocate but often linger beyond them. Reported by 23% of switchers and 3% of persisters, they include: feelings of guilt, failure, or regret; shame, self-blame for their difficulties, and feeling stigmatized as someone who failed or "couldn't make it"; being tired of feeling miserable, bad about one's self, or being too paralyzed to act.

Finally, chronic levels of stress, depression, fear, anxieties and/or guilt, shame, stigma, and self-blame took their toll on 15% of switchers and 7% of persisters in the form of mental or physical health problems, including exacerbation of existing health conditions and clinical diagnoses of depression or anxiety. Nationwide, campus mental health services and offices of student affairs have recently reported rapidly increasing rates of anxiety and depression among college students, and, as we were made aware during our site visits, a number of STEM majors commit suicide each year. As we review in Chap. 10, perfectionism in college students has been extensively studied (though not specifically for STEM majors) in relation to its maladaptive manifestations, such as procrastination, burn-out, anxiety, depression, eating disorders, and suicidal ideation. Curran and Hill's (2019) meta-analysis of studies on perfectionism among college students report a significantly rise in perfectionism over the past 25–30 years. They argue its origins in the increased competitiveness, individualism, and meritocracy that have shaped the neoliberal character of American and other world economies. As we document in Chap. 9, the struggle to belong in the negative, at times-hyper competitive, atmosphere pervading some STEM programs puts students at risk both of ill-health and of switching. As Travers, Randall, Bryant, Conley, and Bohnert (2015) describe, the risks are intensified where students conceal the effort put into achieving high grades, such that their success appears natural and easy, while other students lose face over their more overt struggles. We first noticed this behavior in the original study where young women were demoralized by the apparently effortless success of some young men. Two recent articles describe the rise of "effortless perfectionism" among women on competitive college campuses and the personal and emotional toll of sustaining such standards (Ruane, 2012; Yee, 2003). As we have also reported, for high-performing students with perfectionist tendencies the consequent assumption can be that there is no space between perfection and failure, and that only one very narrow and extremely high standard of achievement leads to success. The personal costs of a STEM educational experience may be disattended as somehow inappropriate to an academic life. However, they are a reminder that the benefits and costs of education are never purely cerebral but involve the whole person.

## *What Enables Persistence?*

Almost half of the persisters in our sample had, at some point, doubted that they had chosen the right path, and 28% had actively considered switching to a non-STEM major. Women were more likely to consider switching than men, as were, African

Americans, and students who entered with low math readiness, but there were no disciplinary differences among persisters who had considered leaving STEM.

Persisters faced the same difficulties in course design, pedagogy, and assessment methods as switchers, and, like switchers, most persisters who had considered switching were responding to grades that they found irrational and disorienting. However, unlike switchers, they were prompted by these experiences to make adjustments, notably in their study habits. As they came to realize that curved grades do not reflect actual learning, high-achieving persisters also began to detach their sense of self-worth from their grades and GPAs. They described this as a long, difficult, but essential transition if they were to stay in their major. It involved accepting that, whether an apparently failing grade represented actual failure depended on how the rest of the class performed. In such crises, persisters differed from switchers in finding ways to adapt their expectations and academic identities, find resources, figure out how to navigate around their difficulties and keep going.

As we found in the first study, for switchers, crises over grades often deepened feelings of isolation and doubts about belonging. When decisions hung in the balance, the scales could be tipped in favor of persistence by the serendipitous intervention of a friend, advisor, or faculty member who normalized the struggle for them. Thus, persistence may depend on the randomness of supportive encounters. Talking to STEM seniors, it was clear that not all their difficulties were resolved. Some still struggled with perfectionist self-demands that they knew kept them at risk.

In describing how they had made it thus far, seniors rarely cited one single approach. Rather, they described an interacting array of individual, social, cultural, and institutional resources that they drew on to survive. These were, broadly, of four kinds. First, they cited personal traits that included maintaining determination and the will to succeed, sustaining intrinsic interest in their discipline and career goals, and a strong belief in oneself. While sheer determination and sustained interest often enabled students to overcome early struggles with low grades and other difficulties, men of color and women of all races and ethnicities also encountered social, cultural, and structural barriers to surmount which required even greater strength, stamina, and determination. Whether students held a privileged or marginalized position within their discipline also determined when the will to persist came into play: students of color and women drew on "grit" to overcome isolation or hostile climates, while more advantaged students used it to combat difficulties with grades or coursework. Altruism was cited by students of color, especially women, in sustaining their interest and motivating them through periods when they felt they did not belong their major.

A second group of adjustments were shifts in behaviors and identities that stressed, accommodating lower grades, developing effective work habits, and maintaining a balanced life style. There were no differences by gender or race/ethnicity in developing effective study and learning habits. Among students who entered under-prepared, whether conceptually or in time management and study skills, persisters managed to develop these capacities more quickly than switchers. As they advanced in their coursework, they learned how to adjust to the workload, expectations, and work habits required, and to adjust their study strategies to the course, the professor, and the material. Contrary to the belief that non-STEM interests are distracting, some seniors (women more often than men) advocated avoidance of total absorption in their major. They had discovered that nurturing wider interests for pleasure and counterbalance improved their STEM disciplinary focus. They engaged in sports, music, art, literature, and languages, or added another major or minor in a non-STEM field.

A third set of strategies focused on seeking out and using social and institutional sources of help and support, both formal and informal, and taking advantage of intellectual and professional out-of-class experiences. An important, common strategy (also reported in the original study) was working with informal peer study groups. These performed several important functions—enhancing students' conceptual understanding, giving social and emotional support, building a sense of belonging and community, and providing guidance and advice about STEM pathways. While some students benefitted from study sessions and supplemental peer instruction organized by departments, the majority of students also found informal peer study groups essential to their survival. Academically oriented peer support (including that found in STEM-related clubs) also supported the retention of underrepresented minority students. Despite their vital role of peer support systems in persistence, most students had to instigate and navigate these relationships for themselves. Students who had difficulty in connecting with a peer study group were at risk. These included students who are marginalized or underrepresented in STEM majors, students with health or emotional disabilities, under-prepared students, and students with outside responsibilities, such as children or full-time jobs. Despite institutional efforts to create a structured system of study groups, we also found (as in the original study) that not all students took advantage of them. An implication arising from these findings is that, discovering why some students do not use formal and informal resources, and experimenting with ways to encourage their greater use, would make a significant contribution to STEM retention, particularly for more vulnerable student groups.

The fourth group of strategies involved figuring out how to play a STEM major as a system in order to increase the likelihood of success. System-playing was noted in *Talking about Leaving* but was far more widespread in this study with persisters proving to be very savvy "consumers" who navigate their pathways through STEM majors in creative and improvisational ways. Seniors described how they carefully researched courses, sections, and faculty members in order to make selections that gave them the best chance of passing courses with good grades. They reviewed online resources and crowd-sourced comments to identify instructors, course sections, or even institutions and departments that would provide the best chances of success. Seniors described these moves as a form of game-playing to maximize their chances of securing a STEM degree. Some students undertook this research because they saw the quality of instruction as vital to their learning and retention, but, more often, students played the game purely to improve their grades. In weed-out courses, beating the class average was sometimes seen as more important to retention or advancement than actual learning or interest, and some seniors had reservations that "gaming their major" was not optimal for learning or long-term success. The circuitous course-taking patterns taken by many STEM students in order to persist and their relocations between majors and sometimes institutions to secure a better fit with interests or career preferences indicate that linear degree paths undertaken at single institutions are no longer seen by many students as the optimal way to secure a STEM degree. We note that the growth and widespread use of game-playing persistence strategies and the highly consumerist approach indicated by our findings also have profound implications for coherence and consistency in STEM degrees.

We deliberately included a sub-set of STEM seniors who had entered with lower math scores than the rest of our interview sample in order to learn how they had managed to persist despite the odds. Students who entered with low math readiness were more likely to consider switching out of STEM during their undergraduate careers. However, compared with high math students, low math students made much greater efforts to enable their own persistence and made use of a wider array of strategies and resources than their high math peers. They were also much more likely to actively manage their schedules in a way that best facilitated their success, including taking "risky" classes at community colleges. They were also more likely to cite their own determination, bolstered by strong parental expectations, as essential to their persistence. This was especially marked among students from first-generation and immigrant families. They also stressed the importance of finding a support system and figuring out how to navigate college STEM courses and coursework in the most advantageous ways. Higher percentages of low math than high math seniors cited the importance of these traits, adjustments, and strategies in all four categories of persistence enablers that we have identified.

We have highlighted the role of legitimating beliefs in explaining the perpetuation of STEM teaching practices with dysfunctional outcomes, and the ways in which such practices and their validations are absorbed by undergraduate and graduate students who carry them forward to another generation. Belief in a meritocratic narrative, which is widespread in STEM disciplines, belies the ways in which courses and institutions are structured in ways that facilitate or impede student success. In keeping with this dominant narrative, we also found that many persisting seniors ascribed their success to talent, intelligence, or self-efficacy while, in the same interviews, also describing their reliance on behavioral adjustments, their use of social and institutional resources, and their ways of navigating STEM majors as systems in calculated ways to secure success. Indeed, most persisters viewed their STEM major as a "game" to play and a challenge to overcome by judicious use of available resources while crediting their success to their personal strengths. However, it was clear from our evidence that personal characteristics alone were never sufficient to secure success. We propose that the danger of the meritocratic narrative that explains persistence in terms of individual interest, effort, and intelligence is that it absolves STEM faculty, departments, and colleges of responsibility for student learning and success. Yet, almost all persisters relied on the kinds of external supports that we have identified in order to succeed. As we concluded in the original study, determination, passion, and a strong will to persist were important, but not sufficient, components of persistence by themselves; we find this still to be true in the current study. This finding alone has implications for the design and teaching of STEM courses because it illustrates the role of institutional systems in students' success.

## *Envisioning the Future: What Institutions Can Do*

As we have reported there is strong collaborative evidence from three of our studies that active and interactive teaching and learning strategies have, to varying degrees, been introduced in all six sample institutions where none were evident 20 years before. Although we cannot generalize from these findings alone, when added to evaluation reports from STEM education reform initiatives (reviewed in Seymour & Fry, 2016), they suggest that credit for the 16% drop in the national switching rate is likely due to the substantial efforts of the STEM education reform community and its funders, to the research that validates their work, and to the disciplinary and professional organizations that promote both of these. Over the last two decades, STEM instructors have collectively created a body of thematic, contextual, research-grounded curriculum and learning materials, an array of classroom-tested, active, interactive, and inquiry-based pedagogies, and learning assessment methods that explore students' depth of understanding, and ability to apply, extend, and transfer their knowledge. This expanding body of tested and discipline-relevant methods and materials has been disseminated to widening circles of instructors through online resources and communities, journal articles, conference presentations, and workshops that offer hands-on exposure to learning theories, research findings, and their classroom applications. With such a body of resources already available, the key question is how best can successful STEM education improvement strategies be taken to scale and sustained? Thus, the emphasis has shifted from the individual and collaborative efforts of STEM instructors to what institutions and departments can do to encourage and sustain all that the STEM education reform community have created.

To address the challenge of transformative change requires consideration of how institutes of higher education (IHEs) work. Their structure has evolved to preserve customary ways of carrying out formal tasks that are evident, for example, in the design of lecture halls, the criteria for departmental funding, and for faculty rewards, tenure, and promotion. Institutional structures privilege traditional teaching methods and do not easily accommodate new ones. This makes the resulting inertia hard to break, involving, as this does, both structural and cultural shifts. Indeed, resistance to change in its educational functioning may be seen as normal in higher education institutions. Thus, those engaged in the process of experimentation with educational improvement have learned that some strategies work better than others. Important among these is that successful teaching reform requires a combined top-down and bottom-up approach. From the top, it requires institutional commitment to the value of research-based instructional strategies (RBISs), shifts in the distribution of funding and rewards, and changes in organizational and physical structures.

Without sufficient top-down buy-in and material support, grass-roots efforts to institutionalize RBIS invariably founder (Seymour, 2001).

In light of this history, it is essential to convince chairs, deans, provosts, and college presidents that high-quality teaching and the scholarship of teaching and learning (SoTL) are important to their mission. That this idea is gaining ground is evident in the spread of institution-sponsored teaching and learning centers, undergraduate research programs, high-school-to-college bridge programs, and women-in-science programs. As it is departmental rather than institutional leaders that have most power to determine matters of curriculum and pedagogy, institutional leverage in these matters may appear to be indirect and marginal (Seymour, 2001; Seymour & DeWelde, 2016). What, then, do institutions have the power to do that contributes to sustained educational improvement that supports greater retention of STEM majors?

The existing institutional rewards system is the main structural deterrent to change for faculty who are otherwise disposed to rethink their teaching. The strategies that would improve their pedagogy are not yet widely embedded in faculty positions and rewards. In accord with Boyer's (1990) proposition that achievements in research and teaching should be judged by parallel criteria, some institutions have extended their criteria for hiring, promotion, and tenure to include evidence of teaching effectiveness and the scholarship of teaching and learning (SoTL) and discipline-based educational research (DBER). The institutional climate for classroom reformers seems to be improving: directors of STEM education initiatives cite tenure successes among their project participants and also a growing trend for new faculty to negotiate career paths that include their innovative teaching and SoTL (Seymour & Fry, 2016).

Institutions also have considerable power to encourage faculty uptake of new curriculum or pedagogy by changing their faculty time allocation policies. Providing release time to allow faculty innovators to do their work is critical to their chances of success. This is especially important for the principal investigators (PIs) of reform efforts who are teaching faculty. Any multi-institution project's administrative efficiency is undermined from the outset unless its faculty PI receives sufficient time from the host institution to organize STEM education reform efforts, particularly those that are multi-institutional in scope.

Strategic use of central resources plays a role both in encouraging and sustaining educational improvements. An effective form of institutional leverage is to offer annual awards to those departments that document educational improvements resulting in desired student outcomes over the previous year. These might include departmental self-study, defining learning objectives for degree programs, aligning course objectives across curricula, the introduction of research-based instructional strategies and assessment of their efficacy, and using data to understand the characteristics of students entering and leaving their majors. The purposive deployment of money, jobs, and resources is also critical in sustaining successful innovations after external funding ends. It entails providing administrative support and funding faculty lines or staff positions that service and support research-based instructional strategies. Administrative and physical structures may also need to be changed to accommodate new ways of teaching, for example, classroom redesign, the addition of technical teaching aids, and provision of service staff. We encountered evidence of such changes at all six of our sample sites.

In support of the institutional changes noted above, we have identified four more specific strategies that have been shown to be critical to the sustained uptake of RBIS pedagogy: (1) providing professional education in teaching methods and learning research for new instructors and graduate TAs; (2) financing and promoting workshops that support existing instructors in hands-on learning of new teaching and student assessment methods; and (3) adopting course assessment tools that provide instructors, departments, and the institution with data on the efficacy of course teaching. (4) Departmental self-study.

1. Durable, nationwide STEM education reform requires development of institution-wide professional education programs that ground the pedagogical knowledge and skills of current and future faculty in learning research. The professional education of graduate students as the new STEM teaching force is essential in enabling departmental transition to research-based teaching methods. Where instructors are developing new forms of pedagogy, it is critical to their success to ensure that their graduate teaching assistants understand both how the new methods work and their basis in learning research (Seymour, Melton, Pedersen-Gallegos, & Wiese, 2005). Institutions can provide funding and resources for teaching and learning centers, and institute professional development programs for the future professoriate—new instructors, post-docs, and graduate teaching assistants. They can also actively leverage departmental understanding and use of these programs.
2. Disciplinary-based workshops have, for two decades, been the main conduits of teaching knowledge and know-how. There is strong evidence that workshops foster the uptake and spread of within- and cross-institution educational reform by drawing in, educating, and enabling new faculty and incorporating them into the change effort. Workshops are effective because participants learn from each other by trying out alternative methods in a relaxed, private, and congenial context that encourages collaboration and builds connections. To work optimally, workshops must be of sufficient duration, offer repeated exposure in a progressive sequence, provide support for new reformers in their departments, use "old hands" as facilitators, and build facilitator capacity among newer recruits (Andrews, 1997; Connolly & Millar, 2006; Hilsen & Wadsworth, 2002). Workshops give participants motivation and skills. They also offer portals to like-minded people whom they might not ordinarily meet—colleagues in different disciplines, administrators, senior faculty, and graduate students. Many professional and disciplinary societies have stepped into this role by mounting workshops as part of annual meetings. Institutions that offer workshops can draw on this expertise, reap these benefits, and build disciplinary and cross-disciplinary communities where colleagues learn from each other how to deploy best practices in their classrooms.
3. Adoption and continued use of RBISs to improve pedagogy depend on faculty being able to get valid, reliable data about their efficacy for student learning

(Dancy & Henderson, 2010). As explained in Chap. 1, in order to learn what students in foundational classes were (or were not) gaining, we used a customized version of the Student Assessment of their Learning Gains (SALG) one-line survey (cf., Appendix C). Originally designed for STEM instructors with NSF support, but widely used by faculty in other disciplines, the SALG is available, free-of-charge, to all instructors. It is also used by departments and institutions as their course evaluation instrument. For example, the Gateways to Completion (G2C) initiative (whose collaborative research was cited in Chap. 7) systematically deploys SALG instruments, and the resulting data help participating institutions to improve student success in courses with historically high DFWI rates.

We encourage consideration of the SALG survey (https://salgsite.net/) by institutions and departments because systematic assessment of what students are gaining from their courses is essential if course improvements are to be sustained. SALG survey templates can be edited to reflect the learning objectives, contents, and methods of any course. Thus, they provide detailed information about what progress students are making toward course learning goals and which pedagogical strategies do, and do not, enable this. Such data can help instructors to make rational, targeted improvements to their course design and pedagogy, demonstrate the value of their RBISs to others, reduce RBIS discontinuation, and document their teaching achievements for tenure and promotion purposes. As a replacement for traditional course evaluation instruments (in which students are asked to assess their instructor), SALG surveys enable departments and colleges to make evidence-based improvements to programs and curricula and gather data for departmental accreditation purposes.

In addition to the 10,000 instructors who regularly been using the SALG for over a decade, the NSF has recommended SALG surveys to evaluate STEM education projects and they are widely used in the STEM education reform community for project evaluation and to gather data for SoTL publications. (The results of SALG surveys have been cited in over 450 scholarly publications.) The SALG-based Undergraduate Research Student Self-Assessment (URSSA) is used by UR directors (including the NSF's REU programs) to track student learning gains in UR experiences. A growing number of departments and institutions have adopted the SALG as their course evaluation instrument and use its group functions to assess curricula, programs, and pedagogical innovations. SALG data are also increasingly accepted by accreditation agencies as prima facie evidence of student learning for program review purposes.

4. We have also pointed to the need for research data to understand why 20% of STEM entrants nationally leave college without a degree and which students are most at risk of such departures. As we have illustrated from our own analysis of data provided by the six participating institutions, it is entirely possible that institutions already have access to data that can answer these questions. Using existing institutional records data, we were able to identify which students are lost from "weed-out" foundation courses. Thus, departments and colleges could also keep track of what students, with what characteristics, enter and leave their

majors and monitor improvements in these patterns as a result of changes in course teaching and assessment methods. It is possibly a task for ethnographic researchers to discern the instructor beliefs that underpin and perpetuate the pedagogy, course design, and grading practices typical of "weed-out" courses. This undertaking would also underpin departmental consideration of more effective, alternative ways to introduce students to important disciplinary content. Recent research on effective transformation (Weaver, Burgess, Childress & Slakey, 2016) points to the effectiveness of departmental self-study combined with faculty learning communities in creating educational improvements at the departmental level.

The most serious implication from our findings is the need to address the chronic pattern of under-preparation of talented students from working-class families of all races and ethnicities who enter undergraduate STEM education from under-resourced schools from across the nation's educational system. To adequately address this problem requires considerable rethinking of state and national policy. However, awareness of this significant cause of losses from STEM majors might also prompt or reinforce ameliorative interventions by higher education institutions and STEM departments. The success of summer bridge programs and of math-readiness assessment and remediation prior to and upon entry encourage such ameliorative remedies until larger problems in the education system can be addressed. It is also possible that institutional leaders can use their platform and communication channels to raise this problem to the consciousness of state and national political leaders. Other topics for such conversations might also be the impact of financial debt on students' degree course and career decisions.

Not only do many of the issues that inhibit STEM persistence lie within the scope of institutions but the strategies by which to address them are already available to the leaders of IHEs and to their departments and colleges. The pioneering work of many STEM instructors has provided the educational means for nationwide improvement in the loss rates of STEM majors. To make further progress in resolving the problems that we have described will require the collaborative engagement of institutional and departmental leaders. It would seem timely to pick up the baton and carry it forward.

## References

Andrews, G. J. (1997). Workshop evaluation: Old myths and new wisdom. In J. A. Fleming (Ed.), *New perspectives on designing and implementing effective workshops* (pp. 71–85). San Francisco, CA: Jossey-Bass.

Boyer, E. L. (1990). *Scholarship reconsidered: Priorities of the professorate*. San Francisco, CA: JosseyBass.

Canning, E. A., Muenks, K., Green, D. J., & Murphy, M. C. (2019). STEM faculty who believe ability is fixed have larger racial achievement gaps and inspire less student motivation in their classes. *Science Advances, 5*(2), eaau4734. https://doi.org/10.1126/sciadv.aau473

Chen, X. (2013). *STEM attrition: College students' paths into and out of STEM fields* (Statistical Analysis Report. NCES 2014-001). National Center for Education Statistics.

Connolly, M. R., & Millar, S. B. (2006). Using educational workshops to improve instruction in STEM courses. *Metropolitan Universities, 17*(4), 53–65.

Curran, T., & Hill, A. P. (2019). Perfectionism is increasing over time: A meta-analysis of birth cohort differences from 1989 to 2016. *Psychological Bulletin, 145*(4), 410–429. https://doi.org/10.1037/bul0000138

Dancy, M., & Henderson, C. (2010). Pedagogical practices and instructional change of physics faculty. *American Journal of Physics, 78*(10), 1056–1063.

Eagan, K., Hurtado, S., Figueroa, T., & Hughes, B. E. (2014). *Examining STEM pathways among students who begin college at four-year institutions*. Washington, DC: National Academy of Sciences.

Ellis, J., Fosdick, B. K., & Rasmussen, C. (2016). Women 1.5 times more likely to leave STEM pipeline after calculus compared to men: Lack of mathematical confidence a potential culprit. *PLoS One, 11*(7), e0157447. https://doi.org/10.1371/journal.pone.0157447

Ferrare, J. J. (2019). A multi-institutional analysis of instructional beliefs and practices in gateway courses to the sciences. *CBE-Life Sciences Education, 18*(2), ar26.

Ferrare, J. J., & Miller, J. (2019). Making sense of persistence in scientific purgatory: A multi-institutional analysis of instructors in introductory STEM courses. *The Journal of Higher Education*. https://doi.org/10.1080/00221546.2019.1602392

Hilsen, L. R., & Wadsworth, E. C. (2002). Staging successful workshops. In K. H. Gillespie (Ed.), *A guide to faculty development: Practical advice, examples, and resources*. Bolton, MA: Anker.

Lee, Y.-G., & Ferrare, J. J. (2019). Finding one's place or losing the race? The consequences of STEM departure for college dropout and degree completion. *The Review of Higher Education, 43*(1), 221–261.

Rask, K. (2010). Attrition in STEM fields at a liberal arts college: The importance of grades and pre-collegiate preferences. *Economics of Education Review, 29*(6), 892–900. https://doi.org/10.1016/j.econedurev.2010.06.013

Ruane, I. W. (2012, July–August). Effortless perfection. The Undergraduate. *Harvard Magazine*, 55–56. Retrieved from http://harvardmagazine.com/2012/07/effortless-perfection

Seymour, E. (2001). Tracking the process of change in U.S. undergraduate education in science, mathematics, engineering, and technology. *Science Education, 86*, 79–105.

Seymour, E., Melton, G., Pedersen-Gallegos, L., & Wiese, D. J. (2005). *Partners in Innovation: Teaching assistants in college science courses*. Boulder, CO: Rowman and Littlefield.

Seymour, E., & DeWelde, C. (2016). Why doesn't knowing change anything? Constraints and resistance, leverage and sustainability. In G. C. Weaver, W. D. Burgess, A. L. Childress, & L. Slakey (Eds.), *Transforming institutions: Undergraduate STEM education for the 21st century*. Purdue, IN: Purdue University Press.

Seymour, E., & Fry, C. F. (2016). The reformers' tale: Determining progress in improving undergraduate STEM education. In G. C. Weaver, W. D. Burgess, A. L. Childress, & L. Slakey (Eds.), *Transforming institutions: Undergraduate STEM education for the 21st century*. Purduc, IN: Purdue University Press.

Travers, L. V., Randall, E. T., Bryant, F. B., Conley, C. S., & Bohnert, A. M. (2015). The cost of perfection with apparent ease: Theoretical foundations and development of the effortless perfectionism scale. *Psychological Assessment. Advance online publication., 27*(4), 1147–1159. https://doi.org/10.1037/pas0000109

Weaver, G. C., Burgess, W. D., Childress, A. L., & Slakey, L. (2016). *Transforming institutions: Undergraduate STEM education for the 21st century*. Purdue, IN: Purdue University Press.

Yee, C. (2003, September 23). Committee unveils Women's Initiative report. *The Chronicle*. Retrieved from http://dukechronicle.com/article/committee-unveils-womens-initiative-report

# Appendix A: Demographic Information on TALR Study Sample Institutions

*Note:* Demographic representation found at: https://nces.ed.gov/ipeds/datacenter

We examined the racial/ethnic and gender makeup of the schools in our study compared to national averages. Most of the institutions in our study had a greater representation of white students and fewer Hispanic and African-American students than is present nationally (Table A.1).

**Table A.1** Racial/ethnic representation at TALR study sample institutions compared to national percentage

| | All | PB1R1 | PB2R1 | PB3R1 | PB4R1 | PR1R3 | PV2R1 |
|---|---|---|---|---|---|---|---|
| African-American | 14 | 2 | 11 | 3 | 2 | 3 | 2 |
| White | 57 | 72 | 50 | 66 | 71 | 67 | 66 |
| Hispanic/Latino | 17 | 9 | 8 | 3 | 9 | 9 | 9 |
| Asian | 6 | 3 | 13 | 8 | 5 | 3 | 5 |
| Native American | 0.8 | <1 | <1 | <1 | <1 | <1 | 0 |
| Two or more races | 3 | 3 | 3 | 3 | 4 | 3 | 8 |



E. Seymour, A.-B. Hunter (eds.), *Talking about Leaving Revisited*,
https://doi.org/10.1007/978-3-030-25304-2

Representation of women was also lower than is seen nationally (Table A.2).

**Table A.2** Percentages of women and men at TALR study sample institutions compared to national percentage

| | All | PB1R1 | PB2R1 | PB3R1 | PB4R1 | PR1R3 | PV2R1 |
|---|---|---|---|---|---|---|---|
| Women | 57 | 51 | 47 | 52 | 44 | 56 | 54 |
| Men | 43 | 49 | 53 | 48 | 56 | 44 | 46 |

Enrollment ranged from ~2000 (PV1R3) to over 34,000 (PB3R1) undergraduates (Table A.3).

**Table A.3** Fall undergraduate enrollment and STEM degrees granted for at TALR study sample institutions

| | All | PB1R1 | PB2R1 | PB3R1 | PB4R1 | PV1R3 | PV2R1 |
|---|---|---|---|---|---|---|---|
| Total enrollment (undergraduate) (2014)a | – | 25,598 | 27,056 | 34,351 | 26,557 | 2050 | 5643 |
| STEM undergraduate degrees granted (2009)b | – | 1139 | 1493 | 1533 | 955 | 101 | 123 |

[a]Dates of data collection
[b]Last available from IPEDs data

Graduation rates were higher at our institutions than the national average, as were averages of graduation rates for racial/ethnic groups at each institution (Table A.4).

**Table A.4** Percentage graduation rates by gender and race ethnicity

| | Total | Men | Women | Native American | Asian | African-American | Hispanic | White |
|---|---|---|---|---|---|---|---|---|
| National | 59 | 56 | 62 | 41 | 73 | 39 | 53 | 63 |
| PB1R1 | 69 | 63 | 67 | 60 | 55 | 59 | 54 | 67 |
| PB2R1 | 85 | 83 | 87 | 88 | 88 | 77 | 80 | 87 |
| PB3R1 | 80 | 78 | 79 | 64 | 71 | 58 | 70 | 82 |
| PB4R1 | 69 | 68 | 72 | 47 | 72 | 57 | 63 | 71 |
| PR1R3 | 88 | 83 | 89 | 60 | 93 | 38 | 89 | 86 |
| PV2R1 | 75 | 76 | 77 | 93 | 76 | 58 | 71 | 78 |

# Appendix B: Description of Quantitative Methods

## Analytical Method Used to Investigate Student Transcript Records

### *Logistic Regression*

In Chaps. 2 and 7 we used logistic regression to predict switching. Logistic regression gives the probability of switching given the values of independent variables. In our study, we use variables such as grades, class level, test scores, and demographic status to predict switching. All six institutions were included in the analysis, and all students who started in STEM and either switched out of or persisted in STEM majors.

The logistic function is:

$$p = 1/\left(1 + e^{-z}\right) \tag{B.1}$$

where $p$ is the probability of switching, $e$ is the natural log (2.718). To find $z$ we use the formula:

$$z = B_1 X_1 + B_2 X_2 + \ldots B_i X_i + c \tag{B.2}$$

where $B_0$ and $B_1$ are logits or the marginal contributions for each variable to the log exponent; $X$'s are the values of the independent variables, and $c$ is a constant.

There are several ways of expressing the predictive power of different variables in the model. Odds are the ratio of the probability of an event over the probability of the event not happening. "Log odds" is the ratio of odds given one value of an independent variable to the odds when that independent variable is increased by one unit. For instance, the log odds of 1.5 indicates that for each one unit increase in a given variable there will be a 150% increase in the *odds* of the occurrence. Values greater than one indicate that an increase in the independent variable is associated



E. Seymour, A.-B. Hunter (eds.), *Talking about Leaving Revisited*,
https://doi.org/10.1007/978-3-030-25304-2

with a greater chance of the outcome (in our case, switching); values less than one are associated with a lower chance of the outcome given the value of the independent variable.

Increases or decreases in the probability of the outcome (Delta-P) can also be evaluated given the values of the independent variables. Holding all other variables constant, the probability value of success can be estimated by increasing each value of the independent variable by one unit while holding all other variables constant. Increases or decreases are evaluated at the average value of a continuous variable, or the difference between 0 and 1 with a dichotomous variable. This is the increase given the central position on the loglinear function but does not reflect increases at either end of the scale.

***Variables Used in Logistic Regression*** We used demographic and academic variables to predict STEM switching. Table B.1 describes the variables and their scales.

**Table B.1** Variables used in prediction of switching

| Variable | Scale/classification | Description/comments |
|---|---|---|
| Gender | 0 = Female, 1 = Male | |
| Underrepresented minority (URM) | 1 = Underrepresented minority<br>0 = Non-URM | URM includes African-American, Hispanic, Native American, Pacific-Islander<br>Non-URM includes all other race-ethnicity |
| ACT/SAT math | −3 to +3 | ACT math and SAT math standardized with population standard deviation (e.g., 100 for SAT). (*X*—mean)/SD. Two standardized scales merged, if students took both tests, average of two standardized scores used |
| DFWI | 0—highest | Number of *D*'s, *F*'s, incompletes and withdrawals. Depending on analysis can be cumulative over all terms, for one term, or for year one |
| DF (SFC) | 0—highest | Number of *D*'s, *F*'s in Severe Foundational Courses |
| IW (SFC) | 0—highest | Number of Incompletes or Withdrawals in Severe Foundational Courses |
| DF (NSFC) | 0—highest | Number of *D*'s, *F*'s in Non-Severe Foundational Courses |
| IW (NSFC) | 0—highest | Number of Incompletes or Withdrawals in Non-Severe Foundational Courses |
| Current class level in data | 3 = Junior<br>4 = Senior<br>5 = Senior+ | Current class level of students in database. Data gathered for all Junior, Seniors, and Senior+ at year of study. Used as covariate to control for bias due to non-completion |
| Grade point average | 0–4 | Average of all letter grades<br>Cumulative includes all letter grades to last term<br>Term includes all letter grades for term |
| Number of terms in data | 1—highest | Highest number of terms enrolled in database. Data gathered for all Junior, Seniors, and Senior+ at year of study. Used as covariate to control for bias |
| Average course difficulty | 0–100 | Average percent of DFWI encountered in coursework |

## Methods Used for Analyzing Responses to the TALR-Student Assessment of Their Learning Gains (SALG) Survey (Tables B.2, B.3, and B.4)

**Table B.2** Number of instructors and students at each institution participating in the SALG survey study

| Institution | Instructors | Students | % Female | % URM | % Declared major |
|---|---|---|---|---|---|
| PB3R1 | 13 | 500 | 56 | 10 | 92 |
| PB2R1 | 7 | 56 | 57 | 9 | 92 |
| PB1R1 | 7 | 110 | 57 | 10 | 78 |
| PV2R1 | 8 | 244 | 51 | 16 | 92 |
| PV1R3 | 6 | 39 | 52 | 16 | 49 |
| PB4R1 | 11 | 478 | 53 | 12 | 97 |
| | 52 | 1427 | 54 | 12 | 91 |

**Table B.3** Class standing of students answering the TALR-SALG survey

| Class standing | *N* | % |
|---|---|---|
| First year | 298 | 24.3 |
| Sophomore | 468 | 38.2 |
| Junior | 313 | 25.5 |
| Senior | 99 | 8.1 |
| Other | 48 | 3.9 |

**Table B.4** Disciplines represented

| Discipline | N | % |
|---|---|---|
| Agriculture | 95 | 8.1 |
| Computer science | 39 | 3.3 |
| Engineering | 238 | 20.3 |
| Life sciences | 426 | 36.3 |
| Mathematics | 23 | 2.0 |
| Physical sciences | 120 | 10.2 |
| Non-stem | 234 | 19.9 |

## Methods Used to Investigate Classroom Teaching Practices

Cluster analyses were conducted in Chap. 8 to categorize teaching methods reported by students in the SALG survey. We used a 2-step cluster analysis using 12 dichotomous variables representing the presence or absence of teaching and learning methods in 28 classes. Cluster analysis used Schwartz's Bayesian clustering criterion. We found two clusters discriminated on the use of clickers in the classroom.

## Methods Used for the "Switch-to" Analyses Investigating What Majors Students Switch to When They Leave Their STEM Major

### *Analysis Used to Investigate Where STEM Students Go*

In Chap. 2 we examined if students switched out of STEM majors to destination non-STEM majors at greater or less than expected frequencies. As mentioned earlier, we categorized STEM and non-STEM majors by discipline, broad subject areas based on the CIP code categories from the National Center for Education Statistics (NCES, 2019). We first found the prior probabilities of STEM and non-STEM students in each discipline at each of the six schools in the study by calculating their relative proportions in the student records of switchers and non-switchers and enrollment information from each institution's websites. The prior probabilities of being in a discipline are denoted $p_{\text{nstem}}$ (probability of non-stem) and $p_{\text{stem}}$ (probability of STEM) and are expressed as proportions.

We then found the cross-tabular observed frequencies ($O_f$) of students switching from STEM to non-STEM subject areas at each institution. These switching patterns are called "pathways" in this analysis.

To find the expected frequency ($E_f$) of students in each cross-tabular cell, we multiplied STEM and non-STEM prior probabilities and the total number of switchers at the institution ($f_I$)

$$E_f = p_{\text{nstem}} * p_{\text{stem}} * f_I \tag{B.3}$$

The resulting expected frequency ratio is the frequency of observed students in each cell divided by the expected number of students.

$$E_{\text{ri}} = O_f \,/\, E_f \tag{B.4}$$

Values of $E_{\text{ri}}$ greater than one indicate that the probability of switching to a non-STEM subject are greater than expected; values less than one mean the opposite, that students in this path are less likely to switch than they would by chance.

Because each institution has both different prior probabilities and different subject areas (e.g., some institutions have agriculture schools and others do not), we calculated expected frequency values with the prior probabilities specific to each institution. We then summed the observed and the expected values within disciplines and across institutions to calculate expected frequency ratios of the five remaining institutions together.

Small cell frequencies, especially at individual institutions made calculating the expected frequency ratio problematic for some cases. Small differences between observed and expected frequencies become exaggerated when small numbers were used. To avoid this we reported ratios only when expected values were greater than 10 cases, a rule of thumb commonly used with proportions (Hopkins, 1998). We also examined large ratios to learn if they were coupled with smaller cell sizes and reported actual frequencies for smaller *N*'s.

We also wanted to know if specific discipline in STEM had greater or less probability of students' switching out of the major than others given the proportion of students in each area at each institution. This comparison was straightforward with expected frequency ratios calculated as:

$$E_{\mathrm{r}} = O_f \,/\, p_{\mathrm{stem}} \qquad \text{(B.5)}$$

Comparing gender, GPA and race/ethnicity used the same methodology described above, but applied cross-tabulation, expected frequencies and expected frequency ratios within these demographic groups.

## References

National Center for Educational Statistics CIP User Site. (2019, May 1). Retrieved from: https://nces.ed.gov/ipeds/cipcode/browse.aspx?y=55.

Hopkins, K.D. (1998) (8th ed.). Educational and Psychological Measurement and Evaluation. Boston: Allyn and Bacon.

# Appendix C: TALR-SALG Survey

## SALG: Class overall

**1. As a result of your work in this class, what gains did you make in your understanding of each of the following?**

| | No Gain | Little Gain | Moderate Gain | Great Gain | Not applicable |
|---|---|---|---|---|---|
| The main concepts explored in this class | ○ | ○ | ○ | ○ | ○ |
| The relationships between the main concepts | ○ | ○ | ○ | ○ | ○ |
| Connecting key ideas from this class with other knowledge | ○ | ○ | ○ | ○ | ○ |
| Applying what I learned in this class in other situations | ○ | ○ | ○ | ○ | ○ |
| Using concepts from this class to address real world issues | ○ | ○ | ○ | ○ | ○ |

Please comment on how your understanding of the subject has changed as a result of this class

**2. How much did the following aspects of this class help your learning?**

| | No gains | a little gain | moderate gain | Great gain |
|---|---|---|---|---|
| The teaching methods used in this class | ○ | ○ | ○ | ○ |
| How class activities, readings, assignments (and labs, if relevant) fit together | ○ | ○ | ○ | ○ |
| The pace of this class | ○ | ○ | ○ | ○ |
| The amount of content presented in this class | ○ | ○ | ○ | ○ |
| The instructor's explanation of how the class activities, reading , labs (if relevant) and assignments related to each other. | ○ | ○ | ○ | ○ |
| The expectation of courteous student-student behavior established by the instructor | ○ | ○ | ○ | ○ |
| The atmosphere of this class | ○ | ○ | ○ | ○ |

Please comment on how (and if) this class helped your learning



E. Seymour, A.-B. Hunter (eds.), *Talking about Leaving Revisited*,
https://doi.org/10.1007/978-3-030-25304-2

## SALG: Class activities

**3. How much did the following aspects of this class help your learning? (Select "not applicable" if you did not do this activity)**

| | Not applicable | No gains | a little gain | moderate gain | Great gain |
|---|---|---|---|---|---|
| Attending lectures | ○ | ○ | ○ | ○ | ○ |
| Participating in discussions during class | ○ | ○ | ○ | ○ | ○ |
| Doing hands on classroom activities | ○ | ○ | ○ | ○ | ○ |
| Participating in group work during class | ○ | ○ | ○ | ○ | ○ |
| Giving class presentations | ○ | ○ | ○ | ○ | ○ |
| Participating in group feedback of student work | ○ | ○ | ○ | ○ | ○ |
| Using clickers or similar tools to participate in class discussions/activities | ○ | ○ | ○ | ○ | ○ |

Describe other class activities that you found helpful to your learning

## SALG: Graded activities

**4. How much did the following aspects of this class help your learning? (Select "not applicable" if you did not do this activity)**

| | Not applicable | No gains | a little gain | moderate gain | Great gain |
|---|---|---|---|---|---|
| Graded assignments | | | | | |
| Writing assignments | | | | | |
| Graded individual projects | | | | | |
| Research papers | | | | | |
| Graded group projects | | | | | |
| Opportunities for in-class review | | | | | |
| The number and spacing of tests | | | | | |
| The feedback on my work received after tests or assignments | | | | | |

## SALG: Support for learning

**5. How much did each of the following aspects of the class help your learning? (Select "not applicable" if you did not do this activity).**

| | Not applicable | No gains | a little gain | moderate gain | Great gain |
|---|---|---|---|---|---|
| Interacting with the instructor during class | ○ | ○ | ○ | ○ | ○ |
| Interacting with the instructor during office hours | ○ | ○ | ○ | ○ | ○ |
| Working with teaching assistants during class | ○ | ○ | ○ | ○ | ○ |
| Working with teaching assistants outside of class | ○ | ○ | ○ | ○ | ○ |
| Working with peers during class | ○ | ○ | ○ | ○ | ○ |
| Working with peers outside of class | ○ | ○ | ○ | ○ | ○ |
| Working with one of the advisors | ○ | ○ | ○ | ○ | ○ |

What other sources of support did you find helpful with this class

## SALG: Attitudes

**6. From your experience in this class, rate the following.**

| | None | a little | moderate amount | a great amount |
|---|---|---|---|---|
| Your enthusiasm for the subject | ◯ | ◯ | ◯ | ◯ |
| Your interest in taking, or planning to take additional classes in this area | ◯ | ◯ | ◯ | ◯ |
| Your interest in discussing the subject area with friends or family | ◯ | ◯ | ◯ | ◯ |
| Your confidence that you understand the materials | ◯ | ◯ | ◯ | ◯ |
| Your confidence you can work in this subject area | ◯ | ◯ | ◯ | ◯ |
| Your willingness to seek help from others (TA, teacher peers) when working on academic problems | ◯ | ◯ | ◯ | ◯ |
| Seeing this as a welcoming, inclusive field in which to start a career | ◯ | ◯ | ◯ | ◯ |
| Confidence that you could succeed in a career in this field | ◯ | ◯ | ◯ | ◯ |

## Major

**7. What is your major [as of current date]? (list all if more than one)**

**8. What is your minor?**

**9. When did you declare your major?**

## 10. I chose this major because:

| | Strongly disagree | Disagree | Agree | Strongly Agree |
|---|---|---|---|---|
| I wanted a career in this field | ○ | ○ | ○ | ○ |
| My major allows me to make a good income after I graduate | ○ | ○ | ○ | ○ |
| My major is necessary to go on to graduate school | ○ | ○ | ○ | ○ |
| My family wants me to pursue this major | ○ | ○ | ○ | ○ |
| I can get a job easily with my chosen major | ○ | ○ | ○ | ○ |
| I have friends who have entered this major | ○ | ○ | ○ | ○ |
| I am good at this subject | ○ | ○ | ○ | ○ |
| This major is easier than others | ○ | ○ | ○ | ○ |
| A career in this major allows me to help others | ○ | ○ | ○ | ○ |

Other (please specify)

## 11. What was your primary reason for choosing this major?

## 12. Agree/Disagree

| | Strongly disagree | Disagree | Agree | Strongly Agree |
|---|---|---|---|---|
| I intended to pursue this major before I started college | ○ | ○ | ○ | ○ |
| It was difficult to choose between this major and other majors | ○ | ○ | ○ | ○ |
| I enjoy studying for courses in my major | ○ | ○ | ○ | ○ |
| After taking this class I am more committed to completing this major | ○ | ○ | ○ | ○ |
| After taking this course I may change my major | ○ | ○ | ○ | ○ |
| I was comfortable asking questions in this course | ○ | ○ | ○ | ○ |
| I was comfortable approaching the teacher out of class | ○ | ○ | ○ | ○ |
| The content in this course was difficult | ○ | ○ | ○ | ○ |
| The way the course was taught made it difficult | ○ | ○ | ○ | ○ |
| This course was very competitive | ○ | ○ | ○ | ○ |
| This course was a "weed out" course for the major | ○ | ○ | ○ | ○ |
| Students worked together in this course | ○ | ○ | ○ | ○ |

## Academic advising

### 13. How often have you used these services or resources since entering college?

| | Never | Sometimes | Often |
|---|---|---|---|
| Meetings or consultations with an academic advisor for my major | ◯ | ◯ | ◯ |
| Meetings or consultations with an academic advisor for undeclared students | ◯ | ◯ | ◯ |
| Special program for student support [need specific name at inst] | ◯ | ◯ | ◯ |
| Tutoring services | ◯ | ◯ | ◯ |
| Services that support disabled students | ◯ | ◯ | ◯ |
| Services that support people of color | ◯ | ◯ | ◯ |
| Services that support international students | ◯ | ◯ | ◯ |
| Services that support women | ◯ | ◯ | ◯ |

Other (please specify)

### 14. I have received helpful advice from my advisor

◯ Strongly Disagree ◯ Disagree ◯ Agree ◯ Strongly Agree ◯ Did not receive any advising

## Student activities

### 15. Since entering the university I have participated in:

| | Yes | No |
|---|---|---|
| Clubs related to my major | ◯ | ◯ |
| Clubs related to my race/ethnicity/nationality | ◯ | ◯ |
| Clubs related to special interests | ◯ | ◯ |
| Clubs related to religion/spirituality | ◯ | ◯ |
| Professional society related to my major | ◯ | ◯ |
| Intramural sports | ◯ | ◯ |
| Team sports | ◯ | ◯ |
| Community service | ◯ | ◯ |
| Volunteer activities | ◯ | ◯ |
| Student government | ◯ | ◯ |
| Research experiences for undergraduates | ◯ | ◯ |
| Fraternity or sorority activities | ◯ | ◯ |
| Study abroad | ◯ | ◯ |
| Internships | ◯ | ◯ |
| Theater productions | ◯ | ◯ |
| Music productions | ◯ | ◯ |
| ROTC | ◯ | ◯ |

Other clubs or organized out of class activities

### 16. Since entering the university I have:

| | Yes | No |
|---|---|---|
| Studied with other students | ◯ | ◯ |
| Tutored other college students | ◯ | ◯ |
| Lived off campus | ◯ | ◯ |
| Commuted to campus from another town/city | ◯ | ◯ |

### 17. I work for pay:

- ◯ 0 hours per week
- ◯ 1 -5 hours per week
- ◯ 6-10 hours per week
- ◯ 11 -20 hours per week
- ◯ More than 20 hours per week

**18. On average I study:**

- 1 - 5 hours per week
- 6-10 hours per week
- 11- 20 hours per week
- 21 -40 hours per week
- More than 40 hours per week

**19. Rate the following:**

| | Poor | Fair | Good | Excellent |
|---|---|---|---|---|
| The overall academic reputation of this school | | | | |
| The overall educational experience I have had at this institution | | | | |
| The instruction I have received at this institution | | | | |
| How this institution treats students of my race/ethnicity | | | | |
| How this institution treats students of my gender | | | | |
| The instruction I have received in this class | | | | |
| My interactions with peers in this class | | | | |
| The use of learning technology in this class | | | | |
| The welcoming atmosphere of this course | | | | |

**20. If you could start over would you go to the same institution?**

- Yes
- No

## Demographic information

**21. What is your gender?**

- ☐ Male
- ☐ Female

**22. What is your race?**

- ☐ African-American
- ☐ Asian-American
- ☐ White
- ☐ Native American
- ☐ Multi-racial

**23. I am...**

- ○ Hispanic
- ○ Non-hispanic

**24. I am a...**

- ○ US Citizen
- ○ International student

**25. I am a...**

- ○ First year student
- ○ Sophomore
- ○ Junior
- ○ Senior
- ○ Other

If other please specify

**26. What is your age? (Please enter a number)**

**27. What is your current grade point average?**

# Appendix D: Persistence Study Interview Protocols

## Switchers

### *Character of the Interviews*

**Aim for a Structured Conversation** I have listed as questions the main issues on which we would like to hear the students' views and experiences. I have sought to place them in an order that may be comfortable to the interviewees—beginning with more concrete matters and working into more sensitive areas. However, you may find that all you need to do is give the student the topic or question and let them talk. In that case, use the questions as prompts for any issues that they do not spontaneously mention. Use whatever issues they spontaneously raise as hooks to shift to different topics. If they get way off-course, do not be afraid to pull them back to what you want them to talk about. Their explanations are gold: encourage them. Their examples are important corroborative evidence—indeed, they support our claim that we are not making this stuff up! But try to steer them away from long rambles.

#### Before You Meet Your Interviewee

1. Make sure that you have added the file name for this interview to your sheet that matches the recording device number for this interview with its filename.
2. Record the demographics of the interview at the start of your recording:

e.g., "White male switcher from *x* to *y* major (if known) at school *Z*; The interviewer is (your name) and the date is."



E. Seymour, A.-B. Hunter (eds.), *Talking about Leaving Revisited*,
https://doi.org/10.1007/978-3-030-25304-2

## Once the Interviewee Arrives

Thank them for coming and get them to sign and date the consent form. Remind them that you will be recording the interview and answer any questions about what will be done with that.

Make sure they understand that no one at their school will know that they have volunteered to give an interview or learn anything about what they say. Stress that their identity will be not be revealed in our reports of findings.

## Explain the Purpose of the Interview

"The question which we would like your help to answer is this: Why do large numbers of students switch from science, computer science, engineering, and mathematics majors—sometimes called STEM majors—into non-STEM majors? We expect that a number of factors contribute to high switching rates. We want you to draw on your own experience in helping us to understand what these factors are, and tell us which of them you see as of greater and lesser importance."

**"I am turning on the recorder and we are recording now."**

1. **Majors (Keep this brief and confirmatory)**

   a. What major did you enter when you came to college (or intend to enter)?
   b. What is your current major?
   c. Did you make any other major changes or add a second major?
   d. When did you make the switch to the major(s) you are now in?

2. **Choice of a STEM major: Tell me something about why you chose your original STEM major.**
   **Prompt for**:

   - What drew you into the sciences?
   - Who or what influenced your choice of a STEM major?
   - How did you learn about what the major involved?

   **Listen for**: HS teachers, role models, family and community, special activities in HS—science clubs, camps, competitions
   **Listen for**: How well-informed are their career aspirations?

3. **Let's talk about your High School Preparation:**

   a. Did you take any of these math and science classes at high school? (HAND THEM A CARD). Can you remember what grade you got?

      - Pre-calculus:
      - Calculus
      - AP math
      - AP calculus
      - College level algebra and trigonometry

- College level or AP Chemistry or Physics?

b. Did you take any college courses while in high school? Which?
c. Have you taken all your college courses at this institution?
d. Did you come into college feeling prepared?
e. Once you started college classes, how well did you find that your high school courses had prepared you for college level work?
f. How were your high school and college science and mathematics courses different?

**Prompt for:**

- in the level of difficulty,
- how much time you had to spend studying,
- what teachers expected of you?

g. What grades were you earning as a senior in high school in science and math? Did your grades change in these courses when you came to college?

- If "Yes," how?
- How did you respond to that?

h. (As relevant) Were there any other ways that you felt under-prepared when you began college level work?

**Listen for and explore**:

- disparities in conceptual level between high school and college,
- math/science courses not taken,
- study skills learned/not learned,
- high school teacher encouragement,
- changes in student's level of interest in science, math, or computer science between high school and present—and their causes.

4. **Student assessments of their learning experiences in STEM (and non-STEM) classes in college.**

a. What kinds of teaching and teachers best help you to understand and master complex material? Why is that?
b. What did you find in particular teaching methods or activities in STEM classes that <u>helped:</u>

   a. your learning?
   b. your motivation?
   c. Were there aspects of teaching methods or class activities in particular STEM classes that <u>made it harder</u> to learn? (Ask them to explain.) What caused you most difficulty?

**Listen for:** more and less traditional forms of teaching. Students may react badly to either.
**Prompts:** If not already discussed, "Tell me something about:"

- the pace at which courses were taught
- amount of material covered ("overload" issue)
- how well/badly different course elements fitted together
- quality of interaction with instructors and TAs; consequences
- learning with peers in class, and study groups outside of class
- discussion sections beyond the lecture ("recitations")

c. Did you have difficulties in understanding the material or doing the work? (If "yes") How did you deal with this?

d. Did any of your STEM courses involve labs? (If yes) How valuable did you find the labs in that/those course(s)?

e. (In majors with lab work) Have you had any undergraduate research experiences? How did your course labs compare with those?

5. **Response to other types of learning experiences:**
   Were there experiences in courses outside of your STEM major that better enabled your learning? Did these experiences influence your switching decision?
6. **Let's talk about tests, assignments, and grading in your STEM classes.**

a. I'm interested in how well the tests or assignments fitted with the course material in your STEM courses?

- Can you give any examples of courses where the tests and the course materials didn't fit well?
- Were there courses where they did fit together well?

b. Did the tests help you to see your progress and understand what you needed to work on?

c. What kinds of grades were you getting?

d. Did the tests and your grades contribute to your switching decision? How important were they in that decision?
**Listen for**:

- problems understanding what the grades meant;
- loss of confidence

e. Was there a weed-out system in some of your STEM courses?

- Which ones?
- How does that system work?
- How did it affect you?

**Probe** for variations by gender and race/ethnicity: "Do weed-out classes affect women (or students in your own ethnic/racial group) differently from the ways they affect white male students?"

- How did you respond to this or deal with it?

7. **Now let's talk about whatever sources of support you have had to help with academic difficulties, or career advice.**

   I want to know something about who you have approached for help or advising:

   a. First, tell me about **official** sources of help and advice you have used and how helpful you found them?

      **If not mentioned, check for:**

      - Faculty, TAs,
      - Advising services—check which advising services worked best for them—within major, non-major related, other.
      - Special programs for women, students of color, first generation students
      - Residential communities (halls, dorms, floors for like majors)
      - Supplemental instruction—tutors

      **Listen for:**

      - Kinds of support that they needed but did not get;
      - Frustrated efforts to get help;
      - "Fork-in-the-road" stories.

   b. Who else has helped you and how important to you was that help?

      **Explore** the significance of <u>informal sources of support</u> (peers, room-mates; clubs, societies (e.g., AWIS, SWE, ROTC); science and engineering dorms), supports outside school (family, community, church), special programs.

      **Listen for:**

      - The significance of <u>out-of-class experiences</u> (undergraduate research, internships, other)
      - The "Do I belong here?" issue.
      - Did they find career role models—where?

      **Women and students of color:**

      - **Explore** their sources of academic and personal support and mentoring. (For some students of color these may be community groups, family, other off-campus supports, e.g., churches)
      - **Ask** if they were helped by special programs or supportive groups/organizations

8. **Culture and Climate Issues**

   A. **Ask all students.**

      a. Were there ways in which the people who taught you behaved that make it hard to feel that you belonged in the major?
      b. Were there ways in which your classmates behaved that made you feel unwelcome?

**Probe:** Did you experience rudeness or hostility

- from other students in your classes?
- your instructors?

If not offered: "Would you give me some examples of this kind of behavior?"

- How did you deal with this kind of behavior?
- Did anyone help you deal with this?

B. **Ask all women and men of color:**

a. Do you see (other) differences between your experiences in your major and those of other classmates that have negatively affected shaped your experiences and your choices?
b. Do you see differences in the climate/environment in the STEM course compared to your non-STEM courses?

**Prompt for** issues that they see as specific to their own racial or ethnic group, and how gender intersects with these.

a. Who and what has helped you address these problems?
b. How would you advise other students like yourself to deal with them?

C. **Ask white male students:**

a. Do you think women or students of color in your major experience any difficulties that could cause them to switch out of their STEM majors?
b. Do you see women or students of color having any advantages in your original STEM major? (If yes, do they seem like unfair advantages?)
c. Did you ever feel that you didn't belong in your STEM major (If yes, ask them to explain.)

9. **Summary question:** Looking back over our discussion so far, what would you say surprised or troubled you most about your educational experiences in your STEM major?
10. **Tell me how you are Financing your college education**

a. **Check for:**

- Sources of support (parents, grants, scholarships, loans, internships)
- Paid work, athletic scholarship (How many hours? Has doing these created stresses or conflicts with academic work?)
- Length of degree: issues?

b. Do the ways in which you are paying for your college education have any bearing on your switching decision?

11. **Let's talk about your Career Plans**

   a. Remind me, what was your career goal when you started college? What prompted you to choose that? (This may be a repetition)
   b. What are you thinking about now as a career direction?
   c. What's been the impact of your switching decision on your career ideas? **Listen for:** The consequences of, and their feelings about, shifts in their career plans related to their switching decision.
   d. What sources of information, career advice, or practical considerations are shaping your ideas about the kinds of career you might pursue?
   e. Are changes in the economic and employment climate a factor in your career thinking?
   f. Do you have any plans for graduate or professional study?
   g. Are you still interested in the sciences (engineering, math, or CS)? Can you see yourself making use of what you have learned in your STEM courses in your future work or in your life?

12. **Let's sum up some of the things we have been discussing**

   a. But first, is there something that I didn't ask about, or that you didn't yet explain to me, that contributed to your decision to switch?
   b. Given all that you have said, what do you see as having contributed **the most** to your switching decision?
   c. Would you put these reasons in order of importance for me?
   d. (If they have not already mentioned this) How important would you say your STEM classroom experiences were in your decision to switch?
   e. **If not already described**: Would you describe how your final decision to switch happened? (They may have explained this earlier.)
   **Listen for**:

      - Last straw events,
      - Push-pull process,
      - Who may have sanctioned or enabled it (women especially),
      - Time spent in studying, study methods,
      - Seeking/not seeking help.

   f. **If not already described**: Was there a particular incident that triggered your decision?
   g. **If not already described**: How did you feel about taking the decision at the time? How are you feeling about it now?
   h. **If not already described**: What kinds of reactions to your decision have you had from other people?
   i. **If not already described**: How would you assess the consequences of your decision—its costs and benefits?
   j. How much would you say that things that you did or didn't do contributed to your switching decision? Are there things that you wish you had known earlier or had done differently?

13. **And this is my Last Request**

- Give some advice to this school, your former department or faculty:
- What could they do if they wanted more people to graduate in your former STEM major?

## Persister Seniors

### *Individual Interviews and Focus Groups*

The persister senior sample is comprised of four sub-sets:

1. STEM seniors of color
2. STEM seniors who entered with less than our Math Readiness indicators

   Members of both of these groups will be interviewed singly.

3. White STEM seniors who will be interviewed in four focus groups of (optimally) five members each. Two focus groups will be all female and two will be all male. Each focus group will contain a mix of seniors from engineering specialties, science disciplines, math, and computer science. Two members of the interviewing team will co-conduct each focus group.

For all groups of persisting seniors, the questions broadly mirror those used for the switcher interviews. The protocol also includes questions and probes with particular relevance for seniors by gender, race/ethnicity, and entering level of math readiness.

#### Before You Meet Your Interviewee

3. Make sure that you have added the file name for this interview to your sheet that matches the recording device number for this interview with its filename.
4. Record the demographics of the interview at the start of your recording:

e.g., "White male physics senior at school *Z*; the interviewer is (your name) and the date is."

#### Once the Interviewee Arrives

Thank them for coming and get them to sign and date the consent form. Remind them that you will be recording the interview and answer any questions about what will be done with that.

Make sure they understand that no one at their school will know that they have volunteered to give an interview or learn anything about what they say. Stress that their identity will not be revealed in our reports of findings.

## Explain the Purpose of the Interview

"The question which we would like your help to answer is this: You are now preparing to graduate in your chosen major. We want you to draw on your experience in helping us to understand what has helped you to persist, what difficulties you found along the way, and how you managed to overcome them."

**"I am turning on the recorder and we are recording now."**

13. **Majors (Keep this brief and confirmatory)**
    a. What major did you enter when you came to college (or intend to enter)?
    b. Are you graduating in the same major: have you changed your major, or added another major or a minor?
14. **Choice of a STEM major: Tell me something about why you chose your original major.**
    **Prompt for**:
    - What drew you into the sciences?
    - Who or what influenced your choice of a STEM major?
    - How did you learn about what the major involved?

    **Listen for**: HS teachers, role models, family and community, special activities in HS—science clubs, camps, competitions
    **Listen for**: What formed career aspirations.
15. **Let's talk about your High School Preparation:**
    a. Did you take any of the following math or science classes at high school? (HAND THEM A CARD). Can you remember what grade you got?
       - Pre-calculus
       - Calculus
       - AP math
       - AP calculus
       - College level algebra and trigonometry
       - College level or AP Chemistry or Physics?

    b. Did you take any college courses while in high school? Which?
    c. Have you taken all your college courses at this institution?
    d. Did you come into college feeling prepared?
    e. Once you started college classes, how well did you find that your high school courses had prepared you for college level work?
    f. How were your high school and college science and mathematics courses different?

**Prompt for:**

- in the level of difficulty,
- how much time you had to spend studying,
- what teachers expected of you?

g. What grades were you earning as a senior in high school in science and math? Did your grades change in these courses when you came to college?
   - If "Yes," how?
   - How did you respond to that?

h. (As relevant) Were there any other ways that you felt under-prepared when you began college level work?

**Listen for and explore**:

- disparities in conceptual level between high school and college,
- math/science courses not taken,
- study skills learned/not learned,
- high school teacher encouragement,

16. **Student assessments of their learning experiences in STEM (and non-STEM) classes in college.**

a. What kinds of teaching and teachers best help you to understand and master complex material? Why is that?

b. What have you found in particular teaching methods or activities in your STEM classes that have <u>helped:</u>
   a. your learning?
   b. your motivation?
   c. Were there aspects of teaching methods or class activities in particular STEM classes that <u>made it harder</u> to learn? (Ask them to explain.) What caused you most difficulty?

**Listen for:** more and less traditional forms of teaching. Students may react badly to either.

**Prompts:** If not already raised, "Tell me something about":

- the pace at which courses were taught
- the amount of material covered ("overload" issue)
- how well/badly different course elements <u>fitted</u> together
- quality of interaction with instructors and TAs; consequences
- learning with peers in class, and study groups outside of class
- discussion sections beyond the lecture ("recitations")

c. Looking back to your first two years, were there times when you had difficulties in understanding the material or doing the work? ( If "yes") How did you deal with this?

d. Did any of your STEM courses involve <u>labs</u>? (If yes) How valuable did you find the labs in that/those course(s)?

e. (In majors with lab work) Have you had any undergraduate research experiences? How did your course labs compare with those?

17. **Response to other types of learning experiences:**

a. In your first two years, were there experiences in courses outside of your major that better enabled your learning?

18. **Thoughts about leaving.** Did you ever consider leaving your major? When was that? What prompted that thought?
19. **Let's talk about tests, assignments, and grading, particularly in your first two years in STEM classes**

a. I'm interested in how well the tests or assignments fitted with the course material in your freshman and sophomore courses?

- Can you give any examples of courses where the tests and the course materials didn't fit well?
- Where there courses where they did fit together?
- Has the "fit" improved as you moved into more senior courses?

b. Did the tests help you to see your progress and understand what you needed to work on?
c. What kinds of grades were you getting in your STEM courses as a freshman and sophomore?
d. Did the tests and grades that you got in these early classes discourage you or make you wonder if you should switch majors? (This may have come up in answer to Q.6)
e. **If "Yes":** What helped you work through that?
f. Was there a weed-out system in some of your introductory courses?

- Which ones?
- How does that system work?
- How did it affect you?

**Probe** for variations by gender and race/ethnicity: "Do weed-out classes affect women (or students in your own ethnic/racial group) differently from the ways they affect white male students?"

- What helped you deal with it?

20. **Now let's talk about what sources of support you have found to help with academic difficulties, or career advice, particularly in your first two years.**

a. Tell me about any **official** sources of help and advice you used as a freshman or sophomore and how helpful you found them?
**If not mentioned, check for:**

- Faculty, TAs,
- Advising services—check which advising services worked best for them—within major, non-major related, other,

- Special programs for women, students of color, first generation students,
- Residential communities (halls, dorms, floors for like majors),
- Supplemental instruction—tutors.

**Listen for:**

- Kinds of support that they needed but did not get;
- Frustrated efforts to get help;
- "Fork-in-the-road" stories.

b. Who or what else has helped you?

**Explore** the significance of <u>informal sources of support</u> (peers, room-mates; clubs, societies (e.g., AWIS, SWE, ROTC); science and engineering dorms, supports outside school (family, community, church), special programs.
**Listen for:**

- The significance of <u>out-of-class experiences</u> (undergraduate research, internships, other)
- The "Do I belong here?" issue.
- Did they find career role models–where?

**Ask Women and students of color:**

- **Ask about** their sources of academic and personal support and mentoring. (For some students of color these may be community groups, family, other off-campus supports (e.g., churches)
- **Ask** if they were helped by special programs or supportive groups/ organizations

21. **Culture and Climate Issues**

A. **Ask ALL seniors.**

a. Again, thinking back to your freshman and sophomore years, were there ways in which your classmates, or the people who taught you, behaved that make it hard to feel that you belonged in the major?
b. Were there ways in which your classmates behaved that made you feel unwelcome?
c. Did you experience rudeness or hostility from other students in your classes or your instructors?

**If Yes:**
If not offered: "Would you give me some examples of this kind of behavior?"

- How did you deal with this?
- Did anyone help you deal with this?
- Did some of that kind of behavior continue into junior and senior years?

B. **Ask all women and men of color:**

   a. Do you see (other) differences between your experiences in your major and those of other classmates that have negatively affected shaped your experiences and your choices?
   b. Do you see differences in the climate/environment in the STEM course compared to your non-STEM courses?

   **Prompt for** issues that they see as specific to their <u>own</u> racial or ethnic group, and <u>how gender intersects</u> with these.

   c. Who and what has helped you address these problems?
   d. How would you advise other students like yourself to deal with them?
   e. Do you mentor younger students?

C. **Ask white male students:**

   a. Do you think women or students of color in your major experience any difficulties that could affect their persistence in your major?
   b. Do you see women or students of color having any advantages in your original STEM major? (If yes, do thee seem like unfair advantages? )
   c. Did you ever feel that you didn't belong in your STEM major (If yes, ask them to explain.)

22. **Tell me how you have Financed your college education**

   a. **Check for:**

      • Sources of support (parents, grants, scholarships, loans, internships)
      • Paid work, athletic scholarship (How many hours? Has doing these created stresses or conflicts with academic work?)
      • Length of degree: issues?

   b. Has the ways in which you have paid for your college education created any difficulties for you?
   c. **If Yes**: How are you addressing them?

23. **Let's talk about your Career Plans**

   a. What are now your career plans?
   b. Do they include graduate or professional study?
   c. Have your career ideas changed over time? What has shaped them?
   d. What sources of information, career advice, or practical considerations have shaped your thinking?
   e. Are changes in the economic and employment climate a factor in your career thinking?

24. **Let's sum up some of the things we have been discussing**

   a. We have focused quite a bit on the earlier years of your major. What would you say were problems or difficulties that have continued for you into your junior or seniors years?
   b. Were there some points along the way when you were not so sure you would make it?
   c. Who or what would you say contributed most to your persistence?
   **Listen for:**

      - Sources of help
      - Things that you did: practices and attitudes
      - Forks in the road: serendipitous outcomes.

   d. Which of these have been the most important?
   e. Are there things that you wish you had known earlier, or done differently that would have made things easier?
   f. If you think about the people that you have known that switched out of your major, what best explains why they didn't make it while you and others did?
   g. Looking back over your education here, how satisfied do you feel with your experience in this major?

25. **And this is my Last Request**

   - Give some advice to this school, your former department or faculty:
   - What could they do if they wanted more people to graduate in your major?

# Appendix E: Persistence Study Participant Samples

Key for institutions participating in the TALR study:

- *PB1R1*: A western state (originally land-grant) university serving a large rural population, with a prestigious engineering research program, and applied science specialties;
- *PB2R1*: A comprehensive, urban north-eastern university with large and diverse STEM undergraduate enrollment;
- *PB3R1*: A large urban, mid-western university with prestige ranking for its STEM research and high production of STEM undergraduates and graduates;
- *PB4R1*: A "flagship" western state university with high reputation for its engineering school and several science departments;
- *PV1R3*: A western liberal arts college with a strong reputation for its science teaching (engineering is not offered);
- *PV2R1*: A small western city university offering engineering, and masters' and doctoral degrees in the sciences (Tables E.1, E.2, and E.3).

**Table E.1** Summary total participants from all TALR study site institutions by type of interview

| Study site | Switchers | Non-switchers, individuals | Focus groups (participants) | Total |
|---|---|---|---|---|
| PB4R1 | 20 | 21 (7 low math) | 7 | 48 |
| PV1R3 | 11 | 17 (12 low math) | 9 | 37 |
| PB3R1 | 19 | 21 (9 low math) | 20 | 60 |
| PB2R1 | 22 | 29 (9 low math) | 16 | 67 |
| PV2R1 | 5 | 24 (11 low math) | 17 | 46 |
| PB1R1 | 19 | 49 (34 low math) | 20 | 88 |
| Total | 96 | 161 (82 low math) | 89 | 346 |



E. Seymour, A.-B. Hunter (eds.), *Talking about Leaving Revisited*,
https://doi.org/10.1007/978-3-030-25304-2

**Table E.2** TALR switcher sample

| Discipline | Type of student | PB4R1 ($n = 20$) | | PV1R3 ($n = 11$) | | PB3R1 ($n = 19$) | | PB2R1 ($n = 22$) | | PV2R1 ($n = 5$) | | PB1R1 ($n = 19$) | | Totals | |
|---|---|---|---|---|---|---|---|---|---|---|---|---|---|---|---|
| | | Male | Female | Male | Female | Male | Female | Male | Female | Male | Female | Male | Female | | |
| Physical Sciences | White students | 2 | 1 | – | 1 | – | 3 | 1 | – | – | – | 1 | 1 | 14 | |
| | Students of color | – | – | – | – | 1 | 1 | – | – | – | 1 | – | 1 | | |
| Life Sciences | White students | 1 | 5 | 3 | 2 | 4 | 1 | – | 2 | 1 | 3 | 1 | 7 | 45 | |
| | Students of color | 1 | 1 | – | 3 | – | 2 | 1 | 6 | – | – | – | 1 | | |
| Engineering* | White students | 3 | 2 | – | 1 | 3 | 2 | 3 | 2 | – | – | 5 | 1 | 25 | |
| | Students of color | 1 | – | – | 1 | – | – | 1 | – | – | – | – | – | | |
| Mathematics | White students | 1 | 2 | – | – | – | – | – | 3 | – | – | – | – | 8 | |
| | Students of color | – | – | – | – | – | – | – | 1 | – | – | | 1 | | |
| Computer Science | White students | – | – | – | – | – | 1 | 1 | – | – | – | – | – | 4 | |
| | Students of color | – | – | – | – | – | 1 | – | 1 | – | – | – | – | | |
| Totals | White students | 7 | 10 | 3 | 4 | 7 | 7 | 5 | 7 | 1 | 3 | 7 | 9 | 70 | 96 |
| | Students of color | 2 | 1 | – | 4 | 1 | 4 | 2 | 8 | – | 1 | – | 3 | 26 | |

*Engineering was offered at five of the six institutions in the study sample

**Table E.3** TALR persister sample

| Discipline | Type of student | PB4R1 (*n* = 28) | | PV1R3 (*n* = 26) | | PB3R1 (*n* = 41) | | PB2R1 (*n* = 45) | | PV2R1 (*n* = 41) | | PB1R1 (*n* = 69) | | Totals | |
|---|---|---|---|---|---|---|---|---|---|---|---|---|---|---|---|
| | | Male | Female | Male | Female | Male | Female | Male | Female | Male | Female | Male | Female | | |
| Physical sciences | White students | 1 | 2 | 1 | 1<br>1[a] | 3 | 2<br>1 | 3 | 1 | 4 | 1 | 1<br>4 | 1 | 38 | |
| | Students of color | – | 1 | – | 1 | 1 | 1 | 1 | – | 1 | 1 | –<br>3 | 1 | | |
| Life sciences | White students | 1<br>2 | 2<br>2 | 1<br>1 | 2<br>3 | 4<br>1 | 4<br>2 | 1<br>1 | 1<br>1 | 2<br>1 | 13<br>6 | 2<br>1 | 8<br>6 | 113 | |
| | Students of color | 1 | 2<br>1 | 1<br>2 | 1<br>4 | 1 | 1 | –<br>2 | 7<br>2 | 1<br>1 | 3<br>3 | 2<br>2 | 3<br>7 | | |
| Engineering[a] | White students | 3 | 2 | 1<br>1 | 1 | 4 | 1 | 5 | 3 | – | – | 3<br>3 | 8<br>1 | 66 | |
| | Students of color | 3<br>1 | 1 | 1 | – | 1<br>2 | 3<br>1 | 5 | 4<br>2 | – | – | 1<br>1 | 2<br>2 | | |
| Mathematics | White students | 1 | – | – | 1 | 1 | 1 | – | 3 | – | 1 | 2 | 1 | 18 | |
| | Students of color | – | –<br>1 | – | – | – | 1<br>1 | – | 1 | 1 | 1 | –<br>1 | – | | |
| Computer science | White students | – | – | 2 | – | – | 1<br>1 | – | – | – | 1 | –<br>2 | –<br>1 | 15 | |
| | Students of color | – | 1 | – | – | 1 | 1<br>1 | 2<br>1 | – | – | – | – | – | | |
| Totals | White students | 8 | 8 | 7 | 9 | 13 | 13 | 10 | 9 | 7 | 22 | 18 | 26 | 150 | 250 |
| | Students of color | 5 | 7 | 4 | 6 | 6 | 9 | 10 | 16 | 4 | 8 | 10 | 15 | 100 | |

[a]Red font = student with low-math readiness score

# Appendix F: Other Problem Iceberg Tables (Tables F.1 and F.2)



E. Seymour, A.-B. Hunter (eds.), *Talking about Leaving Revisited*,
https://doi.org/10.1007/978-3-030-25304-2

**Table F.1** "The problem iceberg" by gender for TAL: Concerns contributing to men's and women's STEM switching decisions, and to the concerns of all men and women STEM switchers, of all men and women STEM persisters, and of all STEM students

| Ranking of concerns in TAL | Significant concerns identified by interviewees | % for whom this factor contributed to switching TAL (%) | % for whom this factor contributed to switching men (%) | % for whom this factor contributed to switching women (%) | % all switcher concerns men (%) | % all switcher concerns women (%) | % all persister concerns men (%) | % all persister concerns women (%) | % concerns of all men (%) | % concerns of all women (%) | % all student concerns (%) |
|---|---|---|---|---|---|---|---|---|---|---|---|
| 16 | Discovery of an aptitude for a non-STEM subject | 10 | 10 | 10 | 11 | 12 | 4 | 5 | 8 | 9 | 8 |
| 9 | Discouraged/lost confidence due to low grades in early years | 23 | 27 | 19 | 32 | 37 | 13 | 12 | 23 | 25 | 24 |
| 1 | Loss of incoming interest and motivation | 43 | 44 | 43 | 62 | 58 | 39 | 32 | 51 | 46 | 49 |
| 6 | Rejection of STEM careers and associated lifestyles | 29 | 20 | 38 | 37 | 50 | 18 | 24 | 28 | 38 | 33 |
| 7 | Shift to a more appealing non-STEM career option | 27 | 27 | 27 | 36 | 30 | 14[a] | 18[a] | 26 | 25 | 25 |
| 12 | Competitive, unsupportive STEM culture makes it hard to belong | 15 | 26 | 4 | 35 | 23 | 8 | 10 | 22 | 17 | 20 |
| 3 | Poor quality of STEM teaching | 35 | 39 | 33 | 92 | 89 | 66 | 80 | 80 | 85 | 83 |
| 13 | Reasons for choice of STEM major prove inappropriate | 14 | 15 | 14 | 74 | 91 | 43 | 36 | 60 | 66 | 63 |
| – | Difficult transition to college | – | – | – | – | – | – | – | – | – | – |
| – | Negative effects of weed-out classes | – | – | – | – | – | – | – | – | – | – |
| 4 | STEM curricular design problems: pace, overload, labs, alignment | 35 | 42 | 29 | 54 | 38 | 40 | 42 | 48 | 40 | 44 |
| 17 | Prefer teaching approach in non-STEM courses | 9 | 8 | 10 | 27 | 22 | 10 | 7 | 19 | 15 | 20 |

| | | | | | | | | | | | |
|---|---|---|---|---|---|---|---|---|---|---|---|
| 2 | Non-STEM major offers better education, holds more interest | 40 | 35 | 46 | 57 | 60 | 25 | 38 | 42 | 50 | 46 |
| 19 | System-playing as means to career goals | 7 | 10 | 4 | 11 | 6 | 4 | 1 | 8 | 4 | 6 |
| 11 | Inadequate high school preparation in subject and study skills | 15 | 15 | 15 | 42 | 40 | 34 | 41 | 38 | 40 | 39 |
| 5 | STEM career options and rewards felt not worth the effort | 31 | 36 | 27 | 48 | 39 | 23 | 17 | 37 | 29 | 33 |
| 14 | Conceptual difficulties with one or more STEM subject(s) | 13 | 9 | 16 | 20 | 33 | 29 | 21 | 24 | 28 | 26 |
| 21 | Problems related to class size | 0 | 0 | 0 | 22 | 17 | 9 | 13 | 16 | 15 | 16 |
| 8 and 15 | Difficulties in seeking and getting appropriate timely help | 24/12 | 20/16 | 29/8 | 68/25 | 84/10 | 42/9 | 43/5 | 56/18 | 66/8 | 65/12 |
| 18 | Time to degree influenced switching decision | 9 | 10 | 7 | 24 | 17 | 26 | 29 | 25 | 22 | 24 |
| 10 | Financial problems with completing STEM degree | 17 | 24 | 11 | 36 | 24 | 23 | 22 | 30 | 23 | 27 |
| 22 | Poor teaching, lab or recitation support by TAs | 0 | 0 | 0 | 20 | 19 | 12 | 9 | 16 | 15 | 16 |
| 20 | Language difficulties with foreign faculty or TAs | 0 | 3 | 3 | 25 | 34 | 20 | 10 | 44 | 56 | 25 |
| 23 | Poor lab/computer lab facilities | 0 | 0 | 0 | 3 | 5 | 1 | 7 | 2 | 6 | 4 |

| 0–25% | 26–50% | 51–75% | 76–100% |
|---|---|---|---|

[a]Issue raised by persisters intending to move into non-STEM field following graduation

**Table F.2** "The problem iceberg" by high-performing students and by gender: Concerns contributing to all male and female high-performing STEM switching decisions, to the concerns of male and female high-performing STEM switchers, overall, to the concerns of male and female high-performing STEM persisters, and to the concerns of all STEM students

| Ranking of concerns in TAL | Significant concerns identified by interviewees | % for whom this factor contributed to switching ($N = 96$) (%) | % for whom this factor contributed to switching high-perform men ($N = 13$) (%) | % for whom this factor contributed to switching high-perform women ($N = 42$) (%) | % all switcher concerns high-perform men ($N = 13$) (%) | % all switcher concerns high-perform women ($N = 42$) (%) | % all persister concerns high-perform men ($N = 34$) (%) | % all persister concerns high-perform women ($N = 47$) (%) | % all student concerns ($N = 346$) (%) |
|---|---|---|---|---|---|---|---|---|---|
| 16 | Discovery of an aptitude for a non-STEM subject | 76 | 85 | 69 | 85 | 69 | 0 | 2 | 25 |
| 9 | Discouraged/lost confidence due to low grades in early years | 61 | 38 | 60 | 62 | 83 | 21 | 26 | 53 |
| 1 | Loss of incoming interest and motivation | 61 | 46 | 67 | 46 | 69 | 12 | 4 | 26 |
| 6 | Rejection of STEM careers and associated lifestyles | 58 | 54 | 64 | 54 | 64 | 3 | 4 | 20 |
| 7 | Shift to a more appealing career option | 54 | 46 | 64 | 46 | 64 | 0 | 2 | 18 |
| 12 | Competitive, unsupportive STEM culture makes it hard to belong | 52 | 54 | 57 | 77 | 88 | 6 | 28 | 53 |
| 3 | Poor quality of STEM teaching | 48 | 46 | 50 | 85 | 98 | 32 | 32 | 78 |
| 13 | Reasons for choice of STEM major prove inappropriate | 48 | 46 | 50 | 54 | 71 | 9 | 9 | 34 |
| – | Difficult transition to college | 43 | 31 | 48 | 92 | 90 | 24 | 28 | 66 |
| – | Negative effects of weed-out classes | 35 | 8 | 21 | 8 | 21 | 3 | 2 | 25 |
| 4 | STEM curricular design problems: pace, overload, labs, alignment | 31 | 46 | 26 | 100 | 83 | 29 | 26 | 65 |
| 17 | Prefer teaching approach in non-STEM courses | 29 | 15 | 29 | 62 | 57 | 3 | 2 | 22 |
| 2 | Non-STEM major offers better education, holds more interest | 26 | 23 | 26 | 23 | 26 | 0 | 2 | 9 |

| | | | | | | | | | |
|---|---|---|---|---|---|---|---|---|---|
| 19 | System-playing as means to career goals | 26 | 23 | 26 | 23 | 26 | 0 | 0 | 8 |
| 11 | Inadequate high school preparation in subject and study skills | 19 | 8 | 24 | 46 | 60 | 21 | 15 | 42 |
| 5 | STEM career options and rewards felt not worth the effort | 17 | 8 | 26 | 8 | 26 | 0 | 0 | 5 |
| 14 | Conceptual difficulties with one or more SME subject(s) | 16 | 23 | 12 | 69 | 76 | 24 | 23 | 53 |
| 21 | Problems related to class size | 15 | 8 | 5 | 38 | 57 | 21 | 11 | 36 |
| 8 and 15 | Difficulties in seeking and getting appropriate timely help | 13 | 15 | 17 | 77 | 79 | 6 | 15 | 45 |
| 18 | Time to degree influenced switching decision | 13 | 15 | 17 | 15 | 17 | 0 | 2 | 4 |
| 10 | Financial problems with completing STEM degree | 10 | 0 | 14 | 62 | 79 | 21 | 21 | 54 |
| 22 | Poor teaching , lab or recitation support by TAs | 3 | 0 | 0 | 0 | 0 | 0 | 2 | 3 |
| 20 | Language difficulties with foreign faculty or TAs | 1 | 0 | 0 | 54 | 45 | 15 | 13 | 29 |
| 23 | Poor lab/computer lab facilities | 0 | 0 | 0 | 0 | 0 | 0 | 0 | 0 |

| | | | |
|---|---|---|---|
| 0–25% | 26–50% | 51–75% | 76–100% |

# Index




E. Seymour, A.-B. Hunter (eds.), *Talking about Leaving Revisited*,
https://doi.org/10.1007/978-3-030-25304-2

## F



## G

**R**

CPSIA information can be obtained
at www.ICGtesting.com
Printed in the USA
BVHW010108111220
595366BV00006B/674

9 783030 253066